DATE			
NO 22 85			
OC 31 '86			
AP 17 '87			
	DISCARDED		

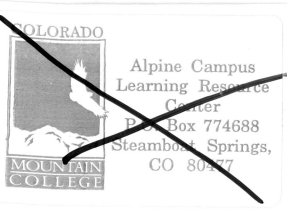

ELECTRICAL
FUNDAMENTALS

SECOND EDITION

ELECTRICAL FUNDAMENTALS

J. J. DE FRANCE

Professor Emeritus
City University of New York

Prentice-Hall, Inc., Englewood Cliffs, New Jersey, 07632

Library of Congress Cataloging in Publication Data
DE FRANCE, J. J. (JOSEPH J.), date

   Electrical fundamentals.
   Includes index.
   1. Electric engineering. I. Title.
TK146.D38 1983   621.3   82-21404
ISBN 0-13-247262-7

Editorial/production supervision and
  interior design: Virginia Huebner
Cover design: Ray Lundgren
Manufacturing buyer: Gordon Osbourne

Printed in the United States of America

10 9 8 7 6 5 4 3 2 1

ISBN 0-13-247262-7

Prentice-Hall International, Inc., *London*
Prentice-Hall of Australia Pty. Limited, *Sydney*
Editora Prentice-Hall do Brasil, Ltda., *Rio de Janeiro*
Prentice-Hall Canada Inc., *Toronto*
Prentice-Hall of India Private Limited, *New Delhi*
Prentice-Hall of Japan, Inc., *Tokyo*
Prentice-Hall of Southeast Asia Pte. Ltd., *Singapore*
Whitehall Books Limited, *Wellington, New Zealand*

To my wife, Margaret

CONTENTS

20. Direct-Current Generators   302

21. Direct-Current Motors   314

22. Kirchhoff's Laws   323

PART II—ALTERNATING CURRENTS

33. Series Circuits—Electronic Aspects   504

34. Parallel and Series–Parallel Circuits   535

35. AC Networks   568

PREFACE

This text is intended for use as a first course in electrical/electronics technology, and is aimed primarily at the technical institute, community, and junior college level. Essentially, this volume is an updated version of the author's earlier *Direct Current Fundamentals* and *Alternating Current Fundamentals*. In addition to combining the two volumes, the content has been modified to reflect the latest advances in the field. Also, in keeping with the recommendations of the I.E.E.E., the SI system (Système Internationale d'Unités) is used for the units, symbols, and abbreviations of physical quantities—except where such use would unduly complicate the communication of information, or differs from industrial practice.

Three outstanding features are claimed for this book. First is the styling —the author has retained the conversational or lecture style that proved successful in the previous volumes. The conversational style is intended to assist the student to a clear, easy understanding, by making the subject "live." The interspersed questions stimulate his or her thinking; the analogies help to connect the subject matter to previously acquired knowledge, or to common sense. Second, the emphasis is on concepts and not on mathematical derivations. However, mathematics is used whenever the material presented has quantitative aspects, and the technician should know how to evaluate. Third are the numerous "small-bit" review questions at the end of each chapter. These can be used for self-evaluation by the student, or for classroom discussion, using the *programmed machine question technique* discussed under "Instructional Notes."

Desirable to the understanding of this volume is a foundation in the principles of algebra, and some knowledge of basic trigonometry and logarithmic functions. Because the calculus is used in some of the derivations, an introduction to the calculus would enhance the students' appreciation of such derivations, but knowledge of the calculus is not required for understanding the electrical theory, or for the solution of problems.

However, in my years of teaching electricity and electronics, I found that many would-be technicians lacked adequate mathematical background. To make up such deficiencies via formal mathematics courses (in arithmetic, algebra and trigonometry), could require two semesters prior to starting the electrical studies. Generally, these students have never been exposed to the technical language of significant figures and powers of ten. This second edition, therefore, starts with an introductory mathematical chapter, wherein scientific notation is explained, and the students are taught how to add, subtract, multiply and divide using powers of ten. They are also given an introduction to proportions and equations, and a short cut to the transposition of equations. This is enough to get them started with electrical studies. Then, throughout the text additional mathematics is introduced as needed.

No attempt is made to cover arithmetic for two reasons. First, it is expected that the student has this basic skill. Second, it is assumed that the student will be using a calculator. Examples for using the calculator in a variety of situations are shown throughout the text, as the need arises.

A major change in this second edition is the switch from electron flow to conventional current flow. When Benjamin Franklin discovered electricity, he described current flow as a "mysterious something" that flows from positive to negative. Many years later, it was learned that the opposite was true, and current flow is due to electron motion—from negative to positive. During the late 1930s, a trend was started to use this true direction of current flow. Unfortunately, it was not picked up. Opposition to changing the old custom was too great. Finally, the I.E.E.E. internationally endorsed the retention of this "conventional" direction of current flow.

With the rapid advance of electronic industries, many electronic technicians and service people find their earnings limited by lack of adequate knowledge. Increased competition, and the advent of more complex circuits has made basic theoretical knowledge indispensable. This volume will be especially valuable to this group.

No book teaches everything about any subject. Much remains for the students to learn on the job. With a sound foundation they can continue their learning, assimilate fresh information, and apply their present knowledge to new developments.

J. J. DE FRANCE

INSTRUCTIONAL NOTES

Many studies of the psychology of learning have shown that effective learning must involve active participation by the learner, and correct responses must be "rewarded." Teaching machines developed in keeping with these basic principles have been very successful. In general, these machines give factual information; ask questions (in small steps) based on this information; elicit some form of response; and finally give or confirm the correct answer.

However, machines have serious limitations. One type, although it allows the student to make any answer, merely states the correct response and continues. Another type restricts the student to a choice of one out of only four answers. If a wrong response is selected, the machine indicates why it is wrong and then allows the student to make another response. When the correct response is given the program advances to the next step.

In a classroom situation, a live instructor can combine the better features of each type of machine. Not only can the student be allowed complete freedom of response, but the instructor can then modify the teaching "program" instantaneously to fit any response.

This text was written with such a teaching technique in mind. In the body of the text, factual information is given and circuit operation is described in more detail than usual, so that the instructor will not have to spend valuable class time lecturing at length to supplement missing items or skimpy treatment. Instead, the lesson time can be spent using a question-discussion-guidance technique, with heavy emphasis on student participation. To help implement this type of lesson, the author has incorporated many review questions at the end of each chapter. These questions follow the text sequence and represent small "bits" of each topic, much in the manner of a teaching machine. Sufficient class time should also be allowed for a satisfying analysis of all problems assigned for homework. If additional time is available, it can be well spent in enriching and motivating each lesson from the instructor's own practical experience.

This system has been tested by the author with several classes, with very gratifying results. Not only were class averages raised, but even more important, the students actually enjoyed these lessons and came to class better prepared to join in the general discussion.

One other feature of this book should be of great assistance to the student and to the instructor. This feature is used in the sections on parallel and series-parallel circuits. To show current distribution among the branches of these circuits, different types of arrows (solid, dotted, dashed, and so on) are used. With this method, it is a simple matter to trace the current distribution even in complex problems. It is recommended that the instructor use the same idea, but with *colored* chalks, for black-board demonstrations.

J. J. De France

0

INTRODUCTION—
MATHEMATICAL REVIEW

No technical textbook can be considered complete unless it includes problem solving in the various theory aspects it covers. So this text on electrical fundamentals contains many problems—both as illustrations and as student exercises. Generally, a student will have had sufficient mathematics to handle "everyday" problems. However, technical calculations tend to use a "language" all their own. Since the average student may not have been exposed to this before, a brief discussion of *scientific notation* is in order. In addition, because students may have forgotten some of their earlier algebra and trigonometry studies, a review of some pertinent aspects of these subjects is given here and throughout the text as the need arises.

SCIENTIFIC NOTATION

Engineering measurements (and calculations) often deal with very large numbers —into the millions of billions, and also with very small numbers—millionths-millionths. This might give the impression that engineering data are extremely accurate—12 places or more. This is not necessarily so. Most data are obtained from instruments with three-digit accuracy.

0-1 Significant figures. If the length of an object is specified as 8 mm, it could be exactly 8 mm—but it is more likely that the object is more than 7.4 mm long or shorter than 8.5 mm long. The figure "8" is really a rounded-off value. (Any value less than 0.5 is dropped, but any value of 0.5 or higher raises the previous digit by one whole number.) Such a measurement has only one *significant figure*. A more accurate measurement could be 8.2 mm. Now we are using two significant figures to show this greater accuracy. This time we can say that the length of the object lies between 8.15 and 8.24 mm. (If the value had been a little less than 8.15, we would have rounded this off to 8.1. Conversely, had the value been 8.25, or a little more, we would have raised it to 8.3.) Had we used a still more accurate measuring device, the recorded length could have been 8.19 mm. This measurement has three significant figures. What are the variation limits now? The length must lie between 8.185 and 8.194.

Of course, significant figures need not contain decimal quantities. For example, the number 427 has three significant figures. (Its variation limits are from 426.5 to 427.4. How about numbers into the thousands, or millions? How are they expressed in three significant figures? Simple—round off, and use as many zeros as are needed to locate the decimal point properly.

Example 0-1

Reduce each of the following numbers to three significant figures.
(a) 1 764 325
(b)   32 678

Solution

(a) Since the fourth digit is less than 5, we make it a zero and add three more zeros to maintain the proper decimal point.

$$1\ 764\ 325 \qquad \text{becomes} \qquad 1\ 760\ 000$$

(b) Since the fourth digit is greater than 5, we raise the third digit to 7, and add sufficient zeros to set the decimal point.

$$32\ 678 \qquad \text{becomes} \qquad 32\ 700$$

Most engineering data are recorded to three significant figures, and calculations are maintained at three significant figures. With slide rules, calculations are automatically maintained at three significant figures, since slide rules are not generally capable of any higher accuracy. Yet, by long-hand, or with a calculator, we can get many more places. Consider, for example, multiplying 324 by 461. The calculator answer is 149 364—or six places. Does this mean greater accuracy! No! Since each of the original data had only three-figure accuracy, the answer is accurate only to three significant figures. So it should be recorded as 149 000.

When dealing with very small numbers—less than 1—such as 0.002 74, the zeros are again used to locate the decimal point—and are not significant figures. Therefore, the number 0.002 74 has only three significant figures. Similarly, 0.000 32 has only two significant figures. On the other hand, the number 0.003 02 has three significant figures. The zero between the three and the two is not used as a decimal locator—and is therefore a significant value. The same applies to the number 60 400. The zero between the 6 and the 4 is not a decimal locator. It is a significant value, and the number has a total of three significant figures.

Now here is a final test. A measurement is recorded as 8.00 mm. How many significant figures does this quantity have? Think. Notice that the decimal point location is not changed whether we include or remove the two zeros. Consequently, they are not decimal locators—they are significant values, and the number 8.00 has three significant figures. Remember, we started this discussion with a measurement of 8 mm. If we wanted to indicate greater accuracy than 7.5 to 8.4 mm, we should have added the significant zeros, and specified it as 8.00 mm.

0-2 Powers of 10.   As was mentioned earlier, at times we must use some very large numbers to express some scientific quantity. For example, in the study of electricity, you will learn that the unit of charge—the coulomb—consists of a total of 6 240 000 000 000 000 000 electrons. Although this number is expressed to three significant figures, it is still a nuisance, and a possible source of error, to carry around all those zeros. So, instead, we express it in *powers of 10* as 6.24×10^{18}. How did we do this?

1. We place the decimal point to just after the first digit. (We normally want the final answer to be a number between 1 and 9, times the proper power of 10.)
2. We count the number of places that the decimal point was shifted to the left (in this case 18 places).
3. We indicate this by affixing "$\times 10^{18}$" after the 6.24.

In other words, the exponent indicates: The full number is bigger than shown, by that many decimal places.

Example 0-2

Express 42 300 000 000 000 in powers of 10.

Solution

1. Place the decimal point after the first digit.

$$4.23$$

2. Count the number of places we shifted the decimal point to the left (13 places).
3. Therefore, expressed in powers of 10, the number is

$$4.23 \times 10^{13}$$

Example 0-3

Convert 1.76×10^{6} to a full conventional number.

Solution

1. Since the *full* number is *bigger* by six places, move the decimal point *back* six places to the right.

2. 1.760000, or 1 760 000.
   1 2 3 4 5 6

When the numbers we encounter are very small—for example, 0.000 000 032 2—a similar technique is used to eliminate the awkward zeros. Again:

1. We place the decimal point to just after the first digit (i.e., 3.22).
2. We count the number of places we shifted the decimal point to the right (eight places for the above case).
3. We affix the power of 10—*but* since this time we made the number look larger than it really is, we use a **negative** exponent. Expressed in powers of 10 the number becomes

$$3.22 \times 10^{-8}$$

Example 0-4

Express 0.000 000 000 002 41 in powers of 10.

Solution

1. Place the decimal point after the first digit.

$$2.41$$

2. Count the number of places of decimal point shift (12 places).
3. Therefore (using a negative coefficient), the number is

$$2.41 \times 10^{-12}$$

Example 0-5

Convert 7.35×10^{-3} to a full number.

Solution

1. Since the exponent is negative, we must make the final number smaller. So we move the decimal point three places to the left.

2. Therefore, 0007.35 becomes 0.007 35.
   3 2 1

Now let us try a combination problem.

Example 0-6

Express 4 375 168 297 in powers of 10 to three significant figures.

Solution

1. Round off. Since the fourth digit is a five, raise the third digit to 8, and add sufficient zeros to set the decimal point properly.

$$4\ 380\ 000\ 000.$$

2. Move the decimal point to just after the first digit, and count the number of places it was moved (nine places).

3. In scientific notation the number becomes

$$4.38 \times 10^9$$

0-3 Addition and subtraction in powers of 10. When the numbers to be added (or subtracted) are expressed in the *same power of* 10, the process is very simple. Merely perform your addition (or subtraction) in the usual manner, and carry the common power of 10 into the final answer.

Example 0-7

Add 3.26×10^3, 5.15×10^3, and 1.37×10^3.

Solution

$$\left.\begin{array}{l} 3.26 \\ 5.15 \\ 1.37 \\ \hline 9.78 \end{array}\right\} = 9.78 \times 10^3$$

Example 0-8

Subtract 3.42×10^{-6} from 8.26×10^{-6}.

Solution

$$\left.\begin{array}{l} 8.26 \\ -\ 3.42 \\ \hline 4.84 \end{array}\right\} = 4.84 \times 10^{-6}$$

Beware, however, when the given numbers are not in the same power of 10. They cannot be added (or subtracted) directly. Such an answer would have no meaning. It is first necessary to convert all numbers to the same power of 10. Then we proceed as in Examples 0-7 and 0-8.

Example 0-9

Add 6.14×10^{-3}, 3.78×10^{-5}, and 2.34×10^{-4}.

Solution

1. Let us convert them all to 10^{-5}.

3.78×10^{-5}	remains	3.78×10^{-5}
6.14×10^{-3}	becomes	614.00×10^{-5}
2.34×10^{-4}	becomes	23.40×10^{-5}

2. Adding, we get 641.18×10^{-5}

3. And rounding to three significant figures, we get

$$641 \times 10^{-5} \quad \text{or} \quad 6.41 \times 10^{-3}$$

Example 0-10
Subtract 8.26×10^{-4} from 3.72×10^{-3}.
Solution

1. Express both numbers in 10^{-4}.

3.72×10^{-3}	becomes	37.20×10^{-4}
8.26×10^{-4}	remains	8.26×10^{-4}

2. Subtracting, we get $\qquad\qquad\qquad\qquad 28.94 \times 10^{-4}$
3. And rounding to three significant figures, we get

$$28.9 \times 10^{-4} \quad \text{or} \quad 2.89 \times 10^{-3}$$

0-4 Multiplying or dividing in powers of 10. The procedure for multiplying in powers of 10 has two simple steps:

1. Multiply the numbers in the usual manner.
2. *Add the exponents* to get the power of 10.

Example 0-11
Multiply 4.26×10^4 by 3.62×10^3.
Solution

1. $\qquad\qquad\qquad\qquad 4.26 \times 3.62 = 15.4212$
2. $\qquad\qquad\qquad\qquad \overset{4}{\vee} + \overset{3}{\vee} = \overset{7}{\vee}$
3. Combining and rounding, we get

$$15.4 \times 10^7 \quad \text{or} \quad 1.54 \times 10^8$$

Example 0-12
Multiply 8.92×10^{-6} by 3.18×10^2.
Solution

1. $\qquad\qquad\qquad\qquad 8.92 \times 3.18 = 28.3656$
2. $\qquad\qquad\qquad\qquad \overset{-6}{\vee} + \overset{2}{\vee} = \overset{-4}{\vee}$
3. Combining and rounding, we get

$$28.4 \times 10^{-4} \quad \text{or} \quad 2.84 \times 10^{-3}$$

To divide numbers that are in powers of 10, the procedure is equally simple. We divide the numbers themselves, in the usual manner, but we *subtract the exponent values* to get the final exponent value.

Example 0-13

Divide 8.92×10^{-6} by 3.18×10^2.

Solution

1. $\dfrac{8.92}{3.18} = 2.805$

2. $10^{-6} - 10^{2} = 10^{-8}$

3. Combining and rounding, we get 2.81×10^{-8}.

This covers the basics of scientific notation. In this text, numerical values will generally be given in this manner, to three significant figures.

BASIC ALGEBRA

As we study electrical theory, we will find that many electrical quantities are interrelated. That is, the value of one quantity is dependent on the value of one, or more, other quantities. These interdependencies are expressed mathematically by proportions or by equations. Let us examine each in turn.

0-5 Proportions.   A proportion shows how one quantity varies with another. A *direct proportion* means that they vary in the same way. If you double one, the other quantity is also doubled. If you triple one, the other is tripled, or if you cut one in half, the other is halved. An everyday example of a direct proportion is the wages earned by an hourly worker. The more hours (H) he works, the more wages (W) he earns. Expressed mathematically, we would write

$$W \propto H$$

where the symbol "$\propto$" means "is directly proportional to." Notice that this does not tell us how much wages he will receive, but only that if you change the number of hours (increase or decrease), the wages will change in the same way.

A basic relation in electricity is Ohm's law, part of which states that the current (I) in a circuit is directly proportional to the voltage (E) applied across the circuit. In mathematical "shorthand" this would be written as

$$I \propto E$$

This means that the value (or amount) of current depends directly on the voltage. If you triple the voltage, the current increases to three times its former value. Conversely, if you reduce the voltage to 90% of its normal value (a brownout), the current will decrease to 90% of its former value.

Another common type of proportion is the *inverse proportion*. In this case the two quantities vary in opposite ways. If you increase one, the other decreases. More specifically, if you double one ($\times 2$), the other is cut in half ($\times \frac{1}{2}$); if you triple one ($\times 3$), the other is reduced to one-third ($\times \frac{1}{3}$). Notice that one variation is the *reciprocal* of the other. An everyday illustration of this is the relation between the time (t) it would take to travel a given distance and the speed (S) at which we travel. If we double our speed, the time is cut in half. Expressed mathematically, this is

$$t \propto \frac{1}{S}$$

This tells us that time is *directly* proportional to the *reciprocal* of speed (which is the same as time is *inversely proportional* to speed).

In electricity, another part of Ohm's law states that the current (I) in a circuit is inversely proportional to the resistance (R) of the circuit, or

$$I \propto \frac{1}{R}$$

From this we can see that if the resistance is increased to say five times its previous value, the current is reduced to one-fifth of its former value.

0-6 Equations. From a proportion, we know by how much one quantity will change compared to the other—but, unless we knew the original value, we will not know what the new value will be. An equation is more specific. It will give us the value. Any proportion becomes an equation if we add other pertinent factors. For example, in the case of the hourly worker mentioned above, we need to add the hourly rate. Now, we can say, the wage he will get is equal to the hourly rate (R), multiplied by the number of hours that he works, or time (t). In algebraic shorthand this is

$$W = R \times t \tag{0-1}$$

This equation is appropriately written to "solve" for wages. For example: "How much will a laborer receive if he works for 8 hours at the rate of $5 per hour?"

To solve:

$$W = R \times t = 5 \times 8 = \$40$$

An equation should always be written in this form—*the thing we are looking for, the "unknown," should be all by itself, on the left-hand side of the "equals" sign.* In the above illustration wages (W) was what we were looking for.

0-7 Transposition of equations. Let us consider a variation of the example above. Suppose that a man wishes to earn $20, and that he can get a part-time job at $5 per hour. How many hours must he work? Notice that the same three factors are involved: wages, rate, and time. Therefore, equation (0-1) must apply. But we are not looking for wages. The "unknown" in this case is "time." The equation should be rewritten with t all alone on the left side of the equality sign. How do we do this?

An equation should be considered as a balance beam—or "the scales of justice." Whatever is on the scale pan on one side is balanced by what is on the pan on the other side. If we wish to change the contents of one side, we can still maintain perfect balance by making exactly the same change to the other side. How does this apply to our problem of rewriting equation (0-1)? We started with

$$W = R \times t \qquad \text{(0-1A)}$$

Let us turn our scale around, left to right.

$$R \times t = W \qquad \text{(0-1B)}$$

Since we did not change either side's content, the equation is still valid—balanced. We want the factor t all alone. If we divide the left side by R, the R's will cancel, $(R \times t)/R$, and we are left with t all alone. Fine, but to maintain the original balance, we must also divide the right side by R. The right side becomes W/R. The revised equation is

$$t = \frac{W}{R} \qquad \text{(0-1C)}$$

This procedure of "rewriting the equation to solve for t is called *transposing* or *transposition*.

Now that we have seen the step-by-step procedure (and *justification*) for transposing an equation, let us look at a simplified method. Just follow this rule: *Whenever we move a quantity from one side of an equation to the other* (i.e., crosses the "equals" sign), *it must do the opposite of what it did before*. Let us apply this technique to equation (0-1):

$$W = R \times t \qquad \text{(0-1A)}$$

Reversing our scales, as before, we get

$$R \times t = W \qquad \text{(0-1B)}$$

To solve for t, the R must cross the "equals" sign. On the left the R is a multiplier, so on the right it should be a divisor.

$$R \times t = W \quad \text{becomes} \quad t = \frac{W}{R} \qquad \text{(0-1C)}$$

This is the same answer we got before using the "long" way.

Suppose, using equation (0-1), that we wanted to solve for R. Such a problem could read: If a man wants to earn $20 in 4 hours, at what hourly rate must he work? As before, we will start with the basic equation,

$$W = R \times t \qquad \text{(0-1A)}$$

and "reverse the scales":

$$R \times t = W \qquad \text{(0-1B)}$$

To solve for R, the t must cross over to the other side. Since it is a multiplier now, it becomes a divisor on the other side, or

$$R \times t = W \quad \text{and} \quad R = \frac{W}{t} \qquad \text{(0-1C)}$$

These equations ($W = R \times t$; $t = W/R$; and $R = W/t$) are not three different equations. They are three forms of the same equation. We use the form that is most appropriate for solving for the unknown quantity. (Remember, the unknown should be on the left of the equality sign, by itself.)

Let us try this shortcut transposing technique on another equation.

$$A = B + C \qquad \qquad \text{(0-2A)}$$

We wish to solve for B. (We know A and C.) The first step, to get the B on the left-hand side, is to "reverse the scales."

$$B + C = A \qquad \qquad \text{(0-2B)}$$

Now the C must cross over to the other side. On the left, it is an "adder." On the right, it must become a "subtractor."

$$B + C = A - C \qquad \text{and} \qquad B = A - C \qquad \text{(0-2C)}$$

Similarly, if we solved equation (0-2a) for C, the B must cross over, and instead of a plus quantity, it becomes a minus quantity, or

$$C = A - B \qquad \qquad \text{(0-2D)}$$

Let us try a third type of equation.

$$A = \frac{B}{C} \qquad \qquad \text{(0-3A)}$$

To solve this equation for B, first we "reverse the scales" and get

$$\frac{B}{C} = A \qquad \qquad \text{(0-3B)}$$

Now the C must cross over. Since it is a divisor on the left side, it becomes a multiplier on the right side, and we get

$$B = A \times C \qquad \qquad \text{(0-3C)}$$

To solve for C, starting with equation (0-3a), the C must cross over—it becomes a multiplier (goes on top).

$$A \times C = B \qquad \qquad \text{(0-3D)}$$

Then the A must cross over, and it becomes a divisor (goes on bottom).

$$C = \frac{B}{A} \qquad \qquad \text{(0-3E)}$$

With familiarity, you can do this in one step: The C and A will crisscross.

$$A = \frac{B}{C} \qquad \text{and} \qquad C = \frac{B}{A}$$

The above equations, using A's, B's, and C's have no practical significance. They were used merely to show how to transpose an equation. Having learned the technique, let us apply it to a real electrical equation. In discussing proportions, we gave you the two parts of Ohm's law, separately ($I \propto E$ and $I \propto 1/R$). Combining these two proportions, we get the equation

$$I = \frac{E}{R} \qquad \text{(0-4A)}$$

In this form it is suitable for solving for current (I), when we know the applied voltage (E) and the circuit resistance (R). If we wanted to find how much resistance we should use to limit the current to some specific value, R belongs on the left side and I must cross over to the right side of the equation. You should recognize this as a crisscross, and

$$I = \frac{E}{R} \qquad \text{becomes} \qquad R = \frac{E}{I} \qquad \text{(0-4B)}$$

Finally, if we wanted to know how much voltage we should apply to the circuit (of known resistance) to get a specific value of current, we must solve the basic Ohm's law, equation (0-4a), for E. The E should be on the left. Therefore, "reverse the scales" ($E/R = I$). Now the E should be alone, so the R crosses over. It was a divisor; it becomes a multiplier (from bottom to top) and we get

$$E = I \times R \qquad \text{(0-4C)}$$

We are now ready for a more complex type of equation, of the form

$$A = \frac{B + C}{D} \qquad \text{(0-5A)}$$

Solving for D is a simple crisscross. The D crosses and goes from bottom (divisor) to top (multiplier). The A crosses, going from top to bottom.

$$D = \frac{B + C}{A} \qquad \text{(0-5B)}$$

To solve equation (0-5a) for B, let us do this in steps. First, move the D across to the left side. It becomes a multiplier.

$$A \times D = B + C \qquad \text{(0-5C)}$$

Now move the C across. It becomes a subtractor.

$$(A \times D) - C = B$$

For the last step, "reverse the scales":

$$B = (A \times D) - C \qquad \text{(0-5D)}$$

Solving for C would be similar to solving for B, and so it need not be shown.

Let us cover just one more type of equation—one containing a squared quantity. For example,

$$A = B^2 \times C \qquad \text{(0-6A)}$$

If we want to solve for C, first we transpose the B^2. In crossing the equals sign, it becomes a divisor, or

$$\frac{A}{B^2} = C \qquad \text{(0-6B)}$$

Now "reversing the scales," we get

$$C = \frac{A}{B^2} \tag{0-6C}$$

To solve equation (0-6a) for B we can use two similar steps: transpose the C, and reverse the scales. This gives us

$$\frac{A}{C} = B^2 \quad \text{and} \quad B^2 = \frac{A}{C} \tag{0-6B}$$

Finally, to solve for B, we must take the square root of both sides, and we get

$$B = \sqrt{\frac{A}{C}} \tag{0-6C}$$

Of course, there are still other "types" of equations. It would be foolhardy to try to give an example of each. However, from these basics, you should now be able to handle most any equation. Just remember that the transposing technique is always the same—when we move a quantity across the "equals" sign, it must do the opposite of what it did before.

PROBLEMS

1. Reduce each of the following numbers to three significant figures.
   (a) 46 327   **(b)** 3983
   (c) 5066   **(d)** 0.062 88
   (e) 24.033   **(f)** 3.507

2. Express each of the above numbers in powers of 10 and three significant figures.

3. Convert each of the following to "full" numbers.
   (a) 3.65×10^4 **(b)** 4.16×10^{-3}
   (c) 7.22×10^8 **(d)** 8.91×10^{-6}

4. Add the following, expressing the answer in powers of 10.
   (a) $9.65 \times 10^4 + 7.33 \times 10^4$
   (b) $4.12 \times 10^{-3} + 2.67 \times 10^{-3}$
   (c) $3.24 \times 10^3 + 8.48 \times 10^2$
   (d) $2.68 \times 10^{-2} + 9.86 \times 10^{-3}$

5. In each part of Problem 4, subtract the second number from the first.

6. In each part of Problem 4, multiply the two numbers.

7. In each part of Problem 4, divide the first number by the second.

part one

Direct Currents

1

ELECTRON THEORY

COLORADO MOUNTAIN COLLEGE
Alpine Campus - L. R. C.
P.O. Box 775288
Steamboat Springs, CO 80477

At one time, students of electricity used to be told: "We don't know what electricity is, but—. We don't know what current is, but—, We don't know how electricity goes through a solid wire, but—."

The electron theory explains these things clearly and simply. In addition, it explains the true meaning of voltage, resistance, insulation, magnetism, induced voltage, and vacuum tubes. Therefore, an understanding of the fundamentals of the electron theory is basic to the understanding of electrical and electronic theory.

Scientists now agree that our universe is fundamentally dependent on two factors, one of which is matter, the other, energy. Let us consider matter first, then energy.

1-1 Composition of matter. Matter is anything that occupies space and has weight (mass). It can exist in any of three forms: solid, liquid, or gaseous. A chemist would subdivide all pure forms of matter into *elements* and *compounds*. An element consists of only one substance. There are over 100 known elements. Some of the most abundant elements are: oxygen (O), silicon (Si), aluminum (Al), iron (Fe), calcium (Ca), sodium (Na), chlorine (Cl), and hydrogen (H).* If we take a bar of iron and break it, slice it, chop it, file it—and finally reduce it to the smallest possible particle of iron—we now have an

---

*The abbreviations in parentheses are the chemical symbols of each element. Some of these abbreviations are obvious, such as O for oxygen, and Al for aluminum. Others, such as Fe for iron, and Na for sodium, come from the Latin *ferrum* and *natrium*, respectively.

atom of iron. In other words, *the atom is the smallest possible subdivision of an element to remain that element.*

Now let us consider the compound. A compound is a chemical combination of two, or more, elements. For example, water is a compound of the elements hydrogen and oxygen; salt is a compound of sodium and chlorine. If we take a drop of water and reduce it to the smallest particle of water, we have a **molecule** of water. *The molecule, then, is the smallest possible subdivision of a compound to remain that compound.* We can break down this molecule of water still further. However, when we do, we will get two atoms of hydrogen and one atom of oxygen—from which we get the familiar appellation H_2O.

1-2 Electron theory of matter.   All matter, according to the electron theory, is fundamentally the same. It consists of three ingredients. Before discussing them, let us note some familiar examples of ingredients. In baking, a baker uses certain ingredients: flour, baking powder, milk, and so on. If he mixes then in a certain proportion, he may have some buns. If he mixes them in another way, he may have a cake or a pie. Let us take another example. A desk is composed of wood, nails, and glue. If we take it completely apart, we have a certain amount of wood, nails, and glue. If we take part of these and put them together in a different arrangement, we may have a chair. If we start with the same ingredients and different proportions and arrangement, we might have a table. Fundamentally, these pieces of furniture are identical; merely the proportions and arrangement of the ingredients are different.

The composition of matter is divided into three ingredients: positrons, neutrons, and electrons. Let us see, in turn, what each of these is. The word "positron" gives us a clue that the positron is a positive charge. (Do not ask at present what a positive charge means. It will be explained later.) This ingredient has very little weight. The neutron, as the name indicates, is neutral; that is, it has no charge. But what is it there for? It supplies practically all the weight of matter. The third ingredient is the electron. Its charge is opposite to that of the positron; that is, it has a negative charge. In turn, it has very little weight. These are our basic ingredients. If we could examine the structure of an atom of iron, we would find that it consists of a specific number of positrons (26), neutrons (56), and electrons (26), arranged in a particular manner. In an atom of copper, some other quantities of positrons, neutrons, and electrons are arranged in a different way. Yet a positron (neutron or electron) from the iron atom is identical to that from the copper atom—and this is true for all elements. They are all made up of the same ingredients. They differ only in the quantities and arrangements of the ingredients. Since the arrangement of the ingredients is so important, let us briefly review the structure of matter.

1-3 The structure of the atom.   Any atom can be considered as having two major components: a central nucleus and one, or more, orbiting electrons. Let us examine each of these in turn.

Nucleus. We find that the atom has, first of all, a nucleus. The nucleus is composed of the positrons and neutrons. All the positrons and all the neutrons which that material has are grouped into one small mass which we call the *nucleus.* As soon as the positron and neutron get together they form what we call the *proton.* One positron plus one neutron equals one proton. Since all known elements have more neutrons than positrons, the positron seldom exists for any length of time by itself. (For this reason many texts omit all reference to the positron as an ingredient and speak only of protons. However, it must be remembered that the proton is a combination of two basic ingredients.) Since the positron has practically no weight, the proton weight is approximately the same as the weight of the neutron alone. What would the charge of a proton be? It consists of a positron, which is positive, and a neutron, which has no charge; therefore, the proton must be positive. The nucleus may be considered as positrons and neutrons, as stated, or as protons plus excess neutrons. Therefore, the nucleus must also have a positive charge. In addition, we find that the nucleus spins on its own axis.

Those of you who have studied chemistry may recall atomic weights and atomic numbers. The word "weight" tells us that atomic weight has something to do with the number of neutrons in the atom. The atomic number corresponds to the number of positrons in the atom (and also to the number of orbiting electrons). For example, iron has an atomic weight of 56, and an atomic number of 26. Therefore, as stated in an earlier illustration, iron has 56 neutrons, 26 positrons, and 26 electrons. Of course, the 26 positrons would join with 26 of the neutrons, forming 26 protons and leaving only 30 extra (free) neutrons.

Electrons. The electrons are the remaining ingredients of our atom. You may have seen or heard that these electrons are arranged in shells around the nucleus. Let us see what that statement means. One electron may be a certain distance away from its nucleus. That electron is rotating around the nucleus, making a ring around it. This ring may be elliptical. At the same time, the plane of the electron shifts a few degrees, so that we get another ring, and then another and another, so that we finally have the electron tracing the path of a complete sphere or shell. We may have some more electrons a greater distance away also shifting their planes or orbits of rotation. They are arranged in a similar fashion, forming a second shell (see Fig. 1-1). The maximum number of shells for any known element is seven. (The shells are labeled alphabetically K through Q, starting with the innermost shell.) Each shell has a definite maximum limit as to the number of electron orbits it can have. This limit is given by

$$\text{maximum orbital paths} = 2n^2 \qquad \qquad \textbf{(1-1)}$$

where n is the shell number. To solve this equation using a calculator, you would enter

$$2, \boxed{\times}, \text{ the value of } n, \boxed{x^2}, \boxed{=}$$

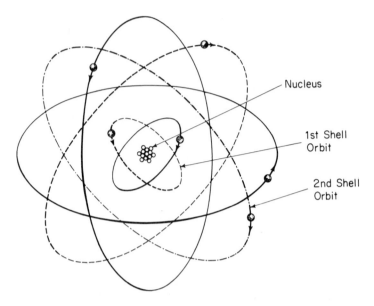

Figure 1-1 Electron orbits and shells.

Obviously, the K or first shell has a maximum electron quota of two, (2×1^2); the second shell, eight, (2×2^2); the third shell, eighteen; and so on.

However, there are two other limitations. The *outermost* shell is filled when it reaches eight electrons; and the next-to-last shell cannot contain more than eighteen regardless of its quota. Some examples follow:

1. Argon (Ar), atomic number 18, has an electron grouping of 2–8–8. The last shell is filled. (Yet, as the third shell, its maximum quota is 18.)
2. Potassium (K), atomic number 19, has four shells, with an electron grouping of 2–8–8–1.
3. Iron (Fe), atomic number 26, has an electron grouping of 2–8–14–2. Notice that since the third shell is no longer the last shell, it can build up toward its quota of 18.
4. Cesium (Cs), atomic number 55, has an electron grouping of 2–8–18–18–8–1. The last electron cannot go into shell 4, because it would then be the next-to-last shell and its maximum is 18; nor can this electron go into shell 5, because that would exceed the last-shell limitation of 8.

The electrons in the last shell are called **valence** electrons, and the electrical (and chemical) properties of a material are dependent on the number of such electrons. A filled last shell (eight valence electrons) produces an inert material. The electrons are tightly bound to their nucleus. Atoms with fewer than four valence electrons tend to give up one or more electrons, and the fewer the valence electrons, the greater this tendency. Conversely, atoms with more than four electrons in their last shell have a tendency to acquire one or more additional electrons. In elements with atomic valences of four, adjacent atoms form into a crystal structure sharing their electrons in **covalent bonds.** Such bonding fills the valence shell, and the material is electrically inert. This creates the **semiconductor** properties that are fundamental to solid-state (transistor) electronics.

What else are these electrons doing? They spin on their own axes and at the same time they revolve around the nucleus. Also, in any one shell, of the number of electrons revolving around the nucleus, some move in a clockwise direction, whereas others revolve in a counterclockwise direction. The number revolving clockwise may or may not be equal to the number revolving counterclockwise. These motions compare very well to our own solar system. The sun is the center of the universe and revolves on its own axis. The planets revolve around the sun and rotate on their own axes. So the picture is very clearly seen when we compare the atom with the solar system (see Fig. 1-1).

There is one more point about the structure of the atom; In a normal atom the number of electrons is equal to the number of positrons. What is the total charge in an atom? It is balanced. There are just as many positive charges as negative, so that a normal atom has no charge.

1-4 Atomic dimensions.   At this point a word about atomic dimensions is in order. Atomic dimensions are necessary to explain such a phenomenon as conductance. (These figures do not have to be memorized, but the relative dimensions are important.)

Consider the electron's weight (mass). If we take 1.1×10^{27} electrons, put them all together, and weigh them, they equal 1 gram.* As to diameter, if we take 3.6×10^{11} electrons and line them up, one alongside the other, we have 1 mm.† The proton is heavier than the electron. This fact is obvious when we recall that the proton contains the neutron, which is the heaviest part of the atom. To obtain a gram's weight, we require only 6.0×10^{23} protons. On the other hand, the proton is smaller than the electron in diameter: 7.2×10^{14} protons equal 1 mm. This time we need 2000 times as many to equal 1 mm. The radius of the orbit from the electron to the nucleus is equal to 10 000 times the diameter of the electron. So if you were asked to state what the major portion of the atom consists of, the only answer would be "empty space."

If a powerful microscope could be obtained to magnify an electron to the size of a basketball, its proton would look like a pea and the space between them would equal about 5 km!‡ When later we consider a piece of "solid" copper wire and say that electrons flow through it, we can see that there is plenty of room for electron motion.

*A gram is about equivalent to the weight of a paper clip.
†A millimeter is about equivalent to the thickness of a dime.
‡A kilometer is approximately 0.6 mile.

1-5 Forces within the atom. You may wonder what prevents the electron, which is minus, from crashing into the proton, or nucleus, which is positive. Why does it stay in a circular path? Why doesn't it fly out? Let us consider the forces in the atom. First, there is a definite force of attraction between the positive nucleus and negative electron. Second, did you ever tie a tin can to a string and whirl it around your head? There is a tendency for the can to fly off. You must pull on the string to keep the can in its circular path. Our electron rotating around the nucleus develops a similar force. This centrifugal force tends to make the electron fly away. So if the electron stays in its orbit, it is obvious that the two forces must be balanced (see Fig. 1-2).

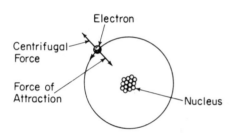

Figure 1-2   Balance of forces within the atom.

Let us consider an electron in the second shell. It is farther away from the nucleus. The force of attraction is weaker, but the electron is rotating at a slower speed. The centrifugal force is also weaker. The two forces are still balanced. The forces in the third and fourth shells must be still weaker; that is, the forces on outer shells decrease.

1-6 Energy. In talking about forces we come to the next fundamental factor of the universe—energy. Energy has been defined as the ability to do work. We cannot see or feel energy itself, but we can notice its effects. At one time it was thought that energy could be created or destroyed, but careful analysis has proven that energy is actually being liberated or converted from one form to another. For example, the sun's rays fall on a lake, causing the water to evaporate. The water rises as vapor to the clouds. The water, because of its high position in the clouds, has *potential* energy. Water falls as rain on the mountaintop and forms a stream; then it comes to a waterfall, where the energy changes to *kinetic* energy as the water falls. This type of energy may be used to turn a waterwheel, which in turn might produce *mechanical* energy that drives a generator. The generator would produce *electrical* energy, which now can be used to light a bulb, which in its turn gives off *light* energy. Some of the *electrical* energy is also converted to *heat*, as you well know if you have ever tried to remove a large bulb from its socket just after switching off the electrical supply. The bulb's energy is similar to the original rays of the sun, being composed of light and heat.

1-7 Charged bodies—electrostatic fields.   Suppose that we have a piece of glass rod. It is composed of atoms. Each atom has the same number of electrons and positrons, and therefore no charge. If we take a piece of silk and rub it on the glass rod, we find that the rod can now pick up pieces of paper. It has ability to do work. In other words, it has energy. Where did this energy come from? We ate breakfast—chemical energy. Our bodies converted this into mechanical energy in our muscles. In rubbing, we converted some of this mechanical energy into friction, or heat energy, but the rest remained as energy in the glass rod. Exactly what has taken place? By rubbing, the silk has removed electrons from the glass rod. The silk now has an excess of electrons, or a negative charge. The glass has a deficiency of electrons, or a positive charge. These charged bodies have energy. The energy is not in the charged body itself but in a field of force surrounding the body. This assertion can readily be proved as follows: Charge your hard-rubber or plastic ball-point pen by rubbing it on a cotton cloth. The pen is now surrounded by an energy field. This field can do work. Tear up some paper into small pieces. The pen can pick up these pieces, but notice that you do not have to touch the paper. If the pen is held close to the paper, the pieces will jump up to the pen. The energy is therefore in a field surrounding the body. This energy is stored in what is called a *dielectric* or *electrostatic field.*

The direction in which the force of this field acts can be found by putting a small test object in the field and noticing the direction in which this test piece moves or tends to move. For example, wind direction at an airfield can be determined by noticing in what direction a windsock tends to move. In our case, the small test object can be either a small positive charge or a small negative charge.

It has been the custom to use a small positive charge as the test object (in spite of the fact that in electrical circuits it is the electron that moves). Therefore, the direction of the field of force is determined by noticing the direction in which this positive test object tends to move. Let us apply this technique to examine the field around a negatively charged body. A positive test piece placed anywhere near this charged body will be pulled into the charged body in a straight line. To represent this field of force, a good diagram must show not only the direction of the force, but also how the strength of the field is affected by the distance from the charged body. From observations, we know that close to the body the power of attraction is strong; the energy is strong. As we move away from the charged body, the field strength gets weaker and weaker. (For example, the charged ball-point pen will pick up pieces of paper close by, but if we move the pen farther away, the pen may sway the pieces of paper but will not pick them up. Obviously, the power of attraction, or strength of field, has decreased.)

This energy field is shown very well in Fig. 1-3 by the radial lines around the negatively charged body. These lines are called *lines of force.* The arrowheads show the direction in which the field acts. The closeness or density of the

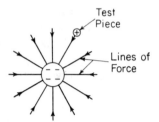

Figure 1-3 Dielectric field of force around a negative charge.

lines indicates the strength of the field. Close to the charged body the lines are close together (strong field), but notice that these radial lines of force get farther apart as they move away from the charged body. The density of the lines is decreasing; that is, the field is getting weaker.

A positively charged body must also have energy. In what direction will it do work? The opposite direction. The positively charged body will repel the test piece. This energy field is shown in Fig. 1-4.

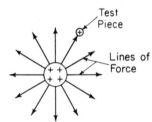

Figure 1-4 Dielectric field of force around a positive charge.

If we anchor two charged bodies close to each other—one positive and the other negative—again using the positive test piece to check the field of force, the negative body attracts it, and the positive body repels it. The field of force is concentrated between the two main bodies (see Fig. 1-5).

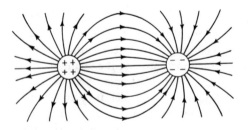

Figure 1-5 Dielectric field of force between two opposite charges.

Now let us examine the energy field between two positively charged bodies (see Fig. 1-6). If we put the test piece at some point between the two charged bodies, it will be pushed away by the first charged body toward the second charged body. But the second charged body will also push it away. The resulting motion is along the line of force as shown. The test piece placed anywhere between the two main bodies will be acted on by two forces and will move along curved lines of force as shown, in Fig. 1-6.

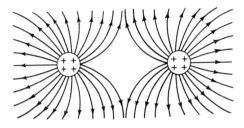

Figure 1-6 Dielectric field between two positive charges.

If the two main bodies are negative charges, the test piece will again be acted on by two forces. It will therefore move along curved lines, as above. However, the test piece will be attracted by the negatively charged bodies; thus the arrows will now point toward the bodies (see Fig. 1-7).

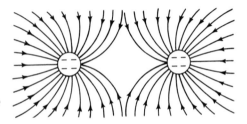

Figure 1-7 Dielectric field between two negative charges.

The magnitude, or strength, of the force between any two charged bodies is directly dependent on the amount of charge on each of the bodies. This, in turn, depends on the number of electrons added to (or removed from) each body. Since, in practical cases, the number of electrons involved is quite large, the unit *coulomb* (Q)* is used. It represents a quantity of 6.24×10^{18} electrons. (Before the electron theory this had been considered as the basic quantity of electrical charge.) We also saw earlier that the fields around the charged body get weaker as we move away from the charge. Therefore, the force between any two bodies must decrease if the distance between them is increased (an inverse relation). Coulomb's law expresses the above relations mathematically:

$$F = \frac{kQ_1Q_2}{d^2} \tag{1-2}$$

where k is a constant of proportionality and depends on the units used.† In the International system (SI), F is in newtons,‡ d is in meters, and $k = 9 \times 10^9$.

*In honor of the eighteenth-century French physicist Charles Augustin de Coulomb (1736–1806).

†Using English units, F is in pounds, d is in feet, and $k = 2.18 \times 10^{10}$.

‡The newton (named in honor of Sir Isaac Newton) is the force required to accelerate a mass of one kilogram at the rate of one meter per second each second. A force of one newton is equal to a force of 0.2248 pound.

Example 1-1

8.6×10^{14} electrons are transferred from body A to body B.

(a) What is the charge on each body?

(b) If the two bodies are separated by a distance of 300 mm, what is the force of attraction (or repulsion) between them?

Solution

(a) 1 coulomb (C) $= 6.24 \times 10^{18}$ electrons. Therefore, the charge on each body is

$$Q = \frac{8.6 \times 10^{14}}{6.24 \times 10^{18}} = 1.38 \times 10^{-4} \text{ C}$$

Using a calculator, this would be entered as.

$$8, \boxed{\cdot}, 6, \boxed{EE}, 1, 4, \boxed{\div}, 6, \boxed{\cdot}, 2, 4, \boxed{EE}, 1, 8, \boxed{=}$$

(b) $d = 300 \text{ mm} = 0.3 \text{ m}$

$$F = \frac{kQ_1Q_2}{d^2} = \frac{9 \times 10^9 \times (1.38 \times 10^{-4})^2}{(0.3)^2} = 1.90 \times 10^3 \text{ N}$$

With a calculator, this would have been entered as

$$9, \boxed{EE}, 9, \boxed{\times}, 1, \boxed{\cdot}, 3, 8, \boxed{EE}, \boxed{+/-}, 4, \boxed{x^2},$$

$$\boxed{\div}, 0, \boxed{\cdot}, 3, \boxed{x^2}, \boxed{=}$$

REVIEW QUESTIONS

1. (a) What is the smallest possible part of a piece of aluminum called?
   (b) Would this also apply to silicon? Why?

2. (a) Can we have an atom of salt? Explain.
   (b) What is the smallest possible amount of salt called?

3. Name the three basic ingredients of all matter.

4. Give the essential characteristic of each of the above ingredients.

5. (a) Which of these three ingredients is (are) found in the nucleus of an atom?
   (b) Name a component found in the nucleus that was not previously mentioned?
   (c) How is this component obtained?

6. What information about atomic structure is obtained from:
   (a) the atomic number of the element?
   (b) the atomic weight?

7. What is meant by the statement that electrons are arranged in "shells" around the nucleus?

8. Helium has an atomic number of 2. How many shells does it have?

9. (a) Except for the first shell, what is the maximum number of electrons that the last shell of an atom can contain?
   (b) What name is given to the electrons in the last shell?

10. (a) In what way (if any) does an electron, or a positron, or a neutron from an oxygen atom differ from the similarly named ingredient of a copper atom?

(b) In what respect does the atom of one element differ from the atom of any other element?

11. What elements (from structure, not by name) are most likely to liberate an electron from their atoms?

12. What is the net charge of a normal atom? Why?

13. (a) Compared to its own size, how far is an electron from its nucleus?

(b) Of what does the major portion of any atom consist?

14. With reference to Fig. 1-2:

(a) What is responsible for the force of attraction?

(b) What causes the centrifugal force?

(c) Why does the electron remain in its orbit?

15. (a) What is energy?

(b) Name five forms of energy.

(c) When electrical energy has been used to light a bulb, is this energy destroyed? Explain.

16. Give an illustration showing how energy goes through at least three transformations.

17. (a) How can a material acquire a negative charge?

(b) How can it acquire a positive charge?

(c) Where is the energy of such charged bodies located?

18. With reference to Fig. 1-3:

(a) What does the circle with the four inscribed little bars sign represent?

(b) What are the radial lines called?

(c) What do they represent?

(d) What do the arrows on these radial lines mean?

(e) Why is the direction inward?

(f) Does this diagram indicate field strength? Explain.

19. Is there a force of attraction, or repulsion, between:

(a) Two neutral bodies?

(b) Two positively charged bodies?

(c) Two negative charges?

20. How is the force of attraction (or repulsion) affected by:

(a) The charge on each body?

(b) The distance between the bodies?

PROBLEMS

1. Gold has an atomic number of 79 and an atomic weight of 197.

(a) How many electrons does it have?

(b) How many protons does it have?

(c) How many *extra* neutrons does it have?

2. Repeat Problem 1 for uranium with an atomic number of 92 and an atomic weight of 238.

3. Argon has an atomic number of 18.
   (a) How many shells does it have?
   (b) What is the number of electrons in each shell?
   (c) Is the last shell filled? Explain.

4. Repeat for zinc with an atomic number of 30.

5. Repeat for silver with an atomic number of 47.

6. Repeat for gold with an atomic number of 79.

7. How many valence electrons does each of the elements in Problems 3 through 6 have?

8. How many coulombs are represented by each of the following quantities of electrons?
   (a) 62.4×10^{18}
   (b) 1.248×10^{19}
   (c) 8.66×10^{16}

9. Draw a diagram showing the energy field:
   (a) Around a negatively charged body.
   (b) Between two positively charged bodies.

10. Calculate the force in newtons between two negative charges of 2.5×10^{-4} and 4.0×10^{-4} C, 600 mm apart.

11. In Problem 10, how many electrons must be removed, or added, to get:
    (a) The 2.5×10^{-4}-C charge?
    (b) The 4.0×10^{-4}-C charge?

12. In Problem 10, at what distance (separation) will the force between the two charges be reduced to 50N?

2

CONDUCTANCE,
RESISTANCE,
AND INSULATION

From Chapter 1, we have seen that every material is composed of atoms. These atoms have a nucleus, and electrons revolving in shells around the nucleus. A force of attraction between the positive nucleus and the negative electrons holds the electrons to their orbits. These forces get weaker on the outer shells. Because of the proximity of the atoms in a material and the elliptical shape of the orbit of some electrons, it happens that an electron in the valence shell may get nearer to the nucleus of an adjoining atom than to its own nucleus. It is atracted more strongly by the second nucleus and it changes its orbit. Such electrons are called *free electrons* (see Fig. 2-1).

2-1 Conductors. Whenever an outside electrical force is applied to a material, it tends to push or pull (or both) the electron out of its orbit; a tug-of-war exists between the nucleus trying to hold the electron and the outside force trying to move the electron. If the holding force of the nucleus is weak, it is easy to make the electrons move through the material. Such a material is called a *conductor*. Another way of looking at this picture is to say that *a good conductor has a large number of free electrons*.

Any pure metal is classed as a good conductor of electricity. However, since the holding force of the nucleus varies one from material to another, some metals are better conductors than others. Based on atomic structures, this means that any element with only one valence electron is therefore a very good conductor: for example, copper, silver, and gold.

Metallic conductors are conveniently grouped into bars, tubes, wires,

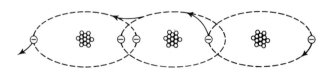

Figure 2-1 Possible path of a free electron.

or sheets. The sheet form is subdivided into sheet, foil, or film, depending on the thickness.

Bus bars are usually solid conductors used to carry very large currents, in the order of hundreds or thousands of amperes. (The ampere, you will learn later, is a unit for measuring current.) They are generally found in large generating plants or substations. When bus bars are used in electronics, they are no larger than about 3 mm square. They are used primarily for self-supporting connections between component parts in transmitting equipment.

Tubes are hollow cylindrical tubing, sometimes oval, usually made of soft copper. At times the tubing is silver-plated for better conductivity. The main use of tubing is again in transmitting equipment, especially for high-frequency circuits.

Sheets are thin plates of material up to 1.5 mm thick. They are made of copper, aluminum, and sometimes of zinc. They are used in shielding materials for radio equipment, as well as for plates in fixed and variable capacitors.

Foils are very thin sheets of tin, lead, aluminum, copper, or other malleable metals. In electronics they are used in the manufacture of paper and mica fixed capacitors and as electrostatic shielding material.

Films of metal are usually deposited on some nonconducting material by electroplating or sputtering. They are used as high-resistance conductors, approaching insulators as a limiting case.

Wires are either single, solid, narrow-diameter rods or several smaller solid wires twisted together. The latter type is called *stranded wire*. It is used where flexibility is required. Where high conductivity is required, wires are made of silver or copper. Where strength as well as good conductivity is required, a copper-clad steel wire is used. Examples of such applications are Signal Corps field-telephone lines and power-transmission lines. Aluminum-clad steel-core wire is also used where weight plays an important factor. When low resistance is required in high frequency circuits (high Q), Litzendraht (German for "covered wire")—commonly called Litz—is used. Litz consists of several strands of wire insulated from each other and connected together only at the ends. Very often the copper wires are tinned to make soldering easier.

Wires are manufactured either bare, as in the form of bus bars, or with various types of coverings, the thickness and type of covering employed depending on the conditions under which the wires are to be used. Some of the common coverings or insulations for wires, and their uses or special properties are as follows:

1. *Enamel:* especially used for winding coils, and for antenna installations
2. *Cotton:* single or double layers: general hookup wire
3. *Silk:* single or double layers: coil winding
4. *Cambric:* high-voltage circuits
5. *Rubber:* general-service cables or outdoor leads
6. *Asbestos:* high-temperature circuits
7. *Cellulose acetate:* flame resistance
8. *Spun glass:* space-saving, moistureproof
9. *Plastic:* high-voltage, moistureproof, space-saving

Insulation on wires often consists of combinations of the above coverings. For hookup wire, the covering is sometimes made so that it can be readily pushed back for convenience in soldering, thus eliminating the need for stripping the wire. This type of wire is called *pushback.* It can be obtained either stranded or solid. Where conductors are to be protected from external magnetic disturbances, they are covered with a metallic braid, in addition to the insulation. Such wires are called *shielded wires.*

Where a number of wires are to be run parallel for certain lengths, they are grouped into a common covering, forming a *cable.*

2-2 Resistors.   When the holding force of a nucleus for its electrons is stronger than in a conductor, it is somewhat more difficult to make the electrons move out of their orbits. Such a material will have relatively few free electrons and is suitable for the manufacturing of resistors. In electronic circuits, resistors are often used deliberately, to reduce the flow of current in the circuit. This action of a resistor is discussed in Chapter 6. Carbon is one of the most widely used resistance materials. It is often desirable to have resistance material in wire form. Since pure metals have a relatively large number of free electrons, they are not suitable for the manufacture of resistance wire. However, it has been found that when metals are alloyed, the atomic structure of each of the component metals is affected. Incomplete electron shells tend to become more nearly completed, so that the number of free electrons is reduced. An alloy is therefore a much poorer conductor than any of the pure metals from which it is made. Examples of resistance wire are constantan, German silver, cast iron, manganin, and nichrome.

2-3 Insulators.   When the holding force of a nucleus for its electrons is very strong, it is almost impossible to make an electron move out of its orbit. Such a material will have practically no free electrons and is suitable for insulation purposes, or electrical isolation of components. Examples of insulating materials commonly used are mica, glass, porcelain, paraffined paper, bakelite, air, oil, and the wire-covering materials mentioned above.

Mica, paraffined paper, or oil is used in the manufacture of fixed capac-itors. Bakelite, steatite, isolantite, and other phenolics are used for coil forms and mounting strips. Porcelain is used for bushings on high-voltage transformers and chokes. Paraffined paper is also used between layers of choke or trans-former windings (see Fig. 2-2).

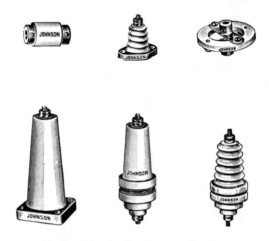

(a) Ceramics for insulators and sockets.
Courtesy E.F. Johnson Co.

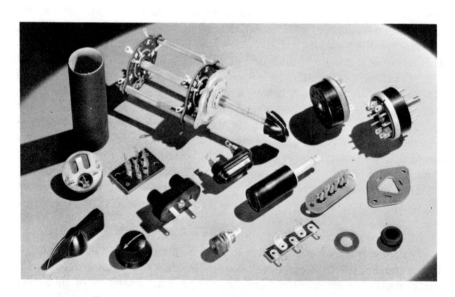

(b) Bakelite and fiber components.
Courtesy P.R. Mallory & Co.

Figure 2-2 Insulation in electronic equipment.

(c) Porcelain mountings for capacitors and switches.
Courtesy Gen Rad Co., P.R. Mallory & Co.

(d) Paper and mica as capacitor dielectrics.
Courtesy Cornell-Dubilier Electronics.

Figure 2-2 (Continued)

2-4 Semiconductors. In Chapter 1 we discussed atomic structure with regard to an isolated atom. When many atoms are combined, other aspects must be considered. For example, in any solid inorganic material, groups of atoms are oriented in a definite orderly manner to form a *crystal* structure. This identical crystal structure is repeated throughout the material. A type of crystal structure of particular importance occurs when *covalent* forces bind the atoms together. This can be seen in materials with four valence electrons, such as silicon and germanium. Each of the four valence electrons of any one atom forms a covalent bond with an electron from one of four adjacent electrons. This sharing of electrons "fills" the last shell, so that such materials have no free electrons.

Figure 2-3 shows this type of structure. In this diagram the circle represents the nucleus *plus all the electrons in the inner shells*. Four valence electrons are shown around each circle. The dashed lines between the circles represent the covalent pair bonds between the atoms. Although this diagram is for a germanium crystal, it can just as well represent the crystal structure of carbon or silicon, since each of these also has four valence electrons. Unfortunately, this diagram conveys a two-dimensional effect, whereas the true crystal structure is three-dimensional, with some of the atoms coming up from the page, and others down.

Based on this crystal lattice structure, with covalent pair bonds, a pure semiconductor should have no conductivity whatever, and therefore theoretically, it should be an insulator. However, this is true only at very low temperatures approaching absolute zero (0°K or −273°C). At room temperature (25°C), there is enough heat energy to break some covalent bonds, creating a few free electrons.* Consequently, these materials are neither good conductors nor good insulators. Electrically, they are classified as **semiconductors**. Germanium and silicon are not the only semiconductor materials. Other materials in this class include elements such as selenium, sulfur, cesium, boron, and most oxides and carbides.

Could semiconductors be classified with resistance materials? Definitely not! To illustrate this point, let us compare the conducting qualities of the semiconductor germanium, the resistance material nichrome (one of the highest-resistance materials in use), and the conductor copper. As a conductor, copper is 60 times better than nichrome, and 600 000 times better than germanium. In other words, with regard to the broad classifications of conductors, semi-

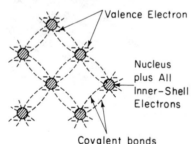

Figure 2-3 Pure germanium, crystal lattice structure.

conductors, and insulators, resistance materials fall under the category of conductors.

2-5 Superconductors. The property called superconductivity was first noted in 1911. It was found that when certain elements are cooled to extremely low temperature all apparent resistance to the flow of current suddenly

*This also creates *holes* or positive current carriers. A more complete discussion is left to texts on semiconductors or transistors. See J. J. DeFrance, *General Electronics Circuits* (2nd ed.), Holt, Rinehart and Winston, New York, 1976, Chap. 1.

disappears. In experimental tests, currents started in a superconducting loop have continued to flow undiminished, for several years—after the external power source is removed. Table 2-1 lists several superconducting elements, and

TABLE 2-1   Superconducting Elements

Element	Temp (°K)	Element	Temp (°K)
Niobium	9.17	Mercury	4.15
Lead	7.23	Tin	3.72
Vanadium	5.03	Indium	3.40
Tantalum	4.39	Aluminum	1.17

the temperature at which they become superconductors. It was further found that combinations of certain elements exhibited this super-conducting quality at higher temperatures. For example, niobium–zirconium and niobium–tin exhibit this effect at around 18°K, while niobium–aluminum–germanium and vanadium–gallium become superconductors at temperatures as high as 20 to 40°K. Although this increase in temperature may seem slight, it does effect a considerable saving in the refrigeration power required to produce these *cryogenic** temperatures. Liquid helium is generally used as the refrigerant, because at these low temperatures it is the only material that does not solidify.

Superconducting wires have played an important role in the production of the very powerful electromagnetic fields required to contain the plasmas of very hot gases used in magnetohydrodynamic (MHD) generating systems, and as shields for the protection of spacecraft from radiation belts in the solar system. Superconducting *thin films* are also being used in switching and storage devices for computer applications.

2-6   Insulation breakdown.

We learned that in an insulator the nucleus holds on very strongly to its electrons. If a small force is applied to such a material in order to push or pull the electron out of its orbit, the electron will not be torn out of its atom. However, the attraction of the nucleus is constant, whereas the outside force can be increased. No matter how good an insulating material may be, it is always possible to increase the outside force to the point where its effect is stronger than the nuclear attraction. Under such circumstances the electrons move out of their orbits; that is, the insulator is now conducting. However, the forces involved are so great that this action is accompanied by mechanical breakage.

Dry wood is normally an insulator. But what happens when lightning strikes a tree? The electrons move—and the tree splits. A fixed capacitor consists of a piece of mica (an insulator) between two plates of metallic conductors.

*Cryogenics is the name given to that branch of science that relates to the production and effects of very low temperatures.

If the force applied to the two plates is strong enough, the mica becomes a conductor. Where conduction takes place, a hole appears in the mica.

Table 2-2 shows the breakdown strength (*dielectric strength*) of some commonly used materials.

TABLE 2-2   Dielectric Strength
(kilovolts per millimeters)[a]

Material	Strength
Air	3.0
Asbestos	10.8
Bakelite	10.0–28.0
Ceramic	8.0
Fiber (commercial)	6.0
Glass	20.0–60.0
Isolantite	12.6
Mica	50.0–225.0
Paper (kraft, dry)	30.0–40.0
Phenolics	19.0–27.0
Porcelain (wet process)	5.7
Rubber	16.0
Varnished cambric	32.0
Vinyl (plastic)	15.8

[a]One millimeter is about the thickness of a paper clip (25.4 mm = 1 in).

REVIEW QUESTIONS

1. With reference to Fig. 2-1:
   (a) To which shell of their atom are the electrons shown normally attached?
   (b) Why may such electrons be attracted to adjacent atoms?
   (c) What are such electrons called?
2. On the basis of the electron theory, explain what constitutes a good conductor.
3. Name five good conductors.
4. Name five types of conductors.
5. On the basis of the electron theory, explain what is meant by a resistor.
6. Why are resistor materials used in electrical circuits?
7. Name five types of resistor materials.
8. On the basis of the electron theory, explain what is meant by an insulator.
9. Name five insulating materials and explain where they are used.
10. With reference to Fig. 2-3:
    (a) What do the circles represent?
    (b) What do the – signs around each circle represent?
    (c) How many valence electrons does each of these atoms have?

(d) What do the dashed arc lines between the circles represent?

(e) How many free electrons are shown in this structure?

(f) At room temperature will there be any free electrons? Explain.

(g) What is this type of material called?

11. Name five elements that may be classified as semiconductors.

12. How does a resistance material compare with a semiconductor with regard to conducting property?

13. **(a)** Can an insulator be made to conduct? Explain.

    (b) What is this effect called?

    (c) Is this a desirable effect? Explain.

14. **(a)** Explain what is meant by dielectric strength.

    (b) Name a material with high dielectric strength.

3

CALCULATION
OF CONDUCTANCE
AND RESISTANCE

In comparing materials to find which are the better conductors, it is not a fair test unless wires of the same cross-sectional areas are used. It is possible for a material of greater cross section to have more free electrons than one of smaller cross section, since the larger contains more atoms. It will therefore seem to be a better conducting material. Before comparing conductances we should compare areas.

3-1 Wire sizes—diameter. Since most commonly used solid wires are round in cross section, the easiest comparison of wire sizes is from a measure of their diameters. Theoretically, wires could be made of any diameter, and so an infinite number of wire sizes could be available. But such helter-skelter production would be highly impractical. Instead, wires are manufactured in selected "standard sizes" only. The first nationally recognized U.S. standard for wire sizes, the *American Wire Gauge (AWG)*, was published in 1912 by the National Bureau of Standards as Circular 31. In the original version, the series started with a wire diameter of 0.005 in, and each successive wire diameter increased by the factor 1.123, so that *for every third wire size larger, the area was doubled.* The prevailing generally used U.S. standard today is an extended range of the original AWG series.

It should be obvious that the AWG sizes are not metric (SI) values. Metric standards for solid conductors are given in the *International Electrotechnical Commission (IEC)* publication IEC 182. It uses a similar system for developing a standard-sizes series, but the "starting" size is 0.100 mm diameter (compared to the U.S. 0.005 in, which is 0.127 mm), and the multiplying factor is 1.122. Therefore, the "selected standard sizes" are not the same in the two systems. This has created problems. Industry in the United States is deeply entrenched in the AWG system, and there is disagreement on how conversion to SI units should be done. Some factions want to keep manufacturing the present wire sizes and connectors, keeping the same AWG identifying numbers but giving diameters and areas in millimeters and square millimeters, respectively. This approach is called "soft metrication." Obviously, this is the cheapest technique. Other factions see "hard metrication"—a full change to the standard SI sizes—as the only final solution, and ask why not change once and for all. Still a third group wants some in-between system. It may be years before this trichotomy is resolved. Meanwhile, if you want to place an order with an electrical supply house for wires and cables, it will have to be in AWG specifications. So in this chapter we must maintain the AWG nomenclature, even as we introduce some SI units.

As mentioned above, the original starting point for wire sizes was a wire of 0.005-in diameter. Since then, even smaller-diameter wires are made—down to about 0.001 in. Rather than work with decimal numbers, a new dimensional unit was introduced—*the mil*—wherein 1 mil = 0.001 in (or 1 in is equal to 1000 mils). The diameter of wires is therefore measured in mils. To change inches to mils, multiply by 1000 (move the decimal point three places to the right). In changing mils back to inches, we would do the opposite (move the decimal point three places to the left).

Example 3-1

(a) 0.012 in = 12 mils   (b) 0.07 in = 70 mils
(c) 0.0035 in = 3.5 mils   (d) 48 mils = 0.048 in

3-2 Wire sizes—area.   Finding the area of a round wire involves finding the area of a circle. You are probably familiar with this formula:

$$\text{area} = \pi r^2$$

where r is the radius.

In terms of diameter,

$$\text{area} = \frac{\pi D^2}{4} \tag{3-1}$$

where D is the diameter. Again we are faced with a laborious calculation. Luckily, we do not need the absolute area of wires. We are interested primarily in the comparison of their areas. This simplifies the calculation.

Let us consider two circles, I and II, of dimaeters D_1 and D_2, respectively:

$$\text{area I} = \frac{\pi D_1^2}{4} \quad \text{and} \quad \text{area II} = \frac{\pi D_2^2}{4}$$

The ratio of areas is

$$\text{area I} \div \text{area II} = \frac{\pi D_1^2}{4} \div \frac{\pi D_2^2}{4}$$

or

$$\text{ratio of areas} = \frac{\pi D_1^2}{4} \times \frac{4}{\pi D_2^2} = \frac{D_1^2}{D_2^2}$$

This relationship means that *to compare areas we merely compare the squares of their diameters!* But diameter squared is not the usual measure of area comparable to square inches or square mils. Therefore, a new name or unit must be given to this quantity: *The area of wires is measured* in **circular mils (cmil).**

To find the area of a wire:

1. Measure the diameter in inches and change this to mils.

2. Square the diameter in mils; *the answer is the area in circular mils.*

Expressed mathematically, this is

$$\text{circular-mil area} = (\text{mils})^2 \tag{3-2}$$

Example 3-2

The diameter of a wire is 0.064 08 in. Find its area in circular mils.

Solution

1.   0.064 08 in $=$ 64.08 mils

2.   Area $= (\text{mils})^2 = (64.08)^2 = 4107$ cmils

3-3 Gauge numbers—wire table. It would be rather cumbersome to specify wires by their diameters or by their areas. Instead, in the AWG system, a number is assigned to each of the selected standard sizes. This numbering is shown in Table 3-1. For example, the wire in Example 3-2 is a No. 14 wire. For wire sizes larger than 0000 (4/0), their area in thousands of circular mils (MCM) is used in place of a gauge number. Such wires are available starting with 250 MCM, and increasing in steps of 50 MCM.

To facilitate the measuring of wire sizes, wire gauges are available (see Fig. 3-1). The wire to be tested is tried in the various slots without forcing. The slots are numbered corresponding to gauge numbers. The size of the wire corresponds to the number of the slot in which it just fits. The circular mil area can then be found from a wire table.

Often it happens that a wire table is not available. A simple approximation is always at your fingertips. Look up the area of a No. 10. You will find that it is 10 380 cmil. As an approximation it is simple to remember that a No. 10 $=$ 10 000 cmil. (This is less than a 4% error. When you consider that commerical resistors are often inaccurate by 5 to 10%, this approximation is permissible.) Now notice the area of a No. 7 and also of a No. 4. You will find that in each case the area is double the previous value. The following rule can be made: For every three numbers below No. 10, double the area; for every three numbers above No. 10, halve the area. For values in between, interpolate.

TABLE 3-1 American Wire Gauge

Gauge No.	Diameter (in)	Area (cmil)	Gauge No.	Diameter (in)	Area (cmil)
0000	0.460 0	211 600	22	0.025 35	642.4
000	0.409 6	167 800	23	0.022 57	509.5
00	0.364 8	133 100	24	0.020 10	404.0
0	0.324 9	105 500	25	0.017 90	320.4
1	0.289 3	83 690	26	0.015 94	254.1
2	0.257 6	66 370	27	0.014 20	201.5
3	0.229 4	52 630	28	0.012 64	159.8
4	0.204 3	41 740	29	0.011 26	126.7
5	0.181 9	33 100	30	0.010 03	100.5
6	0.162 0	26 250	31	0.008 928	79.70
7	0.144 3	20 820	32	0.007 950	63.21
8	0.128 5	16 510	33	0.007 080	50.13
9	0.114 4	13 090	34	0.006 305	39.75
10	0.101 9	10 380	35	0.005 615	31.52
11	0.090 74	8 234	36	0.005 000	25.00
12	0.080 81	6 530	37	0.004 453	19.83
13	0.071 96	5 178	38	0.003 965	15.72
14	0.064 08	4 107	39	0.003 531	12.47
15	0.057 07	3 257	40	0.003 145	9.888
16	0.050 82	2 583	41	0.002 75	7.5625
17	0.045 26	2 048	42	0.002 50	6.2500
18	0.040 30	1 624	43	0.002 25	5.0625
19	0.035 89	1 288	44	0.002 00	4.0000
20	0.031 96	1 022	45	0.001 75	3.0625
21	0.028 46	810.1	46	0.001 50	2.2500

Figure 3-1 American Wire Gauge.

Example 3-3

Without a wire table, find the areas of a No. 4, a No. 5, and a No. 12 wire.

Solution

1. A No. 10 wire $\simeq$ 10 000 cmils. Doubling every three gauge numbers, we get

$$\text{No. } 7 \simeq 20\ 000 \quad \text{and} \quad \text{No. } 4 \simeq 40\ 000 \text{ cmils}$$

2. A No. 5 is smaller than a No. 4 by approximately one-third of the difference between a No. 4 and a No. 7, or

$$40\ 000 - 1/3(20\ 000) \simeq 33\ 300 \text{ cmils}$$

3. For the No. 12 wire, a No. 13 wire $\simeq$ 5000 cmils. A No. 12 is larger than a No. 13, by approximately one-third the difference between the No. 10 and the No. 13, or

$$5000 + 1/3(5000) \simeq 6600 \text{ cmils}$$

3-4 SI wire sizes.   In SI units, the diameter of wires is measured in millimeters (mm), and their true area is calculated, using the full mathematical equation (3-1), for the area of a circle. This is shown in Example 3-4.

Example 3-4

Find the area of a wire 0.710 mm in diameter.

Solution

$$\text{Area} = \frac{\pi D^2}{4} = \frac{\pi(0.710)^2}{4} = 0.396 \text{ mm}^2$$

A table for metric-size conductors is shown in Table 3-2.

3-5 Specific resistance (ρ, rho).   We know that a good conductor has many free electrons; therefore, it offers very little resistance to any external electrical force that tries to make its electrons move through the material. On the other hand, a good *resistance* material is a poor conductor. It has relatively few free electrons, so that much greater difficulty is encountered in making electrons move through the material. Often it is necessary to evaluate these qualities (resistance or conductance) for various materials. A unit for measurement must be used. The unit name for resistance measurements is the *ohms*,* and the unit now used for conductance is the *siemen*.†

When comparing materials to find which are better conductors, two factors must be considered and/or "equalized":

1. *The Areas.* A larger area would contain more free electrons, and would seem to be a better conductor.

*Named after the German scientist Georg Simon Ohm (1787–1854).

†Recently so named, in honor of Ernst Werner von Siemen, a German scientist (1816–1892). Formerly, the unit for conductance was the *mho* (ohm spelled backwards—since conductance is the opposite of ohm).

2. *The Lengths*. It is harder to push or pull electrons through a long wire than through a short one.

TABLE 3-2   **Standard Nominal Diameters and Cross-Sectional Areas of Metric Sizes of Solid Round Wires at 20°C**

Diameter (mm) (R 20 Series)	Area (mm²)	Diameter (mm)	Area (mm²)
18.000	254.5	0.500	0.196
16.000	201.1	0.450	0.159
14.000	153.9	0.400	0.126
		0.355	0.099 0
12.500	122.7		
11.200	98.52	0.315	0.077 9
10.000	78.54	0.280	0.061 6
9.000	63.62	0.250	0.049 1
		0.224	0.039 4
8.000	50.27		
7.100	39.59	0.200	0.031 4
6.300	31.17	0.180	0.025 4
5.600	24.63	0.160	0.020 1
		0.140	0.015 4
5.000	19.63		
4.500	15.90	0.125	0.012 3
4.000	12.57	0.112	0.009 85
3.500	9.898	0.100	0.007 85
		0.090	0.006 36
3.150	7.793		
2.800	6.158	0.080	0.005 03
2.500	4.909	0.071	0.003 96
2.240	3.941	0.063	0.003 12
		0.056	0.002 46
2.000	3.142		
1.800	2.545	0.050	0.001 96
1.600	2.011	0.045 0	0.001 59
1.400	1.539	0.040 0	0.001 26
		0.035 5[B]	0.000 990
1.250	1.227		
1.120	0.985	0.031 5[B]	0.000 779
1.000	0.785	0.028 0	0.000 616
0.900	0.636	0.025 0	0.000 491
		0.022 4	0.000 394
0.800	0.503		
0.710	0.396	0.020 0	0.000 314
0.630	0.312	0.018 0	0.000 254
0.560	0.246	0.016 0	0.000 201
		0.014 0	0.000 154
		0.012 5	0.000 123

Obviously, for a fair comparison of materials, a standard length and a standard cross-sectional area must be used.

Using English units, the standard-size conductor is a piece of wire 1 ft long, and 1 cmil in area. This size wire is called a *circular mil-foot (cmil-ft)*. If we take a standard-size sample of any material and measure its resistance, this resistance will depend only on the type (atomic structure) of the material. This is known as the specific resistance or resistivity of that material. In other words, *the specific resistance of any material is the resistance of a 1 circular mil-foot sample of that material.* A table of resistivity for various materials is shown in Table 3-3.

TABLE 3-3   Specific Resistance (ρ) of Common Materials (at 20°C)

	Ohms per cmil-ft	Ohm-Meters $\times 10^{-8}$
Advance	294	48.9
Aluminum	17	2.83
Brass	42	6.98
Carbon	22 000	3660.
Constantan	294	48.9
Copper (annealed)	10.4	1.73
German silver (18%)	198	32.9
Gold	14.7	2.44
IaIa	300	49.9
Iron (pure, annealed)	61	10.1
Manganin	264	43.9
Nichrome	675	112.
Nickel	60	9.97
Platinum	60	9.97
Silver	9.56	1.59
Tungsten (drawn)	34	5.65

In SI units, the unit for length is the meter (m), and the unit for area is the square meter (m²). This makes the standard-size sample for resistance evaluation, a 1-m cube. Because of its large size, the resistance of this SI "sample" is at times also called *volume resistivity*. The SI values of resistivity are also shown in Table 3-3.

3-6 Calculation of conductance.   *Conductance* (*G*) is the ease with which the electrons can be made to move through a material. The unit for measuring conductance is the siemen (S).

As the area of a wire is increased, we have more atoms in the wire and therefore more free electrons. Thus:

1. *Conductance increases if the area (A) is increased.*

As the length of the wire increases, the average distance between the electrons and the outside force also increases. The effect of the force on the "average" electron will decrease, and it is harder to make the electrons move. (This effect can also be explained by an analogy. Have you ever pulled wires through a conduit? It is obviously harder when the conduit is long.) Thus:

2. *Conductance decreases as the length (l) is increased.*

     If the material used for a conductor has a high specific resistance, it is relatively hard to make the electrons move. Thus:

3. *Conductance decreases as the specific resistance (ρ) is increased.*

     You may have noticed that area and conductance form a direct proportion, whereas the other two factors are in inverse proportion (opposite variations). Summarizing the three statements in a formula, we have

$$G = \frac{A}{\rho \times l} \qquad\qquad (3\text{-}3)$$

where *l* is the length in feet (or meters), and *A* the area in cmils (or square meters).

Example 3-5

Find the conductance of 100 ft of No. 16 copper wire ($\rho = 10.4$; no wire table available).

Solution

1.
$$\begin{aligned}
\text{No. 10 wire} &= 10\,000 \text{ cmils} \\
\text{No. 13 wire} &= 5000 \text{ cmils} \\
\text{No. 16 wire} &= 2500 \text{ cmils}
\end{aligned}$$

2.
$$G = \frac{A}{\rho \times l} = \frac{2500}{10.4 \times 100} = 2.40 \text{ S}$$

Example 3-6

Find the conductance of 200 ft of 0.08-in-diameter aluminum wire.

Solution

1. $\qquad\qquad 0.08 \text{ in} = 80 \text{ mils}$

2. $\qquad\qquad \text{Area} = (\text{mils})^2 = 80 \times 80 = 6400 \text{ cmils}$

3. $\qquad\qquad G = \dfrac{A}{\rho \times l} = \dfrac{6400}{17 \times 200} = 1.88 \text{ S}$

Example 3-7

Find the conductance of 75 m of copper wire, with a diameter of 1.25 mm.

Solution

1. To find the area of the wire in square meters:

    (a) $\qquad\qquad 1.25 \text{ mm} = 1.25 \times 10^{-3} \text{ m}$

    (b) $\qquad A = \dfrac{\pi D^2}{4} = \dfrac{\pi (1.25 \times 10^{-3})^2}{4} = 1.23 \times 10^{-6} \text{ m}^2$

2. $$G = \frac{A}{\rho \times l} = \frac{1.23 \times 10^{-6}}{1.73 \times 10^{-8} \times 75} = 0.946 \text{ S}$$

3-7　Parallel conductance. Suppose that it is necessary to run a wire from point A to point B, 100 ft distant. We thoughtlessly buy a wire and finish the job, only to find that the conductance of this wire is not enough. What can we do to remedy the situation?

1. We could rip out the wire and put in one of better material. But what if we are already using the best material?
2. We could replace it with a wire of larger area—and lose the money already invested in the original wire.
3. We could run another wire alongside the original one and get the same effect as using a larger wire by connecting the wires together at each end.

This last method, connecting the wires in parallel, is the same as increasing the area. The conductance also increases. Thus; the total conductance of a parallel circuit is the sum of the individual conductances:

$$G_T = G_1 + G_2 + G_3 + \cdots \tag{3-4}$$

3-8　Calculation of resistance. A resistor is a material in which the nuclei hold on to their electrons more strongly than in a conductor.

Resistance (R) is therefore a result of the holding force any material has to its electrons. It tends to limit or oppose the flow of electrons through a material. Since it is the opposite of conductance, we can say that

$$R = \frac{1}{G} \quad \text{or} \quad G = \frac{1}{R} \tag{3-5}$$

Furthermore:

1. Resistance decreases as the area is increased.
2. Resistance increases as the length is increased.
3. Resistance increases if the material used has higher specific resistance:

$$R = \frac{\rho \times l}{A} \tag{3-6}$$

Example 3-8

Find the resistance of 400 ft of manganin wire 0.025 in in diameter.

Solution

1. $\quad\quad 0.025 \text{ in} = 25 \text{ mils}$
2. $\quad\quad \text{Area} = (\text{mils})^2 = 25 \times 25 = 625 \text{ cmils}$
3. $\quad\quad R = \frac{\rho \times l}{A} = \frac{264 \times 400}{625} = 169 \ \Omega$

Example 3-9

Find the resistance of 80 m of tungsten wire, 0.8 mm in diameter.

Solution

1. The area of the wire, in square meters, is:

$$0.8 \text{ mm} = 0.8 \times 10^{-3} \text{ m}$$

$$A = \frac{\pi D^2}{4} = \frac{\pi (0.8 \times 10^{-3})^2}{4} = 5.03 \times 10^{-7} \text{ m}^2$$

2.
$$R = \frac{\rho \times l}{A} = \frac{5.65 \times 10^{-8} \times 80}{5.03 \times 10^{-7}}$$

$$= 8.99 \ \Omega$$

REVIEW QUESTIONS

1. What is the listing of standard wire sizes used in the United States called?
2. (a) What unit is used when measuring the diameter of wires in the U.S. system?
   (b) What is the relationship between this unit and the inch?
3. (a) What unit is used to measure the "area" of wires in the U.S. system?
   (b) Knowing the diameter of a circular wire in this system, how do we calculate its "area"?
   (c) Is this a true area? Explain.
   (d) Why can this technique be used without error?
4. With reference to the American Wire Gauge:
   (a) Which wire is thicker, a No. 5 or a No. 20?
   (b) Compare the area of a No. 8 to a No. 11; and to a No. 5.
   (c) Are wires available of diameter larger than 4/0? If so, give an example.
5. In the SI standard-wire system:
   (a) How is the diameter specified?
   (b) Knowing the diameter, how is the area calculated?
6. In the U.S. system:
   (a) What dimension (length and area) of wire is used when stating the specific resistance of a material?
   (b) Give another name for specific resistance.
   (c) What letter symbol is used to represent this quantity?
7. The specific resistance of nickel is given as 60 Ω/cmil-ft? What happens to its specific resistance if:
   (a) The length of the wire is doubled?
   (b) The area of the wire is doubled?
8. On what does the value of specific resistance depend?
9. (a) In the SI system, what dimension (length and area) is used for stating the specific resistance of a material?
   (b) Give another name for this resistance, based on the size of the "sample."
10. Give the name of the unit and the letter symbol used to represent conductance.

11. In evaluating the conductance of a wire, what is the effect of:
    (a) The length of the wire?
    (b) The area of the wire?
    (c) The specific resistance of the wire?
    (d) Connecting two wires in parallel?
12. Give the name of the unit and the letter symbol used to represent resistance.
13. How is the resistance of a wire affected by:
    (a) Tripling its length?
    (b) Doubling its area?
    (c) Changing from copper to nickel?

PROBLEMS

1. Convert the following from inches into mils.
    (a) 0.036   (b) 0.025   (c) 0.08
    (d) 0.18   (e) 0.0063   (f) 0.0226
2. Find the circular mil area of the wires whose diameters are given in Problem 1.
3. What gauge numbers correspond (approximately) to the wires in Problem 2?
4. Find the circular mil area of a wire 0.057 in (in diameter).
5. Repeat Problem 4 for a wire 19.8 mils (in diameter).
6. Repeat Problem 4 for a wire $1\frac{1}{8}$ in (in diameter).
7. Find the diameter of a wire whose area is 133 000 cmils. Express the answer in mils and inches.
8. Repeat Problem 7 for a wire area of 1620 cmils.
9. What is the approximate AWG number for a wire 25.2 mils (in diameter)?
10. Repeat Problem 9 for a wire 0.28 in (in diameter).
11. Find the area of each of the following wires, in SI units.
    (a) 3.15 mm in diameter
    (b) 1.40 mm in diameter
    (c) 0.56 mm in diameter
12. Find the conductance of 50 ft of annealed copper wire. Its diameter is 0.032 49 in.
13. Find the conductance of 125 ft of No. 10 annealed copper wire.
14. Find the conductance of 50 m of annealed copper wire 0.56 mm in diameter.
15. Find the conductance of 10 m of aluminum wire 0.80 mm in diameter.
16. Find the total conductance of the following wires connected in parallel.
    (a) 50 ft of No. 14 annealed copper
    (b) 50 ft of No. 12 aluminum
    (c) 50 ft of No. 18 manganin
17. Find the resistance of the following wires.
    (a) 80 ft of No. 22 constantan
    (b) 200 ft of No. 36 annealed copper
    (c) 30 ft of manganin 0.032 in in diameter
18. Find the resistance (at 20°C) of 900 ft of copper wire $\frac{1}{3}$ in (in diameter).

19. Forty-five feet of aluminum wire has a resistance of 3.5 Ω (at 20°C). Find its diameter in inches.

20. How many feet of silver wire, 0.028 in (in diameter), will be needed to produce a resistance of 2.50 Ω (at 20°C)?

21. Find the resistance of 8000 ft of tungsten wire, 0.062 in (in diameter) (at 20°C).

22. Find the resistance of 300 m of copper wire 8.0 mm in diameter.

23. Fifteen meters of aluminum wire has a resistance of 3.25 Ω (at 20°C). Find its diameter in millimeters.

24. What length of silver wire (in meters) will be needed to produce a resistance of 2.0 Ω at 20°C? The diameter is 0.50 mm.

25. Find the resistance of 800 m of tungsten wire of 1.40 mm diameter at 20°C.

4

ELECTRICAL UNITS AND THEIR RELATION

4-1 Coulomb. In Chapter 1 we found that a normal body has an equal number of positrons and electrons. It therefore is neutral, or has no charge. We can give a body a positive charge by taking away electrons, or we can give it a negative charge by adding electrons to it. The amount of electrons usually involved in practical applications is tremendous. So instead of talking about the individual electrons, we lump a large quantity together and call it a unit charge. The name given to this unit charge is the coulomb. A coulomb represents 6.24×10^{18} electrons.

4-2 Potential. Suppose that we want to give a body a negative charge of 1 coulomb. We have to add 6.24×10^{18} electrons to it. Work must be done, or energy is expended in order to do this. Energy is not created or destroyed. The energy used to charge the body appears in a dielectric field around the charged body. The energy in this field (neglecting frictional heat losses) is equal to the work done in putting these electrons into the body. These extra electrons are pushed away by the electrons normally in the body and by each other. A pressure is therefore created due to the number of electrons crowded together, and by the capacity of the body to hold these electrons.

The production of air pressure is similar. If we take a tank 500 liters* in capacity at normal air pressure and add 1 liter more of air, the increase in pressure is very little. The work required to pump this amount of extra air is small, and the energy in the system is small. However, if we start with a much smaller

*A liter is a cubic decimeter (0.001 m³) and has about the same volume as a quart.

tank and add 1 liter of air, tremendous pressures will be created. Much more work will have to be done this time, and the energy in the system will be much greater. You will notice that the transfer of energy and the pressure created depend *not on the quantity of air put into the tank, but on the work done.* Similarly, the energy around a charged body and the electrical pressures created depend on the work done in charging the body.

Potential energy has been defined in mechanics as energy due to position. A pile driver at the top of its lift has potential energy. If released, it will fall down and do work. The pile driver got its energy from the work that was done in lifting it. The electrons on a charged body have energy due to their position. If "released," they will "fall" back to neutral energy level, and in falling they will do work. The extra electrons have potential energy.

In speaking of the potential energy of a charge we merely say "potential." The earth or ground ($\perp$) is considered as a large reservoir of neutral charge or zero potential. No matter how many electrons have been added or removed from it, the work done is negligible and the charge is negligible. Its potetntial remains zero. As an analogy, consider adding a bucket of water to the Atlantic Ocean; you will realize that the sea level does not change. If we add electrons to a body, we have a negative charge or a negative potential. When we take electrons from a body, we have a positive potential.

Since the electrical potential (energy) gained in putting a charge on a body is dependent on the work done, the unit for evaluating electric potential should reflect the amount of work done for each coulomb of charge moved. Basically, then,

$$\text{electrical potential} = \frac{W(\text{work})}{Q(\text{charge})} \text{ or joules* per coulomb} \qquad \textbf{(4-1)}$$

To simplify electrical calculations, the work done to put a charge on a body, and therefore the electron pressure created, is measured in *volts* (V).†
The potential of a charged body is then given as so many volts plus or minus; plus if electrons were removed and minus if electrons were added as compared to the neutral body or zero potential.

4-3 Potential difference—voltage. Usually, we are interested not in the potential energy at a point but in the difference of potential energy between two points. In electrical work you have probably heard the terms "voltage" or "potential difference"; they are used when referring to the difference in energy levels between two charges.

Now to see if you really understand this discussion, let us start with four neutral bodies, *A, B, C,* and *D* (see Fig. 4-1). From body *A* we take away electrons, and the amount of work done builds up a potential of $+10$ V.

---

*In mechanics, work (in joules) is the product of the force (*F*, in newtons) required to move an object a distance *d* in meters.

†Named in honor of the Italian physicist, Count Alessandro Volta (1745–1827).

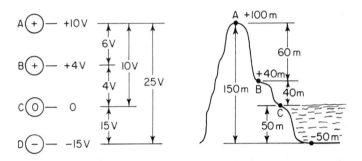

Figure 4-1 Difference of potential.

From body B we also take away electrons, but less work is done. Its potential is $+4$ V. Body C we leave alone. What is its potential? Zero. We add electrons to body D. The work done builds up a potential of -15 V.

What is the voltage between A and B? Voltage means *difference of potential between two points*. The difference between A and B is 6 V. What is the voltage between A and C? Ten volts. What is the potential difference between C and D? Fifteen volts. What is the voltage between A and D? Twenty-five volts: 10 V from $+10$ to 0, and 15 V more from 0 to -15. Notice that the potential of any point is given as plus or minus some value, whereas voltage (potential difference) is just a value without sign.

Now let us consider another body, E. In taking electrons from it we give it a potential of $+10$ V. What is the voltage at point E? The question is improper. How can we give voltage of a point when voltage means the difference of potential *between two points!*

The difference between potential and voltage can be made clearer by this analogy. Point A at the top of the hill has an elevation of $+100$ m above sea level. Point B lower down, has an elevation of $+40$ m. Point C is at sea level or elevation zero. Point D is 50 m below sea level or an elevation of -50 m. A weight placed at point A has potential energy due to the work done in lifting it 100 m. The same weight lower down would have less potential energy because less work is done to put it there. Elevation can be considered similar to potential. Difference of potential or voltage would compare to difference in elevation.

To summarize: voltage (E)* is the difference in potential between two points and is measured in volts, by an instrument called a **voltmeter**. What is a voltmeter? How is it used? For the moment let us merely accept it as an instrument that will measure voltage. The details of its construction and principle of operation will be studied later (see Chapter 16). The important point at this time is how do we connect a voltmeter? A voltmeter has two terminals: one marked positive $(+)$, the other negative $(-)$. Since voltage is the difference in potential between two points, a voltmeter lead should be connected to each of these two

---

*Letter symbols E and V are often used interchangeably. Some texts differentiate, using E for a rise in potential and V for a potential drop.

points—with the positive terminal of the meter to the more positive point and the negative terminal of the meter to the less positive (or more negative) point. Such a connection can be seen in Fig. 4-4, where a voltmeter is used to measure the voltage across the resistor.

4-4   Current (I).   Suppose that we take a piece of copper wire and connect a charged body of -5 V at the top, and a charged body of $+10$ V at the bottom (see Fig. 4-2). What is the voltage across the wire? Fifteen volts. Copper, you will recall, is a good conductor. What do we mean by a good conductor? Electrons can be pulled away from their atoms and can be made to move through the wire. In section A we have some atoms. These atoms contain electrons, which are negative charges. The negative charge of 5 V will repel these electrons. Meanwhile in region B there are more atoms and electrons. The positive charge of 10 V will attract these electrons. The combination of push from the top and pull from the bottom will cause electrons to move from the negative to the plus charge, as indicated. This flow of electrons will continue until the charges are equalized.

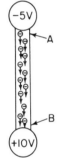

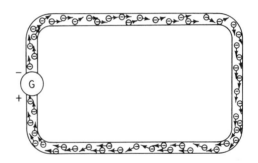

Figure 4-2   Electron flow between two charged bodies.

Figure 4-3   Electron flow in a complete circuit.

Let us apply this principle to a complete circuit (see Fig. 4-3). At G we have some kind of machine that takes electrons from below it and piles them at the top. Since electrons are taken away from the bottom, this point has a positive potential. Since electrons are piled on top, this point has a negative potential. But as soon as this happens, the minus charge pushes electrons away and the plus charge attracts them. The result is a flow of electrons from minus to plus, tending to neutralize the charges. The machine (a generator or battery) maintains the difference in potential, so that we get a continuous flow of electrons, or a *current*.

The word "current" signifies not only that electrons are flowing but also how many are flowing past any given point in the circuit in 1 second. Rather than count the individual electrons, we again use the coulomb. When one coulomb

(6.24 × 10^{18} electrons) flows past any point in a circuit in one second, the current is called one ***ampere*** (A).* If five times that many electrons flow past a given point in one second, the current is 5 A. Although a current of 1 A means that 6.24 × 10^{18} electrons pass a given point in 1 s, it does not signify how fast the electrons are moving. The reference is to the quantity of electrons, not to their speed. Actually, the *forward* velocity of the electrons through a conductor is so slow as to be called a *drift*. For example, in a No. 12 copper wire (often used in house wiring), a current of 1 A causes the electrons to drift through the wire at a velocity just under 0.025 mm/s.

In the above illustrations we have shown that current flow is the motion (or drift) of *electrons* in one general direction. A broader definition is to consider current flow as a direction of "charge" movement. Now the charge could be either negative or positive. Unfortunately, the existence of current was realized before the electron theory was developed. Since the true nature of current was not known, current was assumed to be some mysterious thing that flowed from positive to negative. This convention is in direct contradiction to electron flow, but it is an old established custom. To avoid confusion, this assumed current direction has been called *conventional current*. Many people have been trained and use this old system. Also, in semiconductor device symbols, arrowheads show the direction of conventional current flow. So to maintain this "tradition" it has been internationally agreed to continue the use of conventional current flow.

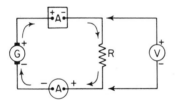

Figure 4-4   Current flow in an electrical circuit.

Now that we know what current means, and the unit in which it is measured, the next point is how do we measure current? The instrument used for this measurement is called an ***ammeter***. The construction and principle of operation of this instrument are postponed to Chapter 16. Its connection into a circuit will be shown here. The ammeter has two terminals, a positive (+) terminal and a negative (−) terminal. If the current to be measured is allowed to flow through the meter, it will indicate the rate of flow of this current. We know that current flows *through* a circuit. Therefore, if we break the circuit and insert the meter at the break, it will measure the circuit current. Two possible connections for an ammeter are shown in Fig. 4-4. Note also that the positive terminal of the meter is connected to the positive side of the break and the negative meter terminal to the negative side of the break.

*Named in honor of the French physicist André Marie Ampère (1775–1836).

4-5   Resistance.   In Chapter 2 we learned that resistance (*R*) is the opposition to the motion of electrons, and is due to the holding force of the nucleus of an atom on its electrons. Resistance depends on the type of material, the cross-sectional area, and the length of the material. Resistance is measured in ohms.

4-6   Ohm's law.   In reading the above, you may have noticed an interrelation between the current (*I*), the voltage (*E*), and the resistance (*R*). Let us analyze this relation more closely. In Fig. 4-4 we see a resistor connected across a generator (*G*) that produces differences of potential. Current (conventional) will flow from the plus to the minus. If the generator is made to work faster, a bigger difference of potential will be produced. The push and pull of the two charges will be greater. As a result more electrons will flow. On the other hand, if the resistance used in the circuit is increased, fewer electrons will be made to move. But the number of electrons moving through is a measure of current; that is:

1. Current *increases* if the voltage is increased.

2. Current *decreases* if the resistance is increased.

This relation can be expressed by formula as

$$I = \frac{E}{R} \qquad \text{(4-2A)}$$

This formula is useful in finding how much current flows in a circuit if the voltage across a circuit and the resistance of the circuit are known.

Example 4-1

A 60-Ω iron is connected across 120-V supply. How much current will it draw?

Solution

$$I = \frac{E}{R} = \frac{120}{60} = 2.0 \text{ A}$$

By calculator we would enter

$$1, 2, 0, \boxed{\div}, 6, 0, \boxed{=}$$

Sometimes it happens that the voltage available is known, and it is desired to find how much resistance is needed to limit the current to some predetermined value. By algebraic transposition,

$$R = \frac{E}{I} \qquad \text{(4-2B)}$$

Example 4-2

The desired bias voltage for a tube is 60 V. What value of resistance should be used if the current in the circuit is 0.08 A?

Solution

$$R = \frac{E}{I} = \frac{60}{0.08} = 750 \ \Omega$$

Calculator entry:

6, 0, $\boxed{\div}$, $\boxed{\cdot}$, 0, 8, $\boxed{=}$

At other times we know the resistance of a piece of equipment, and we must find out how much voltage must be connected across this equipment to get a desired current. In such cases we rewrite the formula as

$$E = I \times R \tag{4-2C}$$

Example 4-3

An 80-Ω relay requires a current of 1.5 A for proper operation. What should the supply voltage be?

Solution

$$E = I \times R = 1.5 \times 80 = 120 \ \text{V}$$

Calculator entry:

1, $\boxed{\cdot}$, 5, $\boxed{\times}$, 8, 0, $\boxed{=}$

4-7 Multiples of electrical units. In measurement of weight using English units, the basic unit is the pound. However, would you measure the weight of a battleship in pounds? No, because the answer would be too unwieldy. You would use a larger unit of weight, the ton. To measure small weight we would use ounces. The ounce, being a *smaller* unit of weight, will give a *larger* numerical answer, avoiding small fractions or decimals of a pound.

The volt, ampere, and ohms are our basic units. Sometimes the quantities involved are too large or too small to be expressed in these basic units. As in the case of measuring weight, new units would give more convenient numbers. However, instead of changing the name of the unit completely, we put a prefix in front of the basic unit. The prefixes used are given in Table 4-1.

The table shows the relation between a prefixed unit and the basic unit. Notice that the largest unit prefix, tera (at the top of the listing), is 10^{12} or 1 million-million time larger than the basic unit. For example, a kilovolt is 10^3 or 1000 times larger than the volt. Conversely, all prefixes with *negative exponents* create units that are *smaller* than the basic unit. A *micro*volt would be 10^{-6} or 1 million times smaller than the volt.

Notice also that the relation between any two adjacent prefixes is 10^3 or 1000 to 1. These are the preferred SI prefixes. To complete the list we should add: hecto (10^2), deca (10^1), deci (10^{-1}), and centi (10^{-2}). Of these four, only the prefix *centi* ($\frac{1}{100}$th or 10^{-2}) has found popular acceptance—and then only in measurements of length [i.e., the *centimeter* (0.01 m)]. Since the use of these additional prefixes is not encouraged in engineering measurements, they have been omitted from Table 4-1.

TABLE 4-1   International System Prefixes

Prefix and Action on Basic Unit		Pronun-ciation	Relation to Basic Unit	Symbol
Divide ↑	tera	ter′a	10^{12} or million-million	T
	giga	jĭ′ga	10^{9} or thousand-million	G
	mega	mĕg′a	10^{6} or million	M
	kilo	kĭl′ō	10^{3} or thousand	k
-Basic Unit ------				
	milli	mĭl′ĭ	10^{-3} or $\dfrac{1}{\text{thousandth}}$	m
	micro	mī′krō	10^{-6} or $\dfrac{1}{\text{millionth}}$	μ
Multiply ↓	nano	năn′ō	10^{-9} or $\dfrac{1}{\text{thousand-millionth}}$	n
	pico	pē′kō	10^{-12} or $\dfrac{1}{\text{million-millionth}}$	p
	femto	fĕm′tō	10^{-15} or $\dfrac{1}{\text{million-billionth}}$	f
	atto	ăt′tō	10^{-18} or $\dfrac{1}{\text{billion-billionth}}$	a

When converting from one multiple to another, the question that sometimes arises is: "Do I multiply or divide by the power-of-10 factor?" To this I say: Use common sense. When changing from a larger to a smaller unit, the final numerical answer must be larger. (For example, recall in English units that 2 tons changed to pounds becomes 4000 pounds.) Conversely, when changing from a smaller unit to a larger unit, the answer must be numerically smaller. (Again, using English units, recall that 32 ounces is only 2 pounds.) If at any time you are confused, think of the problem in terms of tons, pounds, and ounces—and remember: The *smaller the unit, the larger the answer; the larger the unit, the smaller the answer.*

Here is a crutch that can help you in these conversions. Use the "action arrows" in column one of Table 4-1. and multiply or divide as indicated. In each case (whether you multiply or divide) use the difference between exponents as the factor.

Example 4-4

(a) To change 0.05 V to millivolts:

1. We are changing to a smaller unit. The answer must be a larger number.
2. The relation of milli to basic is 10^{-3}.
3. Since the answer must be larger, and the exponent is negative, we *divide* by 10^{-3}.

$$0.05 \text{ V} = \frac{0.05}{10^{-3}} = 0.05 \times 10^{3} \text{ mV} = 50 \text{ mV}$$

(Notice that using the action arrow, we also see that we should divide by the factor 10^{-3}.)

(b) To change 50 mV to microvolts:
1. We are again going to a smaller unit; the answer must be a larger number. Since the exponent is negative, we must divide by the negative exponent.
2. Using the *difference* between exponents, the factor is from 10^{-3} to 10^{-6} or 10^{-3}.

$$50\ mV = \frac{50}{10^{-3}} = 50 \times 10^3\ \mu V = 50\ 000\ \mu V$$

(c) To change 0.5 mA to amperes:
1. Going to a larger unit, the answer must be smaller.
2. The factor from milli to basic is 10^{-3}. Therefore, for a smaller answer we must multiply by the negative exponent.

$$0.5\ mA = 0.5 \times 10^{-3} = 0.000\ 5\ A$$

(Notice that the "action arrow" calls for "multiply.")

(d) To change 0.5 mA to microamperes:
1. The factor is from 10^{-3} to 10^{-6} or 10^{-3}.
2. We are going to a smaller unit; the answer must be larger. (Action arrow calls for divide.) So

$$0.5\ mA = \frac{0.5}{10^{-3}} = 0.5 \times 10^3 = 500\ \mu A$$

(e) To change 2.5 MΩ to ohms:
1. Going to a smaller unit; the answer must be larger.
2. From mega to basic, the factor is 10^6. Therefore,

$$2.5\ M\Omega = 2.5 \times 10^6 = 2\ 500\ 000\ \Omega$$

(f) To change 150 000 Ω to megohms:

$$150\ 000\ \Omega = 150\ 000 \div 10^6 = 0.15\ M\Omega$$

(g) To change 5200 V to kilovolts: The answer should be smaller; the factor is 10^3; therefore,

$$5200\ V = 5200 \div 10^3 = 5.2\ kV$$

(h) To change 0.35 MΩ to ohms: Smaller unit, larger answer, factor 10^6; therefore,

$$0.35\ M\Omega = 0.35 \times 10^6 = 350\ 000\ \Omega$$

These prefixes are used not only with the above units but also with other electrical values that you will learn later.

Precaution. In applying Ohm's law, if any of the given quantities are not in basic units, the *safest* procedure is first to change these quantities to basic units and then to apply Ohm's law. The answer will then be in basic units.

Example 4-5

What voltage is necessary to send a current at 0.5 mA through a resistor of 0.25 MΩ?

Solution

$$E = IR = (0.5 \times 10^{-3}) \times (0.25 \times 10^6) = 0.125 \times 10^3 = 125\ V$$

Calculator entry:

$\boxed{\cdot}$, 5, $\boxed{\text{EE}}$, $\boxed{+/-}$, 3, $\boxed{\times}$, $\boxed{\cdot}$, 2, 5, $\boxed{\text{EE}}$, 6, $\boxed{=}$

Example 4-6

If the supply voltage is 10 V, what value of resistance is needed to limit the current to 120 μA?

Solution

$$R = \frac{E}{I} = \frac{10}{120 \times 10^{-6}} = 0.0833 \times 10^6 = 83\ 300\ \Omega$$

Calculator entry:

1, 0, $\boxed{\div}$, 1, 2, 0, $\boxed{\text{EE}}$, $\boxed{+/-}$, 6, $\boxed{=}$

REVIEW QUESTIONS

1. How is a body given a positive charge? a negative charge?
2. In what unit is the *amount* of charge measured?
3. Two differing bodies are each given a charge of 5 C. Is the work done to charge these bodies the same? Explain.
4. On what does the potential acquired by a charged body depend?
5. When speaking of the potential of a body, or point:
   (a) Is there any other body involved? Explain.
   (b) Is it necessary to specify polarity? Explain.
6. A grounded body is at zero potential. A charge equal to 25 C of electrons is added. What is its potential now? Explain.
7. What is the unit for evaluating electrical potential?
8. How does the term "voltage" differ from potential?
9. With references to Fig. 4-1:
   (a) What is the potential of body A? of body D?
   (b) What is the difference in potential between body A and D?
   (c) What is the voltage between body A and D?
10. (a) What instrument is used to measure voltage?
    (b) How should you connect this instrument to measure the voltage output from a battery? Be specific.
    (c) Repeat part (b) for the voltage across a light bulb.
    (d) State two letter symbols that are used to denote voltage.
11. (a) In its most general sense, what does the term "current" mean?
    (b) In this text, what more specific meaning will be used?
    (c) In what direction does this current flow?
    (d) What is "conventional current"?
    (e) What direction of flow is used with conventional current?
12. (a) What letter symbol is used to denote current?
    (b) What is the unit for measuring current?

(c) What instrument is used for such measurements?

(d) How should you connect this instrument to measure the current flowing through an electric drill?

(e) Why are there polarity marks on this instrument's terminals?

13. What does a current of 3 A mean?

14. With reference to Fig. 4-4:

(a) State two ways by which the current flowing in this circuit could be made to decrease.

(b) What is this relationship called?

15. The current flowing in a circuit is very low. Name two units other than the ampere that could be used for a more convenient numerical answer.

16. The voltage being used for a power transmission line is very high. Name two units other than the basic one that could result in a more convenient numerical answer.

PROBLEMS

1. Forty joules of energy is expended to add 50×10^{18} electrons to an uncharged body.

(a) What charge was transferred?

(b) What is the potential of the body?

2. Starting with two uncharged bodies, A and B, 5 C is transferred from A to B. Thirty joules of energy is expended to remove this charge from body A, and another 50 J is expended in adding it to B. Calculate the voltage between the two bodies.

3. Four bodies (as in Fig. 4-1) have potentials of $+12$, $+8$, $+2$, and -10 V, respectively. Find the voltage between:

(a) A and B

(b) A and C

(c) A and D

(d) C and D

4. Make the following conversions:

(a) 0.025 A to milliamperes and to microamperes

(b) 0.05 mV to microvolts and to volts

(c) 250 μA to amperes and to milliamperes

(d) 150 000 Ω to megohms

(e) 0.47 V to millivolts and to microvolts

(f) 0.65 MΩ to ohms

(g) 0.65 mA to amperes and to microamperes

(h) 370 μV to millivolts and to volts

(i) 0.01 MΩ to ohms

(j) 3 520 000 Ω to megohms

5. A resistor of 80 Ω is connected across a 120-V supply. How much current will flow?

6. What voltage must be connected across a 250-Ω resistor to cause a current of 0.6 A?

7. What should the resistance of a circuit be in order to limit the current in the circuit to 0.8 A if the supply voltage is 220?

8. What should be the value of the bias resistor in a transistor circuit if the current is 25 mA and the voltage across the resistor is 5 V?

9. A resistor of 0.25 MΩ has a current of 0.4 mA through it. What is the voltage across the resistor?

10. In an amplifier circuit, the voltage across the drain load resistor is 180 V. The resistor is 0.05 MΩ. How much current is flowing through the resistor?

11. A milliammeter and a voltmeter are connected as in Fig. 4-4, to read current through the resistor and voltage across the resistor. The readings are 15.0 mA and 4.50 V, respectively. Find the resistance value.

12. A current of 27 μA flows through a 3-MΩ resistor when connected across an unknown voltage source. Find the supply voltage.

13. How much current will flow through a 2700-Ω resistor when a voltage of 1.35 kV is applied across it?

14. What voltage, in microvolts must be applied across a 0.025-MΩ resistor to obtain a current flow of 32 μA through it?

15. The resistance of the field winding of a motor is 95 Ω. Find the current in this winding if a 220-V supply is applied across its terminals.

16. The current in a motor field circuit is 1.8 A when energized from a 120-V supply. Find the resistance of this field circuit.

17. A motor field coil has a resistance of 32 Ω. It has a maximum allowable current rating of 0.75 A. Find the maximum voltage that may be applied to this winding.

18. An automobile headlight draws 3.69 A on high beam and 2.87 A on low beam. The battery supply voltage is 12.8 V. Find the resistance of each filament.

19. The tail light in a car has a resistance of 10.8 Ω when a current of 1.2 A flows through it. Find the voltage of the car battery.

20. Find the current through the dashboard light of an automobile if the resistance of its filament is 11.8 Ω and the battery voltage is 13.1 V.

21. The armature of an automobile starting motor has a resistance of 0.0661 Ω. Find the initial starting current drawn by this motor when connected to the 12.0-V car battery.

22. A generating station sends out 2 kA through a two-wire transmission system that is 5 miles long. The resistance of the line is 0.25 mΩ per 1000 ft. Find the voltage drop in the line.

23. The two-wire 120-V supply line from the electric light company to a consumer's residence has a resistance of 0.07 Ω. If a short circuit occurs at the consumer's residence, find the momentary short-circuit current.

24. A 60-W 120-V lamp draws a current of 500 mA when connected to a 120-V house circuit. Find the resistance of its filament.

25. A toaster has a resistance of 15 Ω. Find the current through the toaster when connected to the 120-V house circuit.

26. An electric lamp is designed to carry a current of 750 mA. At this current, its resistance is 147 Ω. Find the line voltage rating for which it was designed.

27. A relay coil draws a current of 0.03 A when connected to a 12-V battery. Find the resistance of this winding.

28. A relay coil has a resistance of 150 Ω. A current of at least 145 mA must flow through this coil before the relay will close. Find the minimum voltage necessary to operate this relay.

29. What supply line voltage is necessary for an electric furnace having a resistance of 5.1 Ω and rated at 43 A?

30. A voltmeter has a resistance of 0.1 MΩ. Find the current that will flow through this meter when connected across a 440-V line.

5

ELECTRICAL POWER
AND ENERGY

5-1 Power. In mechanics, power has been defined as the rate of doing work. Power is therefore a measure of how fast energy is being developed or consumed. Let us see how this definition applies mechanically. Assume that you are asked to load a truck. The work you must do depends on the weight* of the objects and how high you are lifting them. This work, per unit of weight, depends only on how high you are lifting. The energy you consume *in a given time*, or the power required, depends on how fast you are loading the truck. Are you loading the truck at the rate of 1 object per hour or 100 objects per hour? The more work done per unit of weight (the higher you are lifting) and the greater the number of objects being loaded, the faster the work is being done, or the greater the power being consumed.

The electrical picture is quite similar. The work done per unit weight corresponds to the work done per unit charge. But this is definition of potential and is measured in volts. The number of objects being loaded per unit time corresponds to the number of electrons moving through the circuit per unit time. But this is the definition of current, and is measured in amperes. In other words, *electrical power depends on the voltage across a circuit and the current flowing through the circuit*. The generator develops this power, and the circuit consumes this power. Summarizing:

*The term "weight," as used here (and on page 65), is a unit of *force* and represents the force exerted to support an object against the pull of gravity. Obviously, an object has weight only because of this pull of gravity.

1. Power increases if the voltage is increased.
2. Power increases if the current is increased.

These are both direct relationships. By formula,

$$P = E \times I \tag{5-1A}$$

The unit for measuring power can be derived from equation (5-1A). Since voltage is measured in joules per coulomb, and current is in coulombs per second, then

$$P = E \times I = \frac{\text{joules}}{\text{coulombs}} \times \frac{\text{coulombs}}{\text{seconds}} = \text{joules per second}$$

However, instead of using the derived unit of joules per second, the unit of power has been named in honor of James Watt, the Scottish engineer of steam engine fame.

The power equation as given above, equation (5-1A), is suiatble for finding the power consumed by a circuit when the voltage and current are known. To find how much current an electrical unit is taking, if the voltage and power consumed are known, equation (5-1A) can be transposed by algebra to

$$I = \frac{P}{E} \tag{5-1B}$$

Similarly, to find how much voltage must be applied across an electrical unit if the power and current requirements are known, the formula is rewritten as

$$E = \frac{P}{I} \tag{5-1C}$$

By combination of the power equation and Ohm's law, if two of the quantities in a circuit are known, the other two can be calculated.

Example 5-1

A lamp is rated at 120 V and 60 W.
(a) How much current does it take?
(b) What is the lamp's hot resistance?

Solution

  (a) (Power equation) $I = \dfrac{P}{E} = \dfrac{60}{120} = 0.50 \text{ A}$

  (b) (Ohm's law) $R = \dfrac{E}{I} = \dfrac{120}{0.50} = 240 \ \Omega$

Example 5-2

How much power does a 40-Ω soldering iron consume when connected to a 120-V line?

Solution

In order to calculate power, the current must be known. Therefore:

  1. (Ohm's law) $I = \dfrac{E}{R} = \dfrac{120}{40} = 3.0 \text{ A}$

  2. (Power equation) $P = E \times I = 120 \times 3.0 = 360 \text{ W}$

5-2 Measurement of power. So far, we have seen that power is the product of the current through the circuit and the voltage across the circuit. In Chapter 4 the use of an ammeter and a voltmeter to measure current and voltage, respectively, was shown. Naturally, then, one way of "measuring" power would be by calculation from current and voltage readings. This is an indirect method. Can power be measured directly? In the ammeter, the torque (turning force) that causes the meter to indicate is proportional to the current, while in a voltmeter the torque is proportional to voltage. If another instrument were used, where the torque depended on the simultaneous effects of current and voltage, it would measure power. Such an instrument is the *wattmeter*. The details of construction and principle of operation are left to Chapter 16. For the moment, let us see how it is connected.

The wattmeter has four terminals—two circuits. Two of these terminals, usually larger in size, are from the *current coil* of the wattmeter. The current coil is inserted into a circuit, just like any ammeter. The remaining two terminals are from the *potential coil* and are connected *across* the circuit like any voltmeter. The use of a wattmeter to measure the power taken by a load is shown in Fig. 5-1.

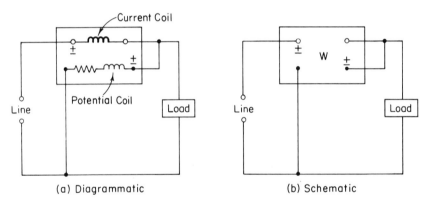

(a) Diagrammatic   (b) Schematic

Figure 5-1 Wattmeter connections. (Note: ± terminal of potential coil may be connected on the line side of the current coil.)

Notice that the ± side of the current coil is connected on the *line* side of the circuit, and the ± side of the potential coil is connected to the same line lead as the current coil. Reversal of either winding will give a backward deflection. It should be noted however that, although the wattmeter indicates power directly, its use in commercial applications is generally limited to ac circuits. The reason for this will be understood when studying power factor of ac circuits.

5-3 Special power formulas. Most power problems can be solved in two steps as in Examples 5-1 and 5-2. To facilitate the solution of these problems, a one-step method can be used. In some cases this one-step method is absolutely necessary. A one-step solution can be obtained by combining Ohm's law and the power equation. Two such combinations follow:

1. From Ohm's law we know that I is equal to E/R. Replacing I in the power equation (5-1A) by its equivalent E/R, we get

$$P = E \times \frac{E}{R}$$

or

$$P = \frac{E^2}{R} \tag{5-2}$$

2. E is equivalent to $I \times R$, or, replacing E in the original power equation, we get

$$P = (I \times R) \times I$$

or

$$P = I^2 R \tag{5-3}$$

Example 5-3

A 25-Ω resistor carries 2.0 A. How much power does it consume?

Solution

$$P = I^2 R = 2.0 \times 2.0 \times 25 = 100 \text{ W}$$

Calculator entry:

2, $\boxed{x^2}$, $\boxed{\times}$, 2, 5, $\boxed{=}$

5-4 Power rating.   Whenever current flows through any electrical equipment, a certain amount of power is consumed because of the resistance of the equipment. The amount of power so consumed can be calculated by $P = I^2 R$. The energy involved appears as heat, which raises the temperature of the equipment. If the temperature rises too much, the insulation on the equipment, or the equipment itself will be damaged. How high will the temperature rise? It depends on the heat being developed ($I^2 R$) and on how fast the equipment can dissipate this heat into the surrounding air. The larger the surface area of the equipment and the freer the circulation of air around this surface area, the more readily the equipment can dissipate the heat. Such equipment would have a higher *power rating*.

Most electrical equipment is marked with a definite voltage and power specification. For example, a small radio set may be rated at 120 V, 85 W. This rating means that when the set is connected across a 120-V supply, it will consume 85 W. Some of the energy will appear as sound from the loudspeaker. The rest appears as heat. If connected to a higher voltage, more current will flow and it will consume more power ($P = E \times I$). But the amount of heat developed ($I^2 R$) will also increase. If the set cannot dissipate the additional heat in any of its components, these components will burn out.

You are probably more familiar with electric-light lamps. No doubt you have noticed the ratings marked directly on the glass, such as 100 W 115 V, 50 W 110 V, or 200 W 120 V. These ratings tell you the amount of power that the lamps will consume when connected to the *rated* voltage. If connected to lower voltages they will not consume the rated power, but neither will they give as much heat and light energy. If connected to higher than rated voltage, they will consume more power and also give more heat and light energy. The extra heat may produce a sufficient increase in temperature to cause the lamp to burn out.

5-5 Power rating of resistors. The power consumed by a resistor is dissipated completely as heat. Two resistors may have the same amount of resistance and yet one may be much larger in physical size than the other. The larger resistor has a greater surface area and can dissipate more heat. It will therefore have a higher *power rating*. With practice you will learn to recognize the approximate power rating of resistors by their physical size (see Fig. 6-4).

Manufacturers of resistors rate their units in ohms and power. The power rating refers to the *maximum* amount of heat the resistor can dissipate without undue rise in temperature. However, these ratings are established for a resistor when mounted one foot away from other units, with free access of air, and at normal room temperatures. When using these resistors in electronic equipment, we crowd them in with hot parts underneath the chassis, add a bottom plate to the chassis, put the chassis in a cabinet, and shove the cabinet against the wall. Is it any wonder why resistors that were calculated accurately as to power rating may burn out?

To avoid such failures, the manufacturer often provides "power derating curves." Such curves (see Fig. 5-2) show the decrease in power rating for a given type of resistor, when operated at various ambient temperatures. Notice that the derating is linear. The temperature at which the derating is started, and the slope of the curve, depend on constructional details. Quite often, the possible ambient temperature under operating conditions is not known. To allow for differences between actual service conditions and the "Free Air Watt Rating" it is general engineering practice to calculate the maximum power that a resistor will have to dissipate, and then order a resistor whose *commercial power rating* is twice to four times this value, depending on the conditions under which this resistor will have to operate.

Example 5-4

A 4000-Ω resistor is connected across a 100-V supply. What resistor should be ordered?

Solution

1.
$$P = \frac{E^2}{R} = \frac{100 \times 100}{4000} = 2.5 \text{ W}$$

Calculator entry:

1, 0, 0, $\boxed{x^2}$, $\boxed{\div}$, 4, 0, 0, 0, $\boxed{=}$

2.   Power rating $= 2.5 \times 2$ or 2.5×4

$= 5$ to 10 W

Order a 4000-Ω 5- to 10-W resistor.

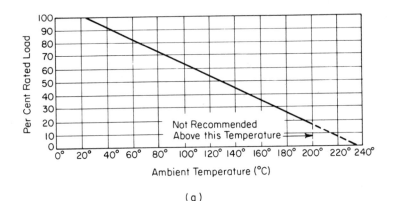

(a)

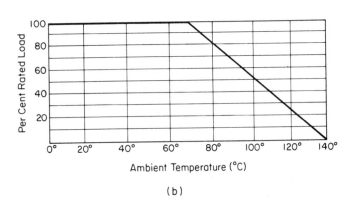

(b)

Figure 5-2   Typical power derating curves. (*Courtesy of Corning Electronics*)

Example 5-5

A resistor is rated at 5000 Ω, 20 W. How much power is it dissipating?

Solution   This is a ridiculous question. If the resistor were just lying on the workbench, it would consume no power at all. The rating of a resistor does not show how much power it will consume but rather the maximum it can safely consume. The actual amount of power consumed depends on the ohms and either the current through it or voltage across it. Using too large a commercial rating means too high a price and too bulky a size. Too low a commercial rating means that the resistor cannot dissipate the heat developed and will burn out. In the above problem, the manufacturer states that this resistor will dissipate 20 W safely under ideal conditions. When this resistor is being used in a circuit, it should not be made to dissipate more than 20 W. In fact, if the resistor is located in a crowded area, it should not be made to dissipate more than 5 to 10 W.

5-6 Electrical–mechanical power conversion. An electric motor takes electrical energy from a supply line and converts it to mechanical energy delivered at its shaft. However, rather than specify (or measure) the total energy converted *in a given period of time*, it is common practice to refer to the *rate* at which the machine takes electrical energy (*power input*) and the *rate* at which it delivers mechanical energy (*power output*). The electrical power input is the product of the voltage applied to the motor and the current drawn by the motor, and is measured in watts. (For high power levels, the kilowatt is used.)

How do we measure the mechanical power output of the motor? At the beginning of this chapter we used the illustration of loading a truck, wherein the mechanical work done depended on (1) the *force* (*f*) needed to lift the load objects (against the pull of gravity—weight); and (2) the height to which the objects were lifted—or *distance* (*d*). In SI units, force is measured in *newtons*, distance in *meters*, and the unit for work (or energy) is $f \times d$ or *newton-meters*. However, to avoid complex units, by internationally agreement the newton-meter was renamed **joule** (J), in honor of the English physicist James Joule (1818–1889). Power, being the rate of doing work, becomes joules per second. This, we saw in our discussion of electrical power, is renamed watts. Notice that the same unit—the watt—is used for both electrical and mechanical power. Since motors deliver mechanical power, their power output—using SI units—should be rated in watts. However, at this time, industry has no plans to change to this SI unit of mechanical power. Instead, the English unit—**the horsepower** (hp)—is still being used. Motors are available from fractional horsepower sizes ($\frac{1}{16}, \frac{1}{4}, \frac{1}{3}$) for small appliances such as fans, mixers, and washing machines, to many hundreds or thousands of horsepower for lifting elevators, driving locomotives, or other industrial applications.

There must be a definite relationship between the electrical power (in watts or kilowatts) taken by the motor and the mechanical power (in horsepower) supplied to the load. This relationship is

$$1 \text{ hp} = 746 \text{ W} \tag{5-4}$$

Example 5-6

(a) Express 15 hp in watts and kilowatts.
(b) Express 12.5 kW in horsepower.

Solution

(a)
$$\text{watts} = \text{hp} \times 746 = 15 \times 746 = 11\ 200 \text{ W}$$
$$\text{kilowatts} = \frac{\text{watts}}{1000} = \frac{11\ 200}{1000} = 11.2 \text{ kW}$$

(b)
$$\text{hp} = \frac{\text{watts}}{746} = \frac{12.5 \times 1000}{746} = 16.8 \text{ hp}$$

5-7 Efficiency (η, eta).   If all the electrical energy taken by a motor were converted to useful mechanical energy at the shaft, the problem of converting from electrical to mechanical units would be simple, as shown in Example 5-6. But this would assume an ideal motor, and no machine, electrical or otherwise, is perfect. Some energy is wasted within the machine and only the remainder of the input is available as useful output. In an electric motor, for example, there are mechanical losses such as those caused by friction, and electrical losses such as those caused by the resistance of its winding. These losses appear as heat, which raises the temperature of the motor, and the allowable temperature rise, in turn, limits the power rating of the machine.

A motor is considered more efficient if a greater portion of the input is converted into useful output. The efficiency of conversion can be obtained from the ratio of energy output to energy input. In common practice, this efficiency is expressed as a percentage, using (the equivalent) *power* ratios:

$$\text{efficiency} = \frac{\text{power output}}{\text{power input}} \times 100 \qquad (5\text{-}5)$$

where power output and power input can be in any power unit, as long as both are expressed in the *same unit*. In general, the larger the motor, the higher the efficiency. Typical values range from over 90% for 100-hp motors, down to approximately 70% for 1-hp machines and even lower for the fractional horsepower sizes.

Example 5-7
A 220-V 15-hp motor has a full-load current rating of 61 A. Find the efficiency of this machine.

Solution

1. Power input $= P = E \times I = 220 \times 61 = 13\ 400$ W
2. Power output (watts) $=$ hp $\times 746 = 15 \times 746 = 11\ 200$ W
3. Efficiency $= \dfrac{\text{output}}{\text{input}} \times 100 = \dfrac{11\ 200}{13\ 400} \times 100 = 83.5\%$

The output of any machine is always less than the input because of the internal power losses, or

$$\text{power input} = \text{power output} + \text{power losses}$$

The efficiency equation could, therefore, be modified by replacing power input by its equivalent, output + losses. Similarly, output could be replaced by its equivalent (input − losses). In other words, whenever data are required concerning power input, power output, efficiency, or power losses, if any two of the above factors are known, the other two can be calculated.

Generators are used to supply electrical power to operate electrical equipment. In turn, these generators are driven (supplied with mechanical power) by some prime mover such as steam engine, gasoline engine, or water turbine. Since the generator supplies electrical power, it is rated in watts (or kilowatts). The prime mover, supplying mechanical power, is rated in horse-power. Here again we need the horsepower-to-watts conversion factor. Also, since the generator has losses, the efficiency term must be included in any conversion problem.

Example 5-8

A 220-V 600-A generator has an efficiency of 92%. Find the horsepower rating needed for the prime mover.

Solution

1. Power output $= E \times I = 220 \times 600 = 132\,000$ W

2. Power input (electrical) $= \dfrac{\text{output}}{\text{efficiency}} \times 100 = \dfrac{132\,000}{92} \times 100$

$$= 143\,500 \text{ W}$$

3. Input required (in hp) $= \dfrac{143\,500}{746} = 192.3$ hp

5-8 Electrical energy (W).   In Chapter 1 we defined energy as the ability to do work, and added that energy is not created or destroyed, but rather converted from one form to another. Once work has been done, energy must have been expended—or more correctly, transformed from one type of energy to another. (It is common, even though erroneous, to speak of energy as being expended, used, or used up. Some justification for such colloquialisms is that insofar as the source of energy is concerned, once work has been done, it has lost or expended energy regardless of in what other form or forms this energy now appears.) The amount of energy used will vary directly with the amount of work done. But we already know that power is the *rate* of doing work. So if we also know over how long a period of time this power rate was used, we can find the energy expended, or

$$W = P \times t \qquad\qquad \textbf{(5-6A)}$$

In electrical units W is the energy in watt-seconds or *joules*, P the power in watts, and t the time in seconds. Since power is the product of voltage and current, equation (5-6A) is more commonly written

$$W = E \times I \times t \qquad\qquad \textbf{(5-6B)}$$

where E and I are voltage and current in volts and amperes, respectively.

Example 5-9

An electric toaster used on a 120-V line draws 5.0 A. It is used on an average of 25 min/day. Find the electrical energy consumed per month.

Solution

1. Total time in seconds $= 25 \times 60$ s/day or $25 \times 60 \times 30 = 45\,000$ s/month

2. $$W = E \times I \times t = 120 \times 5.0 \times 45\,000$$
$$= 27\,000\,000 \text{ W-s or joules}$$

It should be obvious from Example 5-9 that the basic unit of energy, the joule, is not a convenient one. With SI prefixes, we could use kilojoules and megajoules. However, in commecial applications it has been customary to increase the time interval from the second to the hour. So instead of a watt-second (or joule), we now have a *watthour*. Also, by measuring power in kilowatts (instead of in watts), and still retaining the 1-h time interval, we get the unit *kilowatthour*. This is the commonly used unit for electrical energy in industry, and there are no present plans to replace it with the SI unit.

Example 5-10

Find the energy consumed by the toaster in Example 5-9, in watthours, and in kilowatthours.

Solution

1. Time used per month, in hours
$$t = 25/60 \times 30 = 12.5 \text{ h/month}$$

2. Energy consumed
$$W = E \times I \times t = 120 \times 5.0 \times 12.5 = 7500 \text{ Wh}$$
$$= \frac{7500 \text{ Wh}}{1000} = 7.5 \text{ kWh}$$

Now that we know what electrical energy is, and how to calculate the amount of energy consumed in operating a piece of electrical equipment for some given period of time, the question may arise as to the value of this information. To answer this question, let us go back to the source of electrical energy. The electrical utility corporations generate electrical energy. More specifically, they convert the chemical energy from coal, or the potential energy of a waterfall, or energy from a nuclear reactor into electrical energy. Then they distribute this electrical energy to the residential and industrial consumers in the community they service. How much should they charge for this service? The charge should be based not only on the rate at which the energy is consumed (power) but also on the length of time for which this rate is used. For examples, you should pay more for operating a 60-W lamp for 24 h than for operating a 200-W lamp only for 1 h. In other words, the consumer pays for the watthours of energy consumed. In general, the charge per kilowatthour is lower for industrial consumers, but even for residential users the charge per kilowatthour is decreased (after a fixed minimum) as the quantity of energy used increases.

As an illustration, the LILCO (Long Island Lighting Co.) rates effective October 1978 are shown below:

Residential		
First	12 kWh or less	$2.67
Next	78 kWh at	0.0632 per kWh
Next	510 kWh at	0.0527 per kWh
Excess over	600 kWh at	0.0503 per kWh

General Service—Large		
First	6 000 kWh at	0.0381 per kWh
Next	24 000 kWh at	0.0342 per kWh
Excess over 30 000 kWh at		0.0296 per kWh

Example 5-11

Find the cost per month of operating the following electrical equipment, at the residential rates listed above.
(a) 250-W TV receiver for 6 h/day.
(b) 600-W iron for 2 h/day.
(c) 1000-W toaster for 40 min/day.
(d) Three 100-W, two 40-W, and one 200-W lamps for 5 h/day.

Solution

1. Find the total kilowatthours consumed:

(a)	$250 \times 6 \times 30 \div 1000 =$	45.0 kWh
(b)	$600 \times 2 \times 30 \div 1000 =$	36.0 kWh
(c)	$1000 \times \frac{40}{60} \times 30 \div 1000 =$	20.0 kWh
(d)	$3 \times 100 \times 5 \times 30 \div 1000 =$	45 kWh
	$2 \times 40 \times 5 \times 30 \div 1000 =$	12 kWh
	$1 \times 200 \times 5 \times 30 \div 1000 =$	30 kWh
	Total $=$	188.0 kWh

2. First 12 kWh $=$ 2.67
   Next 78 kWh at 0.0632 $=$ 4.93
   Balance $(188 - 12 - 78) = 98$ kWh at 0.0527 $=$ 5.16
                          Total cost $12.76

5-9 Watthour meter.   Using the technique of Example 5-11, any consumer could calculate the cost of the electric energy he consumes. The problem is how he can tell for how long a period each piece of equipment or each light is used? Accuracy would require use of a timeclock on *each* piece of equipment, to record total time of operation per month. With motor loads, accurate timing alone is not enough. The current drawn by a motor (and therefore the power taken) will vary depending on the mechanical load. Now, some form of recording wattmeter would also be necessary. This technique is too involved for accurate cost accounting. (However, it is used for rough estimates.)

An electrical utility corporation must have accurate kilowatthour consumption figures. For this purpose, they use a **watthour meter**. This meter has a current coil and a potential coil, that are connected into a circuit similarly to a wattmeter. The moving element in the meter is free to rotate. The speed of rotation depends on the rate of consumption of energy (power). The *number* of revolutions depends on speed of rotation (power) and time and therefore is a measure of energy. It does not matter whether energy is consumed at a constant rate, or whether the power taken varies erratically during the day. The increase or decrease in speed of the meter reflects these changes in power, and the number of revolutions is automatically proportional to the total energy consumed in any period of time. By gearing the moving element of the watthour meter to a clockwork dial register, the total number of revolutions made can be indicated continuously. In practice, the dials are calibrated to indicate kilowatthours directly. For purposes of billing, the utilities generally take readings of the watthour meters at 1- or 2-month intervals.

REVIEW QUESTIONS

1. (a) Give a definition for electrical power.
   (b) What is the unit commonly used for measuring this power?
   (c) In basic quantities, what should this unit be?

2. (a) Name two electrical quantities that affect the power consumed in a circuit.
   (b) State how the power varies with each of these quantities.

3. (a) What instrument can be used to measure power directly?
   (b) How many terminals does this instrument have?
   (c) Account for these terminals.

4. With reference to Fig. 5-1:
   (a) Why is the current coil inserted (*in series*) into one line?
   (b) Why is the potential coil connected "across" the circuit?
   (c) What is the intent of the $\pm$ coding on one current-coil terminal?
   (d) What is the intent of the $\pm$ coding on one potential-coil terminal?

5. How is equation (5-2) obtained?

6. How is equation (5-3) obtained?

7. (a) Basically, what determines the power rating of any electrical equipment?
   (b) If air circulation by fans is added to a piece of electrical equipment, will this affect its power rating? Explain.
   (c) Vacuum tubes for high-power transmitters are often enclosed in a circulating-water jacket. Why?
   (d) In some applications transistors are mounted on "heat sinks" (large metallic surfaces). Can you see any reason for this?

8. A 50-W 125-V lamp is connected to a 208-V supply. *Explain* what will (probably) happen.

9. A toaster rated at 120 V, 1020 W, is connected to a 110-V supply. Will this affect the rating? Explain.

10. When specifying resistors for use in electrical circuits, what two quantities must be specified?

11. Two resistors—500 Ω, 10 W and 100 Ω, 100 W—are ordered. When they arrive, there are no distinguishing markings. Which is which? Why?

12. Two 120-Ω resistors are each connected across a 120-V line. Resistor *A* has a 5-W rating. Resistor *B* has a 100-W rating. Which unit dissipates more power? Explain.

13. Calculations show that a resistor will dissipate 15 W. What power rating should be ordered? Why?

14. (a) In what respect, if any, is horsepower similar to watts?
    (b) What is the SI unit for mechanical power?

15. From an energy standpoint:
    (a) What is the function of an electric motor?
    (b) What is the function of a generator?
    (c) Is all of the input available as output? Explain.
    (d) How is the efficiency of a machine obtained?

16. (a) What is the SI unit for electrical energy?
    (b) What is the unit commonly used in industry?
    (c) Can this quantity be measured directly?
    (d) How can the energy consumed by a device be estimated?
    (e) Of what practical value is knowing energy consumption?

PROBLEMS

1. An electrical device draws 1.6 A from a 120-V line. How much power does it consume?

2. An electric iron is rated at 80 W, 120 V. How much current does it draw from a 120-V line?

3. A loudspeaker field coil is rated at 18 W, 120 mA. What voltage should be connected across it?

4. (a) What is the hot resistance of a 120-V 300-W lamp?
   (b) Assuming that the resistance remains constant, how much power will it consume when connected to a 110-V line?

5. A 2500-Ω resistor is connected across a 100-V supply.
   (a) How much power does it dissipate?
   (b) What should its commercial power rating be?

6. Draw a schematic diagram showing suitable instruments to measure the current, voltage, and power taken by a toaster from a supply line (show polarity marks).

7. A 5000-Ω resistor is rated at 10 W. Using a safety factor of 4:
   (a) What is the maximum power it should dissipate in a radio chassis?
   (b) What is the maximum voltage that should be connected across it?
   (c) What is the maximum current it should carry?

8. The voltage across a 25 000-Ω resistor is 240 V. What should the power rating of the resistor be?

9. A resistor of 0.05 MΩ is connected in a circuit that carries 40 mA. What should its power rating be?

10. Find the allowable current that can be carried by a 25-W 8000-Ω resistor, using a safety factor of 2.

11. A 7000-Ω resistor carries a current of 0.035 A.
    (a) How much power does it dissipate?
    (b) What should its power rating be?

12. A 60-W 120-V lamp when connected to a 115-V line draws 0.48 A. Find the power dissipated.

13. The heating element of an electric toaster has a resistance of 22 Ω. Find the power dissipated by the toaster when used on a 117-V line.

14. A 60-V arc lamp requires 400 W dissipation for proper operation. How much current does it draw?

15. A generator nameplate shows the rating for current and power as 21.7 A and 5 kW, respectively, but the voltage rating is defaced. Find the voltage rating of this machine.

16. A subway car heater has a resistance of 120 Ω. It is supplied from a 550-V line. Find its power dissipation.

17. Make the following conversions:
    (a) 360 W to kilowatts
    (b) 620 W to horsepower
    (c) 4.2 kW to horsepower
    (d) 5 hp to kilowatts
    (e) 2.85 kW to watts
    (f) $\frac{1}{3}$ hp to watts

18. A 3-hp 120-V motor has an efficiency of 78%. Find the rated current of this machine.

19. Find the efficiency of a 5-hp 220-V motor if the current at rated load is 21.0 A.

20. A 24-V fan motor draws a current of 13 A. Its efficiency is approximately 60%. Find the probable horsepower rating of the motor.

21. A 250-kW generator has an efficiency of 94%. It is to be driven by a steam turbine. Find the horsepower rating needed for the turbine.

22. A 550-V substation generator delivers 727 A to its load when driven by an engine developing 575 hp. Find the efficiency of the generator.

23. A 230-V generator has an efficiency of 90%. How much current will it deliver to its load when the driving power from the prime mover is 50 hp?

24. The motor of Problem 19 is used for 8 h/day. Assuming 21 workdays per month, calculate the cost of operating the motor for 1 month at industrial rate of $0.0342 per kWh.

25. At 5.27 cents/kWh, how much would it cost to operate a 250-W television receiver for 1 month, on an average of 6 h/day?

26. An electric shaver draws 0.06 A from a 120-V line. If the shaving time is 5 min, find the cost per shave at 6.32 cents/kWh.

27. A 20-hp blower motor is used on an average of 8 h/day. If its efficiency is 85%, find the daily cost of operating the blower at the industrial rate of 3.81 cents/kWh.

28. Find the total cost per month for operating the following appliances:

   (a) A 325-W color TV for 5 h/day.

   (b) An 1100-W toaster for 30 min/day.

   (c) A 600-W iron for 3 h/day.

   (d) Four 100-W, three 60-W, and one 200-W lamps for 4 h/day.

The charge for electrical energy is $2.67 for the first 12 kWh (or less), decreasing to 6.32 cents/kWh for the next 78 kWh; to 5.27 cents/kWh for the next 510 kWh.

6

COMMERCIAL RESISTORS

Resistors can be grouped into three classifications, depending on the materials from which they are made: wire-wound, composition, and film. Resistors above 2-W size are usually of the wire-wound type, whereas resistors of 2 W or less are almost entirely of the composition or film type. The exception to this is precision resistors, which may be wire-wound or film, and resistors of less than 500 Ω, which may be of any type regardless of power rating.

6-1 Types of wire-wound resistors.   The basis for wire-wound resistors is the formula

$$R = \frac{\rho \times l}{A}$$

The specific resistance chosen depends on whether high or low resistance is required. The circular mil area of the wire used depends upon the power rating required. Once these two factors have been chosen, a sufficient length of wire is used to obtain the required number of ohms. The resistance wire is wound on an insulated core composed of a material such as fiber, plastic, or ceramic. The wire, if insulated, is close-wound; if it is not insulated, it is space-wound, so that adjoining turns will not come in contact, or short-circuit.

Wire-wound resistors can be classified as follows:

1. *Fixed Resistors.*   These resistors have only two terminals: the start of the winding and the end of the winding. Figure 6-1 shows the schematic diagram used to represent a fixed resistor.

Figure 6-1 Fixed resistor, schematic diagram.

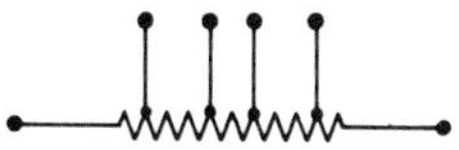

Figure 6-2 Tapped resistor, schematic diagram.

2. Tapped Resistors. Sometimes several values of a fixed resistor are required for a certain purpose. A tapped resistor is used (Fig. 6-2) rather than separate resistors. In the manufacturing, the winding is started as before. When the smallest desired value of resistance is reached, a tap is brought out. Then the winding is continued until the second desired value is reached, and another tap is brought out. This process is continued until the maximum desired resistance is obtained.

3. Semivariable Resistors. The tapped resistor may be excellent in the particular job for which it was designed, but it is almost useless for any other application. You may be able to get a commercially made tapped resistor of the correct total value, but what are the chances of also getting the correct taps? It is impossible for the manufacturer to make tapped resistors to suit every possible situation. However, if the number and location of taps could be adjusted by the user, this problem could be avoided. This adjustability is the principle of the *semivariable* resistor [Fig. 6-4(c)]. The metal band is moved to the correct tap position and is locked in place by tightening the lock screw. These resistors are not intended for use where continuous adjustments are required. They are set once for the particular design and are not changed unless the unit is redesigned.

4. Variable Resistors. When the amount of resistance in a circuit is to be varied frequently, variable resistors are used. The resistance wire is wound on a circular form, bare wire space-wound being used. A metal arm, mounted on a rotating shaft, bears on the top of the winding element [Fig. 6-4(d)]. These resistors were originally made in two types:

(a) *Rheostat.* Two terminals were brought out: one end of the winding, and the movable contact. The amount of resistance in the circuit is the section between the fixed end and the movable contact.

(b) *Potentiometer.* Three terminals were brought out: the two fixed ends of the winding and the movable contact.

In the newer designs all three terminals are brought out. You can then use the same variable resistor as a potentiometer or rheostat. The schematic diagrams of each are shown in Fig. 6-3.

Figure 6-3 Variable resistors, schematic diagram.

Rheostat   Potentiometer

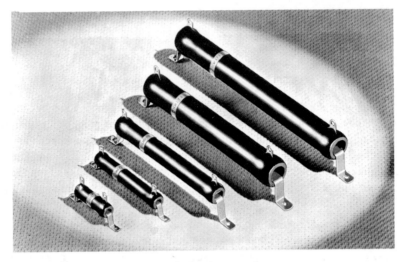

(a) Fixed, wire wound.

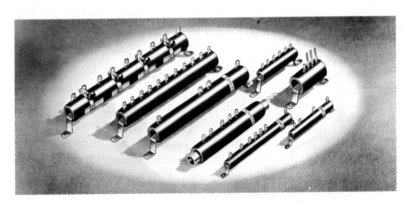

(b) Tapped, wire wound.

(c) Semivariable, wire wound.

Figure 6-4 Commercial resistors. (*Courtesy of Ohmite Mfg Co.*)

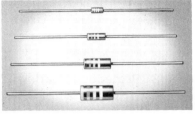

Courtesy Allen-Bradley Co.

(d) Potentiometer, wire wound.　　　　　(e) Fixed, composition.

(f) Potentiometer, composition.

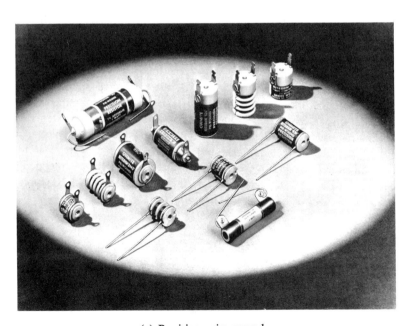

(g) Precision, wire wound.

Figure 6-4 (Continued)

5. *Flexible Resistors.*   In some cases it is convenient to have a resistor wound on a flexible core. These cores are often strips of asbestos. The wire is wound on this core and covered by additional asbestos and a cotton braid as outer covering. More recently, glass fiber cores have been used, and the winding is then protected with an outer braided-glass covering.

6-2 Protective coatings.   In order to protect the delicate wire of a wire-wound resistor from mechanical damage and from electrical short circuits, an insulating coating is applied to the wire. Materials used for such coatings must be good conductors of heat. Otherwise, the heat developed by the electrons as they pass through the wire will not reach the surface. The temperature of the wire will rise too high and the wire will burn out. Materials generally used for this purpose are vitreous enamel, bakelite, baked lava, baked ceramics, cement, spun glass, and fiber. When fiber is used, an outside covering of metal is added to hold it in place. These resistors are called *metal-clad.*

6-3 Composition resistors.   Suppose that it is desired to manufacture a wire-wound resistor of 700 000 Ω. A high resistance material such as nichrome ($\rho = 675$) could be used. In English units (since this is still the commercial practice), $\rho = 675$ Ω/cmil-ft. The smallest area mechanically usable is 10 cmil. The length of wire needed, calculated by

$$R = \frac{\rho \times l}{A}$$

is more than 10 000 ft. Obviously, the price of such a resistor would be prohibitive for general use. Now, carbon has a specific resistance of 22 000 Ω/cmil-ft. By powdering the carbon and mixing it with some insulating binding material, high values of resistance can be made very inexpensively. The proportion of carbon to binding material will determine the resistance obtainable per inch. Composition resistors of any desired resistance value are manufactured by this process for only a few cents each. These resistors are usually coated with some insulating material and baked to a hard finish. This material insulates, protects the unit from mechanical damage, and prevents change in value due to atmospheric moisture absorption. Lead wires (pigtails) are solidly embedded at each end to provide a large contact area and a high "pull strength." A typical construction is shown in the cutaway view, Fig. 6-5.

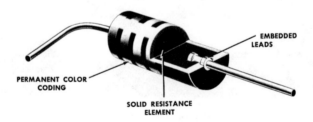

EMBEDDED LEADS

PERMANENT COLOR CODING

SOLID RESISTANCE ELEMENT

Figure 6-5   Fixed composition resistor. (*Courtesy of Allen-Bradley Co.*)

Variable resistors are also made of the composition type. The circular fiber strip is covered with a powdered graphite composition, the rotating arm making contact to this graphite strip [Fig. 6-4(f)].

Composition resistors, although inexpensive, have their drawbacks. One problem is instability. Their resistance value can be permanently changed by overheating. This could happen when soldering the unit into the circuit, or from overload. Oxidation, due to moisture (either from the surface, or in the mix), can affect the resistor. Another problem is noise. Current flow through the mix is not smooth and continuous. Any interruption or random variation in flow produces a noise. Composition resistors should therefore not be used in the early stages of a high-gain amplifier.

6-4   Percent tolerance.   A third problem with composition resistors is that they cannot be manufactured to precise value.* Instead, they are sorted and color-coded according to the measured value after manufacture. If the actual value is within $\pm 5\%$ of the nominal value, a gold band is placed approximately in the center of the body of the resistor. If the actual value falls between ± 5 and $\pm 10\%$, a silver band is put on the body. When the actual value is between ± 10 to $\pm 20\%$, no band is placed on the body. When the error is more than 20%, the resistor is discarded.

6-5   Color code.   In the more expensive wire-wound and film-type resistors, identification data (resistance, tolerance, stock number, etc.) are generally applied using an "ink" composed of ceramic compounds, and fired into the molded coating. This makes the markings an integral part of the resistor. Such techniques are too expensive for a carbon resistor. For these resistors, a four-band color code is generally used: the first three bands to indicate the resistance value, and a fourth band for percent tolerance, as mentioned above. (But recall that if the tolerance is poorer than $\pm 10\%$, the fourth band is omitted.) This technique is shown in Fig. 6-6.

For resistance evaluation each color is assigned a definite value:

Black	0	Green	5
Brown	1	Blue	6
Red	2	Purple	7
Orange	3	Gray	8
Yellow	4	White	9

The first color (first band) designates the first digit. The second color (second band) designates the second digit. The third color (third band) designates the number of zeros to be added. As an illustration, if the first three color bands are yellow–violet–orange, the resistance value is 4-7-000, or 47 000 Ω.

*When the resistance values required must be very accurate, as with instrument shunts and multipliers, precision wire-wound or film-type resistors are used. These resistors can be obtained to an accuracy within 0.1% of their rated value.

For more demanding applications (such as the military), a fifth band is also used. This indicates the reliability (or failure rate) of that type of resistor, in accordance with MIL-R-39008 standards.

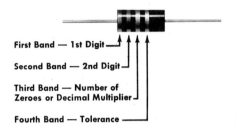

First Band — 1st Digit

Second Band — 2nd Digit

Third Band — Number of
Zeroes or Decimal Multiplier

Fourth Band — Tolerance   **Figure 6-6**   Resistor color coding.

When the resistance value is less than 10 Ω, the above color-coding system is inadequate. To remedy this situation, decimal multipliers are used as the third band. The multipliers are:

Gold	0.1
Silver	0.01

For example, a resistor of 2.7 Ω would have a red first band color, a purple second band color, and a gold third band:

$$2\text{-}7\text{-} \times 0.1 = 2.7 \,\Omega$$

Such a color-coding method gives the resistance value to two significant figures only. Offhand, this rating seems too inaccurate. But remember that the accuracy of the average resistor is at best only within $\pm 5\%$ of the nominal value. Two significant figures are sufficient to express any resistance value to better than the 5% tolerance. The color-coding system is not used for precision resistors.

6-6 Thin-film resistors. In comparing the relative merits of composition versus wire-wound resistors, we find that composition resistors are much cheaper, smaller, and can be obtained in a wide range of values (from approximately 1 Ω to 22 MΩ). However, they have drawbacks of low accuracy, instability, and noise. Thin-film resistors tend to give the advantages of each of the types above, and at a lower price than the wire-wound. These resistors are available in three classifications, depending on the conducting material used.

1. *Carbon Films.* Carbon-film resistors are formed by depositing a thin film of carbon on a ceramic core. For high resistance values, a helical path is cut in the cylinder of the film (see Fig. 6-7). The units are sealed in metal, glass, ceramic, or epoxy coatings to make them impervious to moisture. These units are generally available in $\frac{1}{2}$-W, 1% tolerance ratings, and in resistance values from 10 Ω to 2.5 MΩ.

2. *Metal Films.* For even better accuracy, stability, and lower noise levels, metal-film resistors are available (at a somewhat higher cost). The resistance element is a film of special, nonnoble metal or alloy, deposited on a ceramic core. The film is cut in a spiral to obtain the desired resistance value. Again, suitable encapsulation is provided to protect the unit from air and moisture. Metal-film resistors are generally available in $\frac{1}{8}$- to 1-W ratings; tolerances $\pm 1.0\%$ standard (and up to 0.1% on order); and in resistance values from 10 Ω to 1.5 MΩ.

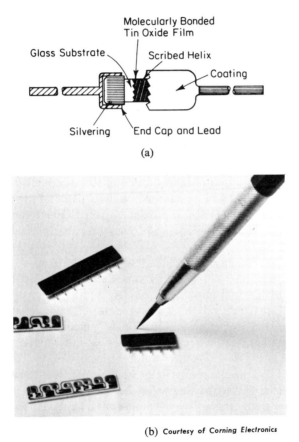

(a)

(b) *Courtesy of Corning Electronics*

Figure 6-7   Tin-oxide resistors.

3. *Tin-Oxide Films.* These resistors are manufactured by fusing tin oxide into the pure surface of freshly drawn, optical quality glass at red heat. The high resistivity of the tin oxide allows the use of thicker films for greater mechanical strength and higher power ratings. Both the tin oxide and glass substrate are chemically inert. Matching of the thermal expansions of the tin-oxide film and the glass eliminates stressing of the film during overloads or thermal

shocking. As a result, these resistors have excellent stability and reliability. A cutaway view of a tin-oxide resistor is shown in Fig. 6-7. Notice the helixing technique used for accurate resistance values.

Tin-oxide resistors are available in power ratings from $\frac{1}{8}$ to 6000 W, and in resistance values from 10 Ω to several megohms. Standard tolerance values range from $\pm\frac{1}{2}\%$ (for the lower power units) to $\pm10\%$. Resistive elements capable of dissipating up to 250 kW, with water cooling, can also be obtained (on special order). For equipment requiring tight packaging of components, resistor networks consisting of from 2 to 10 units on a common substrate are available. Figure 6-7(b) shows an eight-resistor network. Its overall size is 28.1 mm (1.105 in) by 5.72 mm (0.225 in) and 2.79 mm (0.110) in thick.

6-7 Integrated circuits.   Up to this point, the resistors we have discussed are individual *discrete* components. However, for low-power applications, it has been found that by combining the *functions* of passive elements (such as resistors and capacitors) with the active elements (diodes and transistors), into one integrated whole, fantastic reductions in physical size are possible, together with greatly improved reliability. Such combinations are called *integrated circuits*. When these circuits are produced in quantity, the cost is also much lower than that of an equivalent discrete-component circuit.

For example, the LM 565* phase-locked loop (used in FM stereo receivers) comes in a dual-in-line package approximately 60 × 20 mm and 3 mm thick ($\frac{1}{4} \times \frac{3}{4} \times \frac{1}{8}$ in). It contains 4 diodes, 36 transistors, and 27 resistors, and sells at retail for under $2.

Based on manufacturing techniques, integrated circuits fall into two broad classifications: monolithic and hybrid. *Monolithic circuits* are made from a single crystal of material (usually silicon). Resistors (and capacitors), diodes, and transistors are made on the same chip using manufacturing processes common in the semiconductor industry (alloy junctions, diffusion, epitaxial growth, and masking). A cross-sectional sketch of a simple monolithic integrated circuit is shown in Fig. 6-8, together with its schematic.

The monolithic process results in the smallest possible structure and lowest cost. Since there are no "connecting wires" between components, this technique has far better reliability than discrete-component circuits. Unfortunately, there are drawbacks. With regard to resistors, the range of resistance values obtainable is (approximately) from 20 to 25 000 Ω. Also, the resistance values will be affected by temperature much more than with discrete components.

Hybrid circuits use thin-film passive devices interconnected with monolithic active devices. The advantage of this type of circuitry is that optimum design techniques can be used for each type of component. This technique is especially useful where large resistance (and/or capacitance) values are required. Unfortunately, this also results in higher costs and reduced reliability.

---

*National Semiconductor Corporation.

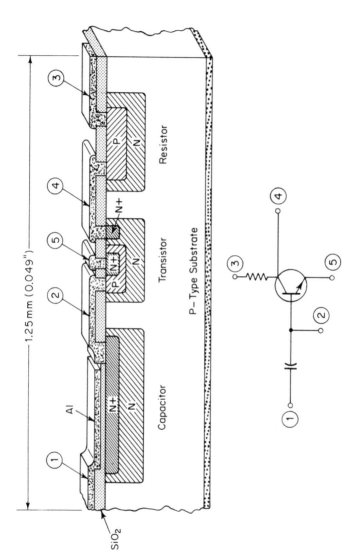

Figure 6-8 Monolithic silicon integrated circuit.

6-8 Temperature coefficient. Have you ever noticed the small gaps between successive lengths of rail on a railroad track? The gaps are there to allow for expansion of the rails when they become hot! When heat is applied to a material, the electrons rotate faster around their nuclei, expanding the electron orbits. The material therefore expands. The increase in length of the rail depends on the increase in temperature and also on the original length. For a given length, the expansion is the same. Therefore, a longer rail will expand more.

Electrically, as a material gets hot and the electrons revolve faster, either of two effects occurs:

1. As the electrons move faster, those electrons moving through the material encounter more and more opposition to their motion. The resistance of such a material will increase with temperature. All pure metals react in this way. These materials have a *positive temperature coefficient.*

2. As the electrons revolve faster, the centrifugal force increases; covalent bonds are broken, producing more free electrons. The resistance of such a material decreases as the temperature increases. Carbon reacts in this way. It has a *negative temperature coefficient.*

When pure metals are mixed together to form alloys, the crystal structure of the atoms of each metal is affected. To explain this effect would require a through knowledge of chemistry and metallurgy and is beyond the scope of this text. At any rate, the result is that when an alloy is heated, the electron structure is practically unaffected and the resistance of the alloy is practically unchanged. By proper choice of the metals in an alloy, and the percentage of each metal, it is possible to obtain an alloy with almost zero temperature coefficient. Examples of such alloys are manganin (84% Cu + 12% Mn + 4% Ni) and Iala (60% Cu + 40% Ni).

The change in resistance of any material depends on the change in temperature and also on the original resistance. The temperature coefficient of any material is therefore the change in resistance *per degree per ohm.* Knowing the resistance (R_1) of a material, at a given temperature (t_1), the change in temperature $(\Delta t = t_2 - t_1)$, and the temperature coefficient (α_1) *at the original temperature* (t_1), it is a simple matter to calculate the change in resistance (ΔR) due to the change in temperature. (The symbol Δ, delta, is used to denote an increment of change.) Expressed mathematically, we have

$$\Delta R = R_1 \alpha_1 \, \Delta t \qquad (6\text{-}1)$$

To find the final resistance (R_2) of the material at a new temperature (t_2) we merely add (*algebraically*) the increment in resistance (ΔR) to the original resistance (R_1), or

$$R_2 = R_1 + \Delta R$$

It should be realized that if the temperature coefficient is negative or if the final temperature is lower than the initial temperature, the increment ΔR will be a negative quantity, and R_2 should be lower than R_1. This explains the need for algebraic addition of R_1 and ΔR.

In Table 6-1, the coefficients as listed, apply only for an initial temperature of 20 degrees Celsius* (approximately normal room temperature) and for temperature changes in Celsius units.

TABLE 6-1   Temperature Coefficients of Resistance

Material	Temperature Coefficient (α_1) at 20°C	Temperature for Inferred Zero Resistance (°C)
Advance	0.000 05	−20 000
Aluminum	0.003 9	−236
Brass	0.002	−480
Carbon	−0.000 5 to −0.003	
Constantan	0.000 008	−125 000
Copper (annealed)	0.003 93	−234.5
German silver (18 %)	0.000 4	−2 480
Gold	0.003 4	−280
Iala	0.000 005	−200 000
Iron (pure)	0.005 5	−180
Manganin	0.000 006	−167 000
Nichrome	0.000 4	−2 480
Nickel	0.006 2	−147
Platinum	0.003	−310
Silver	0.003 8	−243
Tungsten	0.005	−180

Example 6-1

A coil wound with nickel wire has a resistance of 50 Ω when cold (20°C). When current flows through the coil, its temperature rises to 60°C. Find its hot resistance.

Solution A

1. $\Delta t = t_2 - t_1 = 60 - 20 = 40°C$

2. $\Delta R = R_1 \alpha_1 \, \Delta t = 50 \times 0.006\,2 \times 40 = 12.4 \, \Omega$

3. $R_2 = R_1 + \Delta R = 50 + 12.4 = 62.4 \, \Omega$

The mathematics in solution A could have been reduced by combining steps 2 and 3. Replacing the ΔR in step 3 by its equivalent as given in step 2, we get

$$R_2 = R_1 + (R_1 \alpha_1 \, \Delta t)$$

*The SI unit for temperature has been named *Celsius* in honor of the Swedish astronomer Anders Celsius (1701–1744).

and factoring out the R_1 term,

$$R_2 = R_1(1 + \alpha_1 \Delta t) \tag{6-2}$$

Now let us use this combination formula to solve the problem again.

Solution B

1. $\Delta t = t_2 - t_1 = 60 - 20 = 40°C$

2. $R_2 = R_1(1 + \alpha_1 \Delta t) = 50(1 + 0.006\ 2 \times 40) = 62.4\ \Omega$

In most instances either method of solution may be used. Sometimes the step-by-step method will be easier to follow (see Example 6-3 below). There is one case, however, where the combination form must be used. Such a situation arises when the resistance at 20°C is the unknown. A problem will illustrate this point.

Example 6-2

A motor field coil, wound with copper wire, has a resistance of 150 Ω at an operating temperature of 55°C. Find its resistance at 20°C.

Solution

1. $\Delta t = t_2 - t_1 = 55 - 20 = 35°C$

2. $\Delta R = R_1 \alpha_1 \Delta t$

But we do not know R_1, nor can we replace R_1 with R_2, the resistance at 55°C, since $\alpha_1 = 0.003\ 93$ applies only at 20°C. Therefore, this step is impossible. We must use the combination form:

2(a). $R_2 = R_1(1 + \alpha_1 \Delta t)$

   and

$$R_1 = \frac{R_2}{1 + \alpha_1 \Delta t} = \frac{150}{1 + 0.003\ 93 \times 35} = 132\ \Omega$$

In the problem above it was stated that we could not solve for ΔR by using $\Delta R = R_2 \alpha_1 \Delta t$, because R_2 is the resistance value at a temperature of 55°C, while α_1 is the coefficient for a temperature of 20°C. Unfortunately, the value of the temperature coefficient varies with temperature. In the case of copper, for instance, α varies from 0.004 26 at 0°C to 0.003 52 at 50°C. To use the above relationship we would have to obtain the value of α corresponding to 55°C. This complication is avoided by using the combination formula.

How could we solve a problem where neither the initial nor the final temperature is 20°C? One technique would be to include in the table of coefficients, values for each material for many temperatures. Obviously, such a table would be too involved. Another method would be to use a two-step procedure.

For example, assume that we are given the resistance R_2 at some temperature t_2 and we wish to find the resistance R_3 at some other temperature t_3, where neither t_2 nor t_3 is 20°C. First we could evaluate from R_2 at t_2, to R_1 at 20°C, as in Example 6-2. Then we could proceed from R_1 at 20°, to R_3 at t_3 as in Example 6-1. It works, but it is laborious and may be confusing.

Luckily, there is a simpler method. If we plot a curve of resistance versus temperature for any material, over a wide range of temperatures, we would find this plot is a straight line. Now if we extend this line downward in temperature, a point will be reached where the apparent or inferred resistance is zero (see Fig. 6-9). This temperature can be found mathematically. Let us start with any material with a resistance of R_1 at temperature t_1, and a temperature coefficient α_1 at this temperature t_1. In order for the resistance to drop to zero, the change in resistance (ΔR) must equal R_1, or

$$\Delta R = R_1$$

But

$$\Delta R = R_1 \alpha_1 \, \Delta t$$

Substituting R_1 for ΔR, we get

$$R_1 = R_1 \alpha_1 \, \Delta t$$

Now dividing both sides of the equation by R_1 yields

$$1 = \alpha_1 \, \Delta t$$

$$\Delta t = \frac{1}{\alpha_1}$$

The temperature for inferred zero resistance is

$$t = t_1 - \Delta t = t_1 - \frac{1}{\alpha_1} \tag{6-3}$$

Applying the above to copper ($\alpha_1 = 0.003\,93$ at 20°C), we have

$$\Delta t = \frac{1}{\alpha_1} = \frac{1}{0.003\,93} = 254.5°C$$

Since we are starting at a temperature of 20°C and resistance of copper decreases with a drop in temperature, it follows that its resistance would be zero at a temperature of 254.5°C below 20°C or −234.5°C. (This figure should be easy to remember.) In the table of temperature coefficients the temperature of inferred zero resistance for the various materials is also included.

To explain the use of the inferred zero resistance point, let us examine the curve of resistance versus temperature. Such a curve is shown in Fig. 6-9, for copper. (It should be noted that at very low temperatures, the *actual* resistance variation follows the dashed line.) At any temperature such as t_1 or t_2, a right triangle is formed with the resistance value R_1 or R_2 as the altitude and the temperature value $(234.5 + t_1)$ or $(234.5 + t_2)$ as the base. These triangles are similar, and a direct proportion exists between the lengths of their sides, so that for copper

$$\frac{R_1}{R_2} = \frac{234.5 + t_1}{234.5 + t_2} \qquad (6\text{-}4)$$

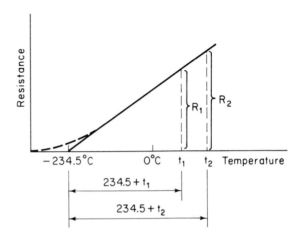

Figure 6-9 Variation of resistance with temperature for copper.

Now with this proportionality, knowing the resistance at *any* temperature, we can find the resistance at any other temperature. Let us repeat Example 6-2 by this method.

Example 6-2 (repeated)

A motor field coil, wound with copper wire, has a resistance of 150 Ω at an operating temperature of 55°C. Find its resistance at 20°C.

Solution (by similar triangles)

$$R_1 = \frac{234.5 + t_1}{234.5 + t_2} \times R_2 = \frac{234.5 + 20}{234.5 + 55} \times 150 = 132 \ \Omega$$

So far we have used this relationship between resistance change and temperature to find the change in resistance or new resistance value. By reversing the analysis, we can also use the same principles to find the change in temperature and new temperature. This technique has an important industrial application. Often it is necessary to find the operating temperature of an electrical component, and it is either impossible or impractical to measure temperature directly by thermometric means. But, if we can measure its cold and hot resistance, we can calculate the operating temperature. One such situation occurs when it is desired to find the operating temperature of a multilayer coil as used in a transformer winding. We could measure the outside surface temperature of the coil with a thermometer. But what is the temperature in the inner layers where the heat generated cannot be dissipated so readily? To make matters worse, many transformer structures are enclosed in hermetically sealed iron cases. Now how can we measure the temperature rise in the windings? I am sure you can see the advantage of calculating temperature from change in resistance.

Example 6-3

The heating element of an electric toaster has a cold resistance (20°C) of 16 Ω. When connected to a 120-V line it draws 6.0 A. The element is wound with nichrome wire. Find the operating temperature of the element.

Solution A (using α at 20°C)

1.
$$R_2 = \frac{E}{I} = \frac{120}{6.0} = 20 \,\Omega$$

2.
$$\Delta R = R_2 - R_1 = 20 - 16 = 4 \,\Omega$$

3. From $\Delta R = R_1 \alpha_1 \, \Delta t$,

$$\Delta t = \frac{\Delta R}{R_1 \alpha_1} = \frac{4}{16 \times 0.0004} = 625°C$$

4.
$$t_2 = t_1 + \Delta t = 20 + 625 = 645°C$$

Solution B (using temperature of inferred zero resistance)

1.
$$R_2 = \frac{E}{I} = \frac{120}{6} = 20 \,\Omega$$

2.
$$\frac{R_1}{R_2} = \frac{2480 + t_1}{2480 + t_2}$$

3.
$$\frac{16}{20} = \frac{2480 + 20}{2480 + t_2}$$

4. By cross multiplying, we obtain

$$16(2480 + t_2) = 20(2480 + 20)$$

$$39\ 700 + 16 t_2 = 50\ 000$$

5.
$$t_2 = \frac{50\ 000 - 39\ 700}{16} = 644°C$$

In the above illustrative problem, we were able to use either method of solution. In general, this is true whenever either t_1 or t_2 is at 20°C. The choice of method is entirely at your discretion. However, when neither starting temperature nor final temperature is at 20°C, the ratio method, using the temperature of inferred zero, is preferable.

The resistance value of commercial resistors is also affected by temperature. However, the temperature coefficient of a resistor may differ appreciably from the coefficient of the material from which it is made, depending on the heat treatment during processing and the method of construction. It is common practice for manufacturers to give the temperature coefficient of each type of resistor in *parts per million* per degree Celsius (ppm/°C). Sometimes, the "per degree Celsius" is omitted, as understood.

Example 6-4

A 200 000-Ω resistor has a temperature coefficient of $+50$ ppm/°C at 25°C. Find its resistance at 60°C.

Solution

1. $$\Delta t = 60 - 25 = 35°$$

2. $$\Delta R = (50)\frac{200\,000}{10^6}(35) = 350\ \Omega$$

3. $$R \text{ at } 60° = 200\,000 + 350 = 200\,350\ \Omega$$

6-9 Special-purpose resistors. Certain resistive components are specifically designed to have a high temperature coefficient so that their resistance varies drastically and nonlinearly with temperature. The more common of these *nonlinear* resistors are discussed below.

1. *Ballast Resistors.* These resistors are used to maintain a constant current flow in spite of changes in line voltage. When the line voltage increases, the current would normally increase. However, the ballast resistor has a high positive temperature coefficient. Even a slight increase in current causes an increase in its resistance, so that the current tends to remain constant at the original value.

2. *Fuse Resistors.* Here again a positive temperature coefficient is used to increase resistance. An increase in current (beyond some predetermined value) will increase the temperature of the resistor, increasing its resistance. The effect is cumulative, and the fuse resistor opens the circuit before damage occurs to other circuit components.

3. *Surge-Limiting Resistors.* These units are used to prevent a high inrush of current when a circuit is first energized. Obviously, they have a high negative temperature coefficient. Their resistance value is high, when cold (room temperature), and drops as they heat up, allowing the current to rise to normal operating value. They are available under a variety of trade names, such as **surgistors**, and **glo-bar**. A typical glo-bar resistor has a value of 250 Ω at room temperature, and only 20 Ω at operating temperature.

4. *Thermistors.* These thermally sensitive devices are made of solid semiconductor material (usually some variety of metallic oxide), and have an extremely high negative temperature coefficient. Thermistor elements are available in bead, rod, disk, probe, or washer forms. They are used in temperature-sensing and temperature-measuring devices, and in a variety of applications wherein changes of temperature are an indirect result. For example, a thermistor bead, suitably mounted in a waveguide is used to measure power in ultrahigh-frequency low-power circuits.

5. *Thyrites.* Another nonlinear resistor, known as a *thyrite*, is made of silicon carbide particles bonded in a ceramic matrix. Although it has a negative

temperature coefficient, the important characteristic of this material is that its resistance decreases sharply as the *voltage* across it is increased. This makes thyrite resistors useful in protecting equipment against line voltage surges as might occur from lightning discharges, or when highly inductive circuits are opened (see pages for Section 19-14 and Fig. 19-8).

REVIEW QUESTIONS

1. Name three classifications of resistors, based on the materials of construction.

2. Explain the distinction between:
   (a) Fixed and tapped resistors.
   (b) Variable and semivariable resistors.
   (c) Rheostat and potentiometer.

3. Compare the merits of wire-wound versus composition types of construction if:
   (a) High precision is required.
   (b) High resistance values (above 1 MΩ) are required.
   (c) High power rating (above 100 W) is required.
   (d) Low cost is desired.

4. With reference to the resistor color code, what is the significance of each of the following?
   (a) Yellow first band
   (b) Red second band
   (c) Orange third band
   (d) Silver fourth band
   (e) Gold third band
   (f) Four bands, left to right, green–black–brown–gold
   (g) Three bands, blue–gray–red. What is the tolerance limits for this resistor?

5. Name three types of thin-film resistors.

6. How do thin-film resistors compare with wire-wound and composition units as to:
   (a) Cost?
   (b) Accuracy (tolerance)?
   (c) Stability?
   (d) Noise level?

7. Which of the thin-film types is available in high-power ratings?

8. State three advantages of monolithic integrated circuit resistors as compared to discrete resistors.

9. Give two limitations of this type of resistor construction.

10. **(a)** Give the unit used to express the temperature coefficient of resistance of a material such as copper.
   (b) Explain what this unit means.
   (c) If a material has a negative temperature coefficient of resistance, what happens when its temperature increases?

11. We plan to operate a dc motor under heavy load for 5 h. There is danger that such operation could overheat the motor field winding. Explain how we can keep a continuous check of the temperature *inside* the winding.

12. A resistor has a rating of -30 ppm/°C. What does this mean?

13. (a) What is a *ballast* resistor used for?
    (b) Explain how it accomplishes this.

14. (a) What is a *surgistor* used for?
    (b) Explain how it accomplishes this.

15. (a) What is a *thermistor*?
    (b) State three uses of this device.

16. (a) What does a thyrite resistor have in common with each of the devices in Questions 13 through 15.
    (b) In what way does it differ?
    (c) What is it used for?

PROBLEMS AND DIAGRAMS

1. How much constantan wire must be used to make a resistor of 10 000 Ω if the gauge number is 38? (Use English units.)

2. Draw symbols for the following.
   (a) Fixed resistor
   (b) Tapped resistor
   (c) Potentiometer
   (d) Rheostat

3. Show the method for color coding the following resistors.
   (a) 240 Ω
   (b) 18 000 Ω
   (c) 350 000 Ω
   (d) 1.6 MΩ
   (e) 400 000 Ω

4. A wire-wound resistor has a resistance of 450 Ω at room temperature (20°C). Under operating conditions, its temperature rises to 160°C. It is wound with nichrome wire. Find its hot resistance.

5. A resistance element, wound with tungsten wire, has a value of 60 Ω at 50°C. Find its resistance at 85°C.

6. Find the resistance of the element in Problem 5 at 20°C.

7. A relay coil (copper wire) has a resistance of 70 Ω at 80°C.
   (a) Find its resistance at 30°C.
   (b) At what temperature will its resistance drop to 60 Ω?

8. The armature resistance of a dc motor (copper wire) rises from 0.054 Ω at room temperature (20°C) to 0.062 Ω after 6 h of operation. If the allowable temperature rise is 30°C, does this machine meet its specification?

9. A transformer winding (copper wire) has a resistance of 12.0 Ω at 35°C, and 14.1 Ω after several hours of operation. Find the temperature of the winding.

10. A 100-W 120-V tungsten filament lamp has a cold resistance (20°C) of 10.5 Ω. Find the operating temperature of the filament when connected to a 120-V supply.

7

SERIES CIRCUITS

7-1 Voltage drop. An electrical circuit with a voltage or difference of potential across it, and electrons moving through it, may very well be compared to a "gravitational circuit" (Fig. 7-1).

A rock at the top of the hill will have an amount of potential energy per pound equivalent to its elevation of 100 m. As the rock falls, its elevation is decreasing and therefore its potential energy is decreasing. The energy is not lost; in this case, it is being converted partly to *energy of motion*, or *kinetic energy*, and partly to heat energy as the rock rubs against the surface of the hill. The loss of potential energy is due to the loss of elevation. When the rock reaches point *B*, it has suffered a loss (drop) of potential energy equivalent to 40 m (per unit weight). It is now at elevation 60 m, and its potential energy per unit weight is equivalent to 60 m.

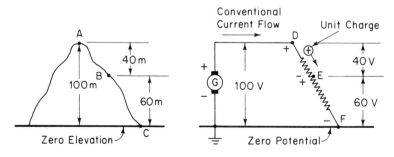

Figure 7-1 Voltage drop and fall in potential energy.

The electrical circuit is similar. Using conventional-current terminology, positive unit charges at point D have potential energy equivalent to 100 V, due to the work done by generator G in "raising" the unit charges to that potential. As the unit charges "drop" toward the zero-potential level through the resistor, they do work. This work appears as the heat energy dissipated by the resistor. Where did this heat energy come from? It is potential energy of the charges, converted to heat. As the unit charges drop to point E, they heat up the portion of the resistor $D–E$, and lose some potential energy. The potential at point E must be less than the potential at point D, or there is a fall in potential (energy) from D to E. Since the potential at any point is measured in volts, we say that there is a drop in volts—or a *voltage drop*—between D and E. This voltage drop occurs because of the work done by the charges "falling down the electrical hill"—or, in other words, by current (I) flowing through some resistance (R). A voltage drop is therefore also called an **IR drop**, and is calculated as the product of (current through the resistance) and (resistance of that portion of the circuit being considered).

7-2 Series circuit. You have often heard the term "series circuit." Just what constitutes a series circuit? If I gave you three pieces of electrical equipment and told you to connect them in series across the line, what mechanical wiring would you have to do? You should connect one terminal of the first unit to the line, the other terminal of that unit to one terminal of the second unit, the other terminal of the second unit to one terminal of the third unit, and the second terminal of this third unit to the opposite line terminal. To summarize:

1. A series connection consists of more than one unit connected in sequence.

Such a "sequential" circuit connection is shown in Fig. 7-2.

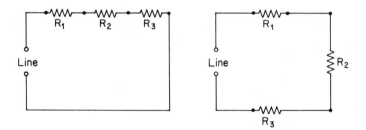

Figure 7-2   Three resistors wired in series.

A *schematic* diagram would be drawn as in either of the diagrams shown in Fig. 7-2. However, in practice, an actual *wiring* diagram would depend on the physical location of the components. As long as the wiring between the units maintains the sequence of connections, the units may be twisted around to any position. The diagrams in Fig. 7-3 are also series circuits.

Now let us analyze these diagrams electrically. Make the upper line

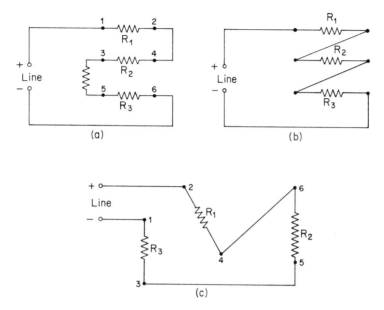

Figure 7-3 Series circuit connections.

terminal positive in potential and the lower terminal negative. Consider a positive unit charge at the positive line terminal. By how many ways can it get to the negative terminals? You will find it must go through the first unit, R_1, then through the second unit, R_2, and finally through the third unit R_3, and to the negative terminal. For an electrical analysis of a series circuit we can then say:

2. There is only one path for the current in a series circuit.

Let us analyze a series circuit by means of an analogy with which you are all familiar. Mathematically, this analogy is not stricly correct in all technical details, but it does present the concept very clearly. At one time or another you have all seen freight trains. In general, the longer the train, the slower it moves; or when high speeds are desired, either a more powerful locomotive or more than one locomotive is used. What does this mean?

1. The greater the pull of the locomotive, the greater the speed, or, conversely,
2. The greater the drag (or resistance) of the freight-car load, the slower the speed.

If the pull of the locomotives exceeds the drag (bearing friction, rolling friction, and wind resistance) of the cars, the train will speed up. When it speeds up, the drag will increase. The speed will become constant when the drag equals the pull. On the other hand, if the drag of the train is increased (as by heavier cars or by pulling up a hill), the train slows down, decreasing the drag until the pull of the engine again equals the drag. In many ways this effect is very similar to an electrical circuit. The pull of the locomotives is the difference in potential, or voltage across the circuit; the drag of the cars is the *IR* drop of the circuit; the speed of the train in miles per hour is the current in the circuit in coulombs per second. Compare statements 1 and 2 above with Ohm's law:

1. The greater the voltage, the greater the current.
2. The greater the resistance, the lower the current.

Also, if the resistance of the circuit is increased, the *IR* drop would tend to increase. The current must "slow down" until the *IR* drop again equals the applied voltage.

Now consider a locomotive and three cars, *A*, *B*, and *C*. If the locomotive were pulling car *A* alone, the drag would be low and the speed would be high. When we add car *B*, the drag is increased, and the train will slow down. Now we add the third car. The drag is increased again, and the locomotive will slow down further. Let us assume that the locomotive speed is now 20 mph. How fast is car *A* moving? At 20 mph. How fast is the second car moving? Also at 20 mph. The third car? Again 20 mph. No matter how many cars there are, all cars must move at the same speed. Then, if the speed changes, all cars will move at this new speed. For a given locomotive pull, what does the speed of the train depend on? The drag of the first car alone? The drag of the second car alone? No! It depends on the sum of the drags of the individual cars. In other words, the speed of the train will change until the sum of the drags of each car equals the pull of the locomotive.

Applying these same principles to a series circuit containing several resistors, we can say:

1. The current in the circuit depends on the total resistance (R_T) of the circuit.
2. The total resistance of the circuit is equal to the sum of the individual unit resistances. Expressed algebraically, we have

$$R_T = R_1 + R_2 + R_3 + \cdots \qquad (7\text{-}1)$$

3. Whatever the circuit current is, that same current flows through each of the units.
4. The applied voltage is equal to the sum of the volt drops (*IR*) of the individual units, or

$$E_T = E_1 + E_2 + E_3 + \cdots \qquad (7\text{-}2)$$

This statement is often given as ***Kirchhoff's law of potentials***: *In any one path of a complete electrical circuit, the sum of the EMFs and of all voltage drops, taken with proper signs, is equal to zero.* Kirchhoff's laws are discussed in more detail in Chapter 22.

Let us apply this information to a problem.

Example 7-1

Three resistors, of 10, 30, and 20 Ω, respectively, are connected in series across a 120-V supply. Find:

(a) The current through each resistor.

(b) The voltage drop across each resistor.

(c) The power dissipated by each resistor and total power.

Solution   The first step is to draw the diagram and label all the parts (see Fig. 7-4).

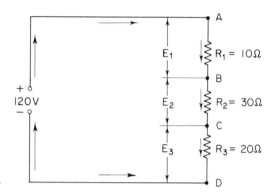

Figure 7-4

(a) Current through each resistor.

1. The current in the circuit depends on the total resistance:

$$R_T = R_1 + R_2 + R_3 = 10 + 30 + 20 = 60 \ \Omega$$

$$I_T = \frac{E_T}{R_T} = \frac{120}{60} = 2.0 \ \text{A}$$

2. The current is the same in all the units:

$$I_1 = I_2 = I_3 = I_T = 2.0 \ \text{A}$$

(b) Voltage drop across each resistor.

3. Applying Ohm's law to *each resistor* to find the individual voltage drops:

$$E_1 = I_1 \times R_1 = 2.0 \times 10 = 20 \ \text{V}$$

$$E_2 = I_2 \times R_2 = 2.0 \times 30 = 60 \ \text{V}$$

$$E_3 = I_3 \times R_3 = 2.0 \times 20 = 40 \ \text{V}$$

4. The applied voltage equals the sum of the individual voltage drops:

$$E_T = E_1 + E_2 + E_3 = 20 + 60 + 40 = 120 \ \text{V (check)}$$

(c) Power dissipated by each resistor and total power.

5. To find the power dissipated in each resistor, apply the power equation to each case:

$$P_1 = E_1 \times I_1 = 20 \times 2.0 = \ \ \ 40 \ \text{W}$$
$$P_2 = E_2 \times I_2 = 60 \times 2.0 = 120 \ \text{W}$$
$$P_3 = E_3 \times I_3 = 40 \times 2.0 = \ \ \underline{\ 80 \ \text{W}}$$
$$\text{Total power} = 240 \ \text{W}$$

6. As a check, apply the power equation to the entire circuit:

$$P_T = E_T \times I_T = 120 \times 2 = 240 \text{ W (check)}$$

If the solution of the problem is done step by step and the proper subscripts are used in each step, the process is very simple.

7-3 Voltage–resistance–power ratios. Notice in the solution of Example 7-1 that the voltage drop E_3 is twice the voltage drop E_1. Also notice that the resistance R_3 is twice the R_1 value. Similarly, $E_2 = 3E_1$ and $R_2 = 3R_1$. In other words, the voltage drops in a series circuit are directly proportional to the respective resistance value. This applies not only to the individual values, but also to the total value ($E_T = 6E_1$, and $R_T = 6R_1$). Expressed mathematically, we have

$$\frac{E_x}{E_1} = \frac{R_x}{R_1} \quad \text{or} \quad \frac{E_x}{E_T} = \frac{R_x}{R_T} \tag{7-3A}$$

By algebraic transposition, this equation can be rewritten as

$$\frac{E_x}{R_x} = \frac{E_1}{R_1} = \frac{E_T}{R_T} \tag{7-3B}$$

But the ratio E_1/R_1 is (by Ohm's law) the current through R_1; and E_x/R_x is the current through component x. This relation merely shows (what we already knew) that the same current flows in each unit of a series circuit. Now, solving equation (7-3B) for E_x, we get

$$E_x = \frac{E_1}{R_1} R_x = \frac{E_T}{R_T} R_x \tag{7-3C}$$

Step 5 of the solution of Example 7-1 also brings out another point. The power dissipated by each resistor is not equal, but rather is directly proportional to their respective resistance values. For example, R_2 (twice R_1) dissipates twice as much power as R_1, whereas R_T, which is six times R_1, dissipates six times more power than R_1.

7-4 General problem-solving technique. There are numerous variations possible in series-circuit problems. It would be impossible to explain how to solve each type; the method of attack varies with each problem. One general rule can be made: *Start with that portion of a circuit where two factors are known.* This method will invariably give you two factors about some other portion. In Example 7-1 the total voltage was given. Therefore, the method used was to find total resistance and total current. Now, let us apply this general technique to another type of problem.

Example 7-2

Three resistors are connected in series across a supply voltage. The drop across R_1 is 40 V, the voltage drop across R_3 is 100 V, the power dissipated in R_2 is 40 W, and the current in the circuit is 2.0 A. Find the values of each resistor, the voltage drop across R_2, and the total voltage.

Solution

1. Draw the diagram and label all known values (see Fig. 7-5).

2. The current is the same in all the units. Therefore,

$$I_1 = I_2 = I_3 = I_T = 2.0 \text{ A}$$

3.

$$R_1 = \frac{E_1}{I_1} = \frac{40}{2.0} = 20 \ \Omega$$

$$R_2 = \frac{P_2}{I_2^2} = \frac{40}{2.0 \times 2.0} = 10 \ \Omega$$

$$R_3 = \frac{E_3}{I_3} = \frac{100}{2.0} = = 50 \ \Omega$$

4.

$$E_2 = \frac{P_2}{I_2} = \frac{40}{2.0} = 20 \text{ V}$$

5.

$$E_T = E_1 + E_2 + E_3 = 40 + 20 + 100 = 160 \text{ V}$$

6.

$$R_T = R_1 + R_2 + R_3 = 20 + 10 + 50 = 80 \ \Omega$$

$$E_T = I_T \times R_T = 2 \times 80 = 160 \text{ V (check)}$$

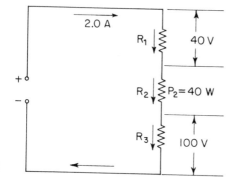

Figure 7-5

To summarize: Start the problem where two factors are known. Ohm's law or the power equation can be applied to any portion of a circuit or to the entire circuit, but *the values used must apply to that portion only, or to the entire circuit only.* Use subscripts on all formulas. Solve the problem step by step. Draw the diagram so you can see what you are doing.

REVIEW QUESTIONS

1. With reference to Fig. 7-1:
   (a) What does the 40-V indication signify?
   (b) What does the 60-V indication signify?
   (c) What causes these voltage effects?
2. Explain *mechanically* why the schematics shown in Fig. 7-2 are series circuits.

3. With reference to Fig. 7-3:
   (a) Which of these three schematics is *not* a series circuit? Explain *mechanically*.
   (b) Trace a path (by the numbers) for the current flow in diagram (a).
   (c) Is there any other path?
   (d) Trace a path for current flow in diagram (c). Is there any other path?
   (e) Explain *electrically* why these are series circuits.

4. With reference to Fig. 7-3(b):
   (a) Which resistance value determines what the line current will be?
   (b) Which resistance value determines what the current through R_2 will be?
   (c) $R_1 = 10\,\Omega$ and $R_3 = 30\,\Omega$. The current I_1 flowing through R_1 is 5 A. What is the current I_3 flowing through R_3?

5. With reference to Example 7-1 and Fig. 7-4:
   (a) What is the voltage drop from A to B? From A to C? From B to D?
   (b) What is the resistance value from A to B? From A to C? From B to D?
   (c) What is the current flowing from A to B? From A to C? From B to D?
   (d) Give a name for the relationship involved in step 4 of the solution.

6. From direct observation of the resistance values in Fig. 7-4:
   (a) Compare the voltage drops across R_2 and R_1.
   (b) Compare the power dissipated by R_1 and R_3.
   (c) Compare the total voltage and the drop across R_2.

7. When solving series circuit problems is the first step, generally, to find the total resistance so that we can then find the total current? Explain.

8. In Example 7-2, since the total current is known, should we use the given 40 V, the given 100 V, or their sum (140 V), to find total resistance by Ohm's law? Explain.

PROBLEMS

1. The following resistors are connected in series across a 120-V supply:
   $R_1 = 25\,\Omega$, $R_2 = 45\,\Omega$, $R_3 = 30\,\Omega$. Find:
   (a) The current in each resistor.
   (b) The voltage drop across each resistor.
   (c) The power dissipated by each resistor.
   (d) The power rating of each resistor.

2. Three resistors are connected in series across a supply line, such that the volt drop in $R_1 = 90$ V, the voltage drop in $R_2 = 50$ V, the power dissipated in $R_2 = 5$ W, and the power dissipated in $R_3 = 6$ W. Find the line voltage and the value of each resistor.

3. A lamp rated at 120 V and 200 W is to be used on a 220-V line.
   (a) What size resistor should be connected in series with the lamp for correct operation?
   (b) What should the power rating of this resistor be?

4. A 120-V 100-W soldering iron overheats while not in use and burns the tips. If the voltage across the iron is dropped to 95 V while the iron is not in use, a good temperature is maintained. What should the rating of a series resistor be for this purpose? (Assume that the resistance of the iron remains constant.)

5. The partial FET amplifier circuit shown in Fig. 7-6 has the following components connected in series across a 38-V supply: (Consider the FET as a resistor, R_i.)

Drain resistor, $R_D = 20\ 000\ \Omega$

Source resistor, $R_S = 2000\ \Omega$

Internal FET resistance, $R_i = $ unknown

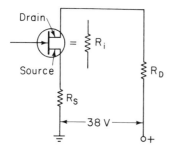

Figure 7-6

The current through the circuit is 1.2 mA. Find:

(a) The potential of the source.

(b) The potential of the drain.

(c) The drain-to-source voltage.

(d) The value of R_i.

(As shown in Sections 4-2 and 4-3, potential means the voltage with respect to ground. The ground is the understood reference point.)

6. A string of eight Christmas tree lights is connected in series. It is designed to operate on a 120-V line. If each lamp has a resistance of 15 Ω, find the line current.

7. A subway system uses banks of six lights in series on a 660-V supply. Each lamp is rated at 40 W.

(a) What is the proper voltage rating for each lamp?

(b) Find the current drawn by each bank of lamps.

8. Another subway system uses 20 lamps in series across a 660-V supply. The lamps are rated at 40 W. Find the resistance of each lamp.

9. Two voltmeters are connected in series across a 550-V line. The resistances of the meters are 22 000 Ω and 40 000 Ω, respectively. What will each meter read? (*Hint:* The "reading" is the same as its voltage drop.)

10. A 300-V (full-scale reading) voltmeter has a resistance of 30 000 Ω. It is connected in series with a 40 000-Ω resistor, and an unknown voltage is connected across the combination. The meter reads 250 V. Find the value of the supply voltage.

11. A motor armature has a resistance of 0.5 Ω. It is to be connected across a 120-V line. What value of series resistor is needed to limit the starting current to 20 A?

12. A motor field circuit draws 1.5 A when connected to the 120-V line. It is desired to reduce the field current to 1.2 A. Give full specifications for the series resistor needed. (Assume a safety factor of 2.)

13. A load having an equivalent resistance of 10 Ω is located 600 m from a 230-V supply source. The resistance of the feeders is 0.33 Ω/km (each wire). Find:

(a) The voltage drop in the feeders.

(b) The voltage at the load.

(c) The power loss in the feeders.

14. A 110-V 5-kW load is located 1200 ft from the supply source. The feeders are No. 2 wire. Find the supply voltage needed for a load voltage of 110 V. (Feeders are copper wire at 25°C.) Use English units.

15. Find the minimum wire size that can be used as feeders to supply a 20-A lighting load from a 120-V supply source when the load is 1800 ft away and the voltage drop must not exceed 5%. (Feeders are copper wire at 25°C.) Use English units.

8

PARALLEL CIRCUITS

You have often heard the term "in parallel," but just what does it mean? If I gave you three pieces of electrical equipment and told you to connect them in parallel, how would you do it?

8-1 Defining a parallel circuit. A parallel circuit consists of the following:

Rule 1. More than one unit, with one terminal of each wired together and connected to one line terminal, and the other terminal of each also wired together and connected to the other line terminal. The simplest way of doing this wiring is shown in Fig. 8-1.

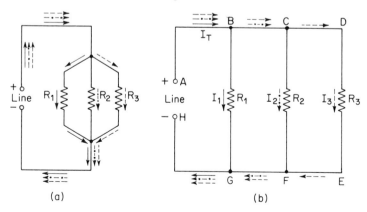

Figure 8-1 Parallel circuit connections.

These resistors are obviously in parallel. However, regardless of the physical placement of the components of a circuit, they will still be connected in parallel if the wiring conforms with the above rule. Close analysis of the circuits in Fig. 8-2 will show that these three resistors are also in parallel.

You will notice that in each case of Fig. 8-2, one end of each resistor is connected to the other resistors and to one line terminal, and that the other end of each resistor is also connected to the other resistors and to the other line terminal. Do not let the physical appearance of a circuit fool you. Trace the connections carefully!

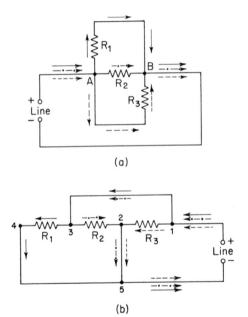

(a)

(b)

Figure 8-2 More parallel circuit connections.

Rule 2. There are as many independent paths as there are units in parallel. In each of these diagrams trace a positive unit charge from the positive line terminal to the negative. How many possible paths does it have? You will notice that the unit charge can go either through R_1 or through R_2 or through R_3.

8-2 Voltage distribution.   Referring to Fig. 8-1(b), points A and B are connected together by a wire. The resistance of the wire is negligible. Regardless of the current through the wire, the voltage drop across it is also negligible. These two points are at the same potential. Similarly, the voltage drop from G to H is also negligible. This makes the difference in potential between B and G the same as that between A and H. But the difference in potential from A to

H is the line voltage (E_T), and the difference in potential from B to G is the voltage across R_1 (or E_1). Therefore, E_1 is equal to E_T. Furthermore, the voltage drops from B to C and from F to G are also negligible, or E_1 is equal to E_2. From similar reasoning, E_2 is equal to E_3. We can then say:

Rule 3. In a parallel circuit the voltage across each unit is the same.

8-3 Current distribution.   Current flows from positive to negative. Referring again to Fig. 8-1(b), all the current in the circuit must come from point A. At point B, part of the current (I_1) goes through R_1. At point C, another part of the current (I_2) goes through R_2. The remainder of the current (I_3) goes through R_3. At point F, I_2 joins with I_3. Then, at point G, I_1 joins with I_2 and I_3. All the current returns to the negative terminal. The current in the wires A to B and G to H contains the three branch components. We can therefore say:

Rule 4. In a parallel circuit the total current (I_T) is the sum of the individual branch currents.

This statement is often given as ***Kirchhoff's law of currents***: *In any electrical network the algebraic sum of the currents flowing to a junction must equal the algebraic sum of the currents flowing away from that junction.* Let us apply this law at junction C of Fig. 8-1(b). The currents flowing away from the junction are I_2 and I_3. Therefore, the current flowing into the junction must equal $I_2 + I_3$. Kirchhoff's laws are discussed in more detail in Chapter 22.

In Section 3-7, we learned that the total conductance is the sum of the individual conductances. But resistance is the opposite of conductance, so that if the conductance is increasing, the resistance is decreasing. Therefore, in a parallel circuit the conductance increases, or

Rule 5. The equivalent resistance of the total circuit, (R_T)* is less than the smallest branch resistance.

Let us see how these rules for a parallel circuit are applied to a problem.

Example 8-1

A 240-Ω, a 600-Ω, and a 400-Ω resistor are connected in parallel across a 180-V line (Fig. 8-3). What is the total current and the equivalent resistance?

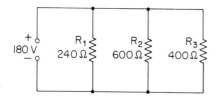

Figure 8-3

---

*We are using the symbol R_T for the equivalent resistance of a parallel circuit, because it is the resistance of the *total circuit*, and matches I_T for total current. Be careful, however, not to confuse it with the total resistance of a series circuit.

Solution

1. The voltage across each unit is the same:

$$E_1 = E_2 = E_3 = E_T = 180 \text{ V}$$

2. The total current is the sum of the individual branch currents. Let us find the branch currents.

$$I_1 = \frac{E_1}{R_1} = \frac{180}{240} = 0.75 \text{ A}$$

$$I_2 = \frac{E_2}{R_2} = \frac{180}{600} = 0.30 \text{ A}$$

$$I_3 = \frac{E_3}{R_3} = \frac{180}{400} = 0.45 \text{ A}$$

$$I_T = \overline{1.50 \text{ A}}$$

3. The equivalent resistance is less than the smallest branch resistance. In this case the equivalent resistance must be less than 240 Ω. How can we find the exact value? The three branches combined are drawing 1.50 A from the 180-V line. What one resistor would do the same thing?

$$R_T = \frac{E_T}{I_T} = \frac{180}{1.50} = 120 \text{ Ω}$$

This one resistor draws the same current as the three resistors in parallel. The single resistance circuit of 120 Ω is equivalent to the three resistors in parallel. R_T must be the equivalent resistance of the circuit.

8-4 Equivalent resistance calculation. It is often necessary to find the equivalent resistance of a parallel circuit, but the voltage across the circuit is not known. We could, of course, assume a voltage and still use the above method. However, a more direct method is preferable. When units are connected in parallel, the total conductance increases, or

$$G_T = G_1 + G_2 + G_3 + \cdots$$

But resistance is the reciprocal of conductance; therefore,

$$\frac{1}{R_T} = \frac{1}{R_1} + \frac{1}{R_2} + \frac{1}{R_3} + \cdots \tag{8-1}$$

Applying this method to the problem above, we substitute the known values of R and get

1.
$$\frac{1}{R_T} = \frac{1}{240} + \frac{1}{600} + \frac{1}{400}$$

Beyond this point, the rest is arithmetic. Some persons handle it by solving each fraction. If you are solving "by hand," and the size of the resistors is around 350 000 Ω, this method results in some inconvenient decimals. Others prefer to find the least common multiple. This method is also awkward if the values are very uneven. The method recommended here is as follows:

2. Pick out the largest known resistance and the unknown resistance ($600R_T$).

3. Multiply both sides by the product $600R_T$:

$$\frac{600R_T}{R_T} = \frac{600R_T}{240} + \frac{600R_T}{600} + \frac{600R_T}{400}$$

4. Simplify:

$$600 = 2.5R_T + 1R_T + 1.5R_T$$

5. Collect your terms:

$$5R_T = 600$$

6. Reduce:

$$R_T = \frac{600}{5} = 120\ \Omega$$

This procedure may seem laborious at first, but with practice you will find that the only steps required are steps 1, 4, and 6. This method can also be used to find one resistor if the equivalent resistance and the other resistor values are known.

Of course, a calculator will simplify the operation tremendously. Solving the equation in step 1 above, we would enter:

$$2, 4, 0,\ \boxed{1/x},\ \boxed{+},\ 6, 0, 0,\ \boxed{1/x},\ \boxed{+},\ 4, 0, 0,\ \boxed{1/x},\ \boxed{=},\ \boxed{1/x}$$

At the equals sign in the calculator entry above we have the sum of the reciprocals, $1/240 + 1/400 + 1/600$, which is $1/R_T$. Then, to get R_T, we need another reciprocal. This accounts for the last entry of $\boxed{1/x}$.

8-5 Special case—equal resistors.   If two equal resistors are connected in parallel, the conductance doubles and the resistance halves. If three equal units are connected in parallel, the conductance is three times the value of one unit, and the equivalent resistance is one-third the value of one unit. In general, the equivalent resistance is equal to the resistance of one unit divided by the number of units.

Example 8-2

A number of 600-Ω resistors are to be connected in parallel. What is the equivalent resistance when 2, 3, 4, or 10 units are used?

Solution

2 in parallel:	$R_T = \frac{600}{2} = 300\ \Omega$
3 in parallel:	$R_T = \frac{600}{3} = 200\ \Omega$
4 in parallel:	$R_T = \frac{600}{4} = 150\ \Omega$
10 in parallel:	$R_T = \frac{600}{10} = 60\ \Omega$

This gives us another technique for *estimating* the equivalent resistance for any combination of resistors in parallel. We had already seen that R_T must be less than the lowest given value. Now, the equal-resistor illustration gives us a *minimum* limit for R_T. It must be more than the lowest given value divided by the number of resistors in parallel. Applying this to Example 8-1, the R_T value must be

1. Less than 240 Ω, but
2. More than 80 Ω ($\frac{240}{3}$)

Use this "by-sight" estimating technique to check any parallel circuit R_T calculation for gross errors.

8-6 Special case—two resistors.   It often happens that only two resistors are connected in parallel. The general method given above can be used. By algebraic transposition, a simpler form of this equation is available:

$$\frac{1}{R_T} = \frac{1}{R_1} + \frac{1}{R_2} < R_T \times R_1 \times R_2$$

$$\frac{R_T R_1 R_2}{R_T} = \frac{R_T R_1 R_2}{R_1} + \frac{R_T R_1 R_2}{R_2}$$

$$R_1 R_2 = R_T R_2 + R_T R_1$$

Factoring yields

$$R_T(R_1 + R_2) = R_1 \times R_2$$

$$R_T = \frac{\text{product}}{\text{sum}} = \frac{R_1 \times R_2}{R_1 + R_2} \qquad \textbf{(8-2A)}$$

When two resistors are in parallel, the equivalent resistance is equal to the product of the two resistances divided by their sum.

Example 8-3

What is the equivalent resistance of a 20-Ω and a 30-Ω resistor in parallel?

Solution

$$R_T = \frac{R_1 \times R_2}{R_1 + R_2} = \frac{20 \times 30}{20 + 30} = 12\ \Omega$$

Calculator entry:

$$2, 0, \boxed{\times}, 3, 0, \boxed{\div}, \boxed{(}, 2, 0, \boxed{+}, 3, 0, \boxed{)}, \boxed{=}$$

On the other hand, if we want some specific resistance value, and the resistor we have is too high a resistance value, we can add a second resistor in parallel to reduce the "total" resistance. For this purpose equation (8-2A) can be rewritten to solve for either branch value, when the other branch and the equivalent resistance are known (see Problem 4).

$$R_1 = \frac{\text{product}}{\text{difference}} = \frac{R_2 \times R_T}{R_2 - R_T} \qquad \textbf{(8-2B)}$$

8-7 Current ratio.   It often happens in a parallel circuit that we know the branch resistances and the total current in the circuit, and we want to find how this current divides among the branches. The general method can always be used as follows:

1. Find the equivalent resistance.
2. All the current flows through the equivalent resistance, and we can find the voltage drop.
3. This is also the voltage across each branch.
4. Knowing voltage and resistance for each branch, we can find the individual currents.

When the circuit has only two resistors, the work can be simplified by algebraic transformations, as follows:

The voltage drop across each branch is the same, or

$$I_1 R_1 = I_2 R_2 \quad \text{but} \quad I_2 = I_T - I_1$$

Substitute:

$$I_1 R_1 = I_T R_2 - I_1 R_2$$

Transpose:

$$I_1 R_1 + I_1 R_2 = I_T R_2$$

Factor:

$$I_1(R_1 + R_2) = R_2 \times I_T$$

$$I_1 = \frac{R_2}{R_1 + R_2} \times I_T \qquad \text{(8-3A)}$$

Similarly, it can be shown that

$$I_2 = \frac{R_1}{R_1 + R_2} \times I_T \qquad \text{(8-3B)}$$

A general rule can therefore be stated as follows:

$$\text{Current divides in the ratio:} \frac{\text{opposite resistance}}{\text{sum of resistances}} \times I_T$$

Example 8-4

A current of 10 A is flowing in a circuit consisting of a 10-Ω and a 40-Ω resistor in parallel (Fig. 8-4). What is the current in each branch?

Solution

$$I_1 = \frac{\text{opposite } R}{\text{sum}} \times I_T = \frac{R_2}{R_1 + R_2} \times I_T = \frac{40}{10 + 40} \times 10 = 8 \text{ A}$$

$$I_2 = \frac{\text{opposite } R}{\text{sum}} \times I_T = \frac{R_1}{R_1 + R_2} \times I_T = \frac{10}{10 + 40} \times 10 = 2 \text{ A}$$

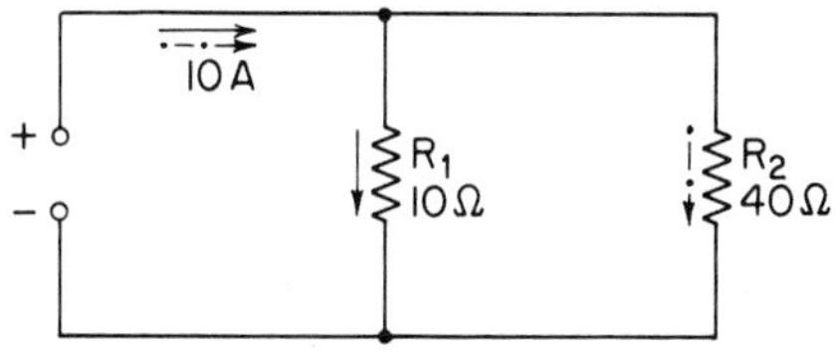

Figure 8-4

8-8 Series versus parallel circuit—comparison. Now that we have discussed both the series and the parallel circuit, a comparison between them would be in order. This is done in the following chart and in Fig. 8-5.

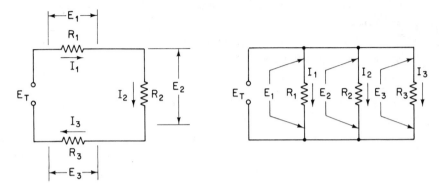

Figure 8-5 Comparison—series versus parallel circuit.

	Series	Parallel
Resistance	The total resistance (R_T): 1. Is greater than any component resistance 2. Is equal to the sum of the individual component resistances 3. $R_T = R_1 + R_2 + R_3 + \cdots$	The equivalent resistance of the total circuit (R_T): 1. Is *less* than the lowest branch resistance 2. Its reciprocal is equal to the sum of the reciprocals of the individual branch resistances
Current	1. Only one path for current to flow 2. The current through each component is the same, regardless of its resistance value 3. $I_T = I_1 = I_2 = I_3 = \cdots$	1. Has as many paths for current flow as there are branches 2. The branch with the lowest resistance has a proportionately higher current flow 3. $I_T = I_1 + I_2 + I_3 + \cdots$
Voltage	1. The component with the highest resistance has a proportionately higher voltage drop across it 2. The sum of the voltage drops across each component is equal to the supply voltage 3. $E_T = E_1 + E_2 + E_3 + \cdots$	1. The voltage across each branch (load) is the same as the supply voltage, regardless of the individual branch resistances 2. $E_T = E_1 = E_2 = E_3 = \cdots$

REVIEW QUESTIONS

1. (a) From *mechanical connections* explain why Fig. 8-2(a) is a parallel circuit.
   (b) By how many paths can a positive charge at junction A reach junction B? Explain.

2. (a) In Fig. 8-2(b), using the numbers, trace one path for current flow from the plus to the minus line terminal.
   (b) Trace a second path.
   (c) Are there any other paths? Explain.

3. With reference to Fig. 8-3:
   (a) What is the voltage across R_1? Why is this so?
   (b) What is the voltage across R_2? Across R_3?
   (c) Make a general statement to cover the above relations.

4. With reference to Fig. 8-1(b):
   (a) What do the three arrows above the wire A–B signify?
   (b) What happens at junction B?
   (c) What happens at junction F?
   (d) State a general rule that applies to any junction.
   (e) What is this rule known as?

5. Five resistors are connected in parallel. Between what maximum and minimum limits will the equivalent resistance fall?

6. A 100-, 200-, 300-, and a 400-Ω resistor are connected in parallel. Between what maximum and minimum limits will the equivalent resistance fall?

7. With reference to Fig. 8-4:
   (a) Which branch (R_1 or R_2) will carry more current?
   (b) What is the ratio of the *branch* currents?
   (c) What portion of the *total* current flows through R_2?
   (d) What portion of the total current flows through R_1?

PROBLEMS

1. Three resistors of 40 Ω, 90 Ω, and 75 Ω, respectively, are connected in parallel.
   (a) Find the equivalent resistance. Make a rough check using maximum and minimum limits.
   (b) If they are connected across a 120-V line, find the current in each resistor and the total current for the circuit.
   (c) Find the power dissipated by each resistor.

2. What is the equivalent resistance of a 25-Ω and a 100-Ω resistor in parallel?

3. What is the equivalent resistance of five 80-Ω resistors in parallel?

4. Solve equation (8-2A) algebraically for R_2.

5. A resistance of 500-Ω is connected across a 220-V supply. How much resistance must be added in parallel to increase the total current to 1.0 A?

6. The total current in a circuit containing an 18-Ω and a 45-Ω resistor in parallel is 5.0 A. What is the current in each resistor?

7. In Problem 1, if the total current is 2.0 A, find:
   (a) The line voltage.
   (b) The current in each resistor.

8. Find the equivalent resistance of a 250-Ω, a 100-Ω, and a 500-Ω resistor connected in parallel. Make a rough check using maximum and minimum limits.

9. Two resistors in parallel have an equivalent resistance of 3250 Ω. One of the resistors is marked as 5000 Ω. What is the second resistance?

10. A circuit has a resistance of 2500 Ω. To reduce the resistance to 1540 Ω, what resistance must be connected in parallel with the 2500-Ω resistor?

11. A circuit has three branches—6 Ω, 10 Ω, and 14 Ω, respectively. If 4.0 A flow through the 14-Ω branch, find the current in the other two branches.

12. A store lighting system consists of 25 lamps of 50 W each, connected in parallel on a 120-V line. Find the equivalent resistance of the system.

13. The series field of a compound motor has a resistance of 0.25 Ω. It is desired to reduce the current flowing through this winding to 80%, by connecting a diverter (resistor) across the winding. Find the resistance required for the diverter.

14. A 0–1-A ammeter has a resistance of 0.12 Ω. A shunt (resistor) of 0.03 Ω is connected across it. What is the total current flowing through this combination when the meter registers 0.5 A?

15. What voltage is required to send 6.0 A through the combination of 14-, 18-, and 8-Ω resistors in parallel?

9

SERIES–PARALLEL CIRCUITS

Do you understand the analysis of series circuits? Of parallel circuits? If you do, a few guiding pointers are all that is necessary now.

1. A series–parallel circuit can consist of any combination of series and parallel elements:
   (a) It may contain a group of two or more units in parallel, connected in series with single units or other parallel groups.
   (b) It may contain a group of two or more units in series, connected in parallel with single units or other series units.
   (c) It may contain combinations of (a) and (b).
2. Start the problem where two factors are known.
3. To the series components apply the series-circuit rules.
4. To the parallel components apply the parallel-circuit rules.

9-1 Equivalent resistance. When the only voltage given in the circuit is the applied voltage (E_T), the only current that can be found is the total current (I_T). This condition makes it necessary for you to calculate the equivalent resistance of the entire circuit (R_T).

Let us analyze a few circuits.

Example 9-1 (See Fig. 9-1)

1. Resistors R_2, R_3, and R_4 are in parallel with each other. They can be replaced by their equivalent resistance (R_A).

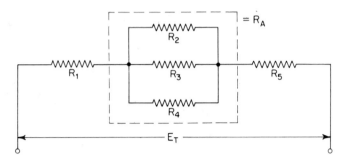

Figure 9-1 Series–parallel circuit.

2. R_1, R_A, and R_5 are in series with each other.

3. $$R_T = R_1 + R_A + R_5$$

Example 9-2 (See Fig. 9-2)

1. R_2 and R_3 are in parallel with each other. They can be replaced by their equivalent resistance (R_A).

2. R_4, R_5, and R_6 are in parallel with each other and can be replaced by their equivalent resistance (R_B).

3. R_1, R_A, R_B, and R_7 are in series with each other.

4. $$R_T = R_1 + R_A + R_B + R_7$$

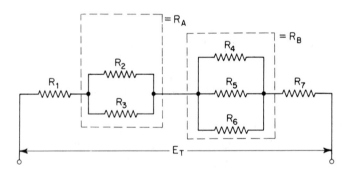

Figure 9-2 Series–parallel circuit.

Example 9-3 [See Fig. 9-3(a)]

1. R_2 and R_3 are in series with each other. Add them. Call this value R_A.

2. R_A is in parallel with R_4. Find their equivalent resistance. Call this value R_B [see Fig. 9-3(b)].

3. R_B is in series with R_6. Add them. Call this value R_C.

4. R_C is in parallel with R_5. Find their equivalent resistance. Call this R_D. [See Fig. 9-3(c).]

5. R_1, R_D, and R_7 are in series with each other. Their sum is R_T.

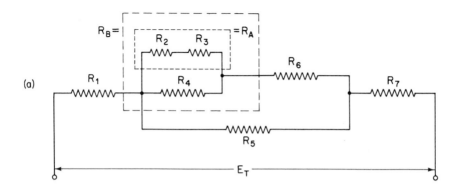

(a)

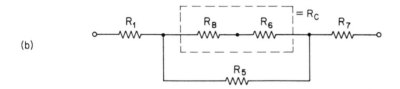

(b)

(c)

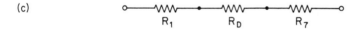

Figure 9-3 Series–parallel circuit.

9-2 Current distribution. In a series–parallel circuit it is important to be able to trace the current distribution in the various parts of the circuit. The important rules to remember are these:

1. In the series elements the current is the same.
2. In the parallel elements the sum of the branch currents is equal to the total current.

For a series–parallel circuit, there is a better way of stating this rule:

2(a). At any junction, the sum of the current flowing into that point is equal to the sum of the current flowing out. (You should recognize this statement as Kirchhoff's law for a parallel circuit.)

Let us analyze a few circuits for currents.

Example 9-4
From Fig. 9-4 it is obvious that

$$I_1 = I_2 + I_3 + I_4 = I_5 = I_T$$

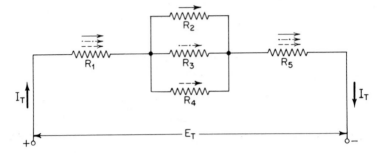

Figure 9-4 Current distribution.

Example 9-5 (See Fig. 9-5)

From Fig. 9-5 we can see that

$$I_T = I_1 = I_2 + I_3$$
$$I_2 + I_3 = I_4 + I_5 + I_6 = I_T$$
$$I_7 = I_4 + I_5 + I_6 = I_T$$

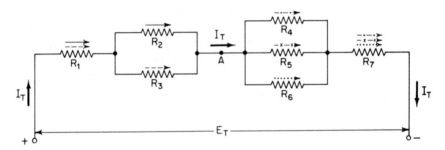

Figure 9-5 Current distribution.

Example 9-6 (See Fig. 9-6)

From analysis of Fig. 9-6, it is obvious that

$$I_T = I_1 = I_2 + I_4 + I_5$$
$$I_2 = I_3$$
$$I_6 = I_3 + I_4$$
$$I_7 = I_6 + I_5 = I_3 + I_4 + I_5 = I_T$$

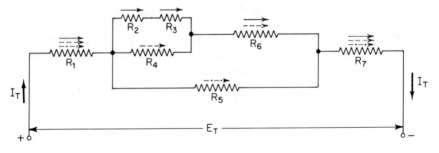

Figure 9-6 Current distribution.

9-3 Voltage distribution. In order to solve series–parallel circuits it is necessary to understand the relation of voltage drops in the various components. The points to remember are these:

1. Voltage is the difference in potential between two points.
2. Voltage drops in the series elements add up.
3. Voltage in the parallel elements remains the same.
4. In adding the voltage drops between any two points, *add the voltage drops for any one path.* This point can be made clearer if you remember that difference in potential is similar to difference in elevation. If a hill is 100 m high and has three paths down to the bottom, each path has an elevation of 100 m. The hill is still 100 m high, not 300!

Let us analyze Fig. 9-7 for voltage distribution.

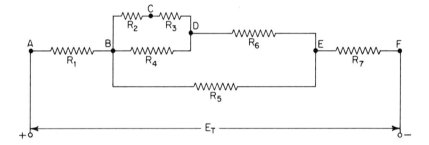

Figure 9-7 Voltage distribution.

Example 9-7

1. The potential difference between B and C is E_2 = voltage drop across R_2.
2. The potential difference between C and D is E_3 = voltage drop across R_3.
3. The voltage between B and D is $E_2 + E_3$. But R_4 is also connected between B and D, in parallel with $R_2 + R_3$ Therefore,

$$E_4 = E_2 + E_3$$

4. The voltage between B and E is $E_2 + E_3 + E_6$ or $E_4 + E_6$. But R_5 is connected between B and E. Therefore,

$$E_5 = E_2 + E_3 + E_6 = E_4 + E_6$$

5. The total voltage is the potential difference between A and F.

$$E_T = E_1 + E_2 + E_3 + E_6 + E_7$$

or

$$E_T = E_1 + E_4 + E_6 + E_7$$

or

$$E_T = E_1 + E_5 + E_7$$

Let us apply all the above relationships to the problem below.

Example 9-8

In Fig. 9-8 find the current through each resistor and the voltage drop across each.

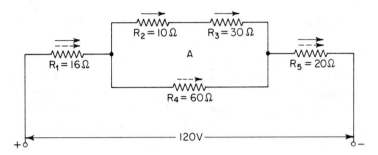

Figure 9-8 Series–parallel circuit.

Solution

1. Since the total voltage is the only known voltage, the equivalent resistance of the entire circuit must be found. Section A consists of R_2 in series with R_3 for a total of 40 Ω; and this 40 Ω is in parallel with R_4, or

$$R_A = \frac{(R_2 + R_3) \times R_4}{(R_2 + R_3) + R_4} = \frac{40 \times 60}{100} = 24\ \Omega$$

$$R_T = R_1 + R_A + R_5 = 16 + 24 + 20 = 60\ \Omega$$

2. $$I_T = \frac{E_T}{R_T} = \frac{120}{60} = 2.0\ \text{A}$$

3. $$I_1 = I_4 = I_5 = I_T = 2.0\ \text{A (series circuit)}$$

4. $$E_1 = I_1 \times R_1 = 2.0 \times 16 = \quad 32\ \text{V}$$

$$E_A = I_A \times R_A = 2.0 \times 24 = \quad 48\ \text{V}$$

$$E_5 = I_5 \times R_5 = 2.0 \times 20 = \quad \underline{40\ \text{V}}$$

$$E_T = E_1 + E_A + E_5 = \quad\quad 120\ \text{V (check)}$$

5. $$E_A = E_4 = E_2 + E_3 = \quad\quad 48\ \text{V}$$

$$I_4 = \frac{E_4}{R_4} = \frac{48}{60} = \quad\quad 0.8\ \text{A}$$

$$I_2 = I_3 = \frac{E_2 + E_3}{R_2 + R_3} = \frac{48}{40} = 1.2\ \text{A}$$

$$I_1 = I_2 + I_4 = \quad\quad \underline{2.0\ \text{A (check)}}$$

6. $$E_2 = I_2 \times R_2 = 1.2 \times 10 = 12\ \text{V}$$

$$E_3 = I_3 \times R_3 = 1.2 \times 30 = \underline{36\ \text{V}}$$

$$E_A = E_2 + E_3 = \quad\quad 48\ \text{V (check)}$$

When the voltage drop across any one portion is given, it is not necessary to find the total resistance. Remember—start where two factors are known. Let us analyze this type of problem using the same circuit as above.

Example 9-9

In Example 9-8 the total voltage is not known. The voltage drop desired across R_2 is 25 V. What should the total voltage be in order to produce this voltage drop?

Solution

$$I_2 = \frac{E_2}{R_2} = \frac{25}{10} = 2.5 \text{ A}$$

$$I_2 = I_3 = 2.5 \text{ A (series circuit)}$$

$$E_3 = I_3 \times R_3 = 2.5 \times 30 = 75 \text{ V}$$

$$E_4 = E_2 + E_3 = 25 + 75 = 100 \text{ V (parallel)}$$

$$I_4 = \frac{E_4}{R_4} = \frac{100}{60} = 1.67 \text{ A}$$

$$I_1 = I_2 + I_4 = 2.5 + 1.67 = 4.17 \text{ A}$$

$$E_1 = I_1 \times R_1 = 4.17 \times 16 = 66.7 \text{ V}$$

$$I_5 = I_3 + I_4 = 2.5 + 1.67 = 4.17 \text{ A}$$

$$E_5 = I_5 \times R_5 = 4.17 \times 20 = 83.4 \text{ V}$$

$$E_T = E_1 + E_4 + E_5 = 66.7 + 100 + 83.4 = 250 \text{ V}$$

Most of the series–parallel circuits that you will encounter as a technician will be of the type shown in Fig. 9-8 and in the problems at the end of this chapter. Such problems are readily solved using the techniques already shown. However, there are complex *networks*—and circuits with multiple sources of power—that result in more than one unknown, and therefore will not respond to the above treatment. Such problems, no matter how complex, can be solved using Kirchhoff's laws and simultaneous equations (or determinants). Special network techniques have also been developed in attempts to simplify solutions. Unfortunately, the analyses and solutions of these problems are often quite involved and laborious. These methods are therefore deliberately delayed until your basics are firmly entrenched (and suitable mathematics have been covered or reviewed). These more advanced chapters, 22 and 23, can be taken up next, if you so desire.

REVIEW QUESTIONS

1. With reference to Fig. 9-1, is R_1 in series with R_2? With R_3? Explain.
2. With reference to Fig. 9-2:
   (a) Is R_2 in series with R_4? Explain.
   (b) Is R_3 in parallel with R_4? Explain.
   (c) How do we get R_T?
3. With reference to Fig. 9-3:
   (a) Is R_1 in series with R_4? Explain.
   (b) Is R_1 in series with R_5? Explain.
   (c) Is R_6 or R_5 in series with R_7?

(d) Is R_2 in parallel with R_4? Explain.

(e) Is R_6 in parallel with R_5? Explain.

(f) Give one combination of parallel resistors.

(g) Give another combination.

(h) How do we get R_T?

4. With reference to Fig. 9-5:

(a) What is the relation between the current I_1 through R_1, and I_T, the total circuit current?

(b) Compare I_7 and I_T.

(c) R_1 has two current arrows above it; R_7 has three current arrows. Compare their values, and account for the number of arrows in each case.

(d) Is there any relation between I_2, I_3, and I_T?

(e) Express Kirchhoff's current relations at location A.

5. In Fig. 9-6, express Kirchhoff's current relations at the junction of R_1 and R_4.

6. With reference to Fig. 9-7:

(a) Express the voltage between B and D in two ways.

(b) Express the voltage between B and E in three ways.

(c) Trace one path (by letters) from the negative to the positive terminal of the power source, and specify the individual voltage drops encountered.

(d) Repeat part (c) for a second path.

(e) Is there a third path? If so, what are the voltage drops encountered?

(f) Does $E_1 + E_4 + E_5 + E_6 + E_7 = E_T$? Explain.

7. In Fig. 9-8:

(a) What is the relation between R_2, R_3, and R_4?

(b) In step 4 of the solution, what does E_A represent?

(c) Explain how the 48 V for E_A is obtained.

(d) Is this (48 V) the voltage across R_2? Explain.

(e) In step 5 account for: $I_2 = I_3 = (E_2 + E_3) \div (R_2 + R_3)$.

8. In Example 9-9, all the individual resistance values are known. (In fact, R_T is the same as in Example 9-8.) Why, then, don't we solve for I_T?

PROBLEMS

1. In Fig. 9-9, find the current through each resistor and the voltage drop across each resistor.

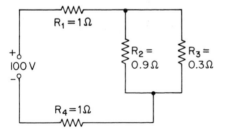

Figure 9-9

2. Using the terminals *A*, *B*, and *C*, two at a time (Fig. 9-10):
   (a) Trace the current flow through the circuits *A–B*, *B–C*, and *C–A*.
   (b) Calculate the resistance between each pair of terminals.

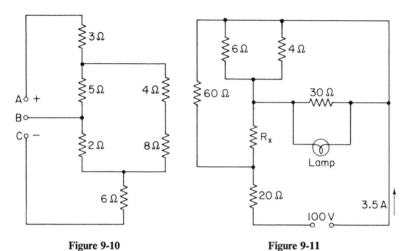

Figure 9-10   Figure 9-11

3. The lamp shown in Fig. 9-11 is rated at 6 V, 0.3 A. Find the value R_x, the current through each resistor, and the voltage drop across each resistor when the lamp is operated at rated value.

4. Find the current through each resistor, the voltage drop across each resistor, and the value of R_x in Fig. 9-12.

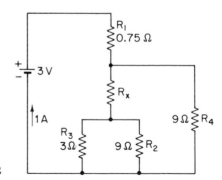

Figure 9-12

5. In Fig. 9-13, find E_T and E_{A-B}.

6. Find the supply voltage, current through each resistor, and voltage drop across each resistor of Fig. 9-14. $I_3 = 2 \times 10^{-2}$ A.

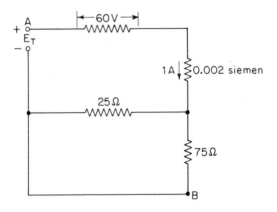

Figure 9-13

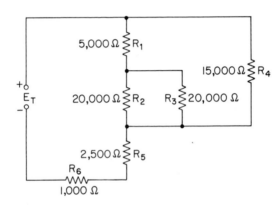

Figure 9-14

7. Find the line current, current through each resistor, and voltage drop across each resistor of Fig. 9-15.

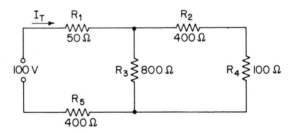

Figure 9-15

8. In Fig. 9-16, the current in R_4 is 10 mA. Mark on the diagram all currents and voltage drops. Find E_T and the value of each resistor.

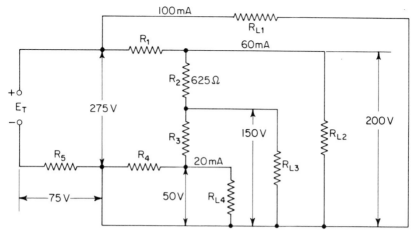

Figure 9-16

9. In Fig. 9-17, mark on the diagram all currents and voltage drops. Find E_T and all resistor values.

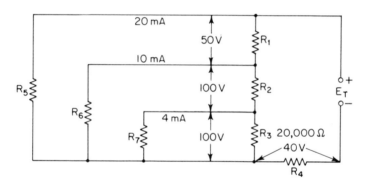

Figure 9-17

10. The current in R_4 of Fig. 9-18 is 10 mA. Find E_T and all resistor values.

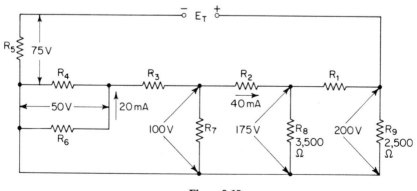

Figure 9-18

11. Find all voltage drops and all currents, including line current, in Fig. 9-19.

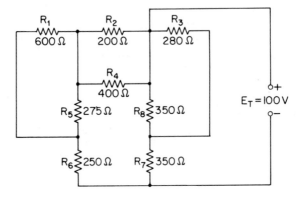

Figure 9-19

12. In Fig. 9-20, find the supply voltage and E_{A-B}.

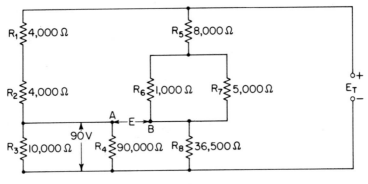

Figure 9-20

13. Find the supply voltage and the value of R_9 in Fig. 9-21.

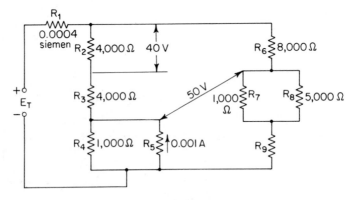

Figure 9-21

10

VOLTAGE DIVIDERS

10-1 Need for voltage dividers. A specific application of the series–parallel circuit is the *voltage divider*. Tapped resistors, semivariable resistors, and potentiometers (Fig. 6-5), or two (or more) series resistors can be used as voltage dividers. The purpose of these circuits is to reduce the voltage fed to some component, or section, below the available power supply value. This is often necessary because the various units or subassemblies in an electronic device give optimum results at widely differing voltages. The power supply is designed for the highest voltage requirement, and this voltage must be reduced for the other requirements. As an illustration, a transistor in the *front end* of a TV receiver may have a recommended operating voltage of only 6 to 12 V, whereas another transistor in the *horizontal deflection* section of the same receiver may need close to 200 V. Meanwhile the *brightness* control, and the *gun* structure of the cathode ray tube may require 300 V or more.* Even within the same transistor widely different voltages are required between the internal elements (*collector, emitter, base*) of the unit. For example, in a "pocket-type" transistor radio, a 9-V battery is used to supply the collector-to-emitter voltage, but the proper emitter–base voltage is less than 1 V.

When such differing voltages are needed, the power supply is designed for the highest voltage requirement, and the voltage divider is connected across this voltage supply. The number of taps or sections used depends on the number of loads supplied. Our problem is to calculate the resistance value and power rating of the voltage-divider sections.

---

*The high-voltage anode of cathode ray tubes generally operate at over 15 000 V; but this voltage is obtained from a separate high-voltage power supply.

10-2 Typical voltage-divider system. Let us apply these ideas to a piece of electronic equipment wherein certain portions require a voltage of 300 V and draw 60 mA; some other portions require 180 V and draw 25 mA; and the remaining portions require 80 V and draw 10 mA. How shall we meet these requirements? This is best shown with the aid of a diagram (see Fig. 10-1). *In voltage-divider calculations, neatness and accuracy of the diagram are of primary importance.* The power supply should develop a 300-V difference of potential. Calling the bottom wire of Fig. 10-1 zero potential, the top wire must

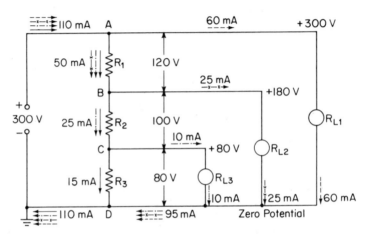

Figure 10-1 Typical voltage-divider circuit.

be at +300 V potential. The first portions of the equipment can be connected to these wires. For simplicity, we will represent these components by a circle and call it R_{L1}. The current in this portion is 60 mA. Mark it on the diagram with its direction. The next portion requires 180 V. Can we connect it directly to the same wires? No, the voltage is too high. But we can use a resistor in series (R_1). The voltage drop across the resistor must be 120 V, so that if we drop 120 V, the potential at the bottom of R_1 will be $300 - 120 = +180$ V. Mark the potential and the voltage drop on the diagram. Now connect the second load section from this wire to the zero potential wire. Represent this second load by another circle and label it R_{L2}. Indicate its current (25 mA), on the diagram. Now starting with the 180-V potential point, we can use another series resistor (R_2) to drop the potential further to 80 V. What should the voltage drop across R_2 be? From 180 volts to 80, or 100 V. Mark this value on the diagram, connect the third load section of the equipment (R_{L3}), and indicate the current (10 mA) on the diagram. To complete the voltage divider, add resistor R_3 from point C to D. We now have a resistor network across the total power supply (A to D).

Let us select R_3 so that a current of 15 mA flows through it, from C to D. (We explain this choice in Section 10-3.) Mark this current alngside R_3. Now let us trace all these currents from start to finish. Using conventional current flow, they all come from the positive supply terminal. Their total value is: $60 + 25 + 10 + 15 = 110$ mA. These four component currents, and their total value, are shown on the upper line wire between the positive supply terminal and point A. Here, at junction A, the current splits. The 60-mA component for R_{L1} continues on to the right and down through R_{L1}. The remaining three components flow down through R_1. What is the value of this current? It is the total current of 110 mA, less the 60 mA for R_{L1}, or 50 mA. This value is shown alongside R_1 with its component arrows. When this 50 mA reaches junction B, the current again divides. The 25-mA component for R_{L2} goes to the right and down through R_{L2}. The remaining 25-mA components continue down through R_2 as shown by the two arrows alongside R_2. Then at junction C, the 10-mA component branches off to the right, and down through R_{L3}, while the 15-mA component flows down through R_3. These four components now combine along the bottom wire, and their total, 110 mA, flows back to the negative supply line terminal. Notice that the currents through R_1, R_2, and R_3, and the voltages across each of these resistors are marked alongside the respective resistors. It is now a simple matter of applying Ohm's law to find the resistance value needed for each:

$$R_1 = \frac{E_1}{I_1} = \frac{120}{0.050} = 2400 \ \Omega$$

$$R_2 = \frac{E_2}{I_2} = \frac{100}{0.025} = 4000 \ \Omega$$

$$R_3 = \frac{E_3}{I_3} = \frac{80}{0.015} = 5330 \ \Omega$$

10-3 Bleeder current.   Notice from Fig. 10-1 that because we added resistor R_3, current will flow from the positive terminal through R_1, R_2, and R_3, and back to the negative terminal even if all the "external" loads were disconnected. This current is called the *bleeder current*. Notice that this current does not flow through the loads; it is merely "bled" from the power supply. As far as the loads themselves are concerned, this current represents a waste of power and should be as small as possible. As far as good regulation (maintaining constant voltages between the taps) is concerned, this current should be as large as possible. As a compromise (between good regulation and high efficiency), good practice recommends that a bleeder current of 10 to 20% of the load be used. In Fig. 10-1, the combined load current is 95 mA, and a bleeder current of 15 mA (about 16%) was used. The bleeder current will increase automatically if the load current should drop. For example, in Fig. 10-1, the total voltage-divider resistance is $R_1 + R_2 + R_3 = 11\ 730 \ \Omega$. If loads R_{L2} and R_{L3} were reduced to zero, the bleeder current would rise to over 25 mA (300/11 730) or almost double its normal value. This will tend to maintain the potentials at B and C at their design values. On the other hand, if the voltage-divider section R_3 were removed, and the above external loads were reduced to zero, what would the potentials at B and C become? There would be no current through R_1 and R_2, and these voltages would rise to the full supply voltage, or 300 V! This could damage the components.

10-4 Power rating of voltage dividers. Voltage dividers are made up either as separate resistors, or as one unit (tapped or semivariable type). When separate resistors are used, the power rating of each resistor is simple to calculate. In the above problem it would be as follows (using a safety factor of 2 to 4):

$$P_1 = E_1 \times I_1 = 120 \times 0.05 = 6 \text{ W; rating} = 12 \text{ to } 24 \text{ W}$$

$$P_2 = E_2 \times I_2 = 100 \times 0.025 = 2.5 \text{ W; rating} = 5 \text{ to } 10 \text{ W}$$

$$P_3 = E_3 \times I_3 = 80 \times 0.015 = 1.2 \text{ W; rating} = 2.4 \text{ to } 5 \text{ W}$$

When a single unit is used, the total resistance must be the sum of the individual step resistances. But the total power rating is *not* the sum of the individual ratings. In the above problem, suppose that we had used a 12 000-Ω resistor rated at 40 W, and that it was 10 cm long. Since the construction is uniform, 1 cm would have a resistance of 1200 Ω and a power rating of 4 W. For resistor R_1 we would need 2 cm for 2400 Ω, but the power rating of a 2-cm piece is only 8 W. This is much too low for R_1. On the other hand, for R_3 (5330 Ω) we need approximately 4.5 cm. This section would have a power rating of 4×4.5 or 18 W, which is more than necessary. The deciding factor, however, is that R_1 will burn out. The power rating for such single units is determined by the I^2R form of the power equation, where R is the total resistance of the steps in the divider and I is the current in the *heaviest* loaded portion. In the above problem the unit must have a total resistance of 2400 + 4000 + 5330 or 11 730 Ω and a power rating of $W = I^2R = (0.05)^2 \times 11\ 730 = 29.3$ W; rating = 60 to 120 W.

10-5 Overvoltage. Sometimes the power supply cannot be designed to deliver the correct maximum voltage. If the voltage is too low, we will not get optimum results from the equipment. Too high a voltage can be easily remedied. Do not use the full voltage directly, but add another step to the voltage divider, between the positive terminal and point A (Fig. 10-1). The full current will flow through this resistor. The voltage drop across this resistor should be the difference between the power-supply voltage and the maximum desired voltage.

10-6 Dual-polarity supplies. In some electronic applications we may need potentials that are negative with respect to ground (zero potential) in addition to the positive potentials of the previous problems. For example, in a field-effect transistor (*FET*) the *gate-to-source* voltage must be opposite in polarity to the *drain-to-source* voltage. The gate is then *reverse-biased* and draws no current.

It should not be inferred from the above illustrations that negative supplies do not deliver current. Whether current is drawn, or not, depends on the type of load connected to the supply. For example, many an integrated circuit not only requires a dual-polarity power source, but also draws current from both the negative and the positive supply. Another illustration that you can appreciate at this time is the TV picture-centering circuit shown in Fig. 10-2(a). A center-tapped potentiometer is connected across a power source. The center tap of the potentiometer is grounded. The deflection coil (horizontal or vertical) is connected as shown. With the variable arm of the potentiometer above center, the top end of the coil is positive, compared to ground, and current will flow *down* through the coil. This, for example, could cause the TV picture to shift to the right. The higher the slide position, the greater the current, and the further the picture will shift. Conversely, with the slide arm below center, the top end of the coil is negative; current will flow up through the coil; the picture shifts to the left.

Figure 10-2(b) shows how the same effect can be obtained using a standard potentiometer. Two equal-value resistors (R_1 and R_2) are used as a fixed voltage divider to produce the ground center tap.

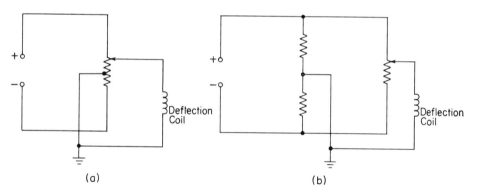

Figure 10-2   Use of ground tap for polarity changes.

Let us now consider a problem with a negative bias included.

Example 10-1

A power supply delivers 400 V to a voltage divider. The equipment requirements are 300 V at 80 mA, 200 V at 40 mA, 90 V at 10 mA, and −70 V at no drain.

Solution   The bleeder current should be between 10 and 20% of the load current. Let us pick 20 mA. The circuit requires a total difference of potential of 300 V from +300 to zero, and 70 V from zero to −70, or 370 V. There is an extra 30 V to be dropped. Draw the diagram and mark all values (see Fig. 10-3).

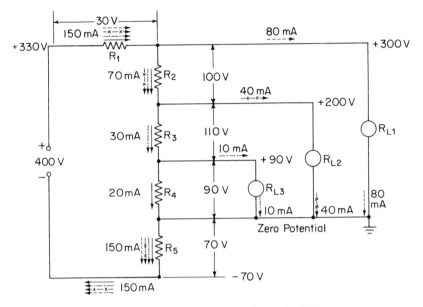

Figure 10-3 Voltage divider with negative bias.

Note the following points:

1. The voltage drops in the voltage divider add up to the total voltage of 400.
2. The voltage drops across each resistor are clearly shown.
3. The current for each load is clearly shown.
4. The total current comes from the positive terminal and returns to the negative terminal of the power supply.
5. The current splits up or adds up at each junction.
6. The negative bias is obtained by having the ground or zero potential as a point between the negative and the positive power-supply terminal.
7. All potentials are figured from the zero potential.
8. All voltage drops are considered as the difference between potentials.

As before, the resistance and power rating for each step can be calculated by applying Ohm's law or the power equation to each step. The current and the voltage drop for each resistor are clearly marked alongside the respective resistors.

In voltage-divider calculations, neatness and accuracy of the diagram are of primary importance.

In all of the above illustrations, the currents drawn by the loads were returned to the negative supply line through the ground, or zero-potential wire. However, when current is also drawn by negative-biased loads, this current flow no longer applies. Only the imbalance current between the positive- and negative-biased drains flows through the ground. This condition is best shown with a problem.

Example 10-2

A power supply delivers 48 V to a voltage divider. The equipment requirements are +28 V at 100 mA, +12 V at 25 mA, and −20 V at 60 mA. The bleeder current is 20 mA. Draw the diagram for a suitable voltage divider, and mark all pertinent voltage values and current distributions.

Solution   (See Fig. 10-4.) The technique used generally follows the previous illustrations, and should need no further explanation—except at junction *F*.

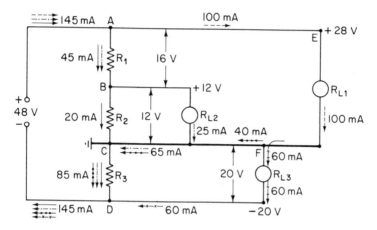

Figure 10-4   Dual polarity with negative and positive loads.

Here the 100 mA from R_{L1} splits, 40 mA returning via the ground wire—as before, but now, 60 mA returns through load R_{L3}. Note that if loads R_{L1} and R_{L3} were balanced (for example, each 100 mA), all of the current from R_{L1} would flow through R_{L3}, and the ground wire current would be only the 25 mA of load R_{L2}.

In the above example, what would the current distribution be if the drain of R_{L3} were 120 mA? The ground wire current would drop to 5 mA, and the other 20 mA from R_{L2} would flow through R_{L3}. What would happen if the drain of R_{L3} were greater than 125 mA? The current through R_2 and R_3 must increase to supply the excess of drain of R_{L3} over $R_{L1} + R_{L2}$.

REVIEW QUESTIONS

1. Why are voltage dividers used?

2. With reference to Fig. 10-1:

    (a) What do the circles marked R_{L1}, R_{L2}, R_{L3} represent?

    (b) What is the potential at point *A*?

(c) How is the 25 mA for R_{L2} obtained?

(d) How is the 80 V across R_{L3} obtained?

(e) How is the +80 V at the top of R_{L3} obtained?

(f) How is the 25 mA through R_2 obtained?

(g) How is the 50 mA through R_1 obtained?

(h) How is the 110 mA on the top wire obtained?

(i) How is the 120 V across R_1 obtained?

(j) How is the 100 V across R_2 obtained?

3. Still with reference to Fig. 10-1:

(a) What is the current flowing through R_3 called? Why?

(b) Account for the current values shown flowing through the voltage-divider resistors R_3 and R_1.

(c) Account for the current values shown along the bottom line wire.

4. (a) In Fig. 10-2(a), what does the resistor symbol represent?

(b) When the potentiometer arm is near the top, in what direction will the current through the coil flow?

(c) Repeat part (b) for the potentiometer arm near the bottom.

(d) Must a potentiometer with a center tap always be used to get this effect? Explain.

5. With reference to Fig. 10-3:

(a) When is the use of resistor R_1 necessary?

(b) What is the function of resistor R_5?

(c) Why is the bottom wire marked with a minus 70 V?

6. With reference to Fig. 10-4:

(a) Account for the 40 mA shown between junctions F and G.

(b) Account for the 85 mA shown flowing through R_3.

(c) If the current through R_{L3} were 80 mA, what would the current be between F and G?

(d) If the current through R_{L3} were 110 mA, What would the current be between F and G?

(e) In what direction would the current in part (c) flow? Explain.

PROBLEMS

1. A power supply delivers 250 V to a voltage divider. The taps are 250 V at 45 mA, 180 V at 30 mA, and 100 V at 10 mA. The bleeder current is 15 mA. Draw a diagram showing all currents and voltages.

(a) What is the resistance of each step of the divider?

(b) What is the power dissipated in each step?

(c) What commercial power rating would you use for each step?

(d) What power rating would you use if a single tapped resistor were employed?

2. Repeat for power-supply voltage of 300 V, and taps of 280 V at 70 mA, 150 V at 30 mA, 90 V at 10 mA, and bleeder current of 15 mA.

3. Repeat for power-supply voltage of 400 V, and taps of 375 V at 100 mA, 300 V at 60 mA, 250 V at 40 mA, 100 V at 25 mA, and bleeder current of 25 mA.

4. Repeat for power-supply voltage of 400 V, and taps of 275 V at 75 mA, 80 V at 20 mA, −90 V at no drain, and bleeder current of 15 mA.

5. Repeat for power-supply voltage of 350 V, and taps of 290 V at 100 mA, 110 V at 20 mA, −60 V at no drain, and bleeder current of 20 mA.

6. Repeat Problem 4, but with the −90-V tap at a current of 50 mA.

7. Repeat Problem 5, but with the −60-V tap at a current of 40 mA.

11

PRIMARY CELLS

A *cell* or a *battery* is a device that converts chemical energy into electrical energy. The terms "cell" and "battery" are often used interchangeably—but incorrectly so. A battery consists of a group of interconnected cells, whereas the so-called flashlight "battery" is only a single cell. Cells (and batteries) are generally grouped into two broad classifications: *primary* and *secondary*.

Primary cells are used until their output voltage falls too low for useful work, and are then discarded. Flashlight cells are an example of this type. On the other hand, when the energy in a secondary cell is exhausted, the chemicals that provided the energy can be restored to their original condition by electrical *recharging* and the cell can be used over and over again. The automobile battery is a prime example of this type of cell. Secondary cells are discussed in Chapter 12.

Primary cells can be further subdivided depending on the chemicals used to provide the electrical output. In this grouping are found the carbon–zinc, the mercury, the alkaline, the "silvercel," and the air cell. However, before discussing each individual type, let us consider a chemical–electronic interaction common to all cells.

11-1 Ionization. When certain chemical compounds are dissolved in water, the molecules break up into atoms, or groups of atoms. In separating, these atoms (or groups) either carry with them extra electrons or leave behind some of their electrons. These particles that are no longer neutral in charge are called *ions*, and the process is called *ionization*. Meanwhile, since the solution contains positive and negative ions, it is capable of conducting electrical current.

Such a solution is called an *electrolyte*. To make electrical contact with the solution, two conducting materials must be immersed in the electrolyte. These are called *electrodes*.

CARBON–ZINC (DRY) CELL

Because of its low initial cost, wide availability, and broad range of desirable features, the carbon–zinc cell is the workhorse of the battery industry. It is also known as the Leclanché cell, so named after Georges Leclanché, who, in 1868, introduced a cell with essentially the same chemical ingredients and reactions as the present cell.

11-2 Construction.   In the familiar round cell, zinc (Zn) serves both as the container and as the negative electrode (see Fig. 11-1). The positive "electrode" is a mix consisting of manganese dioxide (MnO_2), acetylene black (or graphite), zinc chloride, sal ammoniac, and water. Since this powdery mix is not mechanically suited as a positive "terminal," a carbon rod with a large

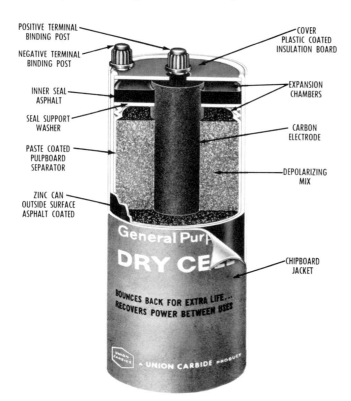

POSITIVE TERMINAL
BINDING POST

NEGATIVE TERMINAL
BINDING POST

INNER SEAL
ASPHALT

SEAL SUPPORT
WASHER

PASTE COATED
PULPBOARD
SEPARATOR

ZINC CAN
OUTSIDE SURFACE
ASPHALT COATED

COVER
PLASTIC COATED
INSULATION BOARD

EXPANSION
CHAMBERS

CARBON
ELECTRODE

DEPOLARIZING
MIX

CHIPBOARD
JACKET

Figure 11-1   Type A or No. 6 dry cell. (Courtesy Union Carbide Co.)

surface area is inserted in the center of this mix, to act as the liaison with the external circuit. (This rod is often mistakenly called the positive electrode.) The carbon is porous enough to allow gases that would accumulate in the cell to escape, while simultaneously preventing leakage of the electrolyte. The electrolyte consists of ammonium chloride (NH_4Cl), zinc chloride, and water, mixed into a gelatinous paste with cornstarch and flour. Since this is not a "wet" liquid, this cell is also called a *dry* cell.

This type of construction has several disadvantages. The zinc electrode forms the case. It can never be utilized fully as an active element. If a hole is eaten through the zinc case, the cell dries up and becomes useless. Also, approximately 40 % of the bulk of the cell is nonproductive chemically. In other words, there is a lot of waste space and the life per dollar of cost is low.

An improvement on this type of construction is the mini-max cell. In this cell the carbon and zinc electrodes are flat disks. The electrolyte mix is sandwiched between the plates. An elastic sealing compound extends around all four sides of the flat cake. Because this seal will expand when necessary, no expansion chamber is needed. Now that the zinc is no longer the sealing case, all of it can be used up chemically. In addition, the carbon "collector" is made comparatively thinner. (It does not wear out.) These features allow for more active mix and smaller nonproductive space. The mini-max type of construction gives more hours of life per unit volume of battery.

11-3 Chemical action. When in solution, ammonium chloride breaks up into NH_4^+ (ammonium) ions and Cl^- (chlorine) ions. The zinc when put into solution gives off positive zinc ions (Zn^{++}). The excess of electrons left in the zinc plate gives the zinc a negative potential. In turn the positive zinc ions repel the positive ammonium ions. These migrate to the carbon rod and take electrons from the carbon rod, becoming neutral ammonia gas (NH_3) plus hydrogen (H). The ammonia gas may remain free and rise to the expansion chamber to combine with the water to form ammonium hydroxide. The carbon rod is then left with a deficiency of electrons, that is, a positive potential. Meanwhile the zinc ions combine with the chlorine ions, producing zinc chloride. In this way chemical energy has been converted into electrical energy because of the difference in potential between the zinc and carbon electrodes.

Regardless of the size of the cell, as long as the electrodes and electrolyte are as given above, the open-circuit voltage (nominally referred to as 1.5 V) may vary from 1.5 to 1.6 V.

The hydrogen, if left free, will gather around the carbon rod, causing detrimental effects, as you will learn later. To prevent this condition, manganese dioxide (MnO_2) is added. This material is abundant in oxygen and recombines with the hydrogen to form Mn_2O_3 and water. The complete analysis may be shown by

$$Zn + 2MnO_2 + 2NH_4Cl \longrightarrow Mn_2O_3 + H_2O + 2NH_3 + ZnCl_2$$

11-4 Internal resistance. Every cell or battery has some internal resistance, due to the resistance of the electrodes, the contact surface between electrodes and electrolyte, and the resistance of the electrolyte itself. To reduce this resistance, a large area of electrodes should be in contact with the electrolyte. Larger cells will therefore have lower internal resistance. For example, the small penlight cell (Type AA) has an internal resistance of approximately 0.3 Ω, while the much larger No. 6 (Type A) has a resistance as low as 0.03 Ω. The internal resistance of a cell increases with use. One of the products from the chemical reaction is hydrogen, which is a poor conductor of electricity. There is a tendency for this hydrogen to form around the carbon rod. The effect is to increase the contact resistance between the carbon and the mix, causing a great increase in internal resistance. Also, as the cell gets older, the mix dries out, further increasing the internal resistance.

The effect of internal resistance is to lower the terminal voltage of a cell. Because of the chemical action, it still generates a potential difference or electromotive force (EMF) of 1.5 V. If no current is drawn from the cell (on open circuit), the terminal voltage is also 1.5 V. But as soon as some electrical equipment is connected across the terminals of the cell, current will flow. This current flowing through the internal resistance of the cell causes a voltage drop. The terminal voltage is then the difference between the generated EMF and the internal voltage drop (see page 141 and Fig. 11-6). When the internal resistance causes an appreciable voltage drop, the cell is useless. The effect of internal resistance will be discussed in more detail later.

From the above, should you test a cell to determine its condition by measuring its open-circuit voltage reading? No! Check its voltage while it is supplying normal current.

11-5 Polarization. The tendency for hydrogen to gather around the carbon electrode, increasing the internal resistance of the cell, is known as *polarization*. How did we prevent it? By adding manganese dioxide (MnO_2), which combines chemically with the hydrogen to form water. However, the prevention is not complete, and with age a cell does polarize. Also, if large currents are drawn from a cell, the chemical action proceeds faster. All the hydrogen formed is not absorbed by the manganese dioxide, and the cell polarizes. A cell that has been rendered useless by polarization will often recover if it is allowed to stand idle (no drain) for a period of time. The manganese dioxide will gradually catch up with the excess of hydrogen and depolarize the cell. However, if the supply of manganese dioxide is exhausted, a polarized cell may seem to recover after a standby period, but the recovery lasts only a few seconds. Once load is applied, the cell polarizes again very rapidly.

11-6 Local action. Electromotive force is produced chemically whenever two dissimilar metals are immersed in a chemical solution. If the zinc contains impurities, small cells are formed between the zinc and its impurities.

This action does not add to the voltage of the cell, but it does use up the zinc. As a result, such *local action* can wear out a cell even though it is not being used. To reduce local action, the zinc is amalgamated ("coated") with mercury so as to present a "pure" surface to the electrolyte.

11-7 Shelf life. You may have noticed that many dry cells carry a warning to put the cell into use before a certain date. The purpose of these instructions is to ensure that you get reasonably fresh cells. From local action and drying of the electrolyte, a cell will become useless in time even though it is still on the retailer's shelf. Shelf life has been defined as the length of time a cell or battery can be stored (at room temperature) and still retain 90% of its original capacity.

11-8 Effect of temperature. Carbon–zinc cells provide optimum performance at normal room temperatures (21°C or 70°F). With decrease in temperature, the chemical activity slows down, and the open-circuit voltage drops (at about 0.0004 V/°C over the range 25 to −20°C). The cells become inoperative at about −30°C. On the other hand, storage at reduced temperatures will increase shelf life. At the other end of the temperature scale, operation or storage at high temperatures is abusive. Temperatures above 50°C (125°F) can cause sudden failure.

11-9 Maximum safe discharge rate. A fresh cell has such a low internal resistance that we will neglect it. The amount of current delivered by a cell will then depend upon the resistance of the load we connect across the cell. Yet we saw above that if too large a current is drawn, hydrogen is formed more quickly, and there is not enough manganese dioxide to absorb it as quickly as it forms. The cell becomes polarized. What is the maximum safe current we can draw from one cell? The answer depends on the amount of manganese dioxide in the cell or in turn on the size of the cell. The familiar No. 6 or Type A dry cell (66.7 × 168 mm) has a maximum safe current rating of 0.25 A. It will give higher currents if the resistance of the load is low enough, but the cell could be ruined by polarization.

11-10 Zinc chloride cells. A variation of the carbon–zinc cell is the zinc chloride or *heavy-duty* cell. In this version, the ammonium chloride is omitted, and only zinc chloride is used as the electrolyte. This improves the electrochemical action, minimizing polarization. Therefore, zinc chloride cells can operate at higher current drains for a considerably longer time than a Leclanché cell of the same size. In addition, the voltage level, under load, holds up longer. However, the increase in chemical activity necessitates a new type, or improved seal, making this heavy-duty cell more expensive.

11-11 Cells in series. We have seen that one dry cell, regardless of size, will produce 1.5 V. What if we need a higher voltage? In series circuits we learned that the voltage drops add up. Let us see what happens if we connect cells in series. Starting with one cell, let us call the negative terminal the zero potential, or reference point. Its positive terminal will have a potential of $+1.5$ V. Tie this terminal to the negative terminal of the next cell. Two points connected by a wire are at the same potential, or the negative terminal of the second cell has a potential of $+1.5$ V. Between this terminal and its own positive there is a difference of potential of 1.5 V, the carbon being more positive. Therefore, the positive terminal of the second cell has a potential of $+3$ V. To connect cells in series, connect the positive terminal of one cell to the negative of the next cell, and so on; the two end terminals will give the total voltage.

Example 11-1

Show dry cells connected to supply 4.5 V.

Solution   At 1.5 V per cell, three cells are needed, as shown in Fig. 11-2.

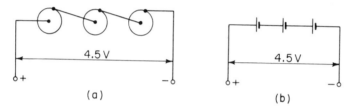

Figure 11-2   Cells in series.

Example 11-2

One dry cell, Type A, has a maximum safe current rating of 0.25 A. What is the maximum safe current that the combination in Example 11-1 can deliver?

Solution   In a series circuit the current is the same in all the units. If each cell has a maximum rating of 0.25 A, it is the same 0.25 A that is flowing through each. So regardless of how many cells are in series, the maximum safe drain is still 0.25 A.

11-12   Cells in parallel.   To connect cells in parallel, all the positive posts are tied together and brought out as the positive terminal and all the negative posts are tied together and brought out as the negative terminal (see Fig. 11-3).

Since the potential difference in any one cell does not add to the potentials created by any other cell, the terminal voltage remains 1.5 V regardless of the

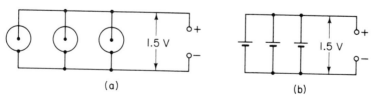

Figure 11-3   Cells in parallel.

number of cells in parallel. But why are cells connected in parallel? Many times
the answer given is "to increase the current." But is this right? Figure 11-4 shows
a 3-Ω bulb connected across a 1.5-V cell. How much current will the bulb take?
$I = E/R = 1.5/3 = 0.5$ A. Where does this current come from? Cell 1. Now
we close the switch, putting cell 2 in parallel. Does the voltage across the bulb
change? No. Does the resistance of the bulb change? No. Does the current
through the bulb increase? No! Do cells in parallel increase the current? No.
The current depends upon the resistance of the unit we connect across the cell!
However, there is a change. When cell 1 was alone, it had to supply the full drain
of 0.5 A. But this is more than its *safe* maximum. The cell will polarize quickly.
By putting cell 2 in parallel, we find that the two cells divide the load between
them. Assuming equally fresh cells, each cell supplies 0.25 A, and the bulb
current is still 0.5 A. We have reduced the drain on the cells to within the safe
maximum rating, thus preventing undue polarization and thereby increasing
the life of the cell. So we see that cells are connected in parallel (1) *to increase
their life* or (2) *to keep the drain on any cell below the safe maximum rating.*

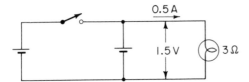

Figure 11-4

11-13 Cells in series–parallel.   Cells are connected in series–
parallel when (1) the voltage required is more than a single cell's voltage, and
(2) the current required is more than the safe maximum rating of one cell.

Example 11-3

A field telephone set requires 6 V at 0.6 A. How should Type A dry cells be
arranged to supply this load?

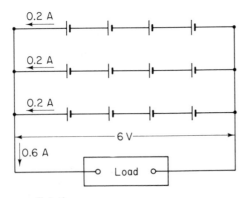

Figure 11-5   Cells in series–parallel.

Solution

1. The voltage required determines the number of cells in series: 6 V requires
   four cells.

2. The current required determines the number of parallel branches. To keep the drain on each cell within 0.25 A, three branches in parallel are required.

3. Each branch in parallel *must* have the same voltage (6 V) or must have four cells in series (see Fig. 11-5).

11-14   Effect of internal resistance—series cells.   In the above cell problems we have completely neglected the internal resistance of the cells. For precise calculations this omission introduces an error. For most practical work this error is negligible, especially if the cells used are fresh and the internal resistance is low or the drain on the cells is low. However, let us now treat these cell problems more accurately.

Figure 11-6 shows a dry cell with its internal resistance (r) connected to an external load. Since it is a dry cell, its EMF is 1.5 V. The load current flowing through the internal resistance of the cell produces a voltage drop (Ir). In this case, the voltage drop is $0.25 \times 0.08 = 0.02$ V. The terminal voltage is reduced by the voltage drop:

$$V = E - Ir = 1.5 - 0.02 = 1.48 \text{ V}$$

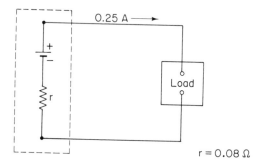

Figure 11-6   Cell with internal resistance.

As you will notice, the error introduced by neglecting the internal resistance is small (less than 1.5%). If the internal resistance or the load current were higher, the error would be correspondingly greater.

When cells are connected in series, the total EMF is the sum of the individual cell EMFs. But the internal resistance of each cell is also in series. Such a group of cells can be replaced by an *"equivalent"* cell whose EMF equals the sum of the individual EMFs and whose internal resistance equals the sum of the individual internal resistances.

Example 11-4

Four dry cells (EMF 1.5) having internal resistances of 0.1, 0.15, 0.05, and 0.2 Ω, respectively, are connected in series across a load of 14 Ω.
(a) What is the terminal voltage of the battery combination?

(b) What is the load current?

(c) What is the voltage across the load?

Solution

1. The first step is to replace the four cells with an equivalent cell. EMF of equivalent cell = 1.5 × 4 = 6 V.

   Internal resistance of equivalent cell = 0.1 + 0.15 + 0.05 + 0.2

   $$= 0.5 \ \Omega$$

   We can use Fig. 11-6 to represent this equivalent cell of 6 V and an internal resistance, r_i, of 0.5 Ω.

2. Before we can find the terminal voltage we must know the load current, so part (b) comes first.

   (b) $$I_T = \frac{E_T}{R_T} = \frac{6}{14 + 0.5} = 0.414 \text{ A}$$

   (a) $$Ir \text{ drp} = 0.414 \times 0.5 = 0.207 \text{ V}$$
   $$V = E - Ir = 6 - 0.207 = 5.79 \text{ V}$$

   (c) Voltage across load must equal cell terminal voltage. Let us check by another method.

   $$E_{\text{load}} = I_L \times R_L = 0.414 \times 14 = 5.79 \text{ V} \quad \text{(check)}$$

11-15   Effect of internal resistance—parallel cells.   When discussing the use of cells in parallel, we pointed out that *if the cells are equally fresh,* they will divide the load among them equally. This statement is true if the cells have equal internal resistance. Again, an equivalent cell can be used to replace cells in parallel. First, cells connected in parallel should be of the same type; that is, they should have equal EMFs. Since they are in parallel, the total EMF is the same as the EMF of any one cell. Also since they are in parallel, their internal resistances are in parallel. Therefore, the internal resistance of the equivalent cell is *less* than the resistance of any one cell. The total internal resistance can be found by the usual formula for resistances in parallel.

Example 11-5

Three cells, EMFs 1.5 V each, are connected in parallel. Their respective internal resistances are 0.1, 0.05, and 0.2 Ω. Calculate (a) the EMF and (b) the internal resistance of the equivalent cell.

Solution

(a)   EMF = 1.5 V (cells in parallel)

(b) $$\frac{1}{R_T} = \frac{1}{R_1} + \frac{1}{R_2} + \frac{1}{R_3} = \frac{1}{0.1} + \frac{1}{0.05} + \frac{1}{0.2}$$

$$R_T = 0.0286 \ \Omega$$

Calculator entry:

$$\boxed{\cdot}, 1, \boxed{1/x}, \boxed{+}, \boxed{\cdot}, 0, 5, \boxed{1/x}, \boxed{+}, \boxed{\cdot}, 2, \boxed{1/x}, \boxed{=}, \boxed{1/x}$$

How will the load divide among parallel cells of unequal internal resistance? As in all parallel circuits, the current will divide inversely as the branch resistances. The fresh cell having the lowest internal resistance will carry the greatest share of the load.

Example 11-6

The above parallel cells are connected to a load of 2.45 Ω.
(a) What is the load current?
(b) What is the current delivered by each cell?
(c) What is the terminal voltage of the battery and the voltage across the load?

Solution   Using the answer from Example 11-5, the three cells can be replaced by an equivalent cell—EMF 1.5 V and internal resistance 0.0286 Ω.

(a) The load current depends on total voltage and total resistance.

$$I = \frac{E_T}{R_T} = \frac{1.5}{2.45 + 0.0286} = 0.605 \text{ A}$$

Calculator entry:

$$1, \boxed{\cdot}, 5, \boxed{\div}, \boxed{(}, 2, \boxed{\cdot}, 4, 5, \boxed{+}, \boxed{\cdot}, 0, 2, 8, 6, \boxed{)}, \boxed{=}$$

(b) Ir drop in equivalent cell $= 0.605 \times 0.0286$
$$= 0.0173 \text{ V}$$

Since the internal resistances of each cell are in parallel, this Ir drop must also be the Ir drop for each individual cell, because of its share of the load current and its own internal resistance. Therefore:

For cell 1: $I_1 r_1 = 0.0173$, $I_1 = \dfrac{0.0173}{0.1} = 0.173 \text{ A}$

For cell 2: $I_2 r_2 = 0.0173$, $I_2 = \dfrac{0.0173}{0.05} = 0.346 \text{ A}$

For cell 3: $I_3 r_3 = 0.0173$, $I_3 = \dfrac{0.0173}{0.2} = 0.086 \text{ A}$

$$I_T = \overline{0.605 \text{ A}} \text{ (check)}$$

(c)   Voltage across load $= I_L R_L = 0.605 \times 2.45 = 1.4823 \text{ V}$

Terminal voltage $= E - Ir = 1.5 - 0.0173 = 1.4827 \text{ V}$ (check)

This example brings out an important point. It is not wise to parallel fresh cells with old or worn cells. If you are in doubt of the condition of a cell do not use it in a parallel group. In Example 11-6 the load current is approximately 0.6 A. Normally, three No. 6 cells in parallel should be sufficient to keep the load on each cell below its maximum safe rating. If they were equal in condition, each cell would be carrying 0.2 A. But notice that cell 2 (the freshest cell) carries 0.346 A. Its life will be shortened!

11-16   Effect of internal resistance—series–parallel cells.   To consider internal-resistance effects on series–parallel cell groupings is no more difficult than the two previous cases; it is merely more laborious. Refer to Fig. 11-5:

1. For each group of four cells in series, the EMF is the sum of the individual cell voltages, in this case, 6 V. In addition, the total internal resistance for each group is the sum of the individual component cell resistances.
2. When paralleling the three branches, the total EMF is the same as for any one branch. The total internal resistance is the equivalent resistance of the three branch internal resistances in parallel.
3. By this method an equivalent cell can be substituted for the entire series–parallel combination.
4. From this equivalent cell the procedure is the same as before:
   (a) Load current is obtained from E_T and R_T (internal + external resistance).
   (b) Ir drop is obtained from load current and equivalent cell resistance.
   (c) Division of load is obtained from Ir drop and internal resistance of each *branch*.
   (d) Terminal voltage and load voltage are obtained from $E - Ir$, or from $I_L R_L$.

11-17 Batteries. To serve applications requiring a higher current and/or voltage than a single cell can supply, manufacturers will package a group of cells (in series, parallel, or series–parallel) to form a battery. Typical battery-operated equipment includes portable transistor radios, lanterns, camera electric-eyes and flash guns, emergency lighting, telephone service, and electrical or electronic instruments. Carbon–zinc batteries are available from 1.5 V (at over 1-A capability) to as high as 3000 V. Some packages will contain two separate groupings—an "A" section providing the high-current, low-voltage

Figure 11-7 Typical "B" battery. (Courtesy Union Carbide Co.)

requirements and a "B" section for the lower-current, higher-voltage needs. A typical B battery, using mini-max cells, is shown in Fig. 11-7.

11-18 Capacity rating.   The capacity of a cell or battery is given in ampere-hours (or milliampere-hours). This rating determines the length of service to be expected from the cell under *normal* conditions. When the current drain × number of hours = ampere-hour rating, the cell's life is finished. Many factors affect this normal expectancy, such as drain on the cells and intermittent or continuous operation. Approximate ratings of commonly used cells are:

Type A (No. 6)	30 Ah
Heavy-duty B	4500 mAh
Medium-duty B	1200 mAh
Standard	450 mAh

11-19 Mercury cell.   The zinc–mercuric oxide cell—more commonly known as the mercury cell—was designed to overcome some of the weaknesses of the carbon–zinc type. Three major advantages of this cell are:

1. Much longer shelf life. They can be relied on to give 3 to 5 years in standby service.
2. Greater capacity for the same volume and weight. This makes them ideal for aircraft and space vehicles, or in small items such as electronic wristwatches.
3. More-constant output voltage with age. These cells are often used as secondary voltage standards in instrumentation and calibration work, supplying an accuracy of better than 1% over several years.

Unfortunately, the mercury cell also has its disadvantages. They are appreciably more expensive in first cost; they have a lower open-circuit EMF (1.350 V)*; and they have a higher internal resistance. In fact, the advantages listed above are realized only at relatively lighter current loads. Except for specially constructed units, these cells suffer a severe loss of capacity at about 4.4°C (40°F).

As to construction, this cell uses a densely pressed structure of red mercuric oxide (HgO) as the positive electrode and depolarizer. (A small amount of graphite is mixed in to improve electrical conductivity.) The negative electrode is a zinc–mercury amalgam (using compressed powdered zinc). These electrodes are separated by an absorbent pad containing the electrolyte—a 40 percent solution of potassium hydroxide (KOH) saturated with zincate ions. A cutaway

---

*Some mercury cells containing a small percentage of manganese dioxide in the depolarizer, have an *initial* voltage of 1.45 V. However, this effect "burns off" during the early stages of operation and the EMF reverts to the 1.35 value.

view of a button cell is shown in Fig. 11-8(a). A typical round-can mercury cell is shown in Fig. 11-8(b). Although it looks like a zinc–carbon cell, the polarity of this mercury cell is reversed—the cap, or top terminal, is negative with respect to the outer steel casing.

Mercury cells and batteries are available in voltages ranging from 1.35 to 97.2 V, and in capacities ranging from 16 mAh to 28 Ah.

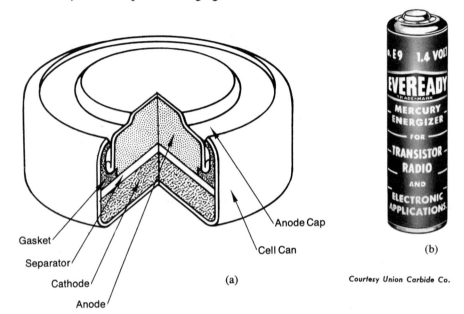

Gasket
Separator
Cathode
Anode

Anode Cap
Cell Can

(a)

(b)

Courtesy Union Carbide Co.

Figure 11-8   Mercury cells.

11-20 Alkaline cell.   The alkaline–manganese dioxide–zinc cell—commonly known as an alkaline cell—represents a major advance in portable power sources over the carbon–zinc cell. This cell is a high-rate source of electrical energy. Its outstanding features result from a combination of unique components and construction methods [see Fig. 11-9(a)]. The cathode, or positive electrode,* is high-density manganese dioxide, which also serves as depolarizer. The outer jacket, of nickel-plated steel, makes contact with the cathode mix and serves also as the current collector. The negative electrode is a zinc anode of extra large surface area. The electrolyte is a high-conductivity potassium hydroxide. This construction produces a cell with very low internal resistance, and a nominal EMF of approximately 1.5 V.

Because of its low internal resistance, the alkaline cell will operate at high

---

*The American convention of electrode nomenclature calls the negative electrode the anode, because the chemical process occurring on discharge is the anodic oxidation of the (negative) metallic electrode, and cathodic reduction of the (positive) oxide electrode. This convention is used by most manufacturers.

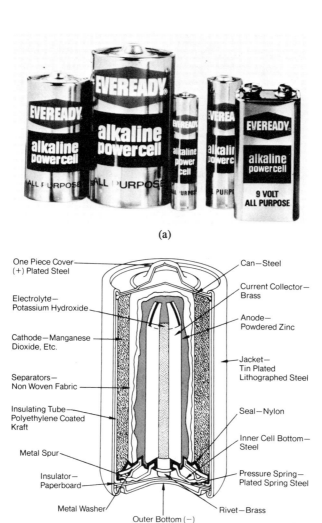

(a)

One Piece Cover—
(+) Plated Steel

Electrolyte—
Potassium Hydroxide

Cathode—Manganese
Dioxide, Etc.

Separators—
Non Woven Fabric

Insulating Tube—
Polyethylene Coated
Kraft

Metal Spur

Insulator—
Paperboard

Metal Washer

Can—Steel

Current Collector—
Brass

Anode—
Powdered Zinc

Jacket—
Tin Plated
Lithographed Steel

Seal—Nylon

Inner Cell Bottom—
Steel

Pressure Spring—
Plated Spring Steel

Rivet—Brass

Outer Bottom (−)
Plated Steel

Courtesy of Union Carbide Corp.

(b) CUTAWAY OF ALKALINE CELL (PRIMARY TYPE)

Figure 11-9 Alkaline cells.

efficiency even with heavy loads. Also, because of improved depolarizing action, these cells can provide continuous operation at high drain without impairment of efficiency and without loss of ampere-hour capacity. A new alkaline cell can provide more than ten times the service of an equivalent carbon–zinc cell.

On light drains (or shelf life), this cell often provides more than twice the service life of the carbon–zinc cell. However, in this respect, it does not equal the mercury cell. With regard to temperature, reasonable service (at lighter loads) can still be obtained down to −40°C.

The initial cost of the alkaline cell falls in between the cost of the carbon-zinc and the mercury cell. The respective operating costs, however, will vary depending on the application. For light, intermittent duty, the carbon-zinc cost is lowest. If a flat-discharge characteristic (constant voltage) is important, the mercury cell cost is lowest. For very light loads, again the mercury cell cost is lowest. In all other applications, the alkaline cell is gaining wide popularity. Typical cells are shown in Fig. 11-9(b).

11-21 Silver oxide cell.   Silver oxide–alkaline–zinc primary cells are a major contribution to miniature power sources. They provide a higher voltage than mercuric oxide cells; offer a flat voltage characteristic; have good low-temperature characteristics; and their internal impedance is low and uniform. The open-circuit voltage of silver oxide cells is 1.6 V, while the operating voltage at typical current drains is 1.5 V. They are available as button cells in capacities of 35 to 210 mAh. They are also available as 6-V batteries, rated at 190 mAh, and at maximum current outputs up to 100 mA.

The silver oxide cell consists of a depolarizing silver oxide cathode, a zinc anode of high surface area, and a highly alkaline electrolyte. The electrolyte is either sodium or potassium hydroxide. A cutaway view of this cell construction is similar to the mercury button cell shown in Fig. 11-8.

Silver oxide batteries are well suited for use in hearing aids, electronic watches, calculators, photoelectric exposure devices, and as reference voltage sources. However, because of the silver content these cells are more expensive than the carbon, mercury, or alkaline cells.

11-22 Recharging of primary cells.   Primary cells are designed for "one-time" use. When their rated capacity has been delivered (and/or their output voltage falls too low for useful work) they should be discarded. Yet home battery chargers that attempt recharging ("rejuvenation" is a better term*) are available on the market. On this matter, the National Bureau of Standards says that for a limited number of cycles, under certain controlled conditions, it is possible to recharge *some* primary cells. However, the conditions are rather restrictive. Recharging of primary cells can be dangerous. Excessive gassing (due to too high a charging rate, or to a defect) can cause a cell to explode.

OTHER PRIMARY CELLS

The cells discussed above are the ones most commonly used. However, for specialized applications several other cells are worthy of mention.

11-23 Lithium cell.   Many varieties of lithium cells have been and are still under experimentation. One common factor is their use of lithium as the negative electrode (anode). Then, depending on the cathode/electrolyte combination used, they can have open-circuit voltages ranging from 1.5 to 4.0 V. The

---

*True recharging recycles the chemical elements in the cells back to their original state. This does not happen here. (Charging methods are covered in Chapter 12.)

1.5-V models would be interchangeable with other primary cells, while in higher-voltage systems, use of the 4.0-V models would require fewer series cells. One manufacturer markets 2.8-V cells, with capacity ratings from 0.5 to 30 Ah. These cells are claimed: to deliver the highest energy per unit weight and volume of any conventional cell; have wide operating temperature range——54 to +74°C (−65 to +165°F); have high discharge rate capacity; and have a shelf life of up to 10 years—even when stored at temperatures of 55°C (130°F). Unfortunately, there is one disadvantage—cost. A typical D-size cell costs over 30 times as much as a carbon–zinc cell. Their use is therefore limited, depending on whether this system offers a cost/performance advantage to the designer and to the consumer.

11-24 Air cell.   Another alkaline cell that has found specialized use is the *air cell*. The electrodes are zinc and carbon, but the carbon electrode is "porous." The electrolyte used is caustic soda (sodium hydroxide, NaOH). Because of the change in electrolyte (as compared to the carbon–zinc cell), the difference in potential developed is only 1.25 V per cell. When the terminal voltage drops to 1 V, the useful life of the cell is ended.

Oxygen to prevent polarization is "breathed" in by the porous carbon "lung" from the air. No manganese dioxide is needed. Since the supply of oxygen is unlimited, greater life is available. The maximum safe current is also increased. However, it should not exceed 0.75 A.

Shelf deterioration is prevented by having the electrolyte as a solid cake. When the cell is put into use, a membrane in the bottom of the filler plug is pierced and pure water is added. The cell remains fresh until the water is added. Air cells have a useful service life up to 3 years.

Whether operation is intermittent or continuous makes no difference in the life of the air cell. The capacity can be accurately stated in ampere-hours. These cells are manufactured in two capacities—300 and 600 Ah. Because of their low cost per ampere-hour, long life, and sustained voltage, these air-depolarized batteries have been particularly successful in such applications as aids -to-navigation markers and buoy lights, railroad signals, and emergency lights.

A "refuelable" zinc–air battery is used by the Marine Corps. The battery can be restored to full activity by replacing the zinc anodes and the electrolyte.

11-25 Seawater battery.   These batteries are designed to produce voltages of from 130 to 250 V and deliver currents up to 580 A, for a time span of 6 to 15 minutes. They were developed to power torpedo and other antisubmarine weaponry. Each cell has a magnesium anode and a silver chloride cathode. The electrolyte is seawater. At launching and during the run of the vehicle, the seawater passes into the battery compartment and out the discharge openings. This seawater flow flushes out the waste reaction products maintaining efficient operation in spite of high discharge rate. The thickness of the cathode plate determines the "run time" of the battery.

11-26 Fuel cells.   A fuel cell is similar to a primary cell in that it produces electrical energy from a chemical reaction, and a "fuel" is consumed. In the carbon–zinc cell, the fuel is the built-in expendable zinc electrode. When this is used up, the cell life is finished. The fuel cell, on the other hand, has an *external* fuel supply, and it continues to produce electrical energy as long as fuel is supplied. The basic reaction in a fuel cell is quite simple. Hydrogen and oxygen are combined to produce water, and in the process they release electrical energy. The cell voltage is approximately 1.0 V.

Reactants for fuel cells are available in a number of forms. Pure hydrogen is commonly stored in compressed-gas cylinders, or it can be produced from hydrides or reactive metals. Hydrogen is also available from hydrocarbon sources such as hydrazine, ammonia, petroleum, propane, and natural gas. Most hydrocarbon fuels are used in indirect oxidation processes. The fuel is first converted to hydrogen which is then oxidized at the fuel electrode. Oxygen can either be supplied as compressed gas in cylinders, or it can be taken directly from the air. Prototype models have been produced using various combinations of fuel sources, as well as liquid, molten, and solid electrolytes. The goal is to produce a cell with direct utilization of organic fuels at relatively low operating temperatures.

The Gemini 5 flight, in 1965, used hydrogen–oxygen fuel cells to deliver 2 kW of peak power. Approximately 200 lb of fuel cells (including gas tanks) did the work of a ton of conventional batteries. Late in 1966, one company demonstrated a General Motors Electrovan powered by 32 hydrogen–oxygen fuel cells, each capable of delivering 1 kW of electrical power, providing the vehicle with a range of about 150 miles. Second-generation fuel cells have shown such promise for high efficiency and negligible pollution that studies are under way for their use in electric power generation.*

11-27 Solar cells.   The photovoltaic effect, which is the basis of all solar cells, was first noticed as far back as 1893, by Edmond Becquerel. In his experiments using a pair of electrodes immersed in a liquid electrolyte, he observed a flow of current when the cell was illuminated with sunlight. However, it was not until growth of semiconductor technology that a practical solar cell became possible. A solar cell is essentially a silicon P-N junction diode.† This consists of a P-type semiconductor in intimate contact with another piece of N-type. Although each piece is electrically neutral, due to added impurity atoms the N-type has extra electron current carriers, while the P-type has a deficiency of electrons or "hole" current carriers. At the junction, some of these opposite-type carriers combine, forming a depletion layer (with no current carriers) and a barrier potential, so that no further carriers can cross over.

---

*The Consolidated Edison Company of New York began construction of a 4.8-MW fuel-cell prototype demonstration plant in New York City in the latter part of 1978. The program is scheduled for testing in late 1982.

†J. J. DeFrance, *General Electronics Circuits* (2nd ed.), Holt, Rinehart and Winston, New York, 1976, Chap. 1.

When energy from a light wave (photons) strikes the cell, it breaks valence bonds within the junction area, and creates electron–hole pairs that cause a potential difference across the cell. This cell voltage is relatively independent of the light level and of the cell size. It is approximately 0.45 V. The current available from the cell, and the power output, do depend on the cell area exposed to the sun. For example, a 57-mm-diameter cell has a power output of approximately 0.25 W, while a 100 mm cell has an output of 1.0 W. Cells can be combined in series to increase the output voltage, and in parallel to increase the output power. Figure 11-10 shows a solar panel array with a peak power rating of 16.5 V, 2.0 A, and 33.0 W.

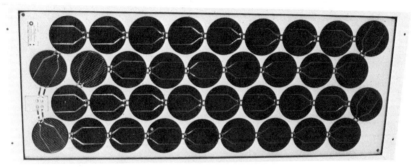

Figure 11-10   Solar battery array. (Courtesy Solar Power Corporation.)

When coupled with a rechargeable battery, solar-cell systems can be used to power many electrical/electronic devices. They are especially useful in remote power supplies such as in communication satellites, radio relay stations, and navigational aids.

Solar energy is also beginning to be used by the electric utilities. A 10-MW solar-thermal power plant was started in California in the early 1982's to supply power to the Southern California Edison system. The facility was developed as a joint effort of the Southern California Edison of Rosemead and the U.S. Department of Energy, with support from the Los Angeles Department of Water and Power and the California Energy Commission. When in full operation, the solar plant is expected to operate on an 8 hour per day schedule during the summer, but only on 4 hours daily in the winter. Thermal storage will supply energy after sundown on a year round basis.

REVIEW QUESTIONS

1. What is the source of the electrical energy output from a cell or battery?
2. What is the distinction between:
   (a) A *cell* and a *battery*?
   (b) A *primary* and a *secondary* cell?

3. Name four types of primary cells.
4. (a) What is an *ion*?
   (b) What is an *electrolyte*?
5. (a) State two functions of the zinc can in the cell shown in Fig. 11-1.
   (b) What is a disadvantage of this type of construction?
6. Name another type of construction that results in a smaller size cell.
7. What is the nominal open-circuit voltage of the carbon–zinc cell?
8. (a) What causes the cell to have internal resistance?
   (b) What is the effect of age on this value?
   (c) How does internal resistance affect the operation of a cell?
9. (a) What is meant by polarization?
   (b) What causes it?
   (c) How is it reduced in the dry cell?
   (d) How is it reduced in the air cell?
   (e) What is its effect?
10. (a) What is meant by local action?
    (b) What causes it?
    (c) What is its effect?
    (d) How is it reduced?
11. (a) What is meant by shelf life?
    (b) In a dry cell, how is the consumer protected against this effect?
    (c) Repeat part (b) for air cell.
12. (a) What determines the maximum safe current of a primary cell?
    (b) What is this value for a Type A dry cell?
    (c) What is the value for an air cell?
    (d) Why does it vary with size of cell?
13. (a) Why are cells connected in series?
    (b) In Fig. 11-2, if each cell has a recommended maximum current rating of 300 mA, what is the recommended maximum current that this group of cells should deliver?
14. (a) When should cells be connected in parallel?
    (b) What is the output voltage available from four carbon–zinc cells connected in parallel?
    (c) When paralleling cells, are their respective ages of any significance? Explain.
15. In battery packaging, what is the significance of an "A" or "B" designation?
16. (a) What are the electrodes in a *mercury* cell?
    (b) What is the open-circuit voltage of this cell?
    (c) Give three advantages of this cell over a carbon–zinc cell.
17. (a) What are the electrodes in an *alkaline* cell?
    (b) What is its open-circuit voltage?
18. Comparing the carbon–zinc, mercury, and alkaline cells:
    (a) Rate them as to first cost.
    (b) Rate them as to shelf life.
    (c) Which is best suited for very light loads?

(d) Which is best suited for light intermittent service?
(e) Which is best suited for heavy loads?
(f) Which is best suited as a voltage reference?
(g) Which is best suited for light-to-moderate loads for long periods of operation?

19. (a) What is the advantage of a silver–zinc cell over the other types noted above?
(b) State two disadvantages of this cell.

20. (a) Can primary cells be recharged?
(b) Give three reasons why recharging is not practical.

21. (a) What is the output voltage from a lithium cell?
(b) How does their shelf life compare with cells discussed previously?
(c) What is their disadvantage?

22. (a) Why may a fuel cell be considered similar to a battery?
(b) How does it differ from a primary cell?
(c) What is the basic principle of the fuel cell?

23. (a) To what class of electronic devices does the solar cell basically belong?
(b) What is the output voltage from this cell?
(c) How does the cell size or strength of sunlight affect this voltage?
(d) On what does the output capacity (current and power) depend?

PROBLEMS AND DIAGRAMS

1. Draw a diagram for dry cells (Type A) connected to supply 4.5 V at 0.8 A. How much current does each cell deliver?

2. A battery-operated receiver requires 6 V and draws 0.06 A. Draw a diagram for dry cells (Type A) to supply this load.

3. A 5-Ω relay requires 0.6 A for correct operation. Draw a diagram for dry cells (Type A) to supply this load. How much current does each cell supply?

4. The open-circuit voltage of a cell (Type A) is 1.50. On short circuit it delivers 6 A.
(a) What is the internal resistance of the cell?
(b) What does this test show with regard to the condition of the cell?

5. The no-load voltage of a cell is 1.52. When the cell is delivering 0.2 A, the voltage drops to 1.50. What is the internal resistance of the cell?

6. Three flashlight cells, EMFs 1.30 V and respective internal resistances 0.2, 0.25, and 0.22 Ω, are connected in series across a 5-Ω bulb.
(a) What are the EMF and internal resistance of the equivalent battery?
(b) What current flows through the circuit?
(c) What is the terminal voltage of the battery and the load voltage across the bulb?

7. Three cells, EMFs 1.4 V each, internal resistance 0.12 Ω each, are connected in parallel across a load of 0.8 Ω.
(a) What are the EMF and internal resistance of the equivalent cell?
(b) What is the load current?
(c) What are the terminal voltage and load voltage?
(d) How much current does each cell deliver?

8. In Problem 1, if each cell has an EMF of 1.5 V and an internal resistance of 0.1 Ω, what is the actual terminal voltage of the battery combination?

9. Four cells are connected in parallel across a load of 0.6 Ω. The EMF of each cell is 1.30 V. The internal resistances are 0.08, 0.10, 0.09, and 0.05 Ω, respectively.
   (a) What is the load current?
   (b) What is the current delivered by each cell?

10. Two batteries, EMFs 18.0 V each and having internal resistances of 0.8 and 1.0 Ω, are connected in parallel across a load. The load current is 4.0 A. Find:
   (a) The load resistance.
   (b) The terminal voltage.
   (c) The current from each battery.

11. The terminal voltage of a cell is 1.38 V, when connected to a load of 6 Ω. The internal resistance of the cell is 0.32 Ω. Find the EMF of the cell.

12. A cell has an open-circuit voltage of 1.42 V and a terminal voltage of 1.38 at a load of 0.25 A. Find the internal resistance of the cell.

13. A bank of 12 cells is connected into three parallel branches, each branch containing four cells in series. Each cell has an EMF of 1.5 V and an internal resistance of 0.1 Ω. When the bank is connected to a load, a current of 0.15 A flows through each cell. Find:
   (a) The equivalent EMF of the bank.
   (b) The equivalent internal resistance of the bank.
   (c) The resistance of the load.
   (d) The terminal voltage of the bank.
   (Neglect the resistance of all interconnecting wires.)

14. The 12 cells of Problem 13 are reconnected to form two parallel branches, with six cells in series in each branch. Repeat parts (a) through (d).

15. A relay having a resistance of 20 Ω requires a current of at least 230 mA for proper operation. It is to be energized from Type A dry cells. Each cell has an EMF of 1.5 V and an internal resistance of 0.1 Ω. How many cells are required and how should these cells be connected?

16. A relay requires at least 28 V across its coil before it will close. The current at this voltage is 600 mA. It is to be operated from Type A dry cells, each cell having an EMF of 1.5 V and internal resistance of 0.1 Ω.
   (a) Find the minimum number of cells required.
   (b) Draw a diagram showing proper connection of these cells.
   (c) Find the actual voltage across the coil.

12

SECONDARY CELLS

When the life of a primary cell is exhausted, the cell is useless. It must be thrown away. A secondary cell, when its ampere-hour capacity is exhausted, can be recharged and used over and over again. Secondary cells are made in several types: lead–acid, nickel–iron (Edison), nickel–cadmium, silver–cadmium, and silver–zinc.

12-1 Lead–acid cell—construction. The *electrodes* consist of a rigid framework made of antimony-lead alloy into which the active material is pasted under pressure. The main ingredient of the active material is lead oxide. The plates are then formed. In this process the positive plates are turned into lead peroxide (PbO_2) and acquire a chocolate-brown color. The negative plates are turned into spongy lead (Pb), a dull slate gray in color. The *electrolyte* used is a dilute solution of sulfuric acid (H_2SO_4), approximately 8 parts of water to 1 part acid.

The capacity of a cell depends on the plate area. To increase the plate area, instead of making the plates larger, a group of positive and negative plates is used. These plates are tied together by connecting straps. These two groups are then sheathed together to form an *element*. It is usual practice to have one more negative plate in an element than positive plates, the two outside plates being negative. To prevent short circuit, or contact between any of the negative and positive plates in the element, *separators* are used. These separators were often made of specially treated grooved porous cedarwood. In more recent construction, separation is done with glass fabrics or fibers, in conjunction with plastic sheaths. In the cover of each cell is an opening, or vent, which provides a place for regular testing of the solution and for replenishing the solution with approved water when necessary. A hard-rubber vent plug should always be screwed firmly in place when the cell is in service, to prevent spilling and spraying of the solution. A hole in the vent plug allows for the escape of gases formed during the charging of the cell. The parts and construction of a cell can be seen in Fig. 12-1.

12-2 Lead–acid cell—chemical action. When a cell is fully charged, the active ingredients are a lead peroxide plate (PbO_2), a spongy lead plate (Pb), and a dilute solution of sulfuric acid ($2H_2SO_4$). The action that takes place can be shown by chemical equation as follows:

$$\overset{Charged}{PbO_2 + Pb + 2H_2SO_4} \rightleftharpoons \overset{Discharged}{2PbSO_4 + 2H_2O}$$

The sulfuric acid ionizes into four positive hydrogen ions and two negative SO_4^- ions. One of the SO_4^- ions combines chemically with the lead plate to produce lead sulfate. The negative charge of the SO_4^- ion gives the lead plate a negative potential. The four positive hydrogen ions take electrons from the lead peroxide plate, becoming neutral, and then combine with the oxygen of the plate, forming water. This plate now has a positive charge of 4. However, the other SO_4^- ion now combines with the lead to form lead sulfate, meanwhile neutralizing two of the positive charges. A net positive charge of 2 is still left on the PbO_2 plate.

In a fully charged cell, the difference in potential between the negative lead plate and the positive lead peroxide is 2.2 V. The voltage at any time on discharge or charge depends on several factors, such as current rate, state of charge or discharge, and temperature. No general average to cover all conditions can therefore be given. In practice it is usually considered as a 2.0-V cell.

12-3 State of charge—specific gravity. Specific gravity of any liquid shows how much heavier that material is in comparison with an equal amount (volume) of water. Pure sulfuric acid has a specific gravity of 1.840. In a lead cell, dilute sulfuric acid is used. As the acid is mixed with water its specific gravity decreases. The dilute sulfuric acid in a fully charged cell varies between 1.275 and 1.300, depending on manufacture and the temperatures at which the cell is to be used. As the cell discharges, water is formed and the sulfate is being taken out of solution by the plates. Both reactions tend to reduce the concentration of acid. Therefore, specific gravity indications can be used to indicate state of charge of the cell. A cell is considered half charged at 1.210 and fully discharged between 1.120 and 1.150.

To measure specific gravity a *hydrometer* is used (see Fig. 12-2). The float in the hydrometer is weighted and calibrated so that when the specific gravity of the acid is 1.300, the liquid-level line will correspond to the 1300 line. As the cell discharges, the electrolyte is diluted and cannot support the

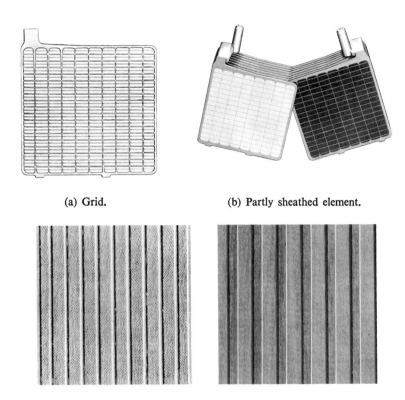

(a) Grid.   (b) Partly sheathed element.

(c) Rubber and wood separators.

(d) Sectional view of three-cell battery.

Figure 12-1 Lead–acid cell construction. (Courtesy of Willard Storage Battery Co.).

(a) Typical hydrometer.

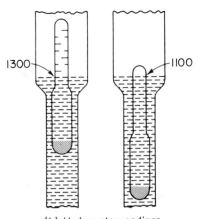

(b) Hydrometer readings.

Figure 12-2 (Photograph Courtesy of Willard Storage Battery Co.).

float as well. The float therefore sinks lower. The top number on the hydrometer is 1100.

Specific gravity readings should not be taken immediately after water has been added. The cell should be allowed to charge or discharge for at least one hour after water is added to allow the water to mix thoroughly. False readings will also be given if the hydrometer is damaged. A moist paper scale or vapor on the inside walls of the float are signs of a leaking and inaccurate hydrometer.

12-4 Capacity rating. The capacity of a battery, given in ampere-hours, depends on the number of plates. Theoretically, a 100-Ah battery can deliver 100 A for 1 h, 50 A for 2 h, 4 A for 25 h, and so on. However, this rating will vary with rate of discharge and temperature. The rated capacity is based on a "normal" discharge rate of 8 h at 28°C (80°F). Greater capacity can be obtained if the discharge current rates are made lower. The *normal current* for a cell is that current which will discharge it in 8 hours; for example, a 100-Ah battery has a normal current rating of $\frac{100}{8} = 12.5$ A. However, unlike the case with primary cells, there is no maximum safe current limitation. The maximum permissible rate of discharge is limited only by temperature. The cell temperature must not exceed 43°C (110°F).

On the other hand, overdischarge may damage the cell. As the cell discharges, the plates change to lead sulfate. This sulfate occupies more space than the active material from which it is formed. To provide for this expansion, the active material is made porous. The size of the pores is reduced as sulfate is formed. Plates are designed to accommodate only a certain amount of expansion, limited to the amount represented by specific gravity readings of 1120 to 1150. Further discharge results in excessive sulfation, which so closes the pores that the cell becomes increasingly difficult to recharge.

12-5 Effect of temperature.   The normal operating temperature for a cell is 28°C (80°F). Temperatures high above or far below this value not only affect the capacity but may also injure the cell permanently.

High temperatures increase chemical activity. Continued or frequent temperatures above 43°C (110°F) will injure the plates and shorten the life of the separators. To reduce likelihood of damage, cells that are to be used in tropical climates should have the specific gravity of their electrolyte reduced. It is common practice to adjust the electrolyte of a fully charged cell to 1.225 as a maximum.

Overheating causes the plates of a cell to buckle. The active material may fall out, or the buckled plates may cause sufficient pressure on the separators to "pinch through" and short-circuit the cell. In addition, high temperatures and increased chemical activity carbonize wooden separators, breaking down their insulating effects.

Low temperature reduces chemical activity, resulting in lower voltage and lower available capacity. At $-18°C$ (0°F) the available capacity of a cell is only 60% of its rated value. In addition, if the cell is discharged, there is danger of the electrolyte freezing. Severe freezing causes expansion and loosening of active material, and soon it falls out of the grids in large pieces. The freezing temperature of the electrolyte varies with its specific gravity. At 1.275 the freezing point is $-29°C(-85°F)$, whereas at 1.100 it is only $-7.2°C$ $(+19°F)$. Obviously, to avoid freezing the cell should be kept charged.

12-6 Charging methods.   On discharge, current flows out of the positive terminal of the cell, through the load, and back into the negative terminal. To charge the cell, merely reverse the direction of current flow! This reversal is accomplished by connecting the negative terminal of the cell to the negative terminal of the charging line, and the positive terminal of the cell to the positive terminal of the charging line. The voltage of the charging line must be higher than the potential difference of the cell, in order to "force" current to flow against the cell's force.

While a battery is being charged, the amount of sulfate in the plates decreases, and the ability of the plate to give up acid is reduced. A cell that is considerably discharged has a large amount of sulfate available and can give up acid at a rapid rate. The rate at which this chemical action proceeds depends on the charging current. High charging currents can therefore be used. As a cell approaches full charge, the amount of sulfate in the plates decreases, and the charging current should correspondingly be decreased. If a high charging rate is maintained, only a portion of this current is used to withdraw acid from the plates. The remainder of the current, by electrolysis, decomposes the water in the solution into oxygen and hydrogen, which are given off in the form of gas. "Gassing" of a cell early during the charging process indicates that the charging current is too high. The action of the bubbles of gas, in escaping from the pores of the plates, has a tendency to wash out and wear away the active material from the plates. This effect is especially noticeable in the positive plate, which softens with use. *Violent gassing should be avoided at all times.* To avoid gassing, the charging rate must be low when full charge is approached. This is known as *finish charge rate.* However, all cells should gas freely when fully charged. If a cell does not gas, it indicates that either the cell is not fully charged or else there is some internal trouble. The temperature of a cell should be watched carefully during charging. If the temperature rises above 110°F, stop charging or reduce the charging rate and allow the cell to cool.

To charge a cell, direct current (dc) must be used. Two methods of charge are in general use. These are known as the *constant-current* and the *modified constant-potential* methods. In the constant-current method, the charging-line voltage is high. Several cells or batteries are connected in series with a rheostat and ammeter to the charging line. The charging current is controlled by the rheostat. The finish rate is recommended. However, a high rate can be used at first and reduced to the finish rate when the cells begin to gas.* This constant-current method is best as far as life of the cell is concerned. Badly sulfated cells require a long, slow rate of charge, continuing for several days. The constant-current method at lower than finish rate is recommended.

The constant-current method has two disadvantages: it requires a long charging time and appreciable power is wasted in the series rheostat. Where speed of charging is important, the modified constant-potential method is used. Chargers for this method are built with voltage control so that the charging-line voltage can be set to a value just above the fully charged EMF of the battery. For example, to charge a 12-V battery, the voltage would be set for approximately 13 V. Each battery to be charged is connected in parallel across the charging line. At the beginning of charge, the current passing through the battery may be high. The battery temperature may rise rapidly and should be watched carefully. As the cells charge, the battery voltage increases, reducing the effective voltage in the circuit. The charging current automatically tapers off. A rule of thumb of the battery manufacturers is that the charging current should not exceed the number of ampere-hours removed from the battery. With a charging-line voltage just above the EMF of the fully charged cell, a small series resistor is inserted to limit the initial charging current to a value not more than the

---

*The *two-rate* method may be either manually or automatically controlled.

ampere-hour rating of the battery. A little more time is required, but the current values are much safer for the battery.

12-7 Maintenance. When a solution, for example, salt water, is left exposed to air, it evaporates. But what is lost—the water, the salt, or both? Careful inspection shows that the salt is left and only the water evaporates. Similarly for the lead–acid cell: the electrolyte evaporates, but only water has been lost. Electrolysis results in further loss of water, in the form of hydrogen and oxygen gas. The level of electrolyte must not be allowed to drop below the top of the separators. Not only does the cell lose capacity but, even worse, the separators are ruined if they are allowed to dry. What should be added? Use only approved or distilled water, and fill to about 1 cm above the separators. If electrolyte has been spilled through carelessness, acid of correct specific gravity should be added. To make this adjustment, bring the cells to fully charged state; then adjust level of liquid and specific gravity to normal, or pour off all the liquid and replace with fresh electrolyte of 1280 specific gravity.

All batteries discharge while standing idle. If a battery is allowed to remain unused for any length of time, especially if originally discharged, the cells will sulfate badly. *A storage battery should not be left discharged for long periods of time.* The lead sulfate changes gradually to an insoluble form. In this condition, it may be impossible to recharge the battery. To prevent this damage, a battery standing idle should be brought to the fully charged state about every 30 days, or kept on trickle charge continuously. When the idle period is very long, this is a nuisance. An alternate method is to bring the cells to full charge and then siphon off all the electrolyte. Flush the cells thoroughly with pure water or fill the cells with water and allow them to stand 12 to 15 h. Pour off the water and the cells will stand indefinitely without injury.

12-8 Maintenance-free batteries. You have probably seen advertisements for "maintenance-free" automotive batteries. These batteries were developed to eliminate the need to add water to maintain cell capacity. This feature was obtained in two ways: first, by designing excess water into the system; and second, by reducing the rate of loss of water.

In the conventional battery, the plates are mounted some distance above the bottom of the case [see Fig. 12-1(d).] With repetitive cycles of charge and discharge, some active material tends to flake or fall out of the grid structures. By allowing space at the bottom, any fallout will lodge below the plate level, and will not short out the cell. In a maintenance-free battery, the plates are enclosed in microporous envelopes. (These are used as separators instead of flat sheets.) Any flaking now remains inside the envelope, and cannot cause a short. The empty space at the bottom is no longer needed. Therefore, the plates are lowered to the bottom of the case, making the electrolyte level appreciably higher above the plates. The cell can now lose much more water without loss of capacity, or damage to the plates.

To give the grid structures rigidity, manufacturers used an antimony alloy in the lead grid structure. In early models, an 11% alloy was used. Unfortunately, antimony produces a chemical reaction that lowers the battery countervoltage. This in turn decreases the battery's resistance to overcharge, promotes gassing (the breaking up of water in the electrolyte into hydrogen and oxygen), and speeds up the loss of water. Several techniques have been used to reduce this effect. The antimony content has progressively been reduced from the 11% to about 7%, then to 4.5%, and most recently to 2.5%. Other manufacturers have replaced the antimony with calcium as a hardener. It too has chemical interactions, but sufficient hardness can be obtained with calcium levels as low as 0.07%. The calcium alloy has its own drawbacks: production difficulties and higher costs. Other alloy materials, such as strontium, are under investigation.

12-9   Sealed lead–acid batteries.   The increasing use of portable, cordless electrical-electronic equipment led to the development (in the mid-1960s) of small, sealed lead–acid batteries. Several new construction features made this "sealed" cell possible:

1. The electrolyte is immobilized. One technique uses a gelling agent to solidify the electrolyte and keep it from flowing. Another technique stores the electrolyte in highly porous separators. In either case, this gelling feature makes it possible to operate such batteries in any position, and minimizes the possibility of leakage.
2. The grids for the electrodes are made of calcium–lead, since this was found to reduce self-discharge. This in turn minimizes the loss of water. (Replacement water cannot be added.)
3. A unique "sealing" method was developed which allows gases to escape but does not allow an electrolyte to leak out, or air to enter.

Sealed lead–acid batteries are available in sizes ranging from 2 to 12 V and with capacities from 1 to 25 Ah. They can be charged, discharged, and stored in any position.

12-10   Edison (nickel–iron) cell—construction.   The case of an Edison cell is made of welded, heavily nickel-plated sheet steel. This construction makes the case unbreakable. The positive plate consists of nickel tubes about 0.6 cm in diameter and 10 cm long. When inserted in the tubes, the active material is in the form of nickel hydrate. After the formation process, this changes to an oxide of nickel (NiO_2). Finely divided pure nickel flakes are mixed in alternate layers with the active material, to improve conductivity. The active material is stamped in at high pressure. Stock for the positive tubes is heavily perforated. The tubes are reinforced, when filled, with eight seamless-steel rings. They are then mounted into steel grids and locked into place under a pressure of 40 tons. These plates are then assembled into groups. The negative plate consists of flat perforated nickel-plated steel pockets filled with finely divided iron. These pockets are mounted into steel grids and assembled into groups in the same manner as the positive plates. The electrolyte used is a 21% solution of caustic potash (potassium hydroxide, KOH) plus added lithium hydroxide.

The completed cell construction can be seen from Fig. 12-3.

Since the 1950s a "sintered plate" has been developed. Instead of using

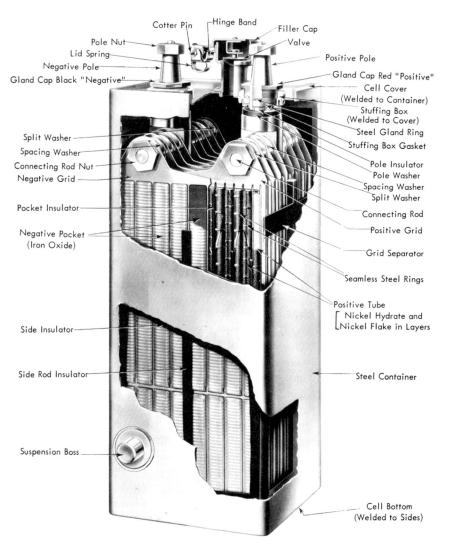

Figure 12-3 Cutaway view of Edison cell. (Courtesy of Thomas A. Edison, Inc.).

"pockets," the active plate material is held in place in the fine pores of a micro-porous sintered nickel plaque with extremely high surface area. The result is a much lower internal resistance. This type of construction is used when high-rate discharge capability is required.

12-11 Edison cell—chemical action. The chemical action in the cell is quite complex, but its nature can be indicated by chemical equation as

$$\text{Fe} + 2\text{KOH} + \text{H}_2\text{O} + \text{NiO}_2 \overset{Charged}{\underset{Discharged}{\rightleftarrows}} \text{FeO} + 2\text{KOH} + \text{H}_2\text{O} + \text{NiO}$$

The potassium hydroxide (KOH) breaks up into negative hydroxide ions and positive potassium ions:

$$2\text{KOH} \longrightarrow 2\text{K}^+ + 2(\text{OH})^-$$

The hydroxide ion moves to the iron plate (Fe), oxidizing the iron. In the process, the ion gives up its excess electrons to the plate, making the plate negative, the hydroxide changing to water (H_2O):

$$\text{Fe} + 2(\text{OH})^- \longrightarrow \text{H}_2\text{O} + \text{FeO} + 2 \text{ electrons}$$

Meanwhile the positive potassium ion moves over to the nickel oxide plate (NiO_2), taking electrons from the plate and neutralizing its own positive charge. The potassium combines with water to form potassium hydroxide and nascent hydrogen. The hydrogen thus formed reduces the nickel oxide to a lower oxide. Since the nickel oxide has lost electrons, it becomes the positive terminal of the battery:

$$2\text{K}^+ + \text{H}_2\text{O} + \text{NiO} \longrightarrow 2\text{KOH} + \text{NiO} - 2 \text{ electrons}$$

The potential difference (EMF, or open-circuit voltage) so created between the iron and nickel oxide plate is approximately 1.40 V per cell.

12-12 State of charge. In the chemical analysis, you saw that the electrolyte remains unchanged and that no water is formed. Hydrometer readings would give no indication as to state of charge, because the specific gravity remains practically constant at 1200. On the other hand, the internal resistance of the cell is comparatively high and increases rapidly as the cell approaches full discharge. Under normal load, the terminal voltage of the cell drops quickly from 1.40 to 1.30 V, then slowly to 1.10, and then rapidly again to 0.9 V. A voltmeter is therefore used to determine the state of charge. The average voltage of an Edison cell, in practice, is considered at 1.2 V. The discharge condition is taken as 1.0 V. The manufacturer has developed an instrument that measures the voltage of a cell as it discharges through a known fixed resistance. From this voltage reading the state of charge of the battery and the current input required to restore the battery to a charged condition are found by referring to a charge-indicator table furnished with the instrument. This *charge test-fork* combines a specially calibrated voltmeter with the known resistor and contact prongs into a single assembly with insulated handle.

12-13 Capacity rating. The capacity of an Edison battery is given in amperehours and depends on the number of plates. This rated capacity is based on a "normal" discharge rate of 5 h. The normal current rate for the cell will then be given as one-fifth the capacity rating.

12-14 Charging methods. To charge an Edison cell or battery, the same procedure is followed as in the lead–acid cell. As the cell approaches full charge, gassing will commence. However, gassing cannot harm the nickel–iron–alkaline battery, so that it is free from finish-rate limitations. Either the constant-current or the modified constant-potential method can be used. The constant-current method as applied to the lead–acid cell requires a long time at the finish rate. In the Edison cell, by increasing the charging-current rate the total time required for charging can be reduced without risk, of injury. The manufacturer of Edison cells recommends the modified constant-potential method. If proper values of charging potentials and series resistances are chosen, it permits an average normal rate without excessively high rates at the start or unnecessarily low rates at the finish. However, neither overcharge nor overdischarge will injure the plates.

12-15 Maintenance. As in the lead–acid cell, evaporation and excessive gassing during charge will result in loss of water from the cell. This loss must be made up periodically by adding pure water to keep the electrolyte level above the plate tops. In rare cases, carbonates either in the water or from excessive carbon dioxide in the air will change the electrolyte into potassium carbonate. Where such conditions exist, the electrolyte should be replaced after approximately 250 complete cycles of charge and discharge. Normally, the solution need not be renewed unless the specific gravity falls below 1.160 after adjusting for solution height and temperature variation.

When the cell is to stand idle for long periods, short-circuit the cells to zero charge and store away in a clean dry place. No further attention is required. Even the full discharge is not necessary. When a cell is to be put to use again, an overcharge at the normal rate is the only preparation necessary.

12-16 Nickel–cadmium battery. The nickel–iron and the nickel–cadmium cells were both first introduced at about the turn of the twentieth century. In the United States, probably because of the cheaper and more readily available raw materials, development of a nickel–iron battery was fostered by Thomas Edison.* In Europe the nickel–cadmium battery was favored. The early nickel–cadmium batteries were vented and used pocket plates—much like the Edison battery (see Fig. 12-3). In the mid-1940s, the nickel–cadmium cell was introduced in the United States. At about the same time, sintered-plate construction was being developed in Germany.

*Therein it got the name Edison cell.

The main advantage of the cadmium cell (over the iron type) is much lower internal resistance. This, in turn, results in more-constant output voltage under load, and allows high-rate discharge. The average voltage on discharge is 1.2 V per cell and at normal discharge rates, this voltage remains very nearly constant until the cell approaches full discharge. Vented cells are available with capacities of over 2000 Ah.

In the uncharged condition the positive electrode of a nickel–cadmium cell is nickelous hydroxide, and the negative cadmium hydroxide. In the charged condition the positive electrode becomes nickelic hydroxide, the negative metallic cadmium. The electrolyte is potassium hydroxide. The overall chemical action can be considered as:

$$\text{Charged} \qquad\qquad\qquad \text{Discharged}$$
$$Cd + 2NiOOH + 2H_2O + KOH \rightleftharpoons Cd(OH)_2 + 2Ni(OH)_2 + KOH$$

Since 1960, technological advances made possible the production of small hermetically sealed rechargeable cells. These are available in three types: button cells (20 to 500 mAh); cylindrical cells (100 mAh to 4.0 Ah); and rectangular cells (1.5 to 23 Ah). The capacity of these cells is based on a 10-h discharge rate. Typical button and cylindrical cells and batteries are shown in Fig. 12-4. Sealed cells should not be charged at rates exceeding the 10-h rate, particularly as they approach full charge. If floating, the trickle-charging current required to keep the battery in a fully charged condition (as in electric toothbrushes), should not exceed the 50-h rate. If these charging rates are exceeded, the gas (oxygen) generated at the nickel electrode will not be absorbed by chemical reaction; pressure will build up and the cell can rupture. To alleviate this condition, recent cell construction includes a safety vent.

These sealed cells are small, convenient packages of high-energy output. They can be recharged for hundreds of cycles giving a long active life. They also have an indefinite storage life. Long idle periods, charged or uncharged, do not cause damage. Because of hermetic sealing, they are leakproof, will operate in any position, and require no maintenance. These cells will operate, with a relatively small loss in capacity, over a temperature range of −20 to +45°C. They can be stored under even wider temperature conditions (−40 to +60°C).

12-17 Silver oxide–zinc.   Where size and weight are of primary importance, this is the preferred power source. The silver–zinc battery is noted for its high energy density, in fact, the highest attainable of any secondary system in use today. It also has a very low internal resistance with a very flat voltage discharge characteristic which is desirable in many high-rate applications. Silver-zinc cells will maintain nearly constant output voltage over a wide range of current rates and can deliver full capacity even at a 15-min rate. It is possible to obtain discharge currents as high as 30 times the rated cell capacity. Silver–zinc rechargeable batteries are available either as vented or sealed cells. Cells have been built with capacities ranging from 0.1 to 30 000 Ah. Some of

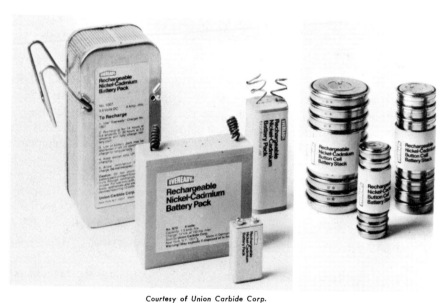

Courtesy of Union Carbide Corp.

Figure 12-4 Sealed, rechargeable nickel–cadmium cells and batteries.

these cells in various sizes and capacities are shown in Fig. 12-5. The silver–zinc system, at the present state of the art, is normally capable of about 100 deep cycles of discharge and charge with an overall activated life of about 2 years. Shallow discharge at moderate rates will give a cycle life of 200 to 300 cycles.

Figure 12-5 Rechargeable silver–zinc batteries. (Courtesy Yardney Electric Corp.).

The nominal open-circuit voltage of silver–zinc cells is 1.86 V. Under load, at moderate rates (i.e., the 1-h rate), the nominal load voltage is 1.50 V.

The high cost of silver, in addition to the high cost of manufacture, has limited application of this system almost entirely to military and aerospace programs as well as very specialized civilian applications, where the emphasis is placed on the performance rather than the economic factor. These include space satellites, missile power, submarine and torpedo propulsion, and some types of portable military communications. Examples of silver–zinc batteries for diverse applications are shown in the accompanying figures. The very large secondary silver–zinc battery shown in Fig. 12-5 is capable of 120 V at 560 A for 6 min. Under these conditions, it will deliver about 50 cycles.

12-18 Silver–cadmium cells. This cell combines the high-energy capability of the silver electrode with the long-life characteristic of the cadmium electrode. On the shelf, it will retain up to 85% of its capacity for periods of 1 to 2 years. In use, it will give a cycle life of over 500 cycles on deep discharge, and over 5000 cycles at 50% discharge. Silver-cadmium cells and batteries will deliver more electrical energy for comparative size and weight than any other type—except the silver–zinc. They have a low internal resistance and can there-

fore supply high discharge rates, and maintain flat discharge voltage under
load. These batteries are available with vented or sealed cells, and in ratings
from 0.1 to 300 Ah. Figure 12–6 shows a 0.25 Ah button cell and a battery
power pack. The battery has a nominal rating of 2.0 Ah, an open-circuit voltage
of 4.2 V, a load voltage of 3.1 V at 2.0 A, and can supply up to 30 A in 5-s
pulses.

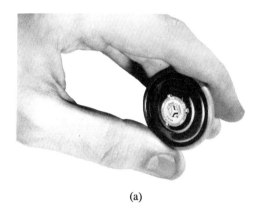

(a)

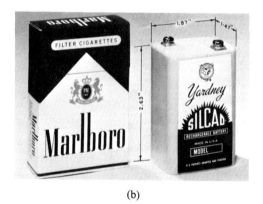

(b)

Figure 12-6   Silver–cadmium button cell and battery. (Courtesy Yardney Electric
Corp.)

12-19 Secondary cells—comparison.   In general, compared to
the lead–acid battery, the alkaline battery has the following advantages:

1. Lighter weight per ampere-hour capacity.
2. Mechanical strength and ruggedness. It can withstand severe vibration
   and shock.

3. Electrical "ruggedness." It can be short-circuited without injury. Even accidental charge in reverse direction will not injure the plates.
4. It requires less maintenance, even for storing away, and can be left discharged without damage.
5. It is not damaged by freezing. Also, freezing temperature does not vary with state of charge.
6. Freedom from corrosive or poisonous acid fumes.
7. It can dissipate heat developed in the cell more easily.

The silver cells, which have the highest output rating per pound, and per cubic inch, have the lowest life, and are, by far, the most expensive.

The nickel–iron (Edison) cell, which is the most rugged, and has the longest life expectancy, has the highest internal resistance, and is not suitable for high-rate discharge.

On the other hand, the lead–acid cell, which has a relatively poor life span, is by far the lowest in first cost. In fact, if cost is considered, this type of battery will put out the highest watt-hours per pound (or per cubic inch) per dollar of first cost. A comparison of the various secondary cells is shown in Table 12-1.

TABLE 12-1   Secondary Cell Voltages

Cell	Electrodes	Electrolyte	Nominal Voltage (V)
Lead–acid	Lead peroxide and Lead	Sulfuric acid	2.0
Edison	Nickel oxide and iron	Potassium hydroxide	1.2
Nickel–cadmium	Nickelic hydroxide and cadmium	Potassium hydroxide	1.2
Silver–zinc	Silver oxide and zinc	Potassium (or sodium) hydroxide	1.5
Silver–cadmium	Silver oxide and cadmium	Potassium hydroxide	1.5

Meanwhile, under experimentation is a variety of batteries for use in electric vehicle propulsion systems. The object is to get higher ampere-hour capacity in a smaller volume and at a lower weight. Among such batteries are a zinc–chlorine, a lithium–metal sulfide, and a sodium–sulfur battery. It is projected that these sources will give three to five times the capacity of an equivalent-size lead–acid battery.

REVIEW QUESTIONS

1. With regard to the lead–acid cell:
   (a) What is the electrolyte?
   (b) Why can the state of charge be determined by measuring the specific gravity of the electrolyte?
   (c) What device is used for this purpose?
   (d) What reading corresponds to full charge? to full discharge?
   (e) What is the correct level of electrolyte?
   (f) In normal operation, if the liquid level falls too low, what should be added? Explain.

2. What is the "normal" terminal voltage of the lead–acid cell?

3. What is the effect of each of the following on a lead–acid cell?
   (a) Low water level
   (b) Overcharging
   (c) High charging rate
   (d) Long idle periods, especially if discharged

4. (a) What is the normal operating temperature of a lead–acid cell?
   (b) What is the effect of too high a temperature?
   (c) Of too low a temperature?
   (d) What changes are made for use in the tropics?

5. (a) What is meant by ampere-hour capacity of a cell?
   (b) What factors affect this capacity?

6. (a) Describe the method used for constant-current charging.
   (b) What is meant by "finish rate"? Why is it used?

7. (a) Describe the method used for modified constant-potential charging.
   (b) What are its advantages and disadvantages?

8. Why are vents provided, particularly for high-capacity cells?

9. Describe two techniques that made "maintenance-free" batteries possible.

10. Describe two features that made a sealed lead–acid cell possible.

11. (a) Give another name for the Edison cell.
    (b) Does it use an acid or alkaline electrolyte? What is the electrolyte?
    (c) Can a hydrometer be used to check the state of charge of this cell? Explain.
    (d) How is the state of charge determined?
    (e) What is the "average" load voltage for this cell?
    (f) What value corresponds to a discharged condition?
    (g) What discharge rate is used in fixing the capacity rating of this type cell?
    (h) What is the advantage of the *sintered plate* construction?

12. (a) What is the advantage of the nickel–cadmium cell over the nickel–iron cell?
    (b) Give a disadvantage.

13. (a) What is the major advantage of the silver–zinc cell?
    (b) Give two disadvantages.
    (c) What is its output voltage?

14. Compare the silver-cadmium cell to the silver–zinc cell, as to:

(a) Output voltage

(b) Energy capability

(c) Life

15. Which type of cell would be best suited for long (many years), unattended operating periods? (Charging is done automatically.) Explain your choice.

16. Why is the lead–acid cell used in automotive batteries?

17. Which type of cell would be best suited for missile use, where payload is of prime importance? Explain your choice.

13

PERMANENT MAGNETISM

Undoubtedly at some time or other each of you has seen and even used a permanent magnet. But have you ever stopped to think what magnetism is? A magnet can lift certain things; it has ability to do work. But this is the definition for energy. Magnetism, then, must be a form of energy. However, we know that energy is not created but is merely transformed from one type of energy to another. What is the origin of this energy we call magnetism? To answer this question let us first consider a mechanical analogy.

A pile driver when raised to the top of its lift has energy due to its elevated position—potential energy. When released it will fall, but in falling it is losing elevation. It must also be losing potential energy. This energy is not lost, because as the pile driver falls it gains speed. The potential energy is converted to *energy of motion*. In mechanics this energy is called *kinetic energy*. The lower the pile driver falls, the lower its potential energy; but it is moving faster, and so the greater its kinetic energy.

13-1   Electron theory of magnetism.   The electrical picture is very similar. A stationary charge has energy due to the position of its electrons. In Chapter 1 we called this energy a dielectric or electrostatic field and said it was a form of potential energy. If this charged body is made to move, some of this potential energy is converted to energy of motion, called *magnetic energy*. In other words, *magnetism is energy due to motion of a charged body* and originates from the dielectric field energy of the stationary charged body. As in mechanics, the faster the charged body moves, the greater the magnetic field and the smaller the dielectric field energy. If a charged body could be made to move with infinite speed, all its energy would be in the magnetic field. The greatest speed we have been able to obtain is the speed of light (3×10^8 m/s). At this speed, the energy of a charged body is stored 50% in its dielectric field and 50% in its magnetic field.

To see how this discussion applies to permanent magnets, let us go back to the structure of matter. All matter is composed of atoms. Each atom has a nucleus, and electrons in concentric shells revolving around the nucleus. The electron is a negative charge, and it is in motion! It must have magnetic energy.

This result seems to indicate that all materials would have magnetic effects. Yet we know that wood is not a magnet and cannot be made into one. Steel sometimes acts as a magnet and at other times it does not. Why?

In the structure of an atom, we say that in each shell some of the electrons revolve clockwise and others revolve counterclockwise around the nucleus. Each group of electrons produces magnetic fields. The strength of the magnetic fields depends on the number of electrons, and the direction of the magnetic fields depends on the direction of rotation. Let us consider the result under three classifications:

1. *Nonmagnetic Material.* The number of electrons revolving clockwise in any one shell is equal to the number revolving counterclockwise. The two magnetic fields produced are equal and opposite. The atom has no external magnetic field, or is considered nonmagnetic.

2. *Magnetic Materials.* The number of electrons revolving in one direction is greater than the number revolving in the opposite direction. The two magnetic fields produced are opposite but *unequal.* Therefore, there is a net magnetic field outside the atom, or the atom is magnetic.

3. *Magnetized Material.* A magnetic material may or may not show any external magnetic properties. We saw above how two electrons or groups of electrons can have magnetic fields in opposite directions and neutralize each other. Similarly, a group of magnetic atoms can have their magnetic fields in such directions as to produce no net magnetic field for the entire material. Such a material is not magnetized. A magnetized material must consist of magnetic atoms and, in addition, the magnetic forces of the majority of its atoms must be in the same direction.

Iron is an example of magnetic material. It has 26 electrons. The first shell has 2 electrons, 1 revolving clockwise, the other counterclockwise. The second shell has 8, 4 revolving in each direction. The third shell is incomplete. The capacity of this shell is 18. It has only 14 electrons, of which 9 revolve in one direction and 5 in the other. The last shell has 2 electrons, 1 revolving in each direction. Notice that the third shell has 4 unbalanced electron rotations. These give the atom its magnetic properties.

Cobalt and nickel are also magnetic materials. They are not so strongly magnetic as iron, cobalt having only 3 unbalanced electron rotations, and nickel

2. Crystal structure or *domains* play an important part in magnetic properties and explain why some alloys are such good magnetic materials. However, a study of domains is beyond the scope of this text.

13-2   Permanent and temporary magnets.   When a magnetic material is placed in a magnetic field, the atoms will align themselves so that their magnetic fields conform to the external field. Some materials will retain the alignment after the external field is removed, others will lose their alignment. *Retentivity* is the property of a material to retain its magnetic alignment. (Retentivity may be considered as due to friction between atoms that makes them hard to align. By the same token, once they are aligned they lose alignment only with difficulty.) *Residual magnetism* is the amount of magnetism left in the material after the external magnetizing field is removed, or it is a measure of the number of atoms that remain in alignment. A material with good retentivity will have a strong residual magnetism. Such a material will make good permanent magnets. Examples of this material are steel, and some of its alloys. Alnico, an alloy of aluminum, nickel, and cobalt, and cobalt–chromium alloys make excellent permanent magnets. Soft iron has poor retentivity, and is a good material for temporary magnets.

Have you ever tried to steer an automobile (without power assist) while it is stationary? Difficult, isn't it? How much easier it is to turn the wheels while the car is in rapid motion! The same is true with atomic alignment. A permanent magnet with good retentivity may lose its magnetism if the atomic motion is violently increased. This effect is caused by abuse such as dropping, banging, or subjecting the magnet to high temperatures.

A magnet is very strong when first made, but over a period of months it will continue to lose its magnetism slowly. After this "aging" period, the magnetic strength remains constant unless the magnet is abused. A fresh magnet cannot be used immediately in commercial applications, because the calibration of the apparatus in which it is used would change radically. To ensure constancy, a magnet must be aged. Artificial means are used to reduce the aging process to 12 h. The magnets are placed in a temperature-controlled chamber at 100°C for about 12 h. In this period of time they will lose all the magnetism they would have lost in natural aging.

13-3   Field patterns.   We have seen above that magnetism is a form of energy. This energy is stored in a magnetic field of force surrounding a magnetized material. The truth of this statement is obvious when we consider that a magnet will pick up some materials, not by touching them but merely by being held close enough to them. How can we represent this field? Energy is the ability to do work. What kind of work can a magnet do?

Before we answer, let us review some similar situations that we discussed previously. The direction and force of wind energy at an airfield is checked with a *test instrument*. It may be a simple windsock or preferably an anemometer.

When studying charged bodies, we learned that they were surrounded by an energy field called a dielectric or electrostatic field. Again we analyzed the reaction of the field on a test instrument. In this case the test instrument was a small positive charge. We represented the distribution of the energy of the field by lines of force, and we saw that a test piece placed in the energy field would move in a direction following a line of force. The density of the lines indicated the strength of the field. (Remember that there are no actual lines in the field. Lines of force are used as means of representing energy fields pictorially. See Chapter 1.)

Since magnetic fields are energy fields surrounding a magnet, the same idea of lines of force can be used to represent pictorially the strength and direction of magnetic fields. This time we call them *magnetic lines of force*. Now the problem is: What shall we use as a test instrument to analyze magnetic fields? An ideal test instrument is a small "north" pole (*a unit pole*). However, a north pole does not exist by itself. You will see later that any magnet has two poles. To distinguish between their reactions, one end is called the north pole, the other the south pole. (But you probably are already familiar with this from your knowledge of magnets.) You may wonder, then, why we use this fictional test instrument. It is the simplest way to explain magnetic field patterns, and is the basis for explaining the relation between the unit of magnetic force and magnetic lines of force.

Let us consider a bar magnet (Fig. 13-1) and see what the energy around this magnet can do to our test instrument, the unit north pole.

If we place the small north pole at point *A*, we find that it will be repelled from the left end of the bar magnet and attracted to the right end, along the dashed path shown. This line shows the direction in which the magnetic forces act on the test instrument. The line is called a *magnetic line of force*. Had the

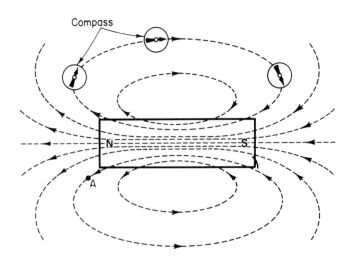

Figure 13-1   Magnetic field around a bar magnet.

unit pole been placed at any other point in the magnetic field around the magnet, it would have been repelled and attracted along some other lines of force, as shown on the diagram. Notice that all the lines of force come out at the left end of the magnet, curve around to the right end, enter the magnet at this point, and continue back to the left end. Magnetic lines of force are closed loops. The point from which they come out is called the *north* pole of the magnet. The point at which they enter is called the *south* pole. In the bar magnet, (Fig. 13-1), the left end is the north pole, and the right end is the south pole. If our test north pole were infinitely small, it would not only be repelled by the north pole of the bar magnet and attracted by the south pole but also would continue through the magnet back to the north pole, to "complete the circuit." From the action of the bar magnet on the test north pole, we can conclude:

1. Like poles repel each other.
2. Unlike poles attract each other.

The strength of the magnetic field is indicated by the density or concentration of the lines of force. Notice in Fig. 13-1 that the magnetic field is strongest close to the poles, and that the strength decreases as we go farther away from these poles. This is common knowledge to anyone who has played or experimented with magnets. Notice also that the lines of force curve away from each other; that is, **lines of force in the same direction repel each other.**

Another way of exploring the magnetic field around a magnet is to use a second bar magnet mounted on a pivot at its center so that it is free to swing in the magnetic field of the first magnet. Such an instrument is the well-known compass needle. The blue end of the compass needle is its "north" pole. If a compass needle is placed in the field of another magnet, it will align itself tangent to the magnetic lines of force of the field (see Fig. 13-1). So if a compass needle is moved around a magnet, it will indicate the direction of the magnetic field, or it will trace the lines of force.

The earth itself has a magnetic field, with its magnetic poles located close to the geographic North and South Poles. Naturally, the pole near the geographic North Pole is called the north magnetic pole. Yet the north pole of the compass needle is attracted to and points to the North Pole of the earth. But this is inconsistent! We just stated that like poles repel each other. Yet the blue end of the compass needle was called the north pole of the needle because it pointed to the North Pole of the earth. Here convention assigned names and terms before the true scientific picture was understood, just as the direction was assigned arbitrarily to current flow before current flow was understood. To avoid confusion, the north pole of the compass needle (and of any magnet) is often called the *north-seeking* pole. Actually, to maintain uniformity of magnetic poles as defined above, the north magnetic pole of the earth is in reality a south magnetic pole.

In a bar magnet, the magnetic field is spread out over a wide area. If we can concentrate this magnetic field into a smaller area, strong magnetic *intensity* can be produced. This can be done by bending the bar magnet into a U shape. In this form it is called a horseshoe magnet. Figure 13-2 shows the magnetic field around a horseshoe magnet. The direction of the field of force is again determined with a small north pole as a test instrument.

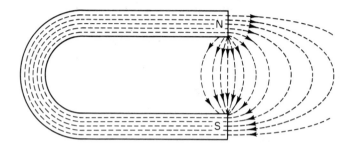

Figure 13-2   Magnetic field around a horseshoe magnet.

The field between two unlike poles facing each other is even more concentrated, as shown in Fig. 13-3. Notice the uniformity of the field in the center. Unless the magnets are anchored, they will be pulled together by the action of the field forces. In other words, **unlike poles attract each other**.

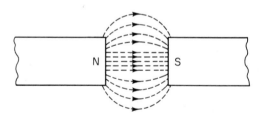

Figure 13-3   Magnetic field between unlike poles.

When two like poles are placed opposite, each pole will repel the small test north pole. The field will then appear as in Fig. 13-4.

From the diagram and from the previous statement that lines in the same direction repel each other, it is obvious that there is a tendency for two north poles to repel each other. If two south poles had been used, the field of force would look the same except that since the south poles would attract the small test pole, the arrows would point into the south poles. The parallel lines of force are again in the same direction and will repel each other; that is, *two south poles repel each other*.

13-4 Magnetic induction.   In the above diagrams you have seen how a magnetic field exists in the space surrounding a magnet. Theoretically, the

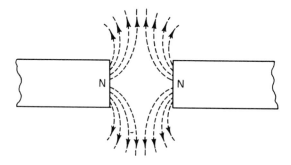

Figure 13-4 Magnetic field between like poles.

lines of force reach out to infinite distance, but the force gets weaker and weaker as the distance increases. Practically, beyond a certain distance the force is so weak that it can be neglected. If a piece of nonmagnetic material, such as wood, is placed in the path of these lines, the lines will pierce through and continue undiverted, just as if the wood were not there. If the material were magnetic, the field pattern would be distorted, as shown in Fig. 13-5.

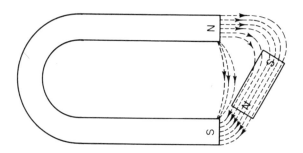

Figure 13-5 Distortion of field by magnetic material.

The explanation is simple. Some of the lines normally would go through the piece of material. Since the material is magnetic, some of its atoms will be aligned with the field. The end where the lines enter becomes the south pole and the end where the lines leave the material becomes the north pole. The material is now partly magnetized.

Now we have two magnets, the original horseshoe magnet and the newly formed magnet. The adjacent poles of these two magnets are unlike! How does this condition affect the energy field? The energy must concentrate between these adjacent poles. Lines of force will be diverted from their normal path and will also pass through the piece of magnetic material. Thus more of the atoms are aligned in the material, and it becomes more strongly magnetized. More of the field energy will concentrate between the adjacent poles. The process is cumulative and continues until all the lines of force go through the material, or until all of the atoms in the material are aligned.

If the material used has good retentivity, it will retain its magnetism when it is removed from the magnetic field. But magnetism is energy. Where did this energy come from? The original magnet! The horseshoe magnet has lost some of its magnetic energy to the piece of material. It therefore becomes weaker.

Whenever a permanent magnet is "unprotected" as above, there is danger that it may gradually lose its strength if magnetic materials of good retentivity are near enough to intercept its field. To prevent this loss, a *keeper* should always be used with a magnet when it is not in use. A piece of soft iron is placed across the poles of the magnet. It will become magnetized, as explained above, and all lines of force will go through it; very few of the lines, if any, will go into the surrounding air (see Fig. 13-6).

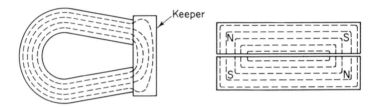

Figure 13-6   Method of "protecting" magnets.

You will recall that soft iron has poor retentivity. When it is removed from the field, will it keep its magnetic alignment? Will it take energy away from the magnetic field? No. Because of its poor retentivity, when taken out of the field it will lose its atomic alignment, and return its energy to the field.

13-5 Magnetic shielding.   Certain components of electronic equipment, such as power transformers and filter chokes, create strong magnetic fields. If these fields are allowed to extend into space, they may react adversely on other components. In a radio receiver, for example, this would produce a bad hum. Magnetic shielding is used to prevent interaction. The shielding can be applied to the offending unit, to keep the magnetic field from piercing through to the outside space. It can also be applied to the reacting unit, to prevent the lines from entering it. In practice either or both methods are used. The principle of magnetic shielding is the same as for a keeper. For use with low-frequency fields, good shields are made of soft iron, and the greater the thickness, the more effective the results. The effect of magnetic shields can be seen from Fig. 13-7.

13-6 Applications to electronics.   Permanent magnets are extensively used in electronic equipment. You may already have seen them in earphones. They are also used in loudspeakers, phonograph pickups and recorders, microphones, and measuring instruments.

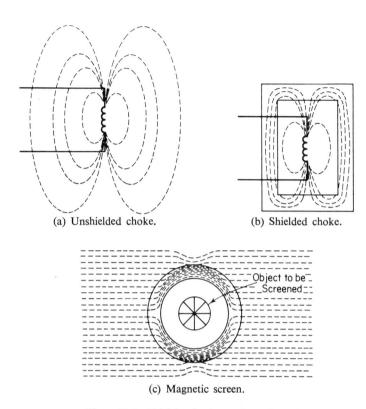

(a) Unshielded choke.

(b) Shielded choke.

(c) Magnetic screen.

Figure 13-7 Shielding in electronic equipment.

REVIEW QUESTIONS

1. **(a)** In what respect is magnetism similar to heat or light?
   (b) What is responsible for this effect called magnetism?

2. By means of the electron theory explain:
   (a) Nonmagnetic materials.
   (b) Magnetic materials.
   (c) Magnetized materials.

3. **(a)** What is meant by retentivity?
   (b) What is it due to?
   (c) What is residual magnetism?

4. What type of material will make good permanent magnets? Why?

5. Why should permanent magnets be aged before they are used in equipment?

6. **(a)** How is the strength and direction of a magnetic field determined?
   (b) How is a magnetic field represented?
   (c) What distinguishes this from a dielectric field?
   (d) How is the strength of the magnetic field indicated?

181

7. With reference to Fig. 13-1.
   (a) What does the N at the left end of the rectangle signify?
   (b) Is this a correct indication? Explain.
   (c) What causes the lines of force to separate as they leave the magnet at N?

8. Is there any advantage in the magnet of Fig. 13-2 over the bar magnet? Explain.

9. Which magnet structure (Figs. 13-1, 13-2, or 13-3) would produce the strongest field? Why?

10. Show by diagram the magnetic field:
    (a) Around a bar magnet.
    (b) Between two unlike poles.
    (c) Between two south poles.

11. (a) In Fig. 13-5, what type of material is represented by the rectangle alongside the horseshoe magnet?
    (b) Explain why the field is distorted.
    (c) Would a block of aluminum, in the same position, cause more or less distortion? Explain.

12. Why should keepers be put on magnets when not in use?

13. How can a device be protected from strong magnetic fields?

14. (a) What is the purpose of the rectangular structure in Fig. 13-7(b)?
    (b) Is bakelite a suitable material for this purpose? Explain.
    (c) What is the purpose of the circular structure in Fig. 13-7(c)?
    (d) What is a good material for this purpose?

14

ELECTROMAGNETISM

We have seen that magnetism is a form of energy due to the motion of electrons. In permanent magnets the force was due to the excess of electrons revolving around the nucleus in one direction. But should magnetic results be restricted to motion of electrons around the nucleus? Shouldn't similar effects be observed when electrons move in one direction through a material?

14-1 Magnetic field around a conductor. Let us take a piece of copper wire. It is a good conductor, but since the number of electrons revolving clockwise around the nucleus of each atom is balanced by the number of electrons revolving counterclockwise, it is nonmagnetic. Now let us connect a battery and rheostat in series with the wire. Electrons will move through the wire from the negative to the positive end, and through the rheostat and battery to complete the circuit. But since electrons in motion produce magnetic fields, there must be a magnetic field around the wire.

Hold the wire vertically and perpendicular to the paper. You will then see the cross section of the wire. Since an arrow is used to indicate direction of current, when current (*conventional*) is moving toward us we will see the point of the arrow. When current is flowing away from us we would see the tail of the arrow, or an X. Using a compass needle, we can examine the magnetic field around a wire carrying current toward us (Fig. 14-1). Place the compass at point *A*. We find that the needle is deflected, and points tangent to the wire and to the left.

Place the compass at point *B*. Again the needle is deflected tangent to the wire. This time it points up. When placed at *C* and *D* it deflects as shown. If we start with the compass farther away from the wire, deflections in the same

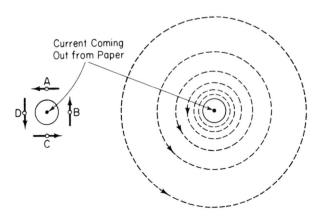

Figure 14-1 Magnetic field around a conductor carrying current toward observer.

direction are obtained, but the reactions are weaker. If we repeat the experiment with the current in the opposite direction, results will be similar but the compass needle will point in the opposite direction (Fig. 14-2).

From these effects we can conclude that:

1. A magnetic field surrounds a wire carrying current.
2. The field gets weaker as we go further away from the wire.
3. The direction of the magnetic field depends on the direction of current.

A simple rule for determining the direction of the magnetic field is: *If the current flow is toward the observer, the field is counterclockwise.*

In many texts you will find a hand rule, called *Fleming's right-hand* rule, for a wire carrying current. This rule states: *"Grasp the wire in the right hand,*

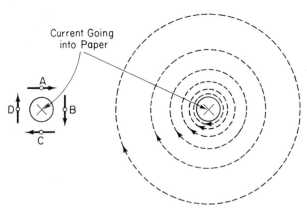

Figure 14-2 Magnetic field around a conductor carrying current away from observer.

with the thumb pointing in the direction of the current (conventional current), and the fingers will curl around the wire in the direction of the magnetic field." This rule is illustrated in Fig. 14-3.

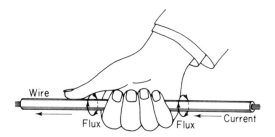

Figure 14-3   Fleming's right hand rule for a current-carrying conductor.

The magnetic field around a wire may be considered as cylindrical with respect to the wire and extending the full length of the wire. The strength of the magnetic field will depend on the amount of electrons flowing through the wire per second, or the current in the circuit. In power plant switchboard layouts, where the current runs into thousands of amperes, the magnetic fields produced by the bus bars may be strong enough to rip these bus bars off the panels unless they are strongly bolted. In electronics, the currents involved are so much lower that the magnetic intensity at any one point along the conductor is small. How can we concentrate the total magnetic field into a smaller space?

14-2 Magnetic field around a coil.   Let us take a long piece of copper wire, wind it around a hollow cardboard cylinder, and connect the terminals to a source of voltage with polarities as shown in Fig. 14-4(a). The current that flows through the wire is indicated by the arrows. To analyze this diagram, let us cut through the center of the coil and examine the cross section

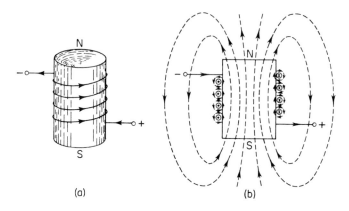

(a)   (b)

Figure 14-4   Magnetic field around a coil carrying current.

[Fig. 14-4(b)]. On the left side of the coil the current would be coming out toward us. We indicate this direction of current by a dot. On the right-hand side the current would be going into the paper or away from us. We indicate this direction of current by an X. Now we can add to the diagram the magnetic field around each cross section of the wire. On the right-hand side the field is clock-wise. For simplicity, let us show only the first ring and break it up into four sections as shown. You will notice that in between adjacent turns the field is in opposite directions. But the two forces are equal in strength, since the current in each turn is the same. The two effects neutralize each other. The net result for the coil is a total force that is downward on the outside of the coil, crossing inward at the bottom and upward inside the coil. Draw this total magnetic field on the diagram (dashed lines). And now, remembering that the magnetic field around each cross section should have been not one ring but a number of concentric rings, we can see that each of these rings would have produced similar total fields. A similar treatment applied to the cross sections of the wires on the left side of the coil will also produce an overall field. Notice that the lines of force come out at the top of the coil and enter at the bottom. This diagram is similar to the field of force around a bar magnet. The top of the coil forms a north pole, the bottom a south pole. But this diagram is the analysis of Fig. 14-4(a). This coil should then be marked with magnetic polarities as shown. The coil forms an electromagnet.

Going through the analysis every time we want to determine magnetic polarity of a coil is a laborious process. A simple rule is as follows: *"Grasp the coil in your right hand so that the fingers curl in the direction the current is flowing; the thumb points to the north pole."* (This is **Fleming's right-hand rule for a coil**, using conventional current.)

Let us apply our rule to three examples as shown in Fig. 14-5.

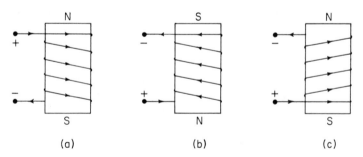

Figure 14-5   Reversal of magnetic polarity.

Notice that from coil (a) to coil (b) the magnetic polarity was reversed by reversing the power supply or battery voltage applied to the coil. Between coils (b) and (c) the magnetic polarity was again reversed by reversing the direction of winding of the coil.

14-3   Factors affecting strength of electromagnets.   Some of you have probably wound or used ready-made electromagnets. Did they have an air core (like the coil above), a wooden core, or some type of iron core? Have you noticed how much weaker the magnet becomes when the iron core is removed? Adding this effect to our previous observations, we can say that the strength of an electromagnetic field depends on:

1. The amount of current in the coil (I)
2. The number of turns of the coil (N)
3. The size and material of the core

The product of the first two factors, amperes × turns, is called ***ampere-turns (NI)***. It is obvious from our previous analysis that the number of lines produced will depend on the ampere-turns of the coil. The third factor, size and material of the core, will bear deeper analysis.

14-4   Simple magnetization curve.   Let us take some copper wire and a hollow bakelite tube, and wind a coil of 100 turns. Now connect this coil through a rheostat and switch to a battery (see Fig. 14-6). Inside the coil put a core of iron.

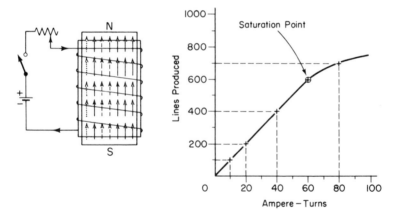

Figure 14-6   Simple magnetization curve.

When the switch is open, there is no current flowing. Therefore, there are no ampere-turns and no lines of force are produced. Plot this point on the curve. Now if we close the switch and adjust the current to 0.1 A, we will have 100×0.1 or 10 ampere-turns. Since current is flowing in the coil, a magnetic field is created. What is the direction of the lines of force produced? Applying the right-hand rule, we find that the north pole is on top, or the direction of the magnetic force is upward. But the core has magnetic atoms. Some of them will be pulled into alignment by the force of the field (first row of dotted arrows). The total number of lines will be the lines due to the coil itself plus the lines added by the weakly magnetized core.

Let us assume that this total number of lines is 100. Plot this point (10 ampere-turns and 100 lines) on the curve. Now double the current in the coil. The number of ampere-turns of the coil is doubled; the number of lines produced by the coil will be doubled; twice as many atoms of the core will be pulled into alignment (first row of solid arrows); the number of lines produced by the core will be doubled and the total number of lines is doubled. Add this point to the curve (20 ampere-turns and 200 lines). Double the current again (0.4 A). The above process is repeated. The core now has four rows of atoms in alignment (the former dotted and solid arrows, and now, two more rows of dashed arrows). Plot this point (40 ampere-turns and 400 lines) also on the curve. Now raise the current to 0.8 A. The ampere-turns will be doubled and the number of lines produced by the coil will be doubled. But all the atoms of the magnetic core will be pulled into alignment without reaching double the previous quantity. The number of lines produced by the core will not be doubled and the total number of lines will be less than double. Let us assume that the number of lines produced is only 700. Add this point (80 ampere-turns and 700 lines) to the curve. Beyond this value of current the number of lines due to the coil will still increase but the core will not add any further lines to this value. The core is *saturated* when all of its atoms have been aligned. You will notice that the curve falls away from the straight line when the core is saturated. The point where the curve bends is considered the saturation point of the core. The curve is called a *saturation, magnetization,* or *B–H curve.*

Had the core been larger in area, it would have had more atoms to be aligned and the strength of the magnet would have been increased further. If a less magnetic material had been used for a core, fewer atoms would have been aligned and a weaker field would have resulted. If a nonmagnetic core had been used, the atoms would be nonmagnetic, no lines would have been added by the core, and the total magnetic strength would have been much smaller.

The picture as presented is not quite accurate. We have assumed that twice as many atoms are aligned by doubling the ampere-turns of the coil until saturation is reached. However, as the core is magnetized it becomes increasingly harder and harder to align the remaining atoms. The curve is not a perfectly straight line up to the saturation point but bends away gradually from the straight line as saturation is approached, and then more sharply beyond the saturation point (see Fig. 14-6).

In many electronic applications saturation of core material is undesirable. To prevent saturation, good design requires cores of large area. This makes the equipment expensive. Cheaper designs introduce air gaps. The effect of these air gaps will be analyzed later.

Electromagnets in electronic applications are often used with alternating currents (ac). What is alternating current? The electrical supply to your home is probably alternating current. The potential between the lines on direct current is always constant. In alternating-current circuits the potential between the wires is rapidly changing. At one instant there is no difference of potential between the lines. Gradually the difference in potential increases to a maximum. Then it decreases again to zero. Then the potential difference increases again to a maximum, but the terminal that was positive is now at a negative potential with respect to the second line terminal. The potential difference then reduces to zero again. This sequence of variations (one cycle) is repeated again and again. Commercial power lines are marked as 60 hertz,* meaning that the above changes occur at the rate of 60 times in one second. The current that flows in an ac circuit will be of the same nature. At one instant no electrons are in motion; the current is zero. Then the current increases to a maximum; the number of electrons flowing from negative to positive line terminals is increasing. Then the current drops to zero. Now the current flowing increases to a maximum again. But this time the electrons are flowing in the opposite direction! Then the current drops to zero again. This sequence completes one cycle of changes for the current in the circuit. A graphical representation of the cyclic changes is shown in Fig. 14-7.

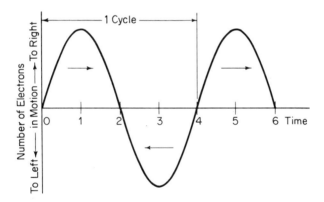

Figure 14-7   Variation of current flow in an ac circuit.

14-5 Hysteresis loss.   When an electromagnet is connected to an ac supply, the current in the coil is changing in direction. As a result the magnetic field must also change in direction. If the magnet has a magnetic core, its atoms must change their alignment to correspond with the changing magnetic field. Let us assume a core of zero retentivity. When the current flows in one direction, the atoms align themselves to correspond. Work is done on these atoms.

---

*The actual unit (dimensionally) is *cycles per second*, which is often colloquially shortened to cycles. More recently this unit was renamed in honor of the German physicist Heinrich Hertz (1857–1894).

The energy so used is stored in the magnetic field produced. As the current drops to zero, the atomic alignment is scattered and the magnetic field "collapses," returning its energy to the electrical circuit. When the current builds up in the opposite direction, the atoms of the core are realigned in the opposite direction. But again the energy required for realignment is returned to the electrical circuit when the current drops to zero. *A core with zero retentivity consumes no power.*

Now consider a core material having retentivity. During the first half-cycle of current flow, the atoms in the core are aligned in one direction. A plot of the magnetic field strength versus magnetizing force would be similar to the magnetization curve of Fig. 14-6. This is repeated in Fig. 14-8 as the curve from *O* to *A*. As the current drops to zero some of the atoms remain in alignment! All the energy stored in the magnetic field is not returned to the electrical circuit. (As you may recall, retentivity was considered as due to frictional resistance between the atoms.) Therefore, when the current—and

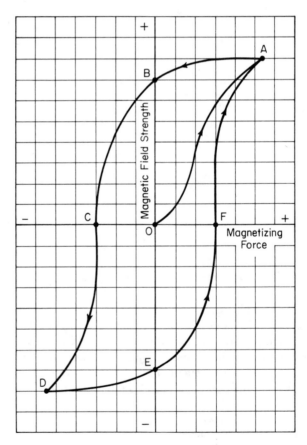

Figure 14-8   Hysteresis curve.

magnetizing force—is zero, a residual magnetic field strength still remains. The magnetization curve is now at *B*. When the current reverses in the second half-cycle, extra work must be done to turn these atoms around against their frictional resistance. This is achieved when the current (and magnetizing force) reaches the level indicated by point *C*. At *D*, the magnitude of the magnetizing force (and current) is the same as at *A*, but opposite in polarity. Once again, as the current drops to zero, retentivity keeps the magnetic field strong (see point *E*), and again a current reversal is needed to reduce the magnetic field strength to zero (at *F*). In completing one cycle of current flow, we have traced out an area *ABCDEF*, proportional to the work done in bringing the core atoms through one cycle of magnetization. This work represents energy lost by the electrical circuit, and appears as heat in the core. The power lost is called *hysteresis loss* (pronounced histerēe′sis).

From the foregoing we can make the following observations concerning hysteresis loss:

1. It is power lost in a magnetic core as heat.
2. It is due to retentivity of the core.
3. It occurs when the magnetizing current changes in direction (in alternating-current circuits) or whenever the core is taken through a cycle of magnetization.

Hysteresis loss is detrimental to the operation of electrical equipment. This is especially true in electronic circuits in which the cyclic changes encountered are not just 60 per second as in power work, but run up to 16 000 or 20 000 for audio frequencies and into millions of cycles per second in radio frequencies. Core material used in such equipment must have extremely low retentivity.

14-6 Relays.   One of the most widespread uses of the electromagnetic principle is in *relays*. These are nothing more than magnetically operated switches. Such switches could have one contact (single pole) or many contacts (double pole, triple pole, etc.). The contacts could be *normally open* and are closed by electromagnetic action, or, vice versa, the contacts could be *normally closed* and are opened by electromagnetic action. An example of this second type is the *circuit breaker* used to protect against overloads or short circuits. The circuit breaker is closed manually to apply power. Should the current in the circuit at any time exceed some preset limit, the electromagnetic action of the relay will cause the circuit breaker contacts to open. Another example of the normally closed relay is the *reverse-current* relay used in the battery-charging circuits of aircraft and automobiles. When the engine is running at above some minimum speed, the generator charges the battery. If the engine is stopped (or running at too low a speed), the generator cannot supply power to the battery, but the battery could discharge—or even be short-circuited—

through the generator. A reverse-current relay will open the circuit between generator and battery and prevent discharge of the battery.

Relays could also be double-acting or *double-throw*, that is, *break* one circuit and simultaneously *make* another circuit. On this basis we could have a single-pole double-throw or multipole double-throw. Obviously, by using sufficient "poles," any combination of normally open, normally closed, or double-throw actions can be obtained.

This simple electromagnetic device is a major factor in our automated technology. In the home, relays are responsible for the clicks and clacks heard during operation of dishwashers, clothes washers, air conditioners, etc. In industry, relays are the "fingers" that control the sequence of operation in the myriad of automated processes.

Regardless of the complexity of an automatic machine; regardless of how many and what type of contacts a single relay may have, the basic action is quite simple. Figure 14-9 shows a relay with normally open contacts. When power is applied to the *solenoid* (current-carrying coil), the soft-iron core is magnetized. The armature is pulled down by magnetic attraction, closing the contacts to the controlled circuit. When the solenoid is deenergized, the spring action causes the contacts to open.

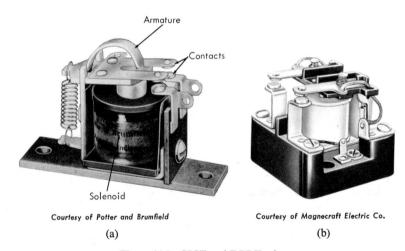

Courtesy of Potter and Brumfield   Courtesy of Magnecraft Electric Co.

(a)   (b)

Figure 14-9   SPST and DPDT relays.

REVIEW QUESTIONS

1. With reference to Fig. 14-1 (left side):
   (a) What does the circle represent?
   (b) What does the black dot in the center represent?
   (c) If the direction of current flow were reversed, how would this be indicated?

(d) What does the arrow at *A* represent?

(e) Why does the arrow point to the left?

2. (a) In Fig. 14-1, right side, what do the concentric circles represent?

(b) Why are broken lines used for these circles?

(c) Why aren't the circles shown with uniform spacing?

3. (a) Why are all directional arrows in Fig. 14-2 opposite to those in Fig. 14-1?

(b) State a rule for determining the direction of these magnetic fields.

4. With reference to Fig. 14-3:

(a) Which hand is used?

(b) In what direction should the thumb point?

(c) What does the curl of the fingers indicate?

5. With reference to Fig. 14-4:

(a) What happens to the magnetic field between the turns of the winding? Explain.

(b) Is the total magnetic field energy for this coil the same, more, or less than would be produced by the same length of wire before coiling? Explain.

(c) What is the advantage of coiling?

(d) What does the N on top of the coil in diagram (a) signify?

(e) Is this a correct designation? Explain.

6. With reference to Fig. 14-5:

(a) Is the N on top of diagram (a) correct? Show why.

(b) Is the S on top of diagram (b) correct? Show why.

(c) Account for the change in magnetism.

(d) The power-supply connections to coils *B* and *C* are the same, yet the magnetic poles are reversed. Which is correct? Account for your answer.

7. State three ways by which a more powerful electromagnet can be obtained.

8. (a) What is meant by saturation?

(b) Will it occur in any air-core electromagnet? Explain.

(c) How can saturation be prevented?

9. With reference to Fig. 14-7:

(a) What two quantities are plotted on this graph?

(b) What does the arrow pointing to the right signify?

(c) What does the arrow pointing to the left signify?

(d) At time 1, what can be said about the magnitude and direction of flow of current?

(e) Repeat part (d) for time instant 2.

(f) Repeat part (d) for time instant 3.

(g) What time interval constitutes "one cycle"?

10. (a) What is hysteresis loss?

(b) When does it occur? Why?

(c) In an audio-frequency transformer, the current reverses rapidly. Is a hard-steel core desirable? Explain.

11. Name three basic parts of any relay.

12. Are a relay's contacts normally open or closed? Explain.

15

MAGNETIC-CIRCUIT CALCULATIONS

In the previous chapters on permanent magnetism and electromagnetism, the topics were treated in a qualitative manner. Magnetic units were deliberately avoided. However, in the design of magnetic circuits, calculations similar to Ohm's law for electrical circuits are often required. This necessitates establishment of quantiative units. Unfortunately, this can create a problem, because there are three systems of units currently in use:

1. *English Units:* still in use by American industries, but being slowly converted to SI.
2. *CGS (Metric) Units:* used mainly by European countries, but also being converted to SI.
3. *SI (Metric) Units:* favored internationally by academic and scientific societies.

Since the trend is to full conversion to SI units, this text will use primarily SI units. However, because the other units may still be encountered in practice, correlation with English and CGS units will also be shown.

15-1 Permeability. In Chapter 14 we saw that the strength of an electromagnet depends on the type of material used for a core. Under similar conditions, certain core materials will establish more lines of force than others; in other words, they will produce stronger magnetic fields than others. *Permeability (μ, pronounced mu)* of any material is a measure of the ease with which its atoms can be aligned, or the ease with which it can establish lines of force.

Materials are rated on a comparative basis. In SI units, the permeability of air (or vacuum) is taken as the standard, and is called the *absolute or free-space permeability* (μ_0). It has a value of

$$\mu_0 = 4\pi \times 10^{-7} \qquad (15\text{-}1)$$

The permeability of any other material is then given as the improvement in strength (or number of lines) produced by that material, as compared to air; and is called the *relative permeability* (μ_r), or

$$\mu_r = \frac{\text{number of lines produced with the material as a core}}{\text{number of lines produced with air as a core}}$$

The relative permeability of commercial core materials varies from less than 50 to more than 25 000. Nonmagnetic materials cannot be magnetized. Since they do not add to the number of lines in the field, permeability of all nonmagnetic materials is the same as air. A few materials when used as cores may actually cause a slight decrease in the number of lines of force, as compared to air. Such materials (e.g., copper, $\mu_r = 0.999\,998$) are called *diamagnetic* materials. On the hand, materials that have permeabilities just slightly greater than 1 (e.g., aluminum, $\mu_r = 1.000\,008$) are known as *paramagnetic* materials.

From the development of the simple magnetization curve (Fig. 14-6), we saw that as a magnetic material approaches saturation, the remaining atoms become increasingly harder to align. But the ease of alignment is a measure of permeability. Is the permeability of a material constant? Unfortunately, no; it decreases as we approach saturation.

15-2 Magnetic units.   In practice, magnetic-circuit calculations are not encountered as frequently as electrical-circuit calculations. There is a tendency to forget the meaning and relationship of magnetic units. However, since magnetic circuits are quite similar in analysis to electrical circuits, magnetic units and relationships are more readily understood and remembered if compared to the corresponding electrical values.

1. In the electric circuit, *current* goes through the circuit. The symbol for current is I, and it is measured in amperes but basically it is the number of electrons per second). In the magnetic circuit, *flux* goes through the circuit. The symbol for flux is ϕ (phi), and it is basically measured in number of lines of force. However, since in practical situations the number of lines involved is quite high, the SI unit for flux groups 10^8 lines together, and this "bunch" is called one *weber* (Wb).*

2. In the electrical circuit, the applied *electromotive force* (*EMF*)—symbol E, measured in volts—is the force that causes current to flow. Correspondingly, *magnetomotive force* (*MMF*)—symbol F_m—is the force that

---

*The SI unit name is in honor of William Edward Weber, a German physicist (1804–1891).

causes flux to flow through a magnetic circuit. As we saw in Chapter 14
(Fig. 14-4), flux lines are created by *current* flowing through a coil, and
the number of lines increases with an increase in current. Therefore, the
basic SI system uses the **ampere** (A) as the basic unit for magnetomotive
force. However, it should be understood that this refers to a coil of one
turn, and therefore represents the magnetomotive force *per turn*. Then,
since a magnetizing coil usually has many turns, the magnetomotive force
for each turn is additive. This makes the effective magnetomotive force
equal to the product of the current through the coil and the number
of turns, or *NI*. The practical *MMF* unit can therefore be taken as the
ampere-turn (**NI**), or

$$F_m = NI \tag{15-2}$$

3. The current in the electrical circuit is limited by the resistance of the
   circuit; the symbol is R; and it is measured in ohms. Correspondingly,
   the number of lines produced in a magnetic circuit is limited by the
   reluctance of the circuit; the symbol is R_m, and there is no specific name
   for this unit. Since reluctance in a magnetic circuit is similar to resistance
   in an electrical circuit, *reluctance can be defined as the opposition of a
   magnetic circuit (or material) to the "flow" or establishment of magnetic
   flux.*

   The resistance of an electrical path depends on the length, area,
   and material of the path. Mathematically, $R = \rho \times l/A$. The reluctance
   of a magnetic path also depends on the same three factors. However, the
   permeability of a material is a measure of how good that material is in
   "conducting" lines of force. In this respect, it is the opposite of specific
   resistance (ρ). Its position in the equation must be reversed, or

   $$R_m = \frac{l}{\mu_r \mu_0 A} \tag{15-3A}$$

   Also, since the permeability of any material, μ, is the product of its
   relative permeability and the permeability constant of free space, then

   $$R_m = \frac{l}{\mu A} \tag{15-3B}$$

   where l, the average length of the magnetic path, is measured in meters,
   and A, the cross-sectional area of the magnetic path, is in square meters.

4. In the electric circuit, the amount of current depends on the voltage (E)
   across the circuit, and on the resistance (R) of the circuit. This relation
   is expressed by Ohm's law as

   $$I = \frac{E}{R}$$

   Similarly, the flux in a magnetic circuit depends on the magnetomotive
   force (F_m), and on the reluctance (R_m) of the circuit. We can therefore
   write an "Ohm's law for a magnetic circuit" as

$$\phi = \frac{F_m}{R_m} \quad \text{or} \quad \phi = \frac{NI}{R_m} \quad\quad \textbf{(15-4)}$$

15-3 Flux density—permeability curves. In Chapter 14 we developed a simple magnetization curve, plotting lines produced (ϕ) against ampere-turns. The relation between these two quantities is given by the flux equation:

$$\phi = \frac{NI}{R_m}$$

If R_m remained constant, doubling the ampere-turns should double the flux; increasing the ampere-turns by five times should also increase the flux by five times, and so on, and the graph would be a straight line. But the graph had a slight curvature up to the saturation point and then bent sharply. Some factor in the above equation must be varying. The variation must be due to change in reluctance.

The reluctance of any circuit is found from equation (15-3B), as

$$R_m = \frac{l}{\mu A}$$

The dimensions of the magnetic circuit (l and A) are not changing. There is only one conclusion possible. The permeability of the magnetic core is changing with the degree of saturation.

We also saw that if the area of the core were increased, we could reduce or delay the saturation effect. The degree of saturation, therefore, depends not on the total number of lines produced, but on how crowded these lines are in the core—or *flux per unit area*. A new unit must be introduced. This unit is *flux density*' symbol **B**. Flux density in SI units is given by the number of webers per square meter, and is given the name *teslas* (T).* Flux density is also referred to as *magnetic induction*. From its definition, it is obvious that flux density can be obtained from

$$B = \frac{\phi}{A} \quad\quad \textbf{(15-5)}$$

Example 15-1

An iron core 460 mm long, and 100×50 mm in cross section, has a total flux of 16×10^{-3} Wb. What is the flux density in the core material?

Solution

1. To find the cross-sectional area in square meters, we must first convert the given dimensions to meters.

$$100 \times 50 \text{ mm becomes } 0.1 \times 0.05 \text{ m}$$
$$A = 0.1 \times 0.05 = 0.005 \text{ m}^2$$

2. $$B = \frac{\phi}{A} = \frac{16 \times 10^{-3}}{5 \times 10^{-3}} = 3.2 \text{ T}$$

*In honor of Nikola Tesla, an American physicist (1857–1943).

Manufacturers of magnetic core materials are well aware of the variation of permeability with flux density. Obviously, giving a permeability "constant" for the various materials would be of little value. Instead, they supply (from test data) curves of permeability versus flux density for any of their core materials.

15-4 Comparison with English and CGS units.
In our discussion so far, we have introduced a variety of parameters for magnetic circuits. All of these were in the preferred SI system. Yet, as mentioned earlier, English and CGS units are still in use in industry. A comparison of the three systems of units will help, if you encounter them in practice.

The permeability of air (or vacuum) in the SI system (μ_0) was given as: $\mu_0 = 4\pi \times 10^{-7}$. Instead, in both English and CGS units, air is assigned a permeability of 1.

Flux (ϕ) in the SI system is grouped together in bunches of 10^8 lines, and this "bunch" is called one weber (Wb). In English units, the unit of flux is the line of force itself, and has no special name, for example, "a flux of 50 000 lines." In the CGS system, the unit of flux is again the line, but it is called a *maxwell** (one maxwell = 1 line).

The unit for magnetomotive force (F_m) in the SI system is basically the *ampere* (A), but remember that this is for a coil of one turn. In a coil of many turns the effective or practical unit of magnetomotive force is the product the current through the turns and the number of turns, or *NI*. This is the same unit as is used in the English system, but the symbol generally used for this parameter is a script F (i.e., $\mathcal{F}$). In the CGS system, again we use the $\mathcal{F}$ to represent magnetomotive force, but the unit is called a *gilbert*† (Gb), where one gilbert = $0.4\pi NI$.

In a magnetic circuit equation the preferred SI symbol for the reluctance of a circuit is R_m. In the English and CGS systems the commonly used symbol is the script R (i.e., $\mathcal{R}$).

When evaluating flux density (B) in SI units, it is a measure of the number of "flux-line bunches" or webers per square meter, and is given the name teslas (T). In the English system, flux density is measured in lines per square inch, and has no special name. In the CGS system, this becomes lines (or maxwells) per square centimeter, and is given the name *gauss*.‡

These three systems of magnetic units are shown for ready comparison in Table 15-1.

15-5 Types of cores.
In the formulas in Table 15-1 we noticed that one way of increasing the flux, and therefore the strength of a magnetic

---

*In honor of James Maxwell, a Scottish physicist (1831–1879).

†In honor of William Gilbert, an English physician and physicist (1544–1603).

‡In honor of Johann Karl Gauss, a German mathematician (1777–1855).

TABLE 15-1   Comparison of Magnetic Units

Magnetic Quantity	SI (MKS) Unit	English Unit	CGS Unit
Flux (ϕ)	Webers 1 weber = 10^8 lines	Number of lines	Maxwells 1 maxwell = 1 line
Magnetomotive force (F_m)	Amperes (A) (per turn)	Ampere-turns (NI)	Gilberts (Gb) 1 gilbert = $0.4\pi NI$
Permeability (air)	$\mu_0 = 4\pi \times 10^{-7}$	$\mu = 1$	$\mu = 1$
Reluctance (R_m)	$R_m = \dfrac{l}{\mu_0\mu_r A}$ $l = $ m $A = $ m^2	$\Re = \dfrac{l}{\mu A}$ $l = $ inches $A = $ inch^2	$\Re = \dfrac{l}{\mu A}$ $l = $ cm $A = $ cm^2
Flux density (B)	Teslas webers/square meter	No name lines/square inch	Gausses maxwells/square cm
"Ohm's law" $\phi = \dfrac{F_m}{R_m}$	$\phi = \dfrac{NI}{R_m}$	$\phi = \dfrac{3.2NI}{\Re}$	$\phi = \dfrac{0.4\pi NI}{\Re}$

field, was to decrease the reluctance of the circuit. In turn this could be done by having a magnetic path of greater permeability (better core material), greater cross section, or shorter length. Let us see how these methods apply to various types of magnetic paths. The simplest type of core is the open core shown in Fig. 15-1.

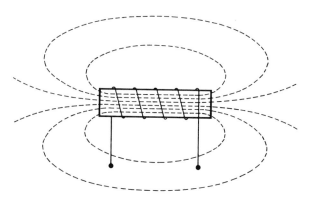

Figure 15-1   Open-type core.

In this type of core, the path consists of two series sections: the comparatively short iron core, plus the air path. Theoretically, the lines extend out to infinity. The average length of path is hard to determine, but it is obviously very long. In addition, the permeability of the major portion of the path is $4\pi \times 10^{-7}$ (air). This type of core has very high reluctance and will be inefficient.

A distinct improvement is obtained if the air path is eliminated. One such type of core is the "closed core" shown in Fig. 15-2.

This type of core is made of thin L-shaped laminations [Fig. 15-2(c)] stacked one behind the other to form the desired thickness of cross section. The cross-sectional area of the magnetic path is shown in Fig. 15-2(b). The magnetizing coil is wound on the "leg" of the core. The resulting flux (except for a very small percentage) will flow through the magnetic path as indicated by the dashed line. Naturally, the lines of force will travel through the entire face and thickness of the laminations. The *average* length of all these lines is the center line shown on the diagram (dashed line). The permeability of this circuit depends on the material used for the laminations. Since no portion of the circuit is air, since the area can be made as large as desired, and since the length is quite finite, the reluctance of such a magnetic path can be made very low.

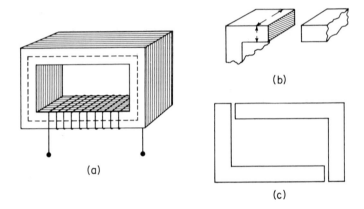

Figure 15-2   Closed core.

In the type of core described above, a small percentage of the lines produced will leak out into the air surrounding the magnetizing coil. To minimize this *leakage flux*, a shell-type core is extensively used. The core is made from a combination of E-shaped and I-shaped laminations, the center leg of the E being twice as wide as the end legs (see Fig. 15-3).

The coil is machine wound on a fiber form and slipped over the center leg. Analysis of the diagram shows that the total flux splits between two identical paths, on each side of the center line. In solving such problems, merely work on one half of the circuit, since the parallel half is the same.

15-6   Mathematical solutions of magnetic circuits.   From the information above we should now be in a position to solve magnetic circuit problems. Suppose that we are given the current and number of turns for the magnetizing coil and the size, shape, and material for the core. We are required

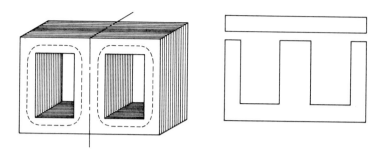

Figure 15-3   Shell-type core.

to find the flux produced. The equation for this is $\phi = NI/R_m$. We can calculate the ampere-turns readily. But what about the reluctance? The permeability of the material, and therefore R_m, depends on the flux density. But we cannot find the flux density until we calculate ϕ! A trial-and-error method must be used. Pick a value for μ and solve for ϕ. Then if the picked value checks with the flux density produced—fine! Otherwise, keep picking! A graphical method of solution avoids this difficulty and is used in practice.

15-7   Graphical solutions.   In the electrical circuit the equation $E = I \times R$ has previously been solved algebraically. The solutions could also have been made graphically, as follows:

1. Draw the coordinates for a graph, labeling the horizontal axis *current* (*I*), and the vertical axis *voltage* (*E*). Graph paper should be used.
2. Pick scales for current and voltage for the range desired.
3. For a given value of resistance (e.g., $R = 10\ \Omega$) plot values of E for the range of currents shown. For example, when $I = 2$, $E = 20$; when $I = 8$, $E = 80$, and so on. A straight line should result. Mark this line $R = 10$.
4. Repeat this procedure for other values of $R(R = 5\ \Omega$ and $R = 20)$.

These curves (Fig. 15-4) could be called E–I curves. Let us use this graph to find how much current will flow in a circuit of 5 Ω resistance if the voltage is 45 V. Locate 45 on the E scale; then run arcoss to the $R = 5\ \Omega$ line and down to the I scale. The answer read on the I scale is 9 A.

To find how much voltage is needed to send 3 A through a 20-Ω circuit, locate 3 on the I scale; then move up to the $R = 20\ \Omega$ line and across to the E scale. The answer on the E scale is 60 V.

The range and accuracy of such a method are limited by the size of the graph. Since the resistance of any given circuit does not change, as evidenced by the straight-line graph, it is much simpler to use the equation mathematically.

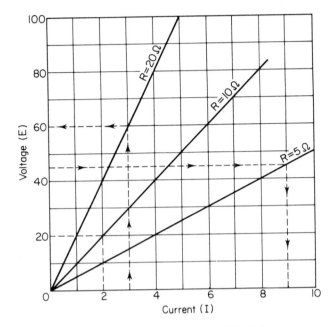

Figure 15-4   Graphical solution of Ohm's law.

15-8 Practical magnetization curve (B–H). In the simple saturation curve, we saw that the saturation point varies with the area of the core. A curve plotted in terms of total flux would apply to only one area of core, and many curves would be needed. If the curve were plotted against flux density instead, it would apply to cores of any cross section, and for any flux values.

Another change in the corrdinates is also necessary. Flux depends not only on ampere-turns but also on reluctance. The reluctance, in turn, depends on the length and area of the magnetic circuit. The area factor is taken care of by the above shift from total flux to flux density ($\phi \div A$). However, the flux produced is still dependent on the length of circuit. Again, many curves would be needed to handle all core lengths. If, on the other hand, the curve is plotted in terms of ampere-turns *per unit length*, it would apply to any magnetic circuit length, and only one curve would be needed.

This introduces another magnetic unit—magnetomotive force per unit length. It is called *magnetizing force*, or *magentic field strength*, and is given the letter symbol *H*. Using SI units, *H* is in *ampere-turns per meter.**

The equation for this new magnetization curve can be obtained from our previous formulas:

---

*In English units, *H* is in *ampere-turns per inch.* The corresponding CGS unit is *gilberts per centimeter*, and is named *oersteds* (Oe) in honor of the Danish physicist Hans Christian Oersted, (1787–1854).

But

$$\phi = NI \div R_m$$

$$R_m = \frac{l}{\mu A}$$

so that

$$\phi = \frac{NI \times \mu A}{l}$$

Dividing both sides by A, we get

$$\frac{\phi}{A} = B = \frac{NI \times \mu}{l}$$

or

$$B = \frac{NI}{l} \times \mu \qquad\qquad (15\text{-}6)$$

This equation is of the same form as the Ohm's law equation, $E = I \times R$, previously plotted. Just as the Ohm's law graph was a plot for resistance in terms of E and I, the magnetization curve is a plot of μ in terms of B and NI/l. With SI units, the coordinates for such curves would be in teslas for the B axis, and in ampere-turns per meter for the H axis. Magnetization curves are often called B–H curves because of the symbols that represent the coordinates used. Sometimes the curves are plotted on semilog paper with H (the x axis) on a logarithmic scale. This method gives greater accuracy at low values of magnetizing force where the curve is almost vertical, and at the same time covers a greater range of values at the higher levels where the curve flattens out.

The graphs in Fig. 15-5 show the B–H curve for cast iron, silicon steel, and wrought iron. More complete charts can be found in handbooks, and in the publications of manufacturers of magnetic core material.

The use of B–H curves can best be shown by an example.

Example 15-2

The core shown in Fig. 15-6 is made of silicon steel. The magnetizing coil has 2000 turns. How much current should be sent through this coil to produce a flux of 3×10^{-3} Wb?

Solution   Since this type of problem is solved by use of the chart, we must first find either NI/l or B. NI/l cannot be found, since I is unknown. B can be found if we first calculate the cross-sectional area.

1. Cross-sectional area (in square meters)

$$A \equiv (40 \times 10^{-3}) \times (80 \times 10^{-3}) = 3.2 \times 10^{-3} \text{ m}^2$$

2. $$B = \frac{\phi}{A} = \frac{3.0 \times 10^{-3}}{3.2 \times 10^{-3}} = 0.94 \text{ T}$$

3. From the curve for silicon steel, corresponding to $B = 0.94$ T, we locate $NI/l = 400$ At/m. This result means that for each meter of length in the magnetic circuit, 400 At are required. To get the total ampere-turns, we must first calculate the length of the magnetic path around the core.

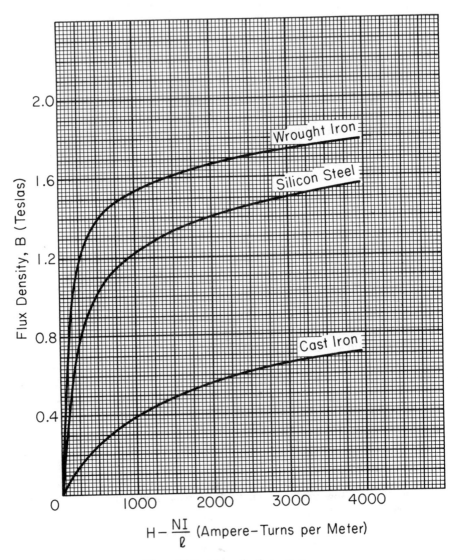

Figure 15-5 Magnetization curves.

4. The average length of this path is the halfway perimeter of the laminations, or

$$l = 120 + 80 + 120 + 80 = 400 \text{ mm} = 0.40 \text{ m}$$

5. The total $NI = 400 \times 0.40 = 160$ At

6. Since ampere-turns is the product of turns × amperes, the current for a given ampere-turns can be found by dividing the NI by the number of turns, or

$$\text{current} = \frac{NI}{N} = \frac{160}{2000} = 0.080 \text{ A} = 80 \text{ mA}$$

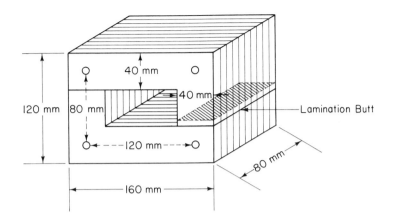

Figure 15-6

Numerous variations of magnetic circuit problems are possible. It may be required to find how many turns the magnetizing coil should have; or how much flux would be produced by a given coil, current, and core; or what area of core should be used to avoid saturation. No rule can be given as to which step comes first, but remember that the *B–H* curves are used in the solution of these problems. What are the coordinates of these curves? Flux density (ϕ/A) and ampere-turns per meter (NI/l). An analysis of the given data should show you which of these factors can be found by calculation.

15-9   Effect of air gap.   In a magnetic circuit it often happens that the magnetic core is not continuous as shown in Fig. 15-6. There may be an air gap in the circuit. The total flux in the iron must also exist in the air gap. This magnetic circuit is similar to a series electrical circuit. A voltage is required to send current through each individual portion of a series circuit, and the total voltage needed must be equal to the sum of the voltages across the individual units. Similarly, a certain amount of ampere-turns are required to send flux through the iron, and additional ampere-turns are required to send flux through the air gap of a magnetic circuit. Since the paths are in series, the total ampere-turns required is the sum of these two. The calculation of the ampere-turns for the iron portion of the circuit has been shown above. It now remains to show how the ampere-turns required for an air gap can be found. Equation (15-6) from which the magnetization curve is plotted is

$$B = \frac{NI}{l} \times \mu$$

Solving for magnetomotive force, we get

$$NI = \frac{B \times l}{\mu}$$

But μ for air is constant at $4\pi \times 10^{-7}$. Therefore, for air, we have

$$NI = \frac{B \times l}{4\pi \times 10^{-7}} \qquad \text{(15-7)}$$

Let us see what effect an air gap would have on the previous example.

Example 15-3

In Example 15-2, suppose that there was an air gap of 0.8 mm. How much current would now be required?

Solution

1. Assuming that the length of iron path has not changed (it is only 0.8 mm less than the previous 400 mm, a negligible distance), the number of ampere-turns required for the iron is, as calculated before, 160 At.

2. In addition, ampere-turns are now required to establish the same flux density in the air gap. For air

$$NI = \frac{B \times l}{\mu} = \frac{0.94 \times 8 \times 10^{-4}}{4\pi \times 10^{-7}} = 598 \text{ At}$$

3.   Total NI for circuit $= 598 + 160 = 758$ At.

4.   Current required $= \dfrac{NI}{N} = \dfrac{758}{2000} = 0.379 \text{ A} = 379 \text{ mA}$

By introducing an air gap of only 0.8 mm, it was necessary to increase the current from 80 to 379 mA—an increase of almost 500%! If the current had not been increased, the flux in the circuit would have dropped considerably. The air gap in the above problem could have been caused by a mere 0.4-mm spacing at each of the butt joints in the L-shaped laminations of the core shown in Fig. 15-6. The possibility of such air gaps is minimized by stacking alternate laminations in the opposite direction. However, this increases construction costs.

You may wonder, why have an air gap in a circuit? In cheap designs it is used to prevent saturation. Since an air gap introduces high reluctance into the circuit, the flux for a given current in a coil is much lower. Reduction in flux means lower flux densities and less chance of reaching saturation. In good design it sometimes happens that an air gap cannot be avoided. If an object must move within a magnetic field, an air gap must be left in the

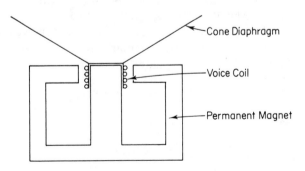

Figure 15-7   Dynamic speaker.

magnetic circuit to allow for such motion. Examples of such cases are loud-speakers, instruments, microphones, and phonograph pickups.

The magnetic circuit of a loudspeaker is shown in Fig. 15-7. Since the voice coil must move within the magnetic field, an air gap must be left in the magnetic core. The air gap should be as small as possible for a strong magnet, and yet large enough to allow the cone to move without rubbing on the sides of the core.

15-10 Tractive force. If we apply Fleming's right-hand rule to the torodial magnetic circuit of Fig. 15-8, we see that flux will flow around the iron core in a clockwise direction, as indicated by the annular arrows. Now, let us cut the ring in half, creating the slightest air gap at each side. Magnetic

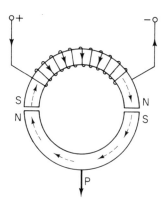

Figure 15-8 Tractive force of an iron-core electromagnet.

poles will be created at each face as shown, and the magnetic pull will try to close these gaps. A physical force, or pull (P), must be applied to maintain these gaps against the magnetic force of attraction. A magnetic flux flows through the core and across the gaps. Electrical energy is used to maintain this flux, and magnetic energy (W) is stored in each air gap. The amount of energy stored depends on the flux density (B) and on the magnetic field strength (H), or

$$(1) \qquad W = \tfrac{1}{2}BH \qquad \text{per unit volume}$$

But, from equation (15-6), $B = (NI/l)\mu$, and by definition $H = NI/l$, so that $H = B/\mu$. We can replace H in (1) by this equivalent, and get

$$(2) \qquad W = \frac{B^2}{2\mu} \qquad \text{joules per cubic meter}$$

Then, if the length of each gap is l meters, and the cross-sectional area of the core is A square meters, the total charge in each gap is

$$(3) \qquad W = \frac{B^2 l A}{2\mu_0} \qquad \text{joules}$$

Now, let us increase the physical force by an amount F', and thereby increase the length of the gaps by an amount Δl. Simultaneously, we will increase the current in the magnetizing coil so as to maintain the same flux density as before. The additional work done is

$$(4) \quad \Delta W = F' \times \Delta l \quad \text{joules}$$

and additional energy is stored in the air gaps.

$$(5) \quad \Delta W = \frac{B^2 \, \Delta l A}{\mu_0} \quad \text{joules}$$

This additional energy must be equal to the additional work done, or

$$(6) \quad F' \times \Delta l = \frac{B^2 \, \Delta l A}{\mu_0}$$

$$(7) \quad F' = \frac{B^2 A}{\mu_0}$$

Notice that this force F' is being exerted against the pull at *both* air gaps. Therefore, the *tractive force* at *each* gap is

$$F_t = \frac{B^2 A}{2\mu_0} \quad \text{newtons} \qquad (15\text{-}7\text{A})$$

Also, since the permeability of air is $4\pi \times 10^{-7}$,

$$F_t = \frac{B^2 A}{8\pi} \times 10^7 \quad \text{newtons} \qquad (15\text{-}7\text{B})$$

Finally, converting this, to find what weight (*mass*) this force can lift (from $F = ma$, where a is the acceleration due to gravity 9.8 m/s²), the tractive weight (mass) is

$$m_t = \frac{B^2 A}{8\pi \times 9.8} \times 10^7 \quad \text{kilograms} \qquad (15\text{-}8)$$

This magnetic pull, or tractive force, can be relatively weak, as in relays, or it may be very powerful as needed for huge commercial lifting magnets. Such magnets can lift weights of many *metric tons.**

Example 15-4

Calculate the pull (force) exerted on a relay armature, if the flux density in the gap is 15×10^{-3} T, and the pole faces are 6.5 mm in diameter.

Solution

1. The air gap area is (in square meters)

$$A = \frac{\pi D^2}{4} = \frac{\pi}{4}(6.5 \times 10^{-3})^2$$

2. $$F_t = \frac{B^2 A}{8\pi} = \frac{(15 \times 10^{-3})^2}{8\pi} \times \frac{\pi(6.5 \times 10^{-3})^2}{4}$$

$$= 2.97 \times 10^{-3} \text{ N}$$

---

*In commercial usage, the megagram (or 1000 kg) is popularly called a metric ton.

REVIEW QUESTIONS

1. (a) What is meant by *permeability*?
   (b) Give the symbol used to represent this quantity.
   (c) What is the permeability of air in SI units?
   (d) What is the relative permeability of nonmagnetic materials? Why?
2. Is the permeability of magnetic materials constant?
3. (a) What is meant by *flux*?
   (b) Give the symbol used to represent this quantity.
   (c) What is the unit for measuring flux in the SI system?
4. (a) What is meant by *magnetomotive force*?
   (b) Give the symbol used to represent this quantity?
   (c) What is the unit for MMF in the SI system?
5. (a) What is meant by *reluctance*?
   (b) Give three factors that determine the reluctance of a magnetic circuit, and tell how the reluctance varies with each.
6. State "Ohm's law" for a magnetic circuit. Using SI units, what is the numerator in this equation?
7. (a) What is meant by *flux density*?
   (b) Give the letter symbol for this quantity.
   (c) Give the name, and/or specific meaning of this unit in the SI system.
   (d) Give another *general* name for flux density.
8. (a) Describe the construction of an open core.
   (b) Repeat part (a) for a closed core.
   (c) Repeat part (a) for a shell-type core.
   (d) What is the advantage of a closed core over an open core?
9. With reference to Fig. 15-2:
   (a) What does the dashed line in diagram (a) signify?
   (b) What do the dimension lines in diagram (b) signify?
   (c) What does diagram (c) represent?
10. With reference to Fig. 15-3:
    (a) Where would the magnetizing coil be placed?
    (b) How many magnetic paths does this circuit have?
    (c) Is the center leg thicker? Explain.
    (d) What is the advantage of this type of core over the core in Fig. 15-2?
11. With reference to Fig. 15-4:
    (a) What does the slope of these curves signify?
    (b) Why is each "curve" a straight line?
    (c) How does the resistance value affect the steepness of these curves?
12. (a) How does the term *magnetizing force* differ from magnetomotive force?
    (b) Give another name for this term.
    (c) Give the letter symbol for this quantity.
    (d) Give the name and/or specific meaning of this unit in the SI system.
13. (a) In solving magnetic circuits containing magnetic cores, why are charts used instead of equations?
    (b) Magnetization curves show the relation between what three factors?

14. With reference to Fig. 15-5:
    (a) Why are the curves *not* straight lines?
    (b) Why not use magnetomotive force (F_m) in ampere-turns as the abscissa?
    (c) Why not use total flux as the ordinate?
15. In Fig. 15-6, explain how the 80-mm and the 120-mm values are obtained for the mean length dimensions?
16. (a) State a disadvantage of an air gap in a magnetic path.
    (b) State two reasons for including an air gap in a magnetic path.
17. (a) In an electric hoist, what two factors determine the lifting power of a magnet?
    (b) How does the tractive effort vary with each?
    (c) Give the equation for this pull.

PROBLEMS

1. A toroid air-core coil (Fig. 15-9), has an inside diameter of 100 mm, and an outside diameter of 150 mm. Find:
   (a) The cross-sectional area of the coil.
   (b) The mean length of the magnetic path.
   (c) The reluctance of the magnetic path.
   (*Note:* Save these answers for Problems 3 and 5.)

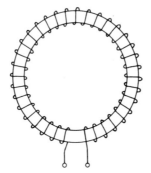

Figure 15-9

2. A rectangular air gap has a cross section of 75 mm × 50 mm, and is 5.0 mm long. Find its reluctance. (Save this answer for Problem 6.)
3. The coil of Problem 1 has 400 turns. Find the current required to produce a flux of 1.0×10^{-5} Wb. (Save this answer for Problem 5.)
4. A relay requires a flux of 5.0×10^{-5} Wb to pull in its armature. The air gap has an area of 30 mm² and is 1.5 mm wide. It is desired to energize this relay by a current of not more than 200 mA. How many turns should the solenoid have? (Save this answer for Problems 7 and 15.)
5. What is the flux density in the toroid of Problem 3?
6. It is desired to establish a flux density of 0.155 T at the air gap of Problem 2.
   (a) What total flux is needed?
   (b) What current will be needed, if an exciting coil of 5000 turns is used?

7. Find the flux density in the air gap of Problem 4.
8. In Fig. 15-6, the dimensions of the core are as follows: height, 150 mm; length, 200 mm; thickness, 40 mm. The laminations are 40 mm wide. A flux of 1.5×10^{-3} Wb is required. From a coil of 500 turns, calculate the current necessary when the core is made of silicon steel.
9. Repeat Problem 8 for wrought iron.
10. Repeat Problem 8 for wrought iron but with an air gap of 6 mm.
11. In Fig. 15-6, the overall size is 100×100 mm. The laminations are 25 mm wide and are stacked 50 mm thick. The core is made of wrought iron and has an air gap of 1.6 mm. The magnetizing coil has 3000 turns. Find the current necessary to produce 2×10^{-3} Wb.
12. In a core as in Problem 11, but without an air gap, the magnetizing coil has 2000 turns and 250 Ω. What flux will be produced if 90 V is applied across the coil?
13. In Fig. 15-3, the center leg is 50 mm wide. All other paths are 25 mm wide. The laminations are stacked 50 mm thick. The overall size of the core is 150 mm $\times$ 75 mm $\times$ 50 mm. The laminations are of silicon steel. How many turns should the magnetizing coil have to produce a flux of 1.4×10^{-3} Wb in the center leg if the current in the coil is 50 mA?
14. Repeat Problem 13 for a center-leg flux of 1.0×10^{-3} Wb and with air gaps of 2.5 mm between the E and I laminations.
15. In the relay of Problem 4, what is the maximum air gap length that will still allow the relay to pull in, when energized with a current of 0.35 A?
16. Calculate the tractive force of the relay in Problem 4.
17. A lifting magnet has a cross-sectional area of 1.4 m² and a total flux of 0.24 Wb.
    (a) Calculate the tractive force.
    (b) What weight (mass) can it lift?

16

DIRECT-CURRENT
MEASURING INSTRUMENTS

The fundamental principle upon which all moving-coil instruments operate is based on the reaction between two magnetic fields. One of these fields is obtained from a permanent magnet, the other from a "moving coil." When current is fed this moving coil, it becomes an electromagnet with north and south poles. The reaction between the two magnetic fields tends to cause the coil to rotate until its north and south poles are adjacent to the opposite poles of the permanent magnet. This is called the *D'Arsonval principle*. The effectiveness of any force to produce rotation is called *torque*. In the D'Arsonval movement, since the flux from the permanent magnet is fixed, the torque developed depends on the current flowing through the moving coil.

16-1 The Weston movement. The Weston movement employs the D'Arsonval principle, and is the basis for practically all dc ammeters and voltmeters. Its construction can be explained with the aid of Fig. 16-1(a). Mounted across the top and bottom of the pole pieces is an insulating non-magnetic strip. In the center of the strips are mounted the upper and lower jeweled bearings and the anchors for the free ends of the coil's springs. The coil is wound on a rectangular aluminum bobbin and a pointer is fixed to the bobbin as shown. A soft-iron core is mounted inside the coil bobbin. This feature reduces the air gap, thereby increasing the magnetic flux and also making for a uniform magnetic field around the coil. Current is fed to the coil through two spiral hair springs. The springs are so adjusted as to provide a restoring torque, which will return the pointer to its zero position when the current in the coil is zero. The bobbin is provided with hardened steel pivots at top and bottom.

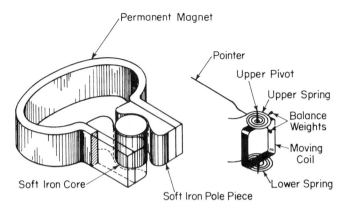

(a) Construction details.

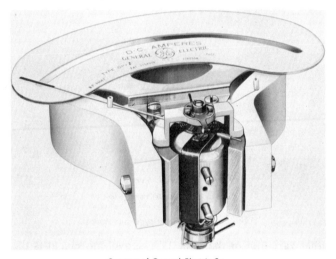

Courtesy of General Electric Co.

(b) Commercial unit.

Figure 16-1 Weston movement.

These pivots rest in the jeweled bearings, supporting the coil so that it is free to turn in the air gap of the magnetic circuit.

In order to keep the movement light in weight, fine wire is used for the coil. Therefore, the coil cannot carry high currents. Weston movements have an operating current range of approximately 10 μA to 30 mA. Currents below 10 μA develop insufficient torque; currents above 30 mA require too large a wire size. As the current range of the movement is lowered, more turns must be used in the coil to maintain a strong field, and thus the resistance of the movement is increased. Resistances for Weston movements spread from as high as 3000 Ω for a 20-μA instrument to 1.2 Ω for a 30-mA range.

16-2 Core and annular-ring magnets. The "external-magnet" movement described above has a drawback in that the instrument reading can be affected by external magnetic fields. Also, these movements cannot be mounted on steel panels without affecting the calibration. With the advent of more powerful magnetic materials (such as alnico), it was possible to redesign the structure and put the magnet in the center, in place of the core. Now, the soft-iron core surrounds the moving coil and magnet, acting as a shield against external fields. Such a structure is shown in Fig. 16-2. Obviously, elimination of the large horseshoe magnet also results in a smaller-size movement. However, because of the smaller magnet—even with the use of high-energy magnets—the sensitivity of core-type movements is limited.

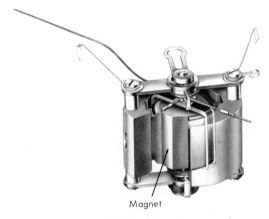

Magnet

Courtesy of Weston Instruments Inc.

Figure 16-2 Core-type movement.

Another variation of the basic D'Arsonval movement is the annular- or bar-ring instrument. This magnetic structure is shown in Fig. 16-3. It combines the external-magnet and core-type movements. It is self-shielding, since the magnet and moving coil are both within a soft-iron ring. Meanwhile, because of the longer magnet length, the sensitivity of this type is comparable with the external-magnet type.

16-3 Taut-band suspensions. In Fig. 16-1, the moving coil was held in place by means of jeweled bearings. No matter how well designed, these bearings will introduce frictional losses. In a movement drawing high power* this friction loss is insignificant. However, in high-sensitivity movements, pivot friction can be serious. The taut-band suspension eliminates this friction loss by eliminating the pivots. Instead, the moving coil is held in place by means of tightly stretched metal ribbons at the top and bottom. The instrument shown in Fig. 16-3 uses a taut-band suspension. The taut band serves three purposes:

1. Provides frictionless suspension for the moving coil

Figure 16-3   Taut-band movement with annular bar-ring.

2. Provides current path to and from the coil
3. Provides the mechanical restoring torque, formerly supplied by the spiral springs

Due to the lack of friction, less driving and restoring torque is now necessary. As a result, for a given full-scale current, the taut-band movement will have fewer turns and a lower resistance. Conversely, for a given coil resistance, these mechanisms can measure smaller currents. Taut-band instruments are available with sensitivities up to 2 μA (and on request, to 1 μA).

It should not be concluded that the taut-band design is always preferable. It is recommended where driving powers of less than 0.05 mW are desired. On the other hand, vibrational stresses can interfere with their operation and extreme shock can make them permanently inoperative. Furthermore, taut-band instruments generally cost more, so that for driving powers above 0.1 mW, the jewel-and-pivot movement would be recommended.

16-4 Ammeters.   We saw above that the moving coil of a Weston movement cannot carry currents greater than 30 mA. Yet often we must measure currents that are much higher. If we put a resistor in parallel with the moving coil, the current going into the instrument will divide, part going through the coil and the rest through the resistor. By proper choice of resistance we can adjust the current division so that exactly one-half, one-fifth, one-tenth, and so on, of the total current goes through the coil. The total current will then be 2, 5, or 10, and so on, times the amount indicated by the galvanometer. This resistor which is connected in parallel with the moving coil, to extend the current range of the instrument, is called a *shunt*.

When an instrument is made with one built-in shunt, its scale is calibrated to read the total current directly. In a multirange ammeter a selection of shunts

---

*All instruments draw power (I^2R) from the circuit, depending on the current through the coil and the resistance of the moving coil.

is available. Such an instrument may have only one scale, in which case the multiplying factor of each shunt range is marked. When the multiplying factor of a shunt is not a decimal factor, it is common to use a separate scale for each range.

16-5 Calculation of shunt values.   To calculate the size of shunt necessary to extend the range of a milliammeter, the parallel-circuit rules are applicable. The simplest way is to use the current-ratio method, as shown in the problem below. A diagram will help to show the current division (see Fig. 16-4).

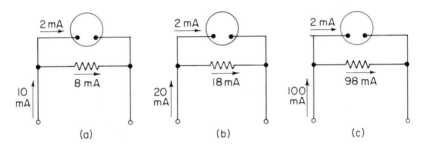

Figure 16-4

Example 16-1

A 0–2-mA instrument has a resistance of 20 Ω. What size shunts are required to extend its range to (a) 10, (b) 20, and (c) 100 mA?

Solution

    (a) 10-mA range: Since 8 mA flows through the shunt, and only 2 mA through the movement, the shunt carries $\frac{8}{2}$, or 4, times more current than the movement. Its resistance must be four times smaller than the coil resistance, or

$$R_s = \tfrac{20}{4} = 5.0 \ \Omega$$

    (b) 20-mA range:

$$\text{Current ratio} = \tfrac{18}{2} = 9$$
$$R_s = \tfrac{20}{9} = 2.22 \ \Omega$$

    (c) 100-mA range:

$$\text{Current ratio} = \tfrac{98}{2} = 49$$
$$R_s = \tfrac{20}{49} = 0.408 \ \Omega$$

In order to combine the three ranges on one instrument, a switching arrangement must be used. Figure 16-5 shows a suitable switching arrangement. In using a multimeter when the current to be measured is not known, it is best to start with the largest range and reduce the range until a sizable deflection is obtained. Notice the wide contact on the switch arm. This *shorting-type* switch protects the movement while switching from one range to another. At no time

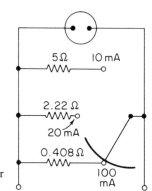

Figure 16-5 Switching arrangement for milliameter.

is the movement unshunted, since the switch makes contact with the next range before opening the circuit of the previous shunt.

16-6   Millivoltmeter and shunt.   Although the technique shown in Fig. 16-5 gives excellent results for multirange milliammeters, it is not suited for high-current service. In such cases, individual, external shunts are used (see Fig. 16-6). For portable instrument use, the shunts are mounted on bakelite bases. For switchboard use, the shunts are bolted directly to the bus bars.

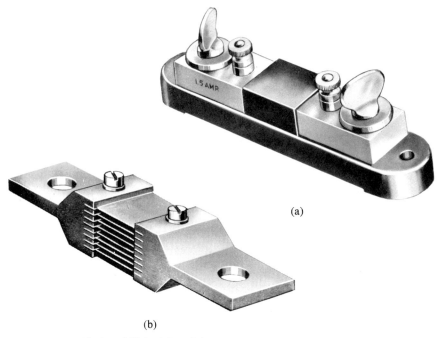

(a)

(b)

Courtesy of Weston Instruments Inc.

Figure 16-6   Ammeter shunts.

Portable shunts are available for current values of 1 to 200 A; switchboard shunts are available up to 8000 A. Notice that the shunt has four terminals. The current to be measured is fed through the shunt using the heavy current-carrying terminals. This keeps the contact resistance low. The voltage drop developed across the inside smaller terminals is applied to the meter through "calibrated" leads. Most shunts are designed for a 50-mV drop at rated current. Obviously, the instrument used with these external shunts will give full-scale deflection on 50 mV. Its scale, however, may be calibrated in amperes to match a specific shunt. For general use with any shunt (portable type), a scale of 0 to 100 is generally used.

16-7 Voltmeter.   The same Weston movement can be used to measure voltages, by adding a resistor in series. The size of the resistor used depends on the current range of the movement and the voltage range desired. The resistor used for this purpose is called a ***multiplier***. Multirange voltmeters can be made by using more than one multiplier either in series or separately, plus suitable switching arrangement.

To calculate the size of multiplier for an instrument, merely apply the series-circuit rules. The method can best be illustrated by a problem.

Example 16-2

A 0–2-mA instrument has a resistance of 20 Ω (Fig. 16-7). What sizes of multipliers are required to convert this to a (a) 10-, (b) 50-, and (c) 100-V voltmeter?

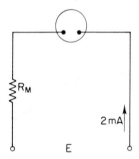

E   **Figure 16-7**

Solution

(a) 10-V range: With 10 V applied across the instrument, the current in the circuit should be 2 mA. Therefore,

$$R_T = \frac{E_T}{I_T} = \frac{10}{0.002} = 5000 \ \Omega$$

Since the total resistance consists of movement resistance and multiplier resistance, the multiplier resistance should be

$$R_m = 5000 - 20 = 4980 \ \Omega$$

(b) 50-V range:

$$R_T = \frac{E_T}{I_T} = \frac{50}{0.002} = 25\ 000\ \Omega$$

$$R_m = 25\ 000 - 20 = 25\ 000\ \Omega$$

In this case the movement resistance is so small compared to the multiplier that the correction can be neglected. The error introduced is only

$$\frac{20}{25\ 000} \times 100 = 0.08\%$$

(c) 100-V range:

$$R_T = \frac{E_T}{I_T} = \frac{100}{0.002} = 50\ 000\ \Omega$$

$$R_m = 50\ 000 - 20 = 50\ 000\ \Omega$$

The three ranges can be combined in one instrument by either separate or series switching, as shown in Fig. 16-8. Notice that in Fig. 16-8(a) the multipliers are in series. The values required for the second and third multipliers are smaller than in the separate resistor method shown at (b). The series method is cheaper, but if one multiplier is defective it affects every succeeding range in addition to its own.

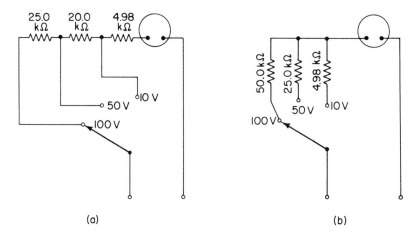

Figure 16-8 Switching arrangement for voltmeter.

16-8 Voltmeter sensitivity.   When measuring the voltage across a component of a circuit, a voltmeter must be connected across or in parallel with the unit. Some of the circuit current will flow through the voltmeter. In order not to unbalance circuit conditions, the voltmeter current should be very small. How small? If the circuit current is 100 A, an additional meter current of 0.1 A (100 mA) would be negligible. On the other hand, if the circuit current is 10 mA, an additional meter current of only 1.0 mA would seriously alter the original circuit conditions. Therefore, the meter current should be small *compared with* *the original circuit current.* A voltmeter is considered more sensitive if it draws less current from the circuit. In other words, the *sensitivity of a voltmeter* varies inversely with the current required for full-scale deflection of its movement. Expressed mathematically,

$$\text{sensitivity} = \frac{1}{I_{fs}}$$

where I_{fs} is the current for full-scale deflection. But

$$I = \frac{E}{R} = \frac{\text{volts}}{\text{ohms}}$$

and its reciprocal is

$$\frac{1}{I} = \frac{R}{E} = \frac{\text{ohms}}{\text{volts}}$$

Therefore, sensitivity of a voltmeter is expressed in **ohms per volt**. From this definition for sensitivity, the sensitivity of a voltmeter using a 0–1-mA movement is

$$\text{sensitivity} = \frac{1}{I} = \frac{1}{0.001} = 1000 \ \Omega/\text{V}$$

A more sensitive movement, 20 μA, used as a voltmeter would have a sensitivity of

$$\frac{1}{I} = \frac{1}{20 \times 10^{-6}} = 50\ 000 \ \Omega/\text{V}$$

This sensitivity figure, ohms per volt, can also be interpreted as the resistance required (ohms) for a 1-V range (per volt). Now we have another method for calculating multiplier resistance. For example, the above 20-μA movement would require 50 000 Ω (movement plus multiplier) for a 1-V range. For a 50-V range the total series resistance should be

$$50\ 000 \ (\text{ohms per volt}) \times 50 = 2\ 500\ 000 \ \Omega$$

Voltmeters are available with sensitivities as low as 100 Ω/V to as high as 200 000 Ω/V. When selecting a voltmeter, be sure to use one of proper sensitivity; otherwise, serious errors in readings will result. Remember that a low-sensitivity meter will give true indication only in high- current, or low-resistance circuits such as found in power applications, while in the high-resistance circuits commonly found in electronics, the higher-sensitivity meters are indispensable. The error introduced by improper voltmeter sensitivity can be readily shown by a problem.

Example 16-3

Find the true voltage across R_2 and R_4 of Fig. 16-9, and the readings of a 100-V voltmeter having a sensitivity of (1) 1000 Ω/V and (2) 20 000 Ω/V.

Solution

1. The true voltage across R_2 and R_4 can be obtained by inspection. Since $R_1 = R_2$ and $R_3 = R_4$, the voltage across R_2 or R_4 must be one-half of the supply voltage or 45 V.

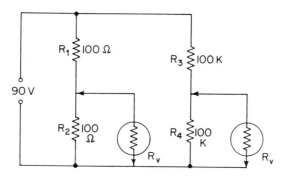

Figure 16-9   Circuit unbalance due to voltmeter.

2. E_2, using the 1000-Ω/V meter,
   (a) Resistance of meter (R_v) = 1000 × 100 = 100 000 Ω
   (b) The parallel resistance of R_2 and R_v equals

$$R_e = \frac{R_2 R_v}{R_2 + R_v} = \frac{100 \times 100\,000}{100\,100} = 99.9\ \Omega$$

Notice that the unbalance in circuit resistance is negligible (0.1 Ω in a total of 199.9 Ω, or less than 0.05%).
   (c) Since voltage divides in proportion to resistance,

$$E_2 = \frac{R_e}{R_1 + R_e} \times E_t = \frac{99.9}{199.9} \times 90 = 44.9775\ \text{V}$$

The error is 0.0225 V in 45 V, or 0.05%.

3. E_2, using the 20 000-Ω/V meter,
   (a) R_v = 20 000 × 100 = 2 000 000 Ω (2 MΩ)
   (b) The 2-MΩ meter resistance will not noticeably affect the resistance of R_2, and therefore the voltage E_2 will be measured as 45.0 V.

4. E_4, using the 1000-ΩV meter,
   (a) The parallel resistance of R_4 and R_v, since each is 100 000 Ω, is R_e = 50 000 Ω. The circuit resistance unbalance is now serious (50 000 out of a new total of 150 000, or $33\frac{1}{3}$%).
   (b) From resistance proportionality, we get

$$E_4 = \frac{R_e}{R_3 + R_e} \times E_t = \frac{50\,000}{150\,000} \times 90 = 30\ \text{V}$$

The error is now 15 V in 45 V, or $33\frac{1}{3}$%.

5. E_4—using the 20 000-Ω/V meter
   (a) The parallel resistance of R_4 and R_v equals

$$R_e = \frac{R_4 R_v}{R_4 + R_v} = \frac{100\,000 \times 2\,000\,000}{2\,100\,000} = 95\,240\ \Omega$$

The unbalance in circuit resistance is not serious (4760 Ω out of a new total of 195 240 Ω, or 2.4%).
   (b) From resistance proportionality, we have

$$E_4 = \frac{R_e}{R_3 + R_e} \times E_t = \frac{95\,240}{195\,240} \times 90 = 43.92 \text{ V}$$

The error is 1.08 V out of 45 V, or 2.4%.

In the above problem, either meter gave true enough indication when the circuit resistance was low. But notice the $33\frac{1}{3}$% error of the low-sensitivity meter in the high-resistance circuit! In fact, even the 20 000-Ω/V meter indication was off by 2.4%. In most electronic applications an error of this amount would be within acceptable limits. The error could of course be reduced by using a meter of even higher sensitivity.

Let us assume that we have to measure voltages in a fairly high resistance circuit, and the only voltmeter available has relatively low sensitivity. Such a situation was seen in Fig. 16-9 for resistors R_3 and R_4 using the 1000-Ω/V voltmeter. Here we saw that the error in our reading was $33\frac{1}{3}$%. Such an error is definitely too much. What can we do? If we know that an error exists we could allow for the extra current drawn by the voltmeter through R_3 and make an estimate at the true voltage. Better yet, using series–parallel circuit theory, we can calculate the true voltage exactly. But such a calculation does take time. Fortunately, another solution to this situation may be possible.

If the voltmeter has a higher voltage range, for example 0 to 1000 V, use it! On this higher range, the voltmeter resistance would be 1000 × 1000, or 1 MΩ. A 1-MΩ voltmeter resistance shunting R_4 would cause much less unbalance in the total circuit resistance. Using the method of Example 16-3, the calculated voltage for E_4 would be 42.6 V, and the error in voltmeter indication would be reduced to only 4.66%. This is far better than the original error of $33\frac{1}{3}$% when using the 100-V range.

At this point you might well wonder, if we do not know the circuit resistance values, how can we know if the shunting effect of the voltmeter is causing serious unbalance in the circuit resistance? That is a good question. A clue to the solution should be seen from the above discussion. Take a reading on a seemingly suitable voltage range. Now shift to a higher voltage range. If the voltage reading *increases*, you obviously have a serious shunting problem. In that case, use the highest voltmeter range that will give you a readable deflection. The correctness of such a procedure can be seen from the discussion on Example 16-3. When the voltage (E_4) was calculated for the 100-V range of the 1000-Ω/V voltmeter, this voltage was 30 V. When the calculation was repeated for the 1000-V scale, the voltage was 42.6 V. (It should be realized that an actual reading of 42.6 V cannot be made on the 1000-V scale. The meter would indicate slightly above the 40-V mark—but below 50 V.) We know from the calculation that the meter was disturbing the original circuit conditions and the unbalance was less when using the 1000-V range.

16-9 Accuracy.   From the above problem and the discussion on sensitivity, you may have drawn a conclusion that low-sensitivity meters are not as accurate as high-sensitivity instruments. Such a deduction is false! The reading obtained with a low-sensitivity voltmeter may be in error, but not because the meter is inaccurate. When a low-sensitivity meter is connected into a high-resistance circuit, the shunting action of the meter changes the circuit resistance and the voltage distribution in the circuit—but the voltmeter indicates the *changed* voltage accurately. In fact, if any relationship is to be made between sensitivity and accuracy, high-sensitivity voltmeters tend to be less accurate. The reason for this may be seen from the discussion that follows.

Accuracy of any indicating instrument (ammeter, voltmeter, wattmeter) depends on the quality of the components used. The tolerance of the resistors used as shunts (for ammeters) or multipliers (for voltmeters) contribute to the meter accuracy. Precision resistors must be used for high accuracy. The magnets, bearings, and springs that supply the restoring torque also contribute to the accuracy of the finished instrument. Finally, where extreme accuracy is required, the scale is hand calibrated to match the actual deflection of the moving coil.

When the manufacturer has done all he can to produce an accurate instrument, there is still another source of potential error—the user. To prevent sticking of the needle against the face of the scale, the needle must of necessity be raised a short distance above the scale (see Fig. 16-1). Because of this spacing, an exact reading can be made only if your eye is directly over the needle. If your eye is to the right, the indication would be read to the left—giving a lower reading. The converse is true, if the eye is to the left. This effect is known as *parallax*. A little experiment will show how serious parallax error can be. Close one eye and hold up one finger approximately one foot in front of the other eye. Now sight on some mark on a wall. Move your head just a small distance to either side and notice how far off the mark your sighting has shifted. Of course, this experiment exaggerates the error, but when you are dealing with accuracies of 1% or better, meter parallax cannot be ignored. To avoid parallax errors, high quality instruments use mirror scales (again see Fig. 16-1). When reading such a meter, your eye is properly aligned when the reflection of the needle in the mirror is directly below the needle itself.

Instruments are available with accuracies ranging from as high as 0.1% to a low of 5%. A point that is not often realized is that *this accuracy figure applies to full-scale deflection*. For example a 10-A meter with 2% accuracy may have an error of 2% of full scale or 0.2 A *anywhere on its scale*. So, when an indication of 1.0 A is read on this meter, the actual current may be 0.8 or 1.2 A, and the actual error at this point on the scale may be as high as 20%. It is therefore desirable, for good accuracy, to use a range that will give readings in the upper half of the meter scale.

16-10 Digital voltmeters.   The above pointer-on-scale instruments can be called "analog" devices—the amount of output (deflection) is continuously variable depending on the input. The best (laboratory) accuracy obtainable with these instruments is $\pm 0.1\%$ *of full scale*. The output value (reading) must be interpreted by the user. Advances in semiconductor technology made practical the development of digital voltmeters. These instruments convert the analog input into digital logic, and display the output directly in decimal numeric values, with from three to seven significant figures. The conversion of the measured value to digital logic involves the use of: *amplifiers, pulse generators, gates, counters* and *clock circuits.* Discussion of these circuits is beyond the scope of this text. Needless to say, the price of a digital voltmeter is appreciably higher than the pointer-and-scale type.

16-11 Wattmeter.   In the Weston movement (see Fig. 16-1) the torque produced was due to the interaction of the magnetic field of the permanent magnet and the electromagnetic field of the moving coil. Since the field of the permanent magnet is fixed, the deflection of the moving coil depends only on the current in this coil. When this type of movement is used in a voltmeter, the moving-coil current depends on the voltage across which the moving-coil circuit (movement plus series multiplier) is connected—and so the deflection is proportional to voltage. But now, in a wattmeter, we want a torque, and resulting deflection, that is proportional not only to the voltage across the circuit but also to the current flowing through the circuit. This can be achieved by replacing the permanent magnet with a second electromagnet. (The basic movement is now called an *electrodynamometer movement.*) This second electromagnet is stationary and consists of a pair of coils, one on each side of the moving coil. Referring to Fig. 16-1(a), these coils would be mounted in the position occupied by the soft-iron pole pieces of the permanent magnet it replaces. The coils are connected in series, so that the magnetic field of both coils are in the same direction. The free end of each coil is brought out to a pair of terminals. The coils are energized by breaking into a circuit and connecting them in series with the circuit. Since the coils carry the line current, they are called *current coils.* Current coils should have low resistance, in order not to reduce the original circuit current, and they should be capable of carrying the full line current.

The moving-coil circuit in a wattmeter is called the *potential coil* and is similar to that of the voltmeters previously discussed. The rest of the mechanical construction (pivots, springs, pointer, scale) follows the Weston movement. Now, since the magnetic field of the stationary magnet depends on circuit current, and the magnetic field of the moving coil depends on circuit voltage, the deflection of this instrument is proportional to the power taken by the circuit. Figure 16-10 shows in diagrammatic form the basic construction of a wattmeter and the proper connection of this meter into a circuit. It should be noted that one terminal of the current coil and one terminal of the potential coil are identified with a polarity marker ($\pm$). Notice that this side of the current coil is connected to the *line* side of the power line and that this side of the potential coil is connected to the *same potential* line wire. If either coil is reversed, the instrument will deflect backwards. Of course, you are probably still wondering why the double polarity marking—plus and minus. This is because the watt-

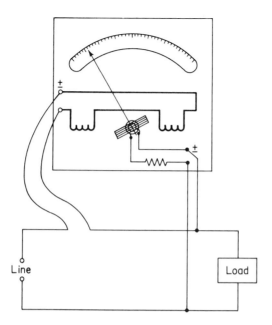

Figure 16-10   Wattmeter: construction and application.

meter, having an electrodynamometer movement, will give true indication on ac as well as on dc circuits, and in any ac circuit the line polarity reverses every half cycle. The important point is not whether the line used is plus or minus, but rather that the marked terminals of each coil are at the same potential at any one instant.

16-12 Ohmmeter—series type.

If the voltage applied to a circuit is low, and/or the circuit has a high resistance, we can use a milliammeter to measure the current flowing through the circuit. Furthermore, if we always use the same value of fixed voltage, the amount of current depends (inversely) on the amount of resistance. Therefore, instead of calibrating the milliammeter to read current, why not mark its scale directly in the ohms required to give the corresponding current? This is the principle of the ohmmeter.

Figure 16-11 shows an ohmmeter circuit made with a 0–1-mA movement and a 3-V battery. With the test prods A and B connected together, the internal resistance is adjusted until the milliammeter reads full scale—1 mA. The *external resistance* is zero. The 1-mA point should be marked 0 on the ohm-meter scale. This adjustment calibrates the ohmmeter. The value of internal resistance required depends on the voltage used and the maximum current range of the movement. In our problem this would be

$$R_i = \frac{E}{I} = \frac{3}{0.001} = 3000 \ \Omega$$

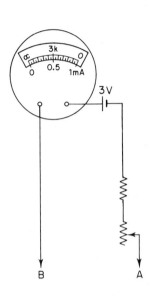

Courtesy of Simpson Electric Co.

(a) Simplified Schematic Diagram   (b) Commercial VOM.

Figure 16-11   Series-type ohmmeter.

You may wonder why part of this resistance is made variable. This variability is to compensate for reduction in the terminal voltage of the battery with age. As the voltage drops, the internal resistance is reduced and the zero setting can still be made.

If we place a resistance of 3000 Ω between the test prods, the total resistance of the circuit has been doubled. The current drops to half scale. This halfway point on the scale should be marked 3000 Ω. This feature is important in the design of series ohmmeters: *The resistance for half-scale deflection is equal to the internal resistance of the ohmmeter circuit.* From this rule it is also obvious that if the battery voltage drops too low and the internal resistance is reduced too much, the calibration of the instrument is in error.

When the resistance between test prods is infinite, the current in the circuit is zero. The zero of the milliammeter scale should be marked infinity on the ohmmeter scale. Notice the inverse nature of the two scales. Theoretically, the ohmmeter has a range of zero to infinite ohms. Practically, this statement is not true. If the resistance to be measured is small compared to the internal 3000 Ω (less than 30), the reduction in current is not noticeable. On the other hand, if the resistor to be measured is too high, very little current will flow through the meter and its deflection cannot be read, so that a limit is placed on the maximum range of the ohmmeter. These limitations to the ohmmeter range can best be shown by a problem.

Example 16-4

A 0–1-mA meter has 50 scale divisions. It is to be used with a 3-V battery as an ohmmeter. Assuming that the instrument can be read to within half a division, what is the maximum and minimum resistance it can measure?

Solution

1. Internal resistance of the ohmmeter must be

$$R_T = \frac{E_T}{I_T} = \frac{3}{0.001} = 3000 \ \Omega$$

2. One-half scale division $= \frac{1}{100}$ of full-scale current $= 0.01$ mA.

3. Maximum resistance measurable corresponds to the minimum current readable:

$$R_{max} = \frac{E}{I_{min}} = \frac{3}{0.01 \times 10^{-3}} = 300 \ 000 \ \Omega$$

4. Minimum resistance measurable corresponds to the maximum current readable. The maximum current is 1 mA, but this value is with zero external resistance. The first noticeable change in current is one-half division less than maximum (1 mA $-$ 0.01 mA, or 0.99 mA):

$$R_{min} = \frac{E}{I_{max}} = \frac{3}{0.99 \times 10^{-3}} = 3030 \ \Omega$$

Since the internal resistance is 3000 Ω, the minimum external resistance measurable is 3030 $-$ 3000 or 30 Ω.

16-13   Changing range of ohmmeter.   The normal usable range of an ohmmeter is at best from $\frac{1}{100}$ to 100 times its internal resistance. To obtain an ohmmeter that will measure lower resistance values, it is necessary to lower the internal resistance. In turn, if the zero calibration is to be possible, either the voltage must be lowered or the current range of the instrument must be increased. The ohmmeter section of the commercial volt-ohm-milliammeter, shown in Fig. 16-11(b), uses a 1.5-V battery for the $R \times 1$ and $R \times 100$ ranges. For the low range, the 50 μA movement is shunted for a full-scale current of 125 mA. The internal resistance—and center-scale resistance—is therefore 12 Ω (1.5 V/125 mA). For the $R \times 100$ range, the meter shunting is reduced to 1.25 mA, full scale. Obviously, with only 1/100 the current, the resistance range is increased by $\times$ 100. When switching to the $R \times 10\ 000$ range, the meter current is reduced to 0.125 mA. This, in itself, would increase the range by a factor of 1000. In addition the battery voltage is increased to 15 V, giving a total change of $\times$ 10 000.

In general, the maximum resistance that can be measured by an ohmmeter increases directly with the voltage of the battery used; the minimum resistance that can be measured decreases in proportion to the increase in current range of the milliammeter. In both cases the calibrating resistance must be changed to correspond.

16-14 Shunt ohmmeter. The shunt ohmmeter is used to measure low values of resistance. Instead of shunting the movement and connecting R_x in series with the calibrating resistor and battery as before, the unknown resistor itself is used as the shunt. A diagram of this type of ohmmeter is seen in Fig. 16-12.

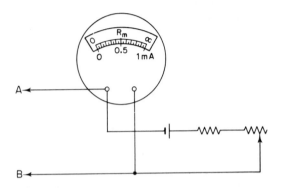

Figure 16-12   Shunt-type ohmmeter.

With the test leads apart, the calibrating resistance is adjusted until the milliammeter indicates full scale. The resistance between the test leads is infinite. The full-scale current point corresponds to infinite resistance of R_x. With the test leads shorted, $R_x = 0$, and all the current supplied by the battery through the calibrating resistance will flow through the test-lead circuit. No current will flow through the movement. Zero current on the milliammeter scale corresponds to zero resistance for R_x. For other values of R_x the current in the circuit divides between R_x and the movement inversely as their respective resistances. When R_x is equal to the resistance of the movement, the current division will be 50% in each. The halfway point on the current scale should therefore be marked with a resistance value equal to the resistance of the moving coil. For a 1-mA movement this value may be as low as 27 Ω. Good accuracy on low resistance can be obtained with this type ohmmeter.

When the proper range is used, ohmmeter accuracy is sufficient for most electronics work. However, when accurate resistors are needed, as in shunts, multipliers, and critical circuits, Wheatstone bridge measurements should be used.

16-15 "Megger" insulation tester. An instrument often used for the measurement of very high resistances is known as a "Megger" (from "megohm"). It is mainly used in power work for the measurement of insulation resistance of cables and machinery windings. Its range is approximately 0 to 50 or 100 MΩ in the 500-V series, or as high as 10 000 MΩ in the 2500-V series. This instrument (see Fig. 16-13) is basically a series-type ohmmeter. As you recall, in order to measure high resistance values, high voltages are required.

Furthermore, when measuring insulation resistances, high voltages are desirable. An insulation that may seem good on low-voltage tests may break down at operating potentials. In the "Megger," a hand-driven magneto (dc generator) is used for supply instead of a battery. A clutch-type drive is used on the hand crank, so that if the crank is turned at above a certain minimum speed, the output voltage from the magneto is constant. Meggers are available in two voltage ratings: 500 to 600 V (common type) and 2500 V. As in the ohmmeter, the current-measuring instrument is calibrated directly in terms of resistance. The resistance to be measured is connected across the output leads (usually supplied with battery clips); the crank is turned until the clutch slips, and the resistance is read directly on the instrument's scale.

Figure 16-13   Megger.

16-16 Wheatstone bridge.   Although the ohmmeter is a quick and convenient instrument for measuring a wide range of resistance values, its accuracy is at best only 5 to 10%. This is quite often sufficient. On the other hand, a commercial *Wheatstone bridge* will measure resistance values from 0.01 to 10 MΩ with excellent accuracy. Such a bridge is a self-contained unit consisting of a 1.5-V cell, a very low range, zero-center current measuring device (a *galvanometer*), and three variable resistors. The principle of operation can be readily explained with the aid of a diagram. In Fig. 16-14, R_x is the unknown resistor. The operating procedure is as follows: Adjust R_1 and R_2 or R_s (or combination) until, with the battery switch closed, *tapping* the galvanometer key will give no deflection of the galvanometer. This operation is called **balancing** the bridge. When the bridge is balanced, no current flows through the galvanometer. This means that there is no difference in potential across the galvanometer, or the potential of point A is the same as the potential of point B! Since no current flows through the galvanometer, all of I_1 flows through R_2, or

$$I_1 = I_2$$

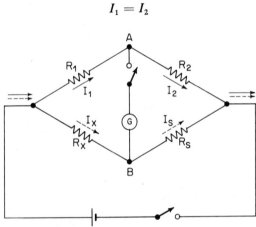

Figure 16-14   Wheatstone bridge.

Similarly,

$$I_x = I_s$$

Points A and B are at the same potential. This can be true only if the voltage drops across R_1 and R_x are equal, or

$$I_x R_x = I_1 R_1 \tag{1}$$

Similarly,

$$I_s R_s = I_2 R_2 \tag{2}$$

Dividing equation (1) by (2) and remembering the current distributions, we get

$$\frac{R_x}{R_s} = \frac{R_1}{R_2}$$

Solving for R_x yields

$$R_x = \frac{R_1}{R_2} \times R_s$$

R_1/R_2 is called the **ratio arm.**

In the commercial bridge (Fig. 16-15) the ratio arm is adjustable by means of a single rotary switch. The fixed ratios obtainable are from 0.001 to 1000 in multiples of 10. R_s is a four-dial *decade box*. Its resistance can be varied from 1 to 10 000 Ω in 1-Ω steps. The resistance setting can be read directly on the dials. The first dial section consists of ten 1-Ω resistors in series. The setting of the dial will indicate how many units are in the circuit. Its range is from 0 to 10 Ω in 1-Ω steps. The second dial section consists of ten 10-Ω units in series. Its

Courtesy of Leeds and Northrup Co.

Figure 16-15   Commercial Wheatstone bridge.

range is from zero to 100 Ω in 10-Ω steps. The third dial section consists of ten 100-Ω units. Its range is from zero to 1000 Ω in 100-Ω steps. The fourth dial, with ten 1000-Ω units, has a range of zero to 10 000 Ω in 1000-Ω steps. The four dial sections are wired in series. The total resistance of the decade box is the sum of the four sections. Reading the dials from large to small sections, if the reading is

$\times$ 1000	$\times$ 100	$\times$ 10	$\times$ 1
6	2	3	5

the resistance is 6235 Ω.

With this type of bridge best accuracy will be obtained when the ratio dial setting is such as to allow the largest value for R_s. That is, use the lowest ratio-dial setting that will still give a balance when R_s is varied.

A variation of the Wheatstone bridge is the **Kelvin bridge**. By using separate potential and current terminals, it can measure resistance values as low as 0.000 01 Ω with better than 0.5% accuracy.

REVIEW QUESTIONS

1. (a) In Fig. 16-1(a), state two functions of the springs.
   (b) What current ranges are suitable for the Weston movement?
   (c) Why is it not suited for lower currents?
   (d) Why is it not suited for higher currents?
2. (a) What is the advantage of a core-type magnet structure?
   (b) What is a disadvantage of this structure?
   (c) Name another type of magnet structure used in meter movements.
   (d) How does it compare with the core type?
3. (a) How can the current range of an instrument be increased?
   (b) What is a multirange milliammeter?
   (c) What precautions are necessary in its construction and use?
   (d) What is the effect of contact resistance in the switch? How can this be reduced?
4. (a) In Fig. 16-5, what is the purpose of the long "arc" on the switch arm?
   (b) What is this type of switch called?
   (c) Can these same resistors (shunts) be used with another movement to provide the same ranges? Explain.
5. With reference to Fig. 16-6:
   (a) With what instrument is this shunt used?
   (b) How are the four terminals of this shunt used?
   (c) This shunt is rated 25 A, 50 mV. It is used with a proper instrument having scale markings of 0 to 100. What is the circuit current when the instrument indication is 25?
6. (a) Can a basic movement (as in Fig. 16-1) be used directly to measure the voltage of a car battery? Explain.
   (b) How is it "converted" into a voltmeter?
   (c) What name is given to the added device?
7. (a) What is meant by sensitivity of a voltmeter?
   (b) Which is preferable, a high or a low value?
8. With reference to Fig. 16-9, it is desired to measure the voltage across R_4. Which of the following is preferable to use?
   (a) A 100-Ω/V voltmeter or a 50 000-Ω/V instrument? Explain.
   (b) The 50-V scale, the 100-V scale or the 300-V scale of a 2000-Ω/V instrument? Explain.
9. When measuring voltages in a series circuit, and the resistance values are unknown:
   (a) How can you tell if the voltmeter is loading the circuit (altering circuit resistance and voltage distribution)?

(b) If there is evidence of loading, how can truer readings be obtained?

10. You have the choice of a 50 000-Ω/V meter of 2% accuracy, or a 100-Ω/V meter of 0.5% accuracy. Which would you use in each of the following cases? Explain your choice.
    (a) Low-resistance circuit.
    (b) High-resistance circuit.

11. **(a)** Why do some meters have mirror scales?
    (b) What is the proper technique in reading this type of meter?

12. **(a)** Name a type of voltmeter that is capable of much higher accuracy than the pointer-and-scale type.
    (b) How is a voltage value "read" on these instruments?

13. **(a)** What type of movement is used in a wattmeter?
    (b) Name the two circuits of this meter.

14. With reference to Fig. 16-11(a):
    (a) What is the purpose of this circuitry?
    (b) How is the instrument "calibrated"?
    (c) What is the unknown external resistance at this time?
    (d) Locate the adjusting knob on the commercial unit, Fig. 16-11(b). What is it called?
    (e) Why is this internal resistance adjustment necessary? Why not use a fixed resistor of proper value?
    (f) Why is the internal resistance only partly variable? Why not completely variable to zero?

15. **(a)** In Fig. 16-11(a), what relation exists between the internal resistance value and the resistance-scale values?
    (b) What determines the choice of battery voltage?
    (c) How can the resistance range be increased?
    (d) How can the accuracy for low-resistance values be increased?

16. With reference to Fig. 16-12:
    (a) Why is this called a shunt-type ohmmeter?
    (b) How is calibration made?
    (c) Is this type used for high- or low-resistance measurements?
    (d) What determines the half-scale calibration?

17. **(a)** What is a megger used for?
    (b) What range of resistances does it cover?
    (c) Why not use the high-resistance range of the ohmmeter?

18. With reference to Fig. 16-14:
    (a) What physical procedure is involved in *balancing* the bridge?
    (b) When is the bridge balanced?
    (c) Why should the galvanometer key be *tapped* (rather than closing a switch) while the bridge is being balanced?
    (d) When the bridge is balanced, what is the relation between R_x and the other resistors?
    (e) In commercial bridges, what is the combination of R_1 and R_2 called?
    (f) What is R_s called?

19. (a) What is the (approximate) range of resistances that can be measured with a "standard" Wheatstone bridge?
    (b) Name a bridge specifically used for low-resistance measurements.

PROBLEMS AND DIAGRAMS

1. A movement having a resistance of 20 Ω gives full-scale deflection with 2 mA. Calculate shunt resistance required to increase the range to:
   (a) 10 mA.
   (b) 20 mA
   (c) 100 mA.
   (d) Draw a diagram for the above three ranges.

2. Using the movement of Problem 1, calculate the multipliers required for the following ranges.
   (a) 2 V   (b) 20 V   (c) 200 V   (d) 500 V
   (e) Draw a diagram for the above four ranges.
   (f) What is the sensitivity of the instrument?

3. A 500-μA movement has a resistance of 300 Ω. Calculate shunts and multipliers required for the following ranges.
   (a) 5 mA, 50 mA, 200 mA
   (b) 5 V, 50 V, 200 V, 500 V
   (c) Draw a diagram for the multimeter.
   (d) What is the sensitivity of the voltmeter?

4. A voltmeter is marked 20 000 Ω/V. Determine its resistance when used on the following ranges.
   (a) 600 V   (b) 200 V   (c) 1 V

5. Draw a diagram showing instruments being used to measure the current in a circuit containing two resistors in series, and the voltage drop across one of the resistors.

6. Two resistors, 10 000 Ω and 6000 Ω, are connected in series across a 240-V supply. A 100-Ω/V voltmeter is used on its 100-V range to measure the voltage across the 6000-Ω resistor
   (a) Find the voltmeter reading.
   (b) Find the true voltage across this resistor.
   (c) Find the voltmeter reading when used on its 600-V range.

7. A transistor is connected in series with a 10 000-Ω resistor across a 30-V supply. The equivalent resistance of the transistor is not known. A 1000-Ω/V voltmeter used on its 15-V scale registers 12 V when connected across the transistor. Find the true voltage across the transistor. (*Note:* Find the voltmeter current, and use series–parallel circuit theory to find transistor current, transistor resistance, and true voltage.)

8. A 0–50-A ammeter has an accuracy of $\pm0.5\%$. Between what limits may the actual current be when the indication is 10 A?

9. Draw a circuit showing proper connections of a wattmeter for measuring power, and indicate polarites of meter windings.

10. A 500-μA movement having a resistance of 310 Ω is to be used in a series ohm-meter. What value of battery voltage and internal resistance should be used if the center of scale should correspond to 60 000 Ω?

11. In Problem 10, if the instrument has 50 scale divisions and a good reading can be obtained for half a division, what are the maximum and minimum resistance markings?

12. The movement of Problem 10 is to be used with a 1.5-V battery as a series ohm-meter. Find:
    (a) The value of internal resistance needed.
    (b) The new center-scale resistance marking.
    (c) The minimum and maximum resistance marking.

13. It is desired to use the movement of Problem 10 in a series ohmmeter with a 1.5-V battery. The center-scale resistance should be 300 Ω.
    (a) What value of internal resistance should be used?
    (b) To what value must the current range of the meter be increased?
    (c) What value of meter shunt resistance is required?

14. Using the instrument of Problem 10 as a shunt ohmmeter, calculate minimum, maximum, and center-scale resistance values.

15. (a) Compare the center-scale resistance value of the series ohmmeter of Problem 13 with the shunt ohmmeter of Problem 14.
    (b) Compare the current drain on the battery for each of the two circuits above.
    (c) What is the advantage of shunt-type ohmmeters for low-resistance measurements?

16. A Wheatstone bridge when balanced (Fig. 16-14) has the following settings: $R_1 = 2000$ Ω, $R_2 = 20\ 000$ Ω, $R_s = 4281$ Ω. Find R_x.

17

INDUCED VOLTAGE

17-1 Principle of induced voltage. In Chapter 13, Permanent Magnetism, we learned:

1. A stationary charge is surrounded by a constant or stationary dielectric field. The energy of this field is entirely a form of potential energy.
2. As the charge moves, the dielectric field moves with it. Motion of the dielectric field converts some of its energy from potential energy to energy of motion or magnetic energy.
3. The faster the charge moves, the faster the dielectric field moves. The magnetic field energy increases; the potential energy decreases.
4. Uniform speed of motion of the dielectric field (constant current) results in a constant magnetic-field strength or "stationary magnetic field."
5. Varying speed of motion of the dielectric field (changing current) results in varying strength of the magnetic field. The lines of force of the magnetic field are expanding and contracting. The magnetic field is "in motion."
6. The energy of the system is constant, the proportion of potential and magnetic energy depending on the speed of the dielectric field.

So far, we have been considering what happens to the magnetic field around a conductor as the potential energy across the conductor is varied. The converse of these effects must also be true:

1. A conductor located in a steady or stationary magnetic field must have a steady or stationary dielectric field. This statement implies that electrons are

not moving through the conductor; or that there is no difference of potential across the conductor.

2. If the conductor is located in a magnetic field of varying strength (a moving magnetic field), in order to preserve the energy balance in the system, the potential energy of the dielectric field must also change in strength. But the potential energy of a dielectric field changes only when the field is in motion. Therefore, the dielectric field must be set in motion by the variations of the magnetic-field strength. Motion of the dielectric field means motion of charged bodies. *The electrons are moving through the conductor.* As the electrons pile up at one end of the wire, this point acquires a negative charge or a negative potential. The opposite end of the conductor has lost electrons; it acquires a positive potential. A difference in potential (voltage) has been established across the wire. This voltage is called an **induced voltage**.

An induced voltage is produced across a conductor when it is located in a varying magnetic field. In effect, the magnetic field must react on the magnetic field around the free electrons, pushing or pulling them to one end of the conductor. The direction in which the electrons move must depend on the direction in which the lines of force of the magnetic field are moving. The polarity of the induced voltage must therefore also depend on the direction in which the magnetic field is moving.

17-2 Lenz's law.*   Lenz's law is used to determine the polarity of an induced voltage. Basically, it is a restatement of Newton's third law of motion (To every action there is an equal and opposite reaction) in electrical terms (see page 238). Let us see how this law of Newton's applies to induced voltages.

The terminals of a coil (Fig. 17-1) are connected to a galvanometer. A magnet is placed above the coil. The magnetic field does not reach the coil.

As we try to push the magnet in Fig. 17-1(a) down, a reaction is produced in the coil, which tries to oppose this motion. Since the magnetic-field strength around the coil is changing, a voltage is induced across the coil. The free electrons in the winding are pushed or pulled to one end. This constitutes a current flow through the coil. But when current flows in a coil, the coil becomes an electromagnet with its own north and south poles. In order to oppose the motion of the permanent magnet, the top of the coil in Fig. 17-1(a) must become a north pole. From Fleming's right-hand rule for a coil (Section 14-2), to get a north pole on top, conventional current must flow from *B* to *A*. Remember, now, that electron flow is opposite to conventional-current flow. Therefore, *electrons* leave point *A*, making it positive in polarity, and pile up at point *B* making it negative in polarity. A voltage has been induced across the coil, and the galvanometer needle deflects to the right, as shown by the dotted arrow. From a casual look at Fig. 17-1(a), it might seem that conventional current is shown flowing from minus to plus. This is not so! The galvanometer is

*Heinrich Friedrich E. Lenz, a Russian physicist (1804–1865).

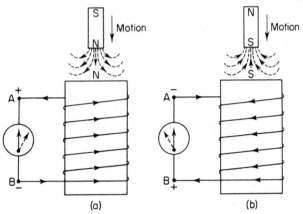

not the battery or source of energy. It is the flow of current (conventional) *to point A* that *makes* point A positive. Now, considering the coil terminals A and B as the *source* of energy, conventional current would flow out of point A—toward a load at the left (not shown)—and back to point B.

If the motion of the magnet is stopped, the magnetic field becomes stationary. The dielectric field also becomes steady. The electrons piled up at terminal B redistribute themselves evenly, and the induced voltage is gone. The galvanometer needle returns to zero.

As the magnet is withdrawn, the magnetic field is moving in the opposite direction. The free electrons in the wire are pushed or pulled to the opposite end of the coil, to point A. Point A becomes negative, point B positive. The galvanometer deflects to the left. The top of the coil becomes a south pole and tries to hold the permanent magnet and prevent it from being pulled out. Again the reaction is opposite to the action.

In Fig. 17-1(b) we are trying to move the south pole of the permanent magnet down. The coil reaction is such that electrons move toward point A, producing a south pole at the top of the coil. The potential of point A becomes negative compared to point B. The galvanometer deflects as shown by the dotted arrow. Again notice that the magnetic reaction is such as to oppose the action producing it.

Mechanical energy is used to move the magnet against the reaction. This energy is converted to electrical energy which appears as the induced voltage. Lenz's law states this effect as follows: *In all cases of electromagnetic induction, the induced currents have such a direction that their reaction tends to stop the motion which produces them.* In the above illustrations, the direction of the currents induced in the coil was such as to produce a magnetic field that tends to stop the motion of the magnet producing the action.

Let us consider another illustration. This time we will move a conductor through a constant magnetic field.

In Fig. 17-2(a) we are trying to push the conductor up, from a point where there is no magnetic field to a point where there is a strong magnetic field. Since the magnetic-field strength around the conductor is changing, a voltage is induced across the conductor. The magnetic reaction resulting from the current induced in our conductor must be such as to oppose the upward motion of the conductor. We know that lines of force in the same direction repel each other, so the lines of force produced around the top of the conductor must be in the same direction as the fixed magnetic field (curved dashed arrow in the diagram). But this direction is clockwise, and a clockwise magnetic field is produced when the current flow is out toward us (see Fig. 14-2). Therefore, we should see the tail of the current arrow, and an X should be put on the cross section of the conductor. Since current (conventional) is moving down, electrons are actually piling up on the near end (top) of this conductor, making this the *negative* terminal.

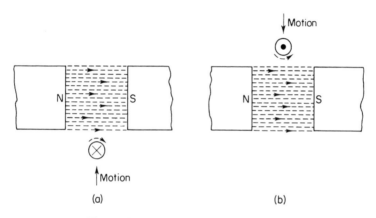

Figure 17-2   Direction of induced voltages.

In Fig. 17-2(b), we are trying to push the conductor down. The magnetic reaction around the conductor must be counterclockwise (in the same direction as the fixed magnetic field). The resulting induced current must be upward, or toward us. The conductor cross section should be marked with a dot. This end of the wire becomes positive in polarity.

Another rule often used to determine the direction of an induced voltage is Fleming's right-hand rule for induced voltage: "Extending the thumb, forefinger, and middle finger at right angles to each other, if the thumb points in the direction of motion of the wire and the forefinger points in the direction of the magnetic field, the middle finger will point in the direction of current flow." Try this rule on the diagrams in Fig. 17-2.

17-3 Methods of cutting lines of force.   In Figs. 17-1 and 17-2, either the magnet was moved with respect to the conductors, or the conductors were moved with respect to the magnetic field. In both cases the magnetic field around the conductor varied in strength. Since the magnetic lines came in contact and passed the conductor, we can say that the lines *linked with* the conductor or that the conductor *cut through* these lines.

There are three methods by which a conductor can cut the lines of force of a magnetic field:

1. *Stationary generating coil:* moving a permanent magnet or dc electromagnet. This principle is used in the larger ac motors and generators.
2. *Stationary permanent magnet or dc electromagnet:* moving the generating coil. This principle is used in all dc motors and generators.
3. *Stationary generating coil and stationary electromagnet:* "moving" the magnetic field by supplying alternating current to the electromagnet winding. This principle is used in *transformers.**

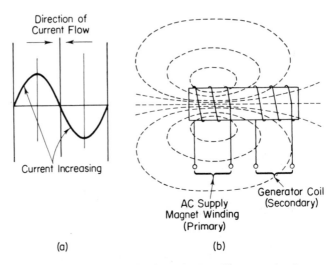

Figure 17-3   Induced voltages by "transformer action."

The first two methods are probably clear from the previous illustrations. The third method may require closer analysis. The current in an ac circuit undergoes four changes (as shown in Fig. 17-3):

1. The current increases from zero to a maximum value. But the strength of an electromagnet depends upon the current in the coil. As the strength of the field increases, the lines of force extend out into space and are "cut" by or link the turns of the generating coil. A voltage is induced in the generating coil.

*See Chapter 36.

2. The current decreases from maximum value to zero. The magnetic field collapses. As the lines of force "shrink," they are cut by or link the generating coil. Again a voltage is induced in the generating coil.

3. The current builds up to a maximum again, but this time it flows in the opposite direction. Again the magnetic field strength increases. The action is similar to change number 1.

4. This current in turn reduces to zero. The magnetic field again collapses. This action is similar to change number 2.

These effects may be seen in Fig. 17-3. Figure 17-3(a) shows the variations in the current, and Fig. 17-3(b) shows the magnetic field.

17-4 Factors affecting magnitude of induced voltage.   If this induced voltage is to be put to practical use, we must know how to control the amount of induced voltage. Let us see what factors are involved. A voltage is induced across a conductor when it cuts a magnetic field. But from the original definition of voltage, we learned that it is equal to the work done in producing the charges that create the difference in potential. In induced voltages, the voltage produced must depend on the work done in cutting the lines of force. The induced voltage is therefore dependent on three factors:

1. Number of turns in the generating coil (N). Each turn may be considered as two conductors in series. Work must be done to make each conductor cut the magnetic lines. The amount of work increases as more turns are used, and the voltage increases directly with the number of turns. Another way of analyzing the relation between voltage and number of turns is to note that a small voltage is induced in each turn. Since the turns are in series, these small voltages are additive. The more turns, the greater the total voltage.

2. The strength of the magnetic field (ϕ). The stronger the field, the greater the number of lines of force existing in the space. More lines will be cut by a conductor; more work must be done to accomplish this result, and hence more voltage will be generated.

3. The relative speed of motion between generating coil and magnetic field (S). When a coil is in motion in a stationary magnetic field, or vice versa, more work must be done in a given time if we make the motion faster. More work done means higher voltage. In the case of the transformer, the magnetic-field motion is caused by the changing current in the magnet winding. The higher the "frequency" of the current wave, the greater the speed of motion and the greater the voltage induced.

Summarizing: The induced voltage depends on the number of turns in the generating coil (N), the strength of the magnetic field (ϕ) and the relative speed of motion (S), or

$$E \propto N\phi S \qquad\qquad (17\text{-}1)$$

17-5 Eddy currents. We have seen that an induced voltage is produced in a conductor when it is cut by a magnetic field. The conductor in the previous examples was a wire, or a coil. But an induced voltage may be produced in any conductor, whether a piece of iron, aluminum, or other material. Sometimes we can make practical applications of this behavior. At other times it causes losses, and methods must be devised to reduce such losses.

In Chapter 15, Magnetic Circuit Calculations, we saw that to produce strong magnetic effects, electromagnets were wound on iron cores. When the current in an electromagnet coil is changing, the magnetic field is either expanding or contracting, and as the flux cuts the turns of the coil, a voltage is induced in the winding. At the same time, the flux is also cutting across the iron of the magnetic circuit. Therefore, a voltage is also induced in the iron! Because of this induced voltage, current actually flows in the iron structure. This current, known as an *eddy current*, is a complete waste of energy and heats up the iron core.

In dc circuits the energy loss is not too serious. An induced voltage is produced only when the circuit is opened or closed. At the ac commercial power frequency of 60 Hz, flux cutting occurs four times in every cycle, or 240 times per second. Although the loss due to one flux cutting is small, as in a dc circuit, it becomes appreciable when you consider that this loss of energy is repeated 240 times in each second. If such an iron-core coil were operated on alternating current for several hours, the loss of energy could be considerable. When you consider that iron-core coils are used in electronics at audio frequencies (20 to 20 000 cycles per second) and at radio frequencies (which often run into megacycles per second), the loss would be terrific and could not be tolerated.

Let us consider the direction of these eddy currents in an iron-core coil. First, remember that according to Lenz's law, the induced voltage produces currents that create a magnetic field in opposition to the originating magnetic field. Applied to this situation, it means that the eddy currents must produce a magnetic field opposite in direction to the magnetic field produced by the coil wound around the iron core. In Fig. 17-4(a) a coil is shown wound around an iron core. The magnetic field produced by the coil on the upper iron structure is to the right. The magnetic field produced by the eddy currents in the same section of iron, when the circuit is closed, must be to the left. In order for this to be true, the eddy currents must be clockwise *across the width of the core*, as shown in the cross-sectional cut of the iron core in Fig. 17-4(b).

Since eddy currents travel crosswise in the iron structure, their magnitude can be reduced to negligible value by increasing the *cross-sectional resistance* of the iron structure. This operation can be performed quite readily. Instead of making the iron structure from a solid mass of iron, it is made in *laminations* (see Fig. 15-2 and 15-3). Each lamination is "insulated" from the others by a

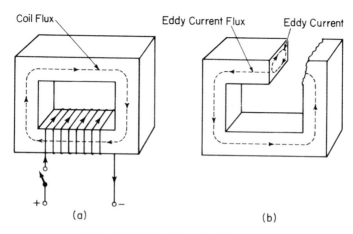

Figure 17-4   Direction of eddy current.

coating of oxide (which has high resistance) and varnish. If such an iron core is taken apart, care must be exercised not to impair the insulation between laminations; otherwise, when the iron core is reassembled the eddy currents will increase and will overheat it. By laminating the iron core, the crosswise resistance is increased, without affecting the reluctance of the iron, in the direction of the magnetic flux.

At radio frequencies, instead of laminating, the iron core is powdered and each granule is insulated from the others to further reduce eddy-current losses.

Another core material used at radio frequencies is a magnetic ceramic or *ferrite*. Commonly used ferrite cores are composed of oxides of nickel and zinc, or of manganese oxide and zinc oxide. Such cores not only provide high permeability, but because they are electrically nonconducting, the eddy-current losses are reduced to very low values. Ferrite cores have found wide application for inductors used at frequencies ranging from several hundred kilohertz to several hundred megahertz.

REVIEW QUESTIONS

1. Since voltage is a form of potential energy, where does the energy represented by an induced voltage come from? Explain.
2. (a) What law in *mechanics* is applicable to induced voltages?
   (b) Give a statement of this "mechanics" law.
   (c) By what name is the induced voltage law known?
3. With reference to Fig. 17-1(a):
   (a) What does the N at the top of the coil signify?
   (b) Is this a correct magnetic polarity? Explain.
   (c) Is the direction of current flow around the coil correct? Explain.

(d) Since current is shown flowing from point B to point A, is the polarity of the voltage A to B correct as shown? Explain.

4. In Fig. 17-1(b), assuming the magnet is just inside the bottom of the coil, and being pulled up:

    (a) By Lenz's law what is the magnetic polarity of the coil?

    (b) In what direction must current flow through the coil to produce such poles?

    (c) What is the potential of point A?

5. With reference to Fig. 17-2(b):

    (a) To what does the word "Motion" and the arrow alongside refer?

    (b) What does the dashed arrow below the conductor signify?

    (c) Is the direction correct? Explain.

    (d) What does the "dot" designation in the wire signify?

    (e) Is this correct? Explain.

6. Show the direction of current flow in the conductor that is being moved in (a), (b), and (c) of Fig. 17-5.

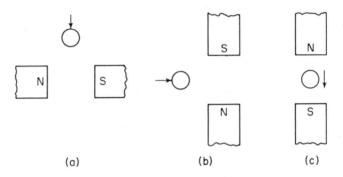

Figure 17-5

7. What is meant by "cutting" lines of force?

8. **(a)** Explain three methods that can be used for cutting the lines of force.

    (b) What practical application is made from each of these methods?

9. Will a transformer work with direct current in the primary winding? Explain.

10. Explain three methods for increasing the magnitude of an induced voltage.

11. What are eddy currents, and how are they produced?

12. **(a)** How does laminating an iron core reduce eddy-current losses?

    (b) Does it affect the strength of the magnetic field? Explain.

13. **(a)** Why are eddy-current losses serious at radio frequencies?

    (b) What precautions are necessary when using iron cores at radio frequencies?

    (c) Name a type of core that minimizes these losses. Explain why.

18

SELF-INDUCTANCE
AND
MUTUAL INDUCTANCE

18-1 Concept of self-inductance. We have seen that if a conductor or coil is cut by a magnetic field, an induced voltage is developed in the conductor or coil. In the previous examples, the flux was produced by some source other than the coil itself. For example, the flux may be produced by a permanent magnet or by another coil forming an electromagnet. Now let us consider just one coil by itself. If this coil is connected to a supply voltage, current will flow through the coil and a magnetic field will be produced around the coil. For a brief instant, just when the line switch to the coil is closed, the current rises from zero to its full Ohm's law value. In turn, the magnetic field grows from zero to its full value and reaches out into space. During that brief instant, the magnetic field is in motion, and it cuts the turns of the coil itself! From Lenz's law, an induced voltage must be produced in the coil itself. This induced voltage is a *counter EMF* in opposition to the line voltage that produced the current in the coil. The induced voltage is a transient phenomenon; it lasts only for the brief instant that the current in the coil is rising to its full Ohm's law value. Once the current reaches this value, it becomes steady. The magnetic field is no longer in motion. The steady magnetic field no longer cuts the turns of the coil, and the induced voltage is zero. The ability of a circuit or component to develop an induced voltage when the current flowing through the circuit is changing is called *self-inductance*. A component having this property is called an *inductor*. Since this induced voltage is developed only when there is a change in current—and since it is a counter EMF opposing the change in current—inductance can also be defined as "the property of a circuit or component to oppose any change in current flow."

245

18-2 Unit of self-inductance. The amount of inductance in any circuit is really a measure of its ability to develop an induced voltage. However, since induced voltage also depends on the speed at which the flux is changing, and therefore the rate at which the current is changing, it would seem that a circuit would have any value of inductance, depending on how fast the current in the circuit is changing. To prevent this ambiguity, a definite rate of change of current must be picked as the standard for the evaluation and comparison of the inductance of various circuits or components. This current value is 1 A/s.

The unit for measuring inductance (L) is the **henry**.* A circuit or component has an inductance of one henry if it develops an induced voltage of one volt when the current is changing at the rate of one ampere per second. Since the variation of current in a circuit is seldom at a uniform rate (as implied above), the evaluation of the henry is preferably given in terms of average values. A circuit or component has an inductance of one henry if it develops an *average* induced voltage of one volt when the current is changing at an *average* rate of one ampere per second, or

$$E_{av} = L \frac{\Delta I}{\Delta t} \qquad (18\text{-}1)$$

where $\Delta I / \Delta t$ is the change in current over a time period of Δt seconds. (The symbol Δ, delta, represents a small change.)

It was noted above that the variation of current with time is not at a uniform rate. However, if we take a very small increment of time (dt),† then, for that small instant of time, the rate of change can be considered uniform. The above relation becomes

$$e = L \frac{di}{dt} \qquad (18\text{-}2)$$

where e is the *instantaneous* induced voltage due to the rate of change of current di/dt *at that instant.*

Example 18-1

The current flowing through a coil rises from 0 to 3 A in 5 s. The average induced voltage is 15 V. Find the inductance of the coil.

Solution

$$L = E \frac{\Delta t}{\Delta I} = 15\left(\frac{5}{3}\right) = 25 \text{ H}$$

Very often the amount of inductance is very low, particularly in high-frequency electronic circuits. Just as with previous units, the prefixes *milli-*, *micro-*, and *pico-* are used with the basic unit, the henry. For example, 10

---

*Named in honor of the American physicist Joseph Henry (1797–1878).

†The symbol d is used in differential calculus to represent an infinitesimal increment.

millihenries (mH) would be equal to 0.01 H; 200 microhenries (μH) would be 0.2 mH, or 0.0002 H (2×10^{-4} H).

18-3   Factors affecting inductance.   We have already explained that inductance is a measure of the ability of a circuit or component to develop an induced voltage. In Chapter 17 we also learned that the induced voltage depends on N, the number of turns in the generating coil; ϕ, the strength of the magnetic field; and S, the speed of cutting. In turn, the speed of cutting is inversely proportional to the time (t) in which the flux cuts the turns, or

$$e \propto N\phi S$$

Replacing S by time, $e \propto N\phi/t$. When studying magnetism, we learned that $\phi \propto NI/R_m$. Using this value for ϕ, we get

$$e \propto \frac{N^2 I}{R_m t}$$

But I/t is the average rate of change of current. Using the standard value of one ampere per second, this term equals unity. Now we can replace e (the induced voltage) by L, or

$$L \propto \frac{N^2}{R_m} \times 1$$

But

$$R_m = \frac{l}{\mu A}$$

Therefore,

$$L \propto \frac{N^2 \mu A}{l} \tag{18-3}$$

From this last equation, we can draw the following conclusions:

1. Inductance of a coil increases as the *square* of the number of turns. Doubling the number of turns in a coil would give four times the inductance; tripling the number of turns, nine times the inductance; and so on. In the opposite direction, halving the number of turns results in *one-quarter* the inductance. Remember this when you have a coil whose inductance is too high. Remove turns slowly, because the inductance decreases rapidly as turns are removed.

2. Inductance increases with permeability of the magnetic circuit. Where high values of inductances are needed, you will find that the coils are wound on iron cores.

3. Inductance increases with the area of the coil. Since the area of a circle is proportional to the square of the diameter, inductance of a coil increases rapidly with increase in the diameter of the coil. Most formulas for inductances will include the term d^2 in place of area.

4. Inductance decreases with the length of coil. This fact is important when selecting the size of wire and type of insulation of the wire used in winding

a coil. Both of these factors determine the *turns per inch* of winding and there-fore the length of the coil. For a given number of turns, the maximum inductance would be obtained by using a fine wire and a thin insulation, such as enamel. On the other hand, when very low values of inductances are desired, the number of turns required would be so few as to make accurate coil production critical. In this case the length of the coil is deliberately increased by *space-winding* the turns on the coil. In better constructions, the coil form is often pregrooved with the desired spacing between turns.

Many practical formulas for the calculation of inductance of coils use the term "form factor." This is the ratio of diameter to length. The first step would then be to calculate the form factor. From tables of form factors, a constant is located and used in the formula for calculating the inductance. The constant in this case would eliminate length (l) from the inductance for-mula, and reduce the area or (diameter)^2 term to diameter (d).

Numerous formulas are in existence for the calculation of various types of inductances. These formulas can be found in any handbook or in specific inductance design books. Discussion of these is beyond the scope of this text.

18-4 Types of commercial inductors. Commercial inductors could be classified in several ways. Depending on type of core, we have air-core, solid magnetic-steel core, and ceramic-core coils. Air-core coils are usually wound on some tubular insulating material as the coil form. These forms may be made of cardboard (treated), fiber, hard rubber, bakelite, steatite, isolantite, or plastic. The better the grade of the insulation, the lower the losses of the coil. Low loss is especially important at high frequencies.

In the larger sizes and for handling high amounts of power, air-core coils are often wound with bare copper wire of large diameter, or with copper tubing. These coils are sufficiently rigid to be self-supporting. No winding form is used; merely a binding-post strip and two or three grooved insulating blocks to preserve uniform spacing between turns of the coil. Smaller coils of low power rating are sometimes made self-supporting by "doping" with some cement and removing the winding form.

In general, air-core and ceramic-core coils find their major application in the radio or electronic field at high frequencies as RF (radio-frequency) chokes and transformers, and IF (intermediate-frequency) transformers. The inductance values of these coils are in the millihenry and microhenry range.

Iron-core chokes are found in two major types, laminated iron cores and powdered iron cores. As explained in Chapter 17, Induced Voltage, the iron is laminated or powdered to reduce the eddy-current losses, which at high fre-quencies would be excessive. Laminated iron cores can be of the closed-circuit or open-circuit type. The closed iron circuit has lower reluctance and is there-fore used for higher inductance values (approximately 0.25 H and higher). Inductors of this type are used mainly at power frequencies (60 Hz) and at audio frequencies (20 to 20 000 Hz). They are used as filter chokes in power supplies, in tone-control or frequency-corrective networks, as audio chokes in amplifiers, in audio and power transformers, and in magnet and relay windings.

Powdered iron cores are of the open-core type. Inductance values of this type of coil are in the millihenry range. Their main application is in the manufacture of RF or IF chokes and transformers.

Commercial inductors can also be classified as fixed or variable. Fixed inductors are designed for one specified value, and the two ends of the winding are brought out to suitable terminals. To change their value would mean redesign. Variable inductors are made in several types as shown in Fig. 18-1.

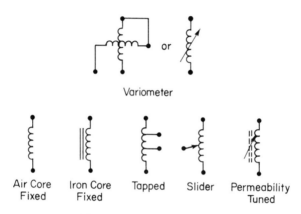

Figure 18-1   Inductor symbols.

1. *Tapped Coils.* When winding the coils, taps are brought out at several points in the winding and connected to suitable terminals. By using one fixed end of the winding and selecting any one of the taps, the inductance value can be changed in steps to suit the circuit requirements. The tap position can be changed by selector switch or plug and jack combination for convenience. Continuous variation of inductance is not possible with this method.

2. *Slider.* This type is quite similar in construction to the slidewire rheostat. When winding the coil, a strip of winding along the side of the coil is left exposed and bare of insulation. A sliding contact makes connection to the exposed section of the turns. The sliding contact and one fixed end of the winding form the two connections to the inductor.

3. *Movable Core.* In an iron-core inductor, the inductance value varies inversely with the reluctance of the circuit. A section of the core is made movable so as to introduce an air gap. As the length of air gap is increased, the reluctance value increases. This principle is used in many RF or IF coils to adjust the coil inductance to the desired value. Circuits employing these coils are often referred to as **permeability-tuned**. This type of coil should be considered as semi-variable, in that the inductance is set once to the desired value and is not intended for continual variation.

4. *Variometer*. The principle for this type of coil will be explained later in this chapter. The coil is made of two sections connected in series. One coil is made to rotate in the field of the other coil. When the angle between the two coils is changed, the inductance of the series coils changes.

Low values of inductances (for use at VHF) can be made using printed-circuit or thin-film techniques. An inductance is obtained by depositing the conducting material in a spiral. Fixed inductors (RF chokes) are also available in molded form that look quite similar to molded resistors (and similarly color-coded).

When drawing electrical circuits containing inductances, symbols are used to represent each of the types of coils listed above. These symbols are shown in Fig. 18-1.

The construction of various types of commercial coils is shown in Fig. 18-2.

18-5 Inductors in series.   Let us consider several coils connected in series but so spaced that the magnetic field of any one does not reach the others. If they are connected to a source of supply, the same value of current will flow through each. A magnetic field builds up around each coil. For a brief instant when the circuit is closed or opened, an induced voltage is developed in each. But these induced voltages are in series, and, as in any series circuits, these voltages add up, or: $e_t = e_1 + e_2 + e_3$, and so on. The magnitude of each induced voltage is directly proportional to the inductance of the respective coils; that is, $e_1 \propto L_1, e_2 \propto L_2, e_3 \propto L_3$, and so on. Similarly, the total induced voltage must be porportional to the total inductance. Replacing each of the induced voltages by their respective inductance values, we get

$$L_T = L_1 + L_2 + L_3 + \cdots \tag{18-4}$$

In words, when inductors are connected in series, the total inductance is equal to the sum of the individual inductances.

18-6 Inductors in parallel.   Inductance in parallel is a little harder to understand than the series coil connection. The stumbling block would be if you forgot how we evaluated the inductance of a coil. Remember, inductance of any circuit is a measure of the voltage induced in the circuit when the current is changing at the *standard rate of one ampere per second*. Let us consider several coils connected in parallel and again so placed that the magnetic field of any one coil does not reach any other. The coils are connected to a source of supply. When the switch is closed, let us assume, for simplicity, that the *total* current rises from zero to one ampere in one second. The current in each coil will also rise from zero to full value in one second. But since the coils are in parallel and the total current rises to one ampere, the current in any one coil is less than one ampere. (The *sum* of the currents in each coil is one ampere, as in resistors in parallel.) The rise in current in each coil in the given one second is from zero to some value less than one ampere. Therefore the rate

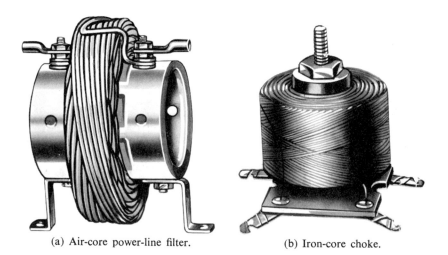

(a) Air-core power-line filter.

(b) Iron-core choke.

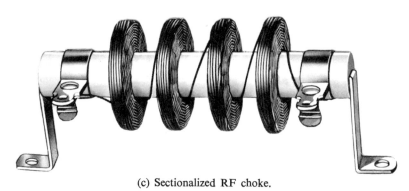

(c) Sectionalized RF choke.

(d) Space-wound shielded RF coil.

(e) Bank-wound (Litz wire) shielded RF coil.

(f) Permeability tuned RF coil.

Courtesy of J. W. Miller Co.

Figure 18-2 Commercial inductors.

Courtesy of E. F. Johnson Co.

(g) Mica-tuned
IF transformer.

(h) Adjust-
able iron core
for permea-
bility tuning.

(i) Self-support-
ing coil.

(j) Plug-in
coil.

(k) Molded thin-film chokes.

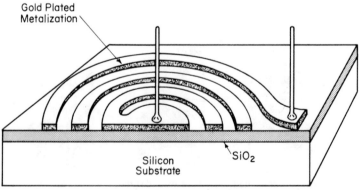

Gold Plated
Metalization

Silicon
Substrate

SiO$_2$

(l) Thin-film spiral inductor.

Courtesy of J. W. Miller Co.

Figure 18-2 (Continued)

of change of current for any one coil is lower than one ampere per second. Yet for the combination we do have the standard rate of one ampere per second! The induced voltage in any one coil is lower than if that coil were carrying the full current in the circuit by itself. The coils are in parallel; voltages in a parallel circuit do not add up. The total inductance is therefore lower than if any one coil were alone, and is less than the smallest value of inductance of any one coil in the combination. This effect, again, is similar to that of resistors in parallel, and the same type of formula applies:

$$\frac{1}{L_T} = \frac{1}{L_1} + \frac{1}{L_2} + \frac{1}{L_3} + \cdots \tag{18-5}$$

This formula is solved by the same method used for resistors in parallel.

Example 18-2

A 30-H, 20-H, and 5-H choke are connected in parallel. What is the total inductance of the combination?

Solution

$$\frac{1}{L_T} = \frac{1}{L_1} + \frac{1}{L_2} + \frac{1}{L_3}$$

$$\frac{1}{L_T} = \frac{1}{30} + \frac{1}{20} + \frac{1}{5} < 30 L_T$$

Using the method shown for solving resistors in parallel (Section 8-4), we take the product of the largest known inductor and the unknown inductor ($30L_T$) as the common numerator, and get

$$\frac{30L_T}{L_T} = \frac{30L_T}{30} + \frac{30L_T}{20} + \frac{30L_T}{5}$$

$$30 = 1L_T + 1.5L_T + 6L_T$$

$$8.5L_T = 30$$

$$L_T = 3.53 \text{ H}$$

Alternatively, using a calculator, we would enter from step 2

$$3, 0, \boxed{1/x}, \boxed{+}, 2, 0, \boxed{1/x}, \boxed{+}, 5, \boxed{1/x}, \boxed{=}, \boxed{1/x}$$

As in the case of resistors in parallel:

1. If the inductances are equal in value, the total inductance equals the inductance of any one coil divided by the number of coils in parallel.
2. If only two coils are in parallel, the following special formula can be used:

$$L_T = \frac{L_1 \times L_2}{L_1 + L_2} \tag{18-6}$$

18-7 Rise of current in an inductive circuit. If a resistor of 20 Ω were connected to a 100-V supply and the line switch were closed, the current would instantly rise to its full Ohm's law value of 5 A. A plot of current versus time would appear as in Fig. 18-3(a). An inductive circuit behaves differently. A coil, being wound with wire, will have some resistance. Let us consider an inductor having a resistance of 20 Ω and connected to a 300-V supply. The Ohm's law value of current is also 5 A. In time the current in the circuit will reach this value. But at the instant that the line switch is closed, the current *tends* to rise to this value. The rate of change of current (from zero toward 5 A) is high, and a large induced voltage is developed. But you will recall that this induced voltage is a counter EMF in opposition to the supply voltage. The *instantaneous* current that flows is*

$$i = \frac{E - e}{R} \qquad (18\text{-}7)$$

where E is the applied voltage, and e is the instantaneous value of the opposing induced EMF.

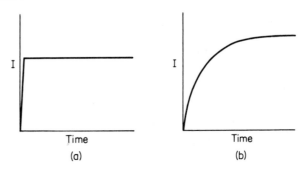

Figure 18-3   Rise of current in resistive and inductive circuits.

Since the counter EMF is large at the start, the net voltage $(E - e)$ is low, and only a small value of current will flow. Let us assume it is 0.5 A. Now the current tends to rise from 0.5 to 5 A—a slightly lower rate than before. The induced EMF drops. This means that the net voltage in the circuit $(E - e)$ is higher. The current rises to a new value. Now we will assume this new value or current is 1.0 A. The current now tends to rise from 1 to 5 A. Again this is a lower rate of change and the induced voltage decreases further, allowing more current to flow. This process continues until finally the induced voltage is zero and the current reaches a steady value, limited only by the coil resistance. Obviously, it will take time for the current to reach its steady Ohm's law value. A typical curve of current versus time for an inductive circuit is shown in Fig. 18-3(b). This is an *exponential* curve.

*Lowercase letters are used to represent instantaneous values.

18-8 Effect of *R* and *L* on rise time.   It can be proven mathematically, and shown experimentally, that the time required for the current in an inductive circuit to reach its Ohm's law value depends on the resistance and the inductance of the circuit. The greater the resistance, the quicker the current reaches its final value (an inverse proportionality); whereas the greater the inductance, the greater the time required before the current reaches its steady value. These effects are shown graphically in Fig. 18-4. A simple analysis will show that these relations are true:

1. The current at any instant was shown to be

$$i = \frac{E - e}{R}$$

Solving for *e*, we get

$$e = E - iR$$

This relationship means that the induced voltage at any instant is the difference between the supply voltage and the *iR* drop due to the coil and circuit resistance. At any given instant, a high-resistance circuit will have a greater *iR* drop and a lower induced voltage. But you just saw that it is this counter EMF that delays the rise of current. Therefore the higher the resistance, the lower the induced voltage and the less time required for the current to reach its final value.

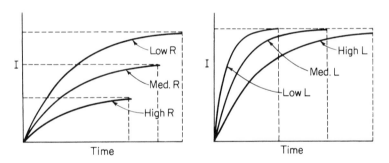

Figure 18-4 Effect of *R* and *L* on current–rise time.

2. You will remember that inductance is a measure of the ability of a circuit (or component) to develop an induced voltage. Under the same conditions, a circuit having a higher inductance will develop a higher counter EMF. At any instant, the net voltage $(E - e)$ will be lower and the current will be lower. As a result, the current rise is slower and it will take more time to reach its final value.

18-9 Time constant (τ, tau).   In the preceding section we considered the rise of current in a qualitative manner, but sometimes it is important to know exactly how long it will take for the current to reach some predetermined value. This can be done either graphically or analytically. Let us first use a mathematical solution.

Starting with equation (18-7), and recalling that the induced voltage (e) is equal to $L\, di/dt$, we get

$$i = \frac{E - L(di/dt)}{R} = \frac{E}{R}\left[1 - \frac{L(di/dt)}{E}\right]$$

Then, solving this differential equation, to eliminate the di/dt, the current at any instant becomes

$$i = \frac{E}{R}(1 - \epsilon^{-Rt/L})* \qquad\qquad (18\text{-}8)$$

Example 18-3

A solenoid has a resistance of 20 Ω and an inductance of 4 H. It is connected to a 40-V dc supply. Find the final current value through the coil, and what percent of this final current value will flow at a time equal to L/R seconds after the coil is energized.

Solution

1. Final current value $= \dfrac{E}{R} = \dfrac{40}{20} = 2.0$ A.

2. Starting with equation (18-8), and substituting L/R for t, we have

$$i = \frac{E}{R}(1 - \epsilon^{-Rt/L}) = \frac{E}{R}(1 - \epsilon^{-1})$$

and since $\epsilon^{-1} = 1/\epsilon$ and $\epsilon = 2.7183$; then $\epsilon^{-1} = 1/2.7183$. Now substituting this value of ϵ^{-1} in the equation for current (i), we get

$$i = \frac{E}{R}\left(1 - \frac{1}{2.7183}\right) = 0.632\frac{E}{R}$$

but E/R is the final current value. Therefore, at $t = L/R$ seconds,

$$i = 63.2\% \text{ of } I_{\text{final}}$$

Notice in the problem above that the last answer is independent of the circuit values (L, R or E). It is therefore a key design point for *any* $R - L$ circuit. This time interval, $t = L/R$, is known as the **time constant (τ, tau)** of the circuit, and represents the time required for the current to reach 63.2% of the final, steady-state, Ohm's law value. (t is in seconds if L is in henries and R is in ohms.)

In pulse work, two other values are important—the time required for the quantity to reach 10% and 90% of its final value. Let us try these conditions on the previous problem.

Example 18-4

In the circuit of Example 18-3, find the time required (in L/R units) for the current to reach 10% and 90% of its maximum (Ohm's law) value.

*Where ϵ, epsilon, is the base for the Naperian system of logarithms and equals 2.7183.

Solution

1. For $i = 0.1\, E/R$: Rewrite equation (18-8) to solve for ϵ,

$$\epsilon^{-Rt/L} = 1 - i\frac{R}{E}$$

Substitute the given value for i:

$$\epsilon^{-Rt/L} = 1 - \left(0.1\frac{E}{R}\right)\frac{R}{E} = 0.9$$

[This is of the form $\epsilon^{-x} = 0.9$, and the unknown value of x (in this case Rt/L) can be found from tables of exponential functions (or Naperian logarithms) or from a calculator.] Using the table of exponential functions (page 000),

$$\frac{Rt}{L} = 0.1$$

and

$$t = 0.1\frac{L}{R} \qquad \text{seconds}$$

If your calculator has natural logarithms, we can use it to solve the above exponential equation. $\epsilon^{-Rt/L} = 0.9$ can be rewritten as a logarithmic equation:

$$\log_\epsilon 0.9 = -\frac{Rt}{L}$$

Calculator entry:

$$\boxed{\cdot}\; , \; 9 \; , \; \boxed{\text{ln}}$$

The display is: -0.105. Therefore,

$$\frac{-Rt}{L} = -0.105$$

and

$$t = 0.1\frac{L}{R} \qquad \text{seconds (rounded off)}$$

2. For $i = 0.9\, E/R$:

$$\epsilon^{-Rt/L} = 1 - \left(0.9\frac{E}{R}\right)\frac{R}{E} = 0.1$$

(a) From the tables

$$\frac{Rt}{L} = 2.3$$

and

$$t = 2.3\frac{L}{R} \qquad \text{seconds}$$

(b) Using the calculator, $\epsilon^{-Rt/L} = 0.1$ becomes $\log_\epsilon 0.1 = -Rt/L$. Calculator entry:

$$\boxed{\cdot}\; , 1, \; \boxed{\text{ln}}$$

The display is: -2.30. Therefore,

$$\frac{-Rt}{L} = -2.30$$

and

$$t = 2.3 \frac{L}{R} \quad \text{seconds}$$

18-10 Universal time-constant curve. Again, notice that the actual circuit values were not used in solving Example 18-4. The above answers therefore apply to any *R-L* circuit, that is, the current rises to 10% of its final value in $0.1\ L/R$ seconds, and to 90% in $2.3\ L/R$ seconds. Using this technique, we can get enough points to plot an accurate curve. Such a curve—a *universal time-constant curve*—is shown in Fig. 18-5. Notice from this curve that for all practical purposes the current can be considered to reach its final value after a time interval of $5\ L/R$ seconds.

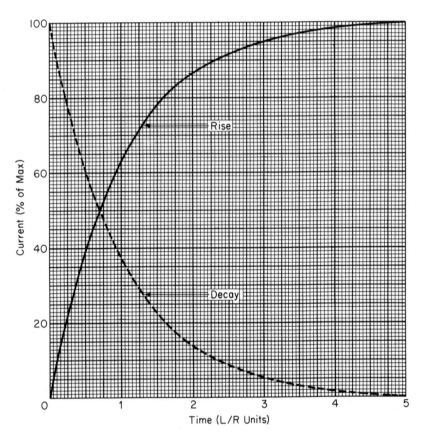

Figure 18-5 Universal time-constant curve.

Example 18-5

A coil, $L = 12$ H, $R = 50\ \Omega$, is connected to a 24-V dc supply. Find the final current through the coil and the time required for the current to reach 50% of maximum. (Use the universal curve.)

Solution

1.
$$I_{final} = \frac{E}{R} = \frac{24}{50} = 0.48 \text{ A}$$

2. For 50% current (from curve), $t = 0.7(L/R)$.

3.
$$t(\text{in seconds}) = 0.7(\tfrac{12}{50}) = 0.168 \text{ s}$$

18-11 Practical time constant.   In the above examples of current rise time we used either a previously prepared universal curve, or exponential tables. However, such aids are not always available. Based on Example 18-5, another technique can be used. Notice that the time for a 50% rise in current is 0.7 L/R. This is also the time for *any* 50% *change*. One such unit (0.7 L/R) caused the current to rise from zero to 50% of maximum. Now from this point a second such time unit (or a total of 1.4 L/R) will cause another 50% change—or from 50% to 75% I_{max} [that is, 50% + ($\frac{1}{2}$ of the remaining 50%)]. Similarly, a third practical unit (or a total of 2.1 L/R) will increase the current from 75% to 75 + ($\frac{1}{2}$ of the remaining 25) or a total of 87.5% of maximum. (Note that a time of 2.1 L/R on the universal curve gives the same answer.) The practical time constant can therefore be used to plot a universal time-constant curve.

The time constant of inductive circuits plays an important role in the design of relays, time-delay switches, and trip coils on automatic controls or protective devices. Where quick-acting circuits are desired, you can see the importance of keeping the inductance as low as possible. Sometimes, resistance is deliberately introduced in the circuit to cut down the time constant or to control the reaction time.

18-12 Decay of current in inductive circuits.   When a circuit containing resistance only is opened, the current immediately drops to zero. In an inductive circuit, when the switch is opened, the current tends to drop to zero. But any change in current produces an induced voltage that tends to keep the current flowing in the circuit. As a result the current drops slowly toward zero. A sketch of current versus time is shown in Fig. 18-6.* The equation for this curve is

$$i = I(\epsilon^{-Rt/L}) \tag{18-9}$$

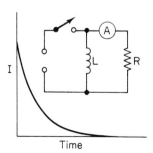

Figure 18-6 Decay of current in an inductive circuit.

*An accurate universal decay curve is shown in Fig. 18-5.

This is also an exponential curve, and can be handled in the same manner as the "rise" curve. The time constant L/R is the time (in seconds) required for the current to *drop* by 63.2% *from* its Ohm's law value. The practical time constant, $0.7L/R$, is the time required for the current to drop by 50% toward zero.

18-13 Induced voltage versus time.

While the current in an R-L series circuit is rising toward its steady-state value, the voltage relations must still follow the basic series-circuit rules. By Kirchhoff's law the total voltage is the sum of the voltage across the resistor and across the inductor (the counter EMF). *At any instant* this relation is $E_T = e_R + e_L$. In terms of current, this can also be expressed as: $E_T = iR + L(di/dt)$. But the current is zero when the circuit is first energized, so that the iR drop is zero and the counter EMF is equal and opposite to the supply voltage. With time, the current increases as per equation (18-8). The voltage across the resistor (iR) must increase in like manner, or

$$e_r = iR = E_T(1 - \epsilon^{-Rt/L}) \tag{18-10}$$

Obviously, the voltage across the coil (counter EMF) must drop, or

$$e_L = E_T - e_R = E_T(\epsilon^{-Rt/L}) \tag{18-11}$$

The drop in induced voltage is also an exponential curve, similar in shape to the current-decay curve. The time constants L/R and $0.7L/R$ can be applied in the same manner as for the current-decay curve; that is, in any time unit of $0.7L/R$, the induced voltage will drop from its previous value by 50%, toward zero.

When an inductive circuit is opened, the immediate value of the induced voltage may be very high, even greater than the supply voltage. Offhand, such value may sound impossible to you; but just think. At the instant the switch is opened, the current tends to drop from its full value to zero. This is the highest rate of change of current, and the induced EMF depending on the value of R (Fig. 18-6), can approach infinity. If the inductance value is high, the induced voltage can reach extremely high values. As the current in the circuit decreases, the rate of change of current is also decreasing (see Fig. 18-5). The induced voltage also decreases toward zero. This curve is identical in shape to the induced voltage curve for the rise of current. The same method can be used to plot this curve.

18-14 Energy stored in magnetic field.

The electrical energy consumed by any circuit in any period of time is equal to the product of the voltage × current × time. When an inductive circuit is first energized, the current through the coil, and the voltage across the coil (the induced voltage), vary with time. Consequently, the power, or rate at which the energy is stored will also vary with time. If we consider a very small interval of time (dt), the energy stored during this infinitesimal time period would be

$$w = ei \, dt$$

Now, if we add up all such time intervals from zero time to infinite time,* we will get the total energy stored in the magnetic field. This summation process (calculus integration) is represented mathematically by

$$W = \int_0^\infty ei \, dt \tag{18-12}$$

Replacing e and i by their respective values from equations (18-8) and (18-11), we get

$$W = \int_0^\infty \left[E(\epsilon^{-Rt/L}) \times \frac{E}{R}(1 - \epsilon^{-Rt/L}) \right] dt$$

This simplifies to

$$W = \frac{E^2}{R} \int_0^\infty (\epsilon^{-Rt/L} - \epsilon^{-2(Rt/L)}) \, dt$$

Integrating, we get

$$W = \frac{E^2}{R} \left[-\frac{L}{R} \epsilon^{-Rt/L} + \frac{L}{2R} \epsilon^{-2(Rt/L)} \right]_0^\infty$$

and

$$W = \frac{E^2}{R} \left[\frac{L}{R} - \frac{L}{2R} \right] = \frac{1}{2} L \frac{E^2}{R}$$

$$W = \tfrac{1}{2}LI^2 \tag{18-13}$$

where I is the steady-state value.

Example 18-6

Find the energy stored in the magnetic field of a solenoid having an inductance of 50 mH and a resistance of 1.6 Ω, when it is connected across a 20-V supply.

Solution

1.
$$I_{\text{final}} = \frac{E}{R} = \frac{20}{1.6} = 12.5 \text{ A}$$

2.
$$W = \tfrac{1}{2}LI^2 = \frac{(50 \times 10^{-3})(12.5)^2}{2} = 3.91 \text{ J}$$

When an inductive circuit is opened, the magnetic field collapses. The energy of this magnetic field must be expended in some manner. At times this energy is put to useful work; at other times this energy can be destructive and even dangerous. The existence of this stored energy can be shown by a simple experiment. A battery, a lamp, and a high inductance are connected as shown in Fig. 18-7.

The lamp should be of higher voltage rating than the battery, so that when the switch is closed, the lamp lights dimly. When the switch is opened, no more current is supplied by the battery. Yet the lamp lights brilliantly for

---

*In an exponential curve, the current theoretically does not reach its steady-state value until infinity.

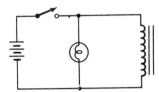

Figure 18-7 Energy stored in an inductive circuit.

a short while! The energy that lights the lamp is supplied by the magnetic field as it collapses. If the circuit is broken quickly, the induced voltage can be very much higher than the battery voltage. Therefore, the lamp burns brighter.

Opening of highly inductive circuits can be destructive to the equipment and dangerous to the operator. For example, the field circuits of large machines and transformers are highly inductive and carry appreciable currents. If these circuits are opened quickly, the induced voltage in the winding can be extremely high—high enough to break down the insulation of the winding, and to break down (ionize) the air at the switch so that a spark jumps across the open switch. Because of the high energy ($\frac{1}{2}LI^2$) stored in the magnetic circuit, the spark will be of sufficient intensity to burn the switch contacts and the hand of the operator. If a voltmeter is connected across an inductive circuit, it should be connected to the supply side of the switch and not directly across the inductive circuit; otherwise, the high CEMF may burn out the instrument.

To prevent such detrimental effects, a discharge resistor is often added in highly inductive circuits, in such a way that when the supply voltage is disconnected, the resistor is shunted across the field. This resistor closes the inductive circuit and prevents the quick decay of current in the inductance, thus limiting the value of the induced voltage. The energy of the magnetic circuit is dissipated gradually as heat in the resistor. Such a connection is shown in Fig. 18-8.

The ignition system used on gasoline engines makes good use of this inductive effect. Energy from a 6-, 12-, or 24-V battery is stored in the ignition coil over a comparatively long period of time. The circuit is then interrupted quickly. The induced voltage developed (approximately 20 000 V) is supplied to the spark plugs in each cylinder and the ensuing spark ignites the fuel.

18-15 Concept of mutual inductance. So far we have discussed how, if the current in a coil was changing, its own flux cut its own turns and an induced voltage was developed in its own winding. This property of a circuit was called self-inductance. Let us now consider two coils close together, their axes in line. If the first coil is connected to a source of voltage, a magnetic field builds up around the first coil. As this field expands to full strength, it cuts not only its own turns but also reaches and cuts the turns of the adjacent coil! An induced voltage is also developed in the second coil. The induced voltage in coil 2 was produced by the flux of coil 1. Now suppose we connect a low resistance across coil 2. The induced voltage will cause a current to flow

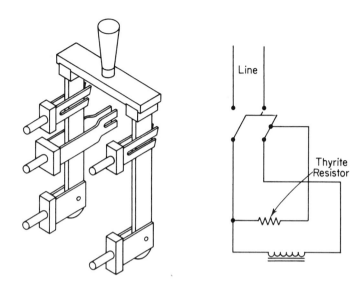

Figure 18-8   Field discharge switch and circuit.

in this second coil. But since the induced voltage is changing, the current will also be changing and the magnetic field produced by the current in coil 2 will also be changing. But this flux from coil 2 reaches and cuts the turns of coil 1. Another induced voltage is developed in coil 1 due to the flux from coil 2. In addition, this flux from coil 2 cuts its own turns and develops an induced voltage in its own winding.

To summarize: Both coils have self-inductance because they had the ability to develop an induced voltage in their own windings as a result of their own flux. In addition, the flux from each coil had the ability to develop an induced voltage in the other coil's winding. This latter effect is known as *mutual inductance* (*M*).

The unit for mutual inductance is also the henry. A circuit has a mutual inductance of one henry when a change of current of one ampere per second in one coil develops an induced voltage of one volt in the second coil.

18-16   Coefficient of coupling.   It is obvious that because of the locations of the two coils just discussed, not all the flux from one coil may reach the second coil. The portion of the flux from one coil reaching the other coil is known as the *coefficient of coupling* (*k*). For example, if *k* between two coils is 0.5, only half the flux from either coil reaches the other coil.

Theoretically, the coupling is never perfect. However, if two coils are wound on a common closed iron core, the coefficient of coupling can be considered as unity. With two air-core coils, or two coils on separate iron cores, the coefficient of coupling depends on the distance between coils and the angle between the axes of the coils. This angular relation has important applications.

When the coils are parallel (and in line), the coefficient k is a maximum. If the coils' axes are at right angles (and in line), k is a minimum. If we desire to prevent interaction between coils, we should locate them at right angles, and as far apart as possible.

Radio coils (RF and IF transformers) have coefficients of coupling of approximately 0.001 to 0.05.

18-17 Coupled coils—series aiding.

Two coils can be connected in series to a supply voltage, and the coils can be so located that there is mutual inductance between them. Also, these coils can be so connected that the flux from each coil at any instant is in the same direction (aiding). If the current in the circuit is changing (supply switch just opened or just closed), the following induced voltages will be developed:

$$e_1: \text{self-induced in coil 1 due to } L_1$$

$$e_2: \text{self-induced in coil 2 due to } L_2$$

$$e_{1,2}: \text{induced in coil 1 by } \phi_2 \text{ due to } M$$

$$e_{2,1}: \text{induced in coil 2 by } \phi_1 \text{ due to } M$$

The total induced voltage in the circuit is

$$e_t = e_1 + e_2 + e_{1,2} + e_{2,1}$$

Since inductance of a circuit is a measure of its ability to develop an induced voltage, it is obvious that the total inductance has increased. Replacing each induced voltage in the above equation by the inductive effect that produced it, we get for total inductance series aiding (L_a):

$$L_a = L_1 + L_2 + 2M \tag{18-14}$$

18-18 Coupled coils—series opposing.

In the above circuit of two coils in series, if the connections to either coil are reversed, the flux from one coil at any instant opposes the flux from the other coil. This connection is called *series opposing*. If the current to this circuit is made to change, as by opening or closing the line switch, the same four induced voltages will be developed as for the series-aiding connection. But this time, since the fluxes from the two coils are in opposite directions, the mutually induced voltages $e_{1,2}$ and $e_{2,1}$ are opposite in polarity to the self-induced voltages e_1 and e_2. The total induced voltage is now

$$e_t = e_1 + e_2 - e_{1,2} - e_{2,1}$$

Replacing each of these induced voltages by the inductive effect that produced them, the total inductance for series opposing (L_o) is

$$L_o = L_1 + L_2 - 2M \tag{18-15}$$

18-19 Calculation of mutual inductance.

The above relations for coupled coils in series can be used for calculating the value of mutual inductance between two coils. There are two simple steps to be performed:

1. Measure the inductance of L_1 plus L_2 in series aiding (L_a).
2. Reverse the connections to one of the coils, and measure the inductance of L_1 plus L_2 in series opposing (L_o).

Now using the equations for L_a and L_o above, and subtracting L_o from L_a, we get

$$L_a - L_o = L_1 + L_2 + 2M - (L_1 + L_2 - 2M) = 4M$$

$$M = \frac{L_a - L_o}{4} \qquad (18\text{-}16)$$

A legitimate question at this point is: How do we measure inductance? This problem will not be discussed in detail here because it involves an ac measurement. But briefly, a Wheatstone bridge is used. The standard resistance is replaced by a standard inductance. The unknown inductance (L_1, L_2, L_a, or L_o) takes the place of the unknown resistance (R_x). The source of supply is alternating current, and earphones or an ac instrument is used in place of the dc galvanometer. The bridge is balanced in the usual manner. Then

$$L_x = \frac{R_1}{R_2} L_s$$

Now, assuming that we can measure the necessary inductances, or that we know their values, let us apply the equation for mutual inductance to a problem.

Example 18-7

When two coupled coils are connected in series, their total inductance measures 80 mH. When the connections to one coil are reversed, the total inductance measures 20 mH. What is the value of the mutual inductance?

Solution   Obviously, the higher inductance (80 mH) must be when the coils were series aiding (L_a). Therefore,

$$M = \frac{L_a - L_o}{4} = \frac{80 - 20}{4} = 15 \text{ mH}$$

18-20   Relation between coefficient of coupling and mutual inductance.   Mutual inductance exists between two coils if variations of flux in one coil produce an induced voltage in the second coil. This means that some flux from coil 1 must reach coil 2. But coefficient of coupling is a measure of the portion of flux that reaches coil 2. Obviously, there must be a direct connection between coefficient of coupling (k) and mutual inductance (M). The greater the coefficient of coupling, the greater the mutual inductance. Let us develop this relation more rigidly. Inductance of any circuit is proportional to the induced voltage it can develop. This statement is equally true for mutual inductance, or

$$M \propto E_{2,1}$$

where $E_{2,1}$ is the voltage developed in coil 2 by the portion of flux 1 reaching coil 2 ($k\phi_1$). But induced voltages are also proportional to the number of turns in the coil, or

$$E_{2,1} \propto N_2(k\phi_1),$$

But

$$\phi_1 \propto N_1$$

Substituting N_1 for ϕ_1, we get

$$E_{2,1} \propto N_2 k N_1$$

But

$$L \propto N^2 \quad \text{or} \quad N \propto \sqrt{L}$$

Therefore,

$$N_1 N_2 \propto \sqrt{L_1 L_2}$$

Replacing $E_{2,1}$ with M, we now have

$$M = k\sqrt{L_1 L_2} \qquad (18\text{-}17)$$

This formula can be used to find M if we know the inductance of each coil and the coefficient of coupling. However, the coefficient of coupling is difficult to measure. If we calculate mutual inductance by the method explained previously, we can now use this equation to solve for k.

Example 18-8

The primary and secondary of an intermediate-frequency transformer each have an inductance of 300 μH. The mutual inductance between windings is 6 μH. What percentage of the flux from one coil reaches the other?

Solution

$$M = k\sqrt{L_1 L_2}$$

$$k = \frac{M}{\sqrt{L_1 L_2}} = \frac{6}{\sqrt{300 \times 300}} = 0.02$$

Two percent of the flux from one coil reaches the other.

18-21 Application of mutual inductance. The basis for all types of transformers is mutual inductance. It applies equally well to power transformers, audio transformers, radio-frequency transformers and inter-mediate-frequency transformers. The detailed theory and application of devices are discussed in a later volume.*†

The variometer is a direct application of mutual inductance. It is used when a continuously variable inductance is desired. As mentioned earlier in this

---

*J. J. DeFrance, *General Electronics Circuits* (2nd ed.), Holt, Rinehart and Winston, New York, 1976.

†J. J. DeFrance, *Communications Electronic Circuits*, Holt, Rinehart and Winston, New York, 1972.

chapter, it consists of two coils, one mounted on pivots inside the other coil. The two coils are connected in series. When the axes of the two coils are parallel, the coefficient of coupling between the two coils is a maximum; the mutual inductance is also a maximum. If the flux from both coils is in the same direction, the coils are in series aiding and the total inductance is $L_1 + L_2 + 2M$. When the movable coil is rotated so that they are at right angles, the coefficient of coupling is zero, and the total inductance is $L_1 + L_2$. When the coil is rotated another 90° (180° total), the coils' axes are again parallel, but this time their fluxes are opposing and the total inductance is $L_1 + L_2 - 2M$. With unity coupling and $L_1 = L_2$, the maximum variation is from $4L$ to zero. A variometer is shown in Fig. 18-9.

Courtesy of General Radio Co.

Figure 18-9   Commercial variometer.

REVIEW QUESTIONS

1. (a) In a dc circuit can the turns of a coil cut its own flux? Explain.
(b) Compared to the applied voltage, what is the result of such action?
(c) What is this property of a coil called?
(d) Name the unit used in evaluating this property.

2. What is the effect of each of the following on the inductance of a coil?
(a) Removing half of the number of turns.
(b) Space-winding a coil, using the same number of turns.
(c) Vibration introducing an air gap in a closed iron-core coil.

3. With reference to Fig. 18-1:
(a) In the upper-right symbol, what does the arrow through the "loops" signify?
(b) In the lower row, second from left, what do the vertical parallel lines signify?
(c) In the lower right, what do the dashed vertical lines signify? What does the arrow through these lines signify?

4. What does the term *permeability tuning* mean?

5. (a) Name four types of variable inductances.
   (b) If smooth continuous variation is desired, which type is preferable?
6. Explain why the inductance of two coils in parallel is less than the inductance of either coil.
7. With reference to Fig. 18-3:
   (a) What type of circuit produces the effect shown in diagram (a)?
   (b) To what type of circuit does diagram (b) apply?
8. With reference to Fig. 18-3(b):
   (a) What prevents the current from rising rapidly to its maximum value?
   (b) What determines this maximum value?
   (c) What circuit parameters determine how long it will take for the current to reach a steady value?
   (d) How does an increase in supply voltage affect this timing?
9. (a) What conclusion can be drawn from Fig. 18-4(a)?
   (b) What conclusion can be drawn from Fig. 18-4(b)?
10. (a) How is the *time constant* of an *L-R* circuit determined?
    (b) What information do we get from this time constant?
    (c) How much time is required in *any L-R* circuit for the current to reach 10% of its final value? to reach 90%? to change by 50%?
11. With reference to Fig. 18-5:
    (a) What unit is used for the current axis?
    (b) What unit is used for the horizontal axis?
    (c) Why aren't these axes plotted in terms of amperes and seconds? Explain.
12. In the circuit of Fig. 18-6, the supply voltage is 20 V, and a *line current* of 3 A (steady-state value) flows after the switch is closed.
    (a) When the switch is opened, does the *line current* immediately drop to zero?
    (b) What current is indicated by the graph?
    (c) With the switch open, what is the voltage *applied* across the coil?
    (d) Why doesn't the coil current drop to zero immediately?
    (e) In L/R seconds what can be said about this current?
    (f) In $0.7L/R$ seconds, what can be said about the counter EMF of the coil?
    (g) What can be said about this EMF in $1.4L/R$ seconds?
    (h) With respect to time constant, how long will it take for the counter EMF and current to drop to zero?
13. An *L-R* series circuit is connected across a dc supply voltage.
    (a) Does the resistor take energy from the supply?
    (b) What happens to this energy?
    (c) Does the coil take energy from the supply?
    (d) What happens to this energy?
    (e) When the circuit is opened, what happens to the energy that the resistor has taken?
    (f) What happens to the energy that the coil has taken?
14. With reference to Fig. 18-7, and a battery voltage of 45 V:
    (a) When the switch is first closed, what is the voltage across the lamp?
    (b) What is the induced voltage across the coil?

(c) After a time interval of at least $5L/R$ units, what is the voltage across the lamp? induced in the coil?

(d) When the switch is first opened, is there any voltage across the lamp? Explain.

(e) Will this voltage be 45 V, more, or less? Explain.

15. In Fig. 18-8, what is the purpose of the resistor?

16. (a) How much *mutual inductance* does a single 5-H winding have?

(b) What conditions must exist in order to have mutual inductance?

(c) With regard to mutual inductance, what does *coefficient of coupling* mean?

17. With regard to mutually coupled coils:

(a) What is the maximum possible coefficient of coupling?

(b) How can such coupling be obtained?

(c) What is a typical value of coupling between two air-core RF coils?

(d) How can the coupling between two coils be reduced to a minimum?

18. With regard to mutually coupled coils:

(a) What does *series aiding* mean?

(b) What does *series opposing* mean?

(c) If two coils are connected in a series-opposing relation, how can this be changed to series-aiding?

19. Two identical mutually coupled coils have a coupling coefficient of 1.0. They can be rotated through series aiding to series opposing. What is the total inductance range obtainable?

PROBLEMS

1. (a) An induced voltage of 1 V is developed in a coil when the current changes at the rate of 1 A/s. What is the inductance of the coil?

(b) In part (a), what would the inductance be if the induced voltage were (1) 5 V; (2) 0.02 V; (3) 0.0005 V?

2. (a) Repeat Problem 1(a), with the induced voltage 10 V and the current changing at the rate of 2 A/s.

(b) Repeat for an induced voltage of 0.2 V and a current rate of change of 5 A/s.

3. A coil of 100 turns has an inductance of 150 μH. What will its inductance be if 20 turns are removed?

4. For the coil in Problem 3, how many turns should be removed to obtain an inductance of 120 μH?

5. (a) Calculate the inductance of a 5-H, 10-H, and 15-H coil in series.

(b) Repeat for $L_1 = 1.5$ H, $L_2 = 160$ mH, and $L_3 = 0.8$ H.

6. Three coils, 5 H, 10 H, and 15 H, are connected in parallel.

(a) *By inspection*, between what maximum and minimum limits must the total inductance value lie?

(b) Calculate the actual value.

7. Repeat Problem 6(a) and (b) for $L_1 = 1.5$ H, $L_2 = 160$ mH, and $L_3 = 0.8$ H.

8. A coil has an inductance of 4 H. An inductance of 2.8 H is desired. What value inductor must be added in parallel to obtain the desired final inductance?

9. Repeat Problem 8 for an original inductance value of 15 mH and a desired value of 5.6 mH.

10. Calculate the time constant for each of the following combinations.
    (a) $L = 4$ H and $R = 20\ \Omega$
    (b) $L = 80$ mH and $R = 150\ \Omega$
    (c) $L = 20$ mH and $R = 20\ k\Omega$
    (d) $L = 160$ mH and $R = 2.2\ M\Omega$

11. A coil has an inductance of 3.5 H and a resistance of 25 Ω.
    (a) What is the time constant of this combination?
    (b) If a time constant of 0.05 s is desired using the same coil, how can this be obtained?

12. A relay having a resistance of 10 Ω and an inductance of 20 H is connected across a 50-V supply. If a current of approximately 3 A is required for the relay to operate, how long after the switch is closed will the relay close?

13. A relay for 24-V operation requires 0.25 A to operate. The final current through the relay shall not exceed 0.5 A. The relay shall not close until 2 s after it is energized. What must the R and L of the relay be?

14. Plot a curve of current versus time for a circuit having an inductance of 10 H and a resistance of 50 Ω connected to a 200-V supply. What will the current be at the end of 0.5 s? [Use equation (18-8).]

15. Repeat Problem 14 using the practical time-constant relation.

16. Two coils of 30 H each are connected in series. The mutual inductance between the coils is also 30 H. Find k, L_a, and L_o.

17. A 32-H coil is connected in series with an 8-H coil; $k = 0.8$. Find L_t:
    (a) When aiding.
    (b) When opposing.

18. Two 20-H coils are connected in series to give a total inductance of 60 H.
    (a) What is the coefficient of coupling?
    (b) What is the total inductance if one coil is reversed?

19. A 15-H coil is connected in series with an unknown inductance. The total inductance is 70 H. When one coil is reversed, the total inductance is 30 H.
    (a) What is the inductance of the unknown coil?
    (b) What is the coefficient of coupling?
    (c) What is the mutual inductance?

20. Calculate the energy that would be stored in the magnetic field of:
    (a) The relay in Problem 12.
    (b) The relay in Problem 13.
    (c) The inductor in Problem 14.

DIELECTRIC CIRCUITS
AND
CAPACITANCE

19-1 Review of charged bodies. At this point it would be wise to review Chapter 1, particularly Section 1-7, Charged Bodies—Electrostatic Fields. You will recall that a charged body may be either positive or negative, depending on whether it has a deficiency or excess of electrons. Work was done in order to obtain a charged body. The energy expended was not lost; it was transformed into an energy field surrounding the charged body. This fact was proven by a demonstration that the charged body had ability to do work.

The energy field surrounding such a charged body was called an *electrostatic* or *dielectric field*. We represented this field by lines of force where

1. The direction of the lines of force (arrows) showed the direction in which a positive test piece in the field would move.
2. The density of the lines of force indicated the strength of the energy field.

A quick review of Chapter 4 is also in order, particularly Sections 4-1, 4-2, and 4-3. In these sections we showed that a tremendous number of electrons was added or removed to produce a charged body. To avoid large numbers, we spoke of the amount of charge on a body in terms of coulombs. A charge of one coulomb would mean that 6.24×10^{18} electrons had been added or removed.

We also saw that when a neutral body was charged, it acquired a *potential* with respect to ground. The more work done to charge the body, the greater its potential. The potential could be positive or negative depending on whether electrons were removed or added. The potential of the charged body was measured in volts. For any two unequally charged bodies, not only did each have a potential with respect to ground, but there was also a *difference of potential or voltage between them*. This difference of potential was measured in volts.

19-2 The dielectric circuit. Although from tradition we have been using "conventional-current" flow, here is an example wherein analysis from electron flow gives a clearer explanation of the action. Figure 19-1 shows two parallel metallic plates separated by an air space and connected to a source of supply. The instant the switch is closed, the following phenomena take place simultaneously:

1. Electrons flow from the negative terminal of the battery to the bottom plate, *A*. Since air is an insulator, these electrons cannot go farther. They pile up on this plate giving it a negative charge.

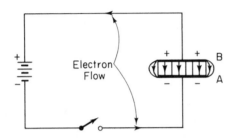

Figure 19-1 Dielectric circuit. (Note: The dielectric field shown is based on the conventional "positive test piece.")

2. The negative charge on plate *A* produces an electrostatic field around it. The direction of this energy field—using the conventional "positive test piece"—is from plate *B* downward to the negative plate *A*.
3. Electrons flow from plate *B* to the positive terminal of the battery, due to the attracting force of the battery (positive terminal) and the repelling force of the dielectric field of plate *A*. Plate *B* acquires a positive charge.
4. When plates *A* and *B* become charged, one negative and the other positive, a difference in potential builds up between them. This difference in potential is in opposition to the battery voltage. As the charge on the plates increases further, a point is soon reached where the difference in potential developed between the plates equals the battery voltage. No more electrons can flow; the circuit has reached a static state.

Let us examine this final state. Whether the switch is opened or closed makes no difference now; no current is flowing. The plates are charged, one negative and the other positive. An energy field exists between the two plates as shown in Fig. 19-1. Because the plates are parallel and of opposite charge, the energy is concentrated between the two plates instead of spreading out in all directions as shown in Chapter 1. The total strength of the field is represented by the total number of lines of force or *dielectric flux*. The Greek symbol ψ (psi) is used to represent this flux. (This idea of using lines of force or flux to represent the energy in a field was used before, in magnetic circuits.)

If, in the above illustration of two parallel plates, we were to increase the supply voltage, additional electrons would flow to plate *A*, increasing its charge. Also, more electrons would be attracted from plate *B*, leaving it with a greater positive charge. This process would continue until the difference in potential built up between the plates again equaled the applied voltage. Since the plates have a greater charge, and more electrical energy was used to charge the plates, it follows that there is now more energy in the dielectric field. In terms of flux, this means that the number of lines of force produced has increased. The flux produced is therefore directly proportional to the applied voltage:

$$\psi \propto E$$

19-3 Elastance (S).   Using Fig. 19-1 again, let us now decrease the distance between the plates. With the plates closer together, the reaction of the energy field of each plate on the other is increased. Plate *A* will repel electrons from plate *B* with greater force. In turn, plate *B* will attract electrons to plate *A* more strongly. The charge on each plate is increased, and the flux produced is increased. We have increased the flux without increasing the applied voltage! This means that flux is inversely proportional to the distance between plates, or

$$\psi \propto \frac{1}{d}$$

This relationship is readily explained by comparison with the magnetic circuit. The flux produced in a magnetic circuit depends not only on the applied magnetomotive force but also on the reluctance of the path. If we decrease the length of the magnetic circuit, we decrease the reluctance of the path, and more flux is produced. In other words, we have decreased the "opposition" of the circuit to the production of flux.

Just as the magnetic circuit has a property of opposition to the production of flux (reluctance), so does the dielectric circuit. (Remember also that in the electric circuit we have opposition to the flow of current, resistance.) In the dielectric circuit this opposition to the production of dielectric flux is called *elastance* (*S*).

Again, comparison of the three circuits will make evaluation of elastance easier. Resistance increases with the length of the conductor. Reluctance increases with the length of the path. Elastance also increases with the length of the path or with the distance between the two plates. Resistance decreases with increase in the area of the conductor. Reluctance decreases with increase in the area of the magnetic path. Similarly, elastance decreases with increase in the area of the dielectric path. In our illustration above, the elastance would decrease if the area of the plates were increased. Expressed by formula, we have

$$S \propto \frac{d}{A}$$

19-4 Dielectric constant (K).   You will recall that the resistance of a conductor depends on the type of material as well as on its dimensions. We took care of this difference by a constant, called specific resistance, for each material. In the magnetic circuit, reluctance also depended on the type of material. The constant used in this case was the permeability of the material. Elastance of any dielectric circuit also depends on the material of the path. Again, a constant is used to evaluate the effect on flux from using various materials. This constant is called *dielectric constant* (K). It is a measure of how good a material is for the production of dielectric flux.

The dielectric constant for air is taken as unity. For any other material, the constant will be more than 1, depending on how many more lines of force would be produced if the material were substituted for air as the path between the plates.

$$K = \frac{\text{flux produced with material as dielectric}}{\text{flux produced with air as dielectric}}$$

The dielectric constant of a material will vary depending on the processing method in its manufacture. Full data can be found in any electrical handbook. Some common values are given in Table 19-1.

TABLE 19-1 Dielectric Constants

Bakelite	4.5– 5.5	Paper	2.0–3.5
Ceramics		Paraffin	2.1–2.5
(Low K)	5.0–570	Plastics	2.1–4.5
(High K)	600–10 000	Porcelain	5.7–7.0
Fiber	2.5– 5.0	Rubber	2.0–3.0
Glass	4.4–10.0	Water	81
Mica	2.5– 8.7	Wood	2.5–7.7
Oil	2.2– 4.7		

If the material used has a higher dielectric constant, the elastance of the path is decreased. Combining this in the formula for elastance, we get

$$S \propto \frac{d}{A \times K} \tag{19-1}$$

In the illustration of the dielectric circuit (Fig. 19-1), the material used between the plates was air. If a slab of bakelite were inserted between the plates so as to fill the air space completely, the elastance of the path would be reduced to approximately one-fifth. Since the "opposition" of the circuit is decreased, the flux will be increased, or

$$\psi \propto \frac{1}{S}$$

Combining this relation with the previous relation between flux and EMF, we have Ohm's law for the dielectric circuit:

$$\psi = \frac{E}{S} \tag{19-2}$$

19-5 Dielectric strength. In Chapter 2, when discussing insulators, we pointed out that if a sufficiently high external force (voltage) is placed across an insulator, its electrons will be torn from their nuclei, and that this action is accompanied by mechanical breakage of the material (insulation breakdown). Then followed a table of the breakdown strength (*dielectric strength*) of some commonly used materials. The table is repeated as Table 19-2, with some additions.

TABLE 19-2   Dielectric Strengths (kV/mm)

Material	Strength
Air	3.0
Asbestos	10.8
Bakelite	10.0– 28.0
Ceramic	8.0
Cloth (varnished)	17.0
Fiber (commercial)	6.0
Glass (commercial)	20.0– 60.0
Glass (electrical)	80.0–330
Isolantite	12.6
Mica	50.0–225
Oil	12.0– 16.0
Paper (kraft, dry)	30.0– 40.0
Phenolics	19.0– 27.0
Porcelain (wet process)	5.7
Pressboard	5.0– 29.0
Rubber	16.0
Varnished cambric	32.0
Vinyl (plastic)	15.8
Wood (paraffined)	4.5

Such tables are usually given in kilovolts per millimeter. In all cases they are voltage per unit thickness. Obviously, the thickness of the dielectric determines its breakdown potential. The thicker the dielectric, the greater the voltage that can safely be applied across the unit.

For example, fiber is rated at 6.0 kV/mm. If we were to apply 15.0 kV across a piece of fiber 1 mm thick, it would break down electrically and also mechanically, and cause a short circuit. A piece 2 mm thick can withstand 6.0 kV across each millimeter of thickness, or 12.0 kV total. (A piece 2 mm thick can be considered as two pieces, each 1 mm thick, in series.) But this piece will also break down. On the other hand, a piece 3 mm thick will withstand the applied voltage with a slight margin to spare.

You may have noticed that all the dielectrics listed are also good insulators. All dielectrics must be good insulators. The better the insulating qualities, the greater the dielectric strength. However, the dielectric *strength* must not be confused with dielectric *constant*. There is no relation between the two terms.

Dielectric constant is a measure of the effectiveness of a material to produce dielectric flux (similar to permeability in a magnetic circuit). A material may have a high dielectric constant and yet not be able to withstand high voltages, because of low dielectric strength.

19-6 Relation between flux and charge.   In any given dielectric circuit, we have shown that the flux is directly proportional to the difference in potential between the two plates. This difference in potential is due to the excess or deficiency of electrons on opposite plates. But the number of electrons added or removed from either plate is the charge on the plate (Q), measured in coulombs. So we have a three-way relationship: Flux is proportional to voltage, voltage is proportional to charge, and flux is proportional to charge,

$$\psi \propto E \propto Q$$

19-7 Capacitance (C).   Elastance has been defined as the opposition of a circuit (or component) to produce dielectric flux. In practice, it is more convenient to speak in terms of the ability of a circuit (or component) to produce flux. Since flux is proportional to charge, and since it is far easier to measure charges accurately, we can replace flux with charge. The property of a circuit or component to hold a charge is called *capacitance* (C). Also, since charge is proportional to voltage, capacitance can also be defined as *the property or ability of a circuit or component to resist or oppose any change in voltage.*

From this brief discussion, it should be obvious that capacitance is just the opposite of elastance, or

$$C = \frac{1}{S}$$

Earlier in this chapter we developed the formula for elastance. For capacitance, invert this formula, or

$$C \propto \frac{AK}{d}$$

The complete formula is

$$C = \frac{8.85AK}{10^{12}d} \tag{19-3}$$

where K is the dielectric constant, A is in square meters, d is in meters, and C is in farads (this unit will be explained in the next section).

Expressed verbally:

1. Capacitance increases if the area of the plates (A) is increased.
2. Capacitance increases if the dielectric constant (K) is increased (by changing the material between the plates).
3. Capacitance decreases if the spacing between plates (d) is increased.

Example 19-1

A capacitor has plates of 2 cm $\times$ 2 cm, separated by a paper dielectric ($K = 3$) 0.1 mm thick. Calculate its capacitance.

Solution

1.   Plate area $= (2 \times 10^{-2})(2 \times 10^{-2}) = 4 \times 10^{-4} \, \text{m}^2$

2.   Distance between plates $= 0.1 \, \text{mm} = 0.1 \times 10^{-3} \, \text{m}$

3. $C = \dfrac{8.85AK}{10^{12}d} = \dfrac{8.85 \times 4 \times 10^{-4} \times 3}{10^{12} \times 0.1 \times 10^{-3}} = 106.2 \, \text{pF}$

Capacitance exists between any two conductors separated by a dielectric. The conductors do not have to be plates; they may be wires, grids, or conductors of any shape. The dielectric may be air or any other material that is an insulator. Obviously, in any two-wire cable there will be capacitance between the two wires. The wires themselves are the conductors, and the insulation between the wires is the dielectric! Following the same reasoning, there is also capacitance between any two turns of a coil! In vacuum tubes there are metal elements separated by air; again, there is capacitance between these elements. The capacitance between the elements of a transistor is even greater, because the semiconductor material has a higher dielectric constant. Where such capacitance effects are detrimental, careful design must be used to reduce these unwanted capacitances to a minimum. One simple way would be to increase the spacing between the conductors, but this simple remedy may not always be permissible.

When the property of capacitance is desirable, components are built having large plate area, small spacing between plates, and using dielectrics with high constants. Components deliberately designed to have capacitance are known as *capacitors*.

19-8 Unit of capacitance.   Ohm's law for the dielectric circuit was developed above as

$$\psi = \frac{E}{S}$$

Since then we have shown that charge is proportional to flux and that capacitance is the reciprocal of elastance. Replacing flux and elastance by their equivalents, we get

$$Q = CE \qquad\qquad (19\text{-}4)$$

This equation can be interpreted in several ways:

1. For any given capacitor, the charge on the plates will increase if the applied voltage is increased; or—the greater the charge put on the plates, the higher the EMF developed across the plates.

2. For any given voltage (either applied or developed across the plates), the larger the capacitor, the greater the charge it will take.

3. For any given charge on the plates, the smaller the capacitor, the larger the voltage developed across the plates, or the larger the voltage that must be applied.

This formula, $Q = CE$, is the basis for evaluation of capacitance. If Q is in coulombs and E is in volts, the capacitance of the component is in *farads*.* On a dc basis we can state this in another way. A component has a capacitance of one farad if a charge of one coulomb is produced on the plates when one volt is applied across the plates. In an ac circuit, a variation of this concept gives a clearer picture. The voltage in an ac circuit is continuously changing. If E changes by one volt, in a circuit having a capacity of one farad, the charge on the plates will increase or decrease by one coulomb. If the voltage is changing at an average rate of one volt per second, then the charge must be changing at the rate of one coulomb per second. But one coulomb per second is one ampere! Therefore, *the capacitance of a component is one farad if one ampere flows through the circuit when the applied voltage changes at the rate of one volt per second.*

To manufacture a capacitor of one farad would be highly impractical— the physical size of the unit would be tremendous. For example, the capacitor in Example 19-1 has a capacitance of only 106 pF. To increase this to 1 F, the plate area would have to be increased by 10^{10}. Each plate dimension would increase from 2 cm $\times$ 2 cm to 2000 m $\times$ 2000 m! In practice, the capacitors used have much lower capacitance. Again subunits are used. The usual capacitances encountered in power and electronic work are in the order of microfarads (μF) or even picofarads (pF).

Example 19-2

A capacitor of 15 μF is connected across a 200-V supply. What is the charge on the plates?

Solution

$$Q = CE = 15 \times 10^{-6} \times 200 = 0.003 \text{ C}$$

Example 19-3

A voltage of 400 produces a charge of 50 μC on the plates of a capacitor. What is the capacitance value?

Solution

$$C = \frac{Q}{E} = \frac{50 \times 10^{-6}}{400} = 0.125 \ \mu\text{F}$$

19-9 Working voltage.   When selecting a capacitor for a particular application, it is obvious that the proper value of capacitance must be specified. In addition it is also important to specify a voltage rating. If the capacitor is to be used in a very high voltage circuit, a dielectric having high dielectric strength, such as mica, is desirable. Furthermore, the dielectric should be sufficiently thick so as to prevent breakdown. This not only raises the cost of the unit but also results in a bulkier unit. On the other hand, if the capacitor is to be used in a low-voltage circuit, the above quality of construction is not necessary, and

*Named in honor of the English physicist and chemist, Michael Faraday (1791–1867).

would be poor engineering from an economical point of view. It would be preferable to use a thin paper dielectric. For this reason, capacitors are manufactured with various grades and thicknesses of dielectric, and the maximum continuous voltage that may be applied across them is specified as the *working voltage dc* (WVDC). It is sometimes desirable to use an extra margin of safety (particularly if high surge voltages may be encountered), and select a working voltage rating that is higher than the circuit voltage.

19-10 Capacitors in parallel.   In Fig. 19-2(a) three capacitors are shown connected in parallel to a source of supply. As in any parallel circuit, the voltage across each component is the same, and is equal to the supply voltage.

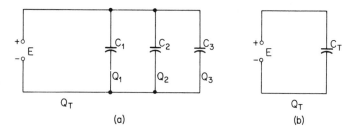

Figure 19-2 Capacitors in parallel.

Each capacitor becomes charged. Let us call these charges Q_1, Q_2, and Q_3, respectively. The total charge supplied by the source must be

$$Q_T \text{ (parallel)} = Q_1 + Q_2 + Q_3 \tag{19-5}$$

Applying the equation $Q = CE$ to each capacitor separately, we get

$$Q_1 = C_1 E$$
$$Q_2 = C_2 E$$
$$Q_3 = C_3 E$$

But from Fig. 19-2(b), one capacitor of capacitance C_T would also take a charge:

$$Q_T = C_T E$$

Substituting these CE values for each of the charges in equation (19-5) yields

$$C_T E = C_1 E + C_2 E + C_3 E$$

Since E is the same for all capacitors it can be canceled from both sides of the equation, or

$$C_T \text{ (parallel)} = C_1 + C_2 + C_3 + \cdots \tag{19-6}$$

When capacitors are connected in parallel, the total capacitance is the sum of the individual capacitances.

19-11 Capacitors in series.   When capacitors are connected in series, the elastance of each component is in series. The total elastance is increased. Therefore, the total charge that is produced on the combination is reduced to a value smaller than would have beeen produced on any one capacitor if it were alone. Since charge is proportional to capacitance, it means that the total capacitance of the combination is *less than that of the smallest capacitor.* Let us analyze this conclusion from a mathematical viewpoint.

Figure 19-3 shows three capacitors connected in series to a supply voltage. Here again, an electron flow analysis is better than a conventional-current analysis. Starting at the negative terminal of the supply, electrons flow to the right plate of capacitor C_3, charging it negatively. Let us call this charge Q. The dielectric field produced will repel electrons from the left plate of C_3 to the right plate of C_2. The left plate of C_3 becomes positively charged, the right plate of C_2 negatively charged. The number of electrons moving from the left plate of C_3 to the right plate of C_2 must be the same as originally went into the right plate of C_3. The charge on these plates must also be Q. In exactly the same way, the left plate of C_2 becomes positive and the right plate of C_1 negative. Again, the charge involved is Q. Also, Q electrons are repelled from the left plate of C_1 to the positive terminal of the voltage supply.

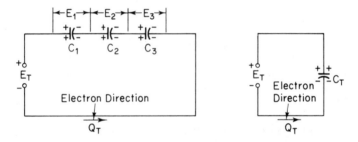

Figure 19-3   Capacitors in series.

From this discussion you can see that the charge on each capacitor is the same, and is equal to the total charge supplied from the power source:

$$Q_T \text{ (series)} = Q_1 = Q_2 = Q_3 \qquad (19\text{-}7)$$

As the three capacitors acquire a charge, a difference in potential is developed across their plates. Let us call these voltages E_1, E_2 and E_3, respectively. But in any series circuit the total voltage is equal to the sum of the voltages across each component, or

$$E_T = E_1 + E_2 + E_3$$

But we also know that the voltage developed across any capacitor when charged is

$$E = \frac{Q}{C}$$

Replacing each of the voltages by their respective Q/C values,

$$\frac{Q_T}{C_T} = \frac{Q_1}{C_1} + \frac{Q_2}{C_2} + \frac{Q_3}{C_3}$$

Since the charge is the same for all units, it cancels from both sides of the equation, and we have

$$\frac{1}{C_T}\text{(series)} = \frac{1}{C_1} + \frac{1}{C_2} + \frac{1}{C_3} + \cdots \tag{19-8}$$

(You will recognize this type of formula. The solution is the same as for resistors in parallel.)

When capacitors of *equal* capacitance are connected in series, the total capacitance is equal to the capacitance of any one capacitor divided by the number of capacitors. For any two capacitors in series, the special formula (product divided by sum) can be used. Again, this formula is the same as for resistors in parallel.

Because of the similarity between formulas for capacitors in series or parallel, and for resistors in parallel or series, respectively, no demonstration problems of such simple nature will be shown. However, series capacitors do present some interesting problems in terms of charges and voltage across each capacitor.

Example 19-4

A capacitor of 8 μF is connected in series with a capacitor of 4 μF across a 600-V supply. What is the voltage across each capacitor, and the charge on each capacitor?

Solution   The voltage across each capacitor can be found from $Q = CE$. But first we must know the charge. For a series circuit the charge on each is the same as the total charge, Q_T. To solve for Q_T we must find C_T.

1.
$$C_T = \frac{C_1 \times C_2}{C_1 + C_2} = \frac{8 \times 4}{8 + 4} = 2.67 \ \mu F$$

2.
$$Q_T = C_T \times E_T = 2.67 \times 10^{-6} \times 600$$
$$= 1.6 \times 10^{-3} \ C$$

3.
$$Q_1 = Q_2 = Q_T = 1.6 \times 10^{-3} \ C$$

4.
$$E_1 = \frac{Q_1}{C_1} = \frac{1.6 \times 10^{-3}}{8 \times 10^{-6}} = 200 \ V$$

$$E_2 = \frac{Q_2}{C_2} = \frac{1.6 \times 10^{-3}}{4 \times 10^{-6}} = 400 \ V$$

This problem brings out one important point: The smaller capacitor has a greater voltage across it. In many electronic power supplies, capacitors are connected in series when the supply voltage is greater than the voltage rating of the capacitors. In such cases either the capacitors must have the same capacitance, or else the lower-capacity unit must have a proportionally higher voltage rating. In the above example, assuming that each capacitor were rated at 350 V, at first thought it might seem that the two in series should be good for 700-V service. Yet even on 600 V, the 4-μF capacitor would break down, since it is rated at 350 V and is operating with 400 V across it.

Example 19-5

A variable capacitor of maximum capacitance 350 pF, minimum 15 pF, is connected in series with a 0.0005-μF fixed capacitor. What is the capacitance range of the combination?

Solution

1. $$0.0005 \ \mu F = 500 \ \text{pF}$$

2. $$\text{Max. } C_T = \frac{C_{max} \times C_2}{C_{max} + C_2} = \frac{350 \times 500}{850} = 206 \ \text{pF}$$

3. $$\text{Min. } C_T = \frac{C_{min} \times C_2}{C_{min} + C_2} = \frac{15 \times 500}{515} = 14.6 \ \text{pF}$$

Notice that although the maximum capacitance of the variable capacitor was reduced drastically, practically no change was made to the minimum capacitance. This method of changing the capacitance range of a variable capacitor is often used in oscillator circuits of superheterodyne receivers.

19-12 Losses in capacitors.   Losses in capacitors can be divided into three categories:

1. *Resistance Losses.* Since conductors are used to make a capacitor, and since no conductor is perfect, it is obvious that any capacitor must have resistance. This resistance is quite low, but if a capacitor is charged and discharged often, as happens on high-frequency ac circuits, this loss cannot be neglected. Also, at high frequency, resistance of conductors can be much higher than the dc Ohm's law value (see Effective Resistance, Section 28-2). At any rate, the resistance of the conductors cannot be neglected as a contributing factor to the losses of a capacitor. This is a series resistance effect.

2. *Dielectric Leakage Losses.* Another source of losses is leakage through the dielectric. No dielectric is a perfect insulator, and no matter how high the insulation resistance, some small current will flow through the material when high voltage is applied across the plates of the capacitor. This leakage current multiplied by the applied voltage represents a distinct power loss. Insulation resistance is a shunt resistance effect.

3. *Dielectric Hysteresis.* A third loss is dielectric hysteresis. When a capacitor is connected to an ac supply, the electrostatic field between the plates reverses its polarity with every alternation of the supply voltage. Each time the field reverses, the electrons in the dielectric material are strained first in one direction and then in the opposite direction. Energy is used in displacing these electrons from their normal atomic paths. This loss is quite similar to hysteresis loss in magnetic circuits. As in magnetic circuits, this loss becomes quite appreciable at high frequencies. Dielectric materials for capacitors used on high frequencies must be carefully chosen for low losses.

The effect of all these losses will cause a capacitor to heat up. Where sufficient power is available, these losses can be tolerated as long as the heating is not excessive. But in electronics, when dealing with weak RF signals, these losses must be reduced to a minimum. A measure of the quality of a capacitor (how low its losses are compared to its capacitive effect) is given by the term *dissipation factor* (*DF*), and is the ratio of *capacitive reactance** to *effective shunt resistance*. For an ideal capacitor the losses are zero, the effective shunt resistance is infinite, and the dissipation factor is zero. The dissipation factor for commercial capacitors may range from 0.025 to 0.0001.

19-13 Types of commercial capacitors. Capacitors may be classified as fixed, adjustable, and variable. Let us consider each in detail.

1. *Fixed Capacitors.* In fixed capacitors the capacitance value is fixed in the manufacture of the unit and cannot be changed. The majority of capacitors used in industry are of this type. Many types of construction are found, depending on the capacitance desired, voltage rating desired, and the amount of leakage permissible. Fixed capacitors are generally named after the type of dielectric used in their construction. The more common types are listed below, and are shown in Fig. 19-4.

(a) *Paper.* A paper-dielectric capacitor of tubular construction consists of aluminum foil plates and a kraft paper dielectric rolled together and impregnated with wax or resin to exclude moisture. Sometimes they are mounted in metal cases. They can be used wherever extremely low leakage is not important. Ratings of these capacitors commonly range from 0.0001 to 4.0 μF, and from 200 to 600 V. A variation of the paper capacitor is the oil-filled capacitor (*Dykanol, Hyvol*). These capacitors are generally of a higher capacitance value (1.0 μF and up), with voltage ratings from 400 to 5000. They are usually mounted in metal cases, and, especially for the higher-voltage ratings, the terminals are brought out through stand-off ceramic insulators. Their main application is in high-voltage circuits such as power supplies and transmitters.

(b) *Plastic Film.* Since the 1950s, a variety of plastic films have been used as dielectric materials for capacitors. Each type of film has some specific advantage over another. In general, use of such films produces a capacitor with lower losses and better resistance to moisture as compared to the paper dielectric. However, for a given capacitance and voltage rating, plastic-film units are either bulkier or costlier than paper types. For some applications, dual dielectric capacitors of polystyrene film and kraft paper are used, taking advantage of the merits of each dielectric to ensure reliability, although at a higher cost.

(c) *Mica.* This type of capacitor is used mainly in the RF circuits of receivers and transmitters. Mica is one of the best natural insulators known.

---

*Capacitive reactance is discussed in Section 29-3.

(a) Paper dielectric units.

Courtesy Cornell Dubilier Electronic Corp.

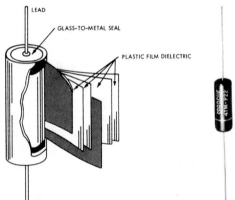

(b) Plastic film units.

Courtesy Sprague Electric Co.

(c) Mica units (molded and dipped).

Courtesy Sprague Electric Co. and Aerovox.

Figure 19-4 Commercial fixed capacitors.

(d) Electrolytic capacitors.

Courtesy Cornell Dubilier Electronic Corp.

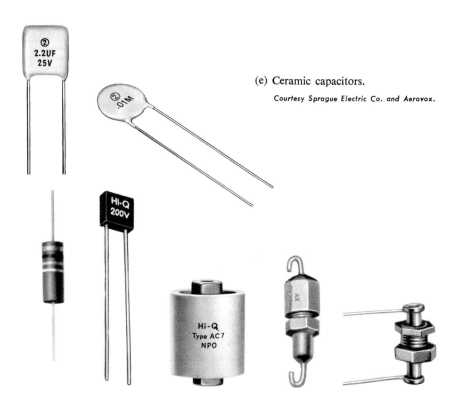

(e) Ceramic capacitors.

Courtesy Sprague Electric Co. and Aerovox.

Figure 19-4 (*Continued*)

Consequently, capacitors with mica dielectric have exceptionally low losses, and can be made with high voltage ratings. For transmitter service, mica capacitors are available with voltage ratings as high as 30 000 V and for currents in excess of 100 A at radio frequencies. These high-power units are generally encased in large ceramic cylinders. Because of high cost, mica capacitors are seldom found with capacitance values greater than 0.05 μF. A special construction—firing a thin layer of silver directly on the surface of the mica—produces capacitors with excellent stability, close tolerances, and repeatable, linear temperature-capacitance variations. These *silvered-mica* units are used in frequency-selective (tuned) circuits, and particularly for temperature (drift) compensation (TC).

(d) *Ceramic.* One broad class of ceramic dielectric materials (TC) produces capacitors with excellent stability, very high insulation resistance, and predictable, linear capacitance change with temperature. The temperature coefficient (in ppm/°C) and the dielectric constant of the ceramic are dependent on the chemical composition, and range from P120 ($+120$ ppm) at a K of 5 (approximately), to N5250 (-5250 ppm) at a K of 570. Capacitors of this type are used in frequency-selective circuits, and for drift compensation.

When temperature effects and dielectric losses are not a prime consideration, a high-K ceramic (with K values as high as 10 000) can be used to produce high capacitance values in a minimum of space—but at a sacrifice in stability as compared to the TC ceramics. Ceramic dielectrics are also suitable for high-voltage applications.

A variation of the "standard" ceramic capacitor is the *semiconductor* type, with capacitance values as much as 100 times greater than with conventional dielectrics. However, this semiconductor material cannot withstand high voltages, and these capacitors are limited to voltage ratings of 25 V (or less). Another limitation of this type is caused by low insulation resistance, which restricts their use to low-impedance circuits. They find their major application in bipolar transistor circuits, where low voltage and low impedance is often no handicap.

(e) *Glass.* The use of glass as a capacitor dielectric first began commercially in the early 1950s. Capacitors of this type are characterized by extremely low losses, excellent stability and reliability. Temperature coefficients range from NPO (0 ppm) to P140. Glass capacitors are available in capacitance values up to 0.01 μF and with voltage ratings up to 6000 V. They are used primarily in RF applications.

(f) *Electrolytic.* These capacitors are used mainly where high capacitance values are required, and leakage current is not important. They are seldom rated at higher than 450 V. Common capacitance values are from 4 to 200 μF in the higher voltage ratings, and even up to 6000 μF in the low voltage ratings. For computer applications, capacitance values as high as 200 000 μF are available.

Electrolytic capacitors may be made using either aluminum or tantalum as the base-oxidizable metal. In an aluminum unit, one plate is aluminum, which may be coiled, wound, etched, and or pleated to give greater surface area for a given total size. The other "plate" is an electrolytic solution, commonly of ammonia, boric acid, and water. The dielectric is a film of oxide formed on the aluminum plate when current passes through the unit. This film is approximately 0.0001mm thick. The metal can of this type of capacitor (or a second metal plate), makes contact with the electrolyte and forms one of the terminals. The other terminal is the aluminum plate.

These capacitors depend on chemical action within them to produce the oxide film, which acts as a dielectric. Current must flow through the unit to maintain this film. Once this film is formed, the capacitor has a high resistance, but still some current does continue to flow. Their leakage current is high compared to that of the other types previously discussed. Unfortunately, the leakage current often increases with aging of the unit. This is especially true if the capacitor is not in use, since the electrolytic film tends to disintegrate. The amount of leakage is therefore an indication of the quality of the capacitor. When using such a basis for comparison, it should be realized that a larger capacitance unit, or a higher voltage rating unit, would normally have higher leakage. Standards published for electrolytic capacitors show the acceptable leakage current limits as

Working Voltage	Leakage Current (mA)
0– 50	$0.2 + 0.01 \times C$
50–350	$0.3 + 0.02 \times C$
350–475	$0.4 + 0.02 \times C$

On this basis, a 20-μF, 300-V capacitor would be acceptable if the leakage current did not exceed $0.3 + (0.02 \times 20)$ or 0.7 mA.

In order to maintain the dielectric film, electrolytic capacitors must be connected into a circuit with proper observation of polarity. Therefore the terminals of these capacitors are marked POSITIVE and NEGATIVE. The applied voltage must be connected plus to plus and minus to minus. If this precaution is not observed, the current flowing through the unit will be opposite in direction to the current that formed the film; chemical action will be reversed; the dielectric oxide is destroyed and the unit is short-circuited.

These capacitors cannot normally be used on ac circuits. However, special nonpolarized electrolytic capacitors are available for use on alternating current. Essentially, these special capacitors consist of two normal units connected in "series reversed" so that one unit is always acting as a capacitor and prevents the other unit from being "unformed" by high reverse currents.

In some constructions, very high capacitance values are obtained with small physical size, by using cotton gauze coated with aluminum threads as the plate. This construction results in very large surface area for a given strip size.

The principal advantage of electrolytic capacitors is that they are much smaller in size for a given capacitance than any other type. For electronic work, 50 μF in a paper capacitor would be prohibitive in size. Another advantage is the lower cost of these units for a given capacitance. However, they also have drawbacks that limit their application: high leakage current and lower voltage ratings. Also, the oxide films deteriorate in time, especially when not in use, and they must be replaced comparatively often.

Tantalum oxide has a higher dielectric constant, and is also less affected by temperature as compared to aluminum oxide. Consequently, tantalum capacitors have higher capacitance ratings for a given volume, better temperature stability, and excellent shelf life. However, they are appreciably more expensive, and are available only at lower voltage ratings.

2. *Adjustable Capacitors.* Adjustable capacitors are used whenever some specific capacitance value is desired and the value is not obtainable accurately in the fixed type; or wherever it is necessary to adjust the capacitance of a circuit from time to time. In range, they are available from as low as 3 to 25 pF or as high as 600 to 1000 pF. Quite often these capacitors are used in parallel (or in series) with a fixed or variable capacitor to "trim" the circuit to some exact capacitance value. Adjustable capacitors are therefore more generally known as *trimmers* or *padding capacitors*.

Adjustable capacitors are available in two types: mica-tuned or air-tuned, depending on the dielectric used. The air-tuned type is constructed from two sets of metal plates (brass, cadmium-plated brass, or aluminum). One set of plates, spaced and connected in parallel, is mounted rigidly to an insulating framework (*stator*). The second set of plates, similarly constructed, is mounted on a shaft with bearings (*rotor*). As the rotor is turned, its plates mesh with the stator plates, increasing the total capacitance. The shaft of the rotor is cut short and is slotted for screwdriver adjustment. A locking nut fits over the shaft, to lock the setting of the plates when the correct capacity has been set.

Mica-tuned capacitors are also constructed from two sets of metal plates, but they are separated by mica dielectric. One end plate (top or bottom) is made of phosphor bronze, a springy material. By means of a screw through this plate and the insulating framework, pressure can be applied to the entire assembly, decreasing the distance between plates and increasing the capacitance. The higher-capacitance trimmers are usually mica-tuned.

3. *Variable Capacitors.* Variable capacitors are used wherever the capacitance of a circuit has to be varied often. In construction they are quite similar to the air-tuned adjustable capacitor. The shaft of these capacitors is made to extend a short distance beyond the framework, so that a dial or knob can be put on the end of the shaft to vary the capacitance. In automobile and aircraft radio, flexible shafts or electric motors may be connected to the shafts for remote control. Where the capacitances of several circuits are to be varied simultaneously, variable capacitors are made in several sections with the rotors all on the

same shaft. These units are called ganged capacitors. Figure 19-5 shows commercial units of the adjustable and variable types.

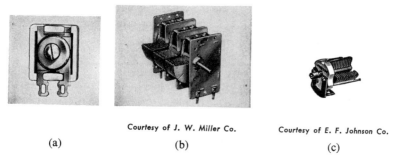

Courtesy of J. W. Miller Co.   Courtesy of E. F. Johnson Co.

(a)   (b)   (c)

Figure 19-5   (a) Mica compression trimmer capacitor. (b) Three-gang variable capacitor. (c) Air-tuned trimmer capacitor.

For use in high-voltage circuits, the spacing between the plates of a variable capacitor is increased. Sometimes the entire unit is enclosed in a sealed container and filled with oil.

19-14   Thin-film and integrated circuitry.   Fixed capacitors can also be made using the thin-film technique described in Section 6-6 for resistors. A capacitor is obtained by depositing successive layers of conducting material, dielectric material, and another conducting layer. Such a unit is shown in Fig. 19-6. Using an *active* substrate, such as silicon, and selective masking techniques, an integrated circuit containing various circuit elements can be produced. A simple electronic circuit of this type (and its equivalent schematic) was shown in Fig. 6-8. However, since large capacitance values cannot be obtained by this method, circuits are often redesigned using additional transistors, to accomplish the capacitive effect.

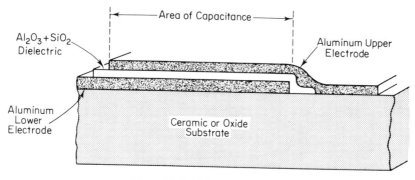

Figure 19-6   Thin-film capacitor.

19-15 Color-coding systems for capacitors. Originally, capacitor ratings were stamped directly on the unit or on a paper wrapper around the unit. But, when the color-coding scheme proved so successful for resistors, manufacturers decided to use a similar system for certain types of capacitors. The color code as used with tubular paper capacitors, mica capacitors, and ceramic capacitors is shown in Appendix 4.

19-16 Charging current—capacitive circuit. Figure 19-7 shows a capacitor and a resistor connected through a switch to a source of dc power. There is no charge on the capacitor and therefore no difference in potential between its plates. When the switch is first closed (time = 0), the only limitation to the flow of current is the resistance R. The current will immediately rise to its maximum value $I = E/R$. But just as soon as the switch is closed and current starts to flow, electrons from the negative terminal of the supply accumulate on the lower plate of the capacitor, making it negative. At the same time, electrons from the upper plate are attracted to the positive terminal of the power source, leaving this plate positive. There is now a difference in potential between the plates of the capacitor, *opposing* the applied voltage.

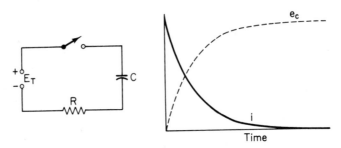

Figure 19-7   Charging current, capacitive circuit.

The net voltage in the circuit is now reduced, but a slightly lower current continues to flow, increasing the charge and the counter EMF (CEMF) of the capacitor. This process is cumulative: The charge on the capacitor increases; the CEMF of the capacitor increases; the charging current decreases. With time, the current continues to decrease. But because of the slower flow of current, the charge on the capacitor and its CEMF rise more and more slowly. Since the CEMF rises more slowly, the opposition to current flow builds up more and more slowly. Therefore, the current decreases but at a slower and slower rate. This is an exponential curve. Eventually, the CEMF of the capacitor equals the applied voltage, and the current drops to zero.

19-17 Effect of R and C on charging time. In Fig. 19-7 let us assume that the value of resistance R is increased. The result is that when the switch is first closed, a lower value of initial current ($I = E/R$) will flow. The charge on the capacitor will build up more slowly; the CEMF builds up more slowly; the current drops more gradually. The curves are still exponential in shape, but it will take more time for the CEMF of the capacitor to equal the line voltage, and for the current to drop to zero. *The charging time increases if the series resistance (R) is increased.*

Let us consider what happens if the capacitance value is increased. If the resistance R is unchanged, the initial charging current is the same as for our first illustration. The charge on the plates, Q, will also be stored at the same rate. But a larger capacitor can hold more charge. Also, from $Q = CE$, the potential difference built up across the plates is $E = Q/C$. A larger capacitor requires more charge to build up the same CEMF. More charge means more time. Since the CEMF builds up more slowly, the current drops to zero more slowly. *The greater the capacitance of a circuit, the longer the charging time will be.*

19-18   Time constant (τ)—capacitive circuit.   From the preceding section, we have seen on a qualitative basis that the charging time of a capacitive circuit increases if the resistance or the capacitance is increased. They are both direct proportions, or

$$\text{charging time } t \propto RC$$

We have also seen that the current curve is exponential, and decays (drops) with time. This is similar to the current-decay curve in the inductive circuit. Notice also that the capacitor CEMF, or voltage-rise curve is an exponential curve similar to the current-rise curve in the inductive circuit. Therefore, the time constant concept applies equally well to a capacitive circuit. The time constant, τ (tau), for a capacitive circuit is

$$\tau = RC \tag{19-9}$$

where τ is in seconds if R is in ohms and C is in farads. (A more practical twist of these units is to use R in megohms and C in microfarads.)

Now let us develop the charging relationships mathematically. In Fig. 19-7, it is obvious that at any instant, the sum of the voltage drop (iR) across the resistor, and the voltage built up across the capacitor (e_c) must equal the supply voltage (E), or

$$E = iR + e_c$$

But the voltage across a capacitor [as given by equation (19-4)] depends on the charge on its plates and on the capacitance value. In turn, the accumulated charge depends on the current flow, and on the length of time this current flows ($Q = It$). But as shown in Fig. 19-7, the current flow is not steady. Therefore, to find the charge at any instant, it is necessary to obtain the sum total of the charges produced by an infinite number of instantaneous current values (i) flowing for an infinitesimal time period (dt), over the full time period t. This is an integration process, and is expressed mathematically by

$$E = iR + \frac{1}{C} \int_o^t i \, dt$$

Using the calculus to solve this equation for i,

$$i = \frac{E}{R} \epsilon^{-t/RC} \qquad\qquad (19\text{-}10\text{A})$$

But E/R is the *initial* or maximum value of current. Therefore,

$$i = I_m \epsilon^{-t/RC} \qquad\qquad (19\text{-}10\text{B})$$

The voltage across the resistor, at any instant, is the product iR, or

$$e_R = E(\epsilon^{-t/RC}) \qquad\qquad (19\text{-}11)$$

And since $E = iR + e_C$, the voltage built up across the capacitor, at any instant, is $E - IR$, or

$$e_C = E(1 - \epsilon^{-t/RC}) \qquad\qquad (19\text{-}12)$$

Example 19-6

A resistor of 100 000 Ω, and a capacitor of 8 μF are connected in series across a 500-V supply. Find the circuit current, and the voltage across the capacitor at the end of 2 s.

Solution

1.
$$RC = 100\,000 \times 8 \times 10^{-6} = 0.8$$

2. To find current:

   (a)
   $$i = I_m \epsilon^{-t/RC} = I_m \epsilon^{-2/0.8} = I_m \epsilon^{-2.5}$$

   (b) From the table of exponential functions, $\epsilon^{-2.5} = 0.0821$. Or, using a calculator instead of tables, we change $\epsilon^{-2.5}$ to $1/\epsilon^{2.5}$, and enter

   $$2, \boxed{\,\cdot\,}, 5, \boxed{\epsilon^x}, \boxed{1/x}$$

   The display is 0.820 849.

   (c)
   $$I_m = \frac{E}{R} = \frac{500}{100\,000} = 5.00 \text{ mA}$$

   (d)
   $$i \text{ (at 2 s)} = 5.00 \times 0.0821 = 0.41 \text{ mA}$$

3. To find capacitor voltage:

   $$e_C = E(1 - \epsilon^{-t/RC})$$
   $$= 500(1 - 0.0821) = 459 \text{ V}$$

Example 19-7

Find the circuit current and the voltage across the capacitor in Fig. 19-7, in percent of maximum values at a time instant $t = RC$.

Solution

1. To find the current value,
   $$i = I_m \epsilon^{-t/RC}$$

   at $t = RC$,
   $$\frac{-t}{RC} = -1$$

and

$$i = I_m \epsilon^{-1}$$

From the table, $\epsilon^{-1} = 0.3679$, and

$$i = 36.8\% \text{ of } I_m$$

2. To find the capacitor voltage,

$$e_C = E(1 - \epsilon^{-t/RC}) = E(1 - \epsilon^{-1})$$
$$= E(1 - 0.3679) = 63.2\% \text{ of } E$$

A comparison of the above results, with the similar discussion of R-L circuits (and particularly Example 18-3, page 256), should make it obvious the the product RC determines the *time constant* (τ, tau) for an R-C circuit, and represents the time required for:

1. The current to drop *by* 63% from its maximum value.
2. The voltage across the capacitor (*CEMF*) to rise to 63% of its maximum value.

Using the logarithmic equations, (19-10) and (19-12), and expressing time in RC units, it is possible to calculate for current and capacitor voltage values (in percent of maximums) for any fraction or multiple of RC units, and then plot a *universal* time constant curve. Such a curve is shown in Fig. 19-8, and can be used with any R or C values, and any value of line voltage. Notice that these curves are identical to those shown in Fig. 18-5 for L-R circuits.

Example 19-8

Using the universal curve, and the circuit values of Example 19-6 ($R = 100\,000$ Ω, $C = 8.0\ \mu F$, $E = 500$ V), find:
(a) The time constant of the circuit.
(b) The circuit current and capacitor voltage at the end of 2.0 s.
(c) How long it will take to charge the capacitor fully.

Solution

(a) $$\tau = RC = 100\,000 \times 10^{-6} \times 8 = 0.8 \text{ s}$$

(b) $$I_m = \frac{E}{R} = \frac{500}{100\,000} = 5.00 \text{ mA}$$

$t = 2$ s corresponds to $2/0.8$ or 2.5 RC units. From the universal curve at $t = 2.5RC$, $i = 8.0\%$, and $e_C = 92\%$ of maximum.

$$i = 5 \times 0.08 = 0.4 \text{ mA}$$
$$e_C = 500 \times 0.92 = 460 \text{ V}$$

(c) Full charge requires 5 RC time units, or

$$5 \times 0.8 = 4.0 \text{ s}$$

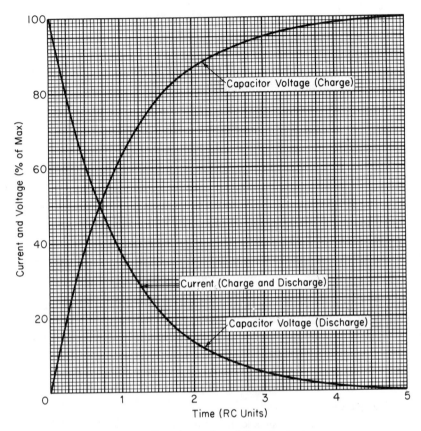

Figure 19-8 Capacitor universal time-constant curves.

In pulse work, dealing with *R-C* circuits, two time intervals of major design importance are the time required for the voltage to rise to 10% and to 90% of the supply value. These key points, as evaluated for the *R-L* circuit, apply equally well now, They are:

1. Time for 10% rise = 0.1*RC*
2. Time for 90% rise = 2.3*RC*

In Examples 19-6 and 19-8, we used the table of exponential functions and the universal time-constant curve to evaluate current and capacitor voltage. If these aids are not available, we can again use the *practical* time constant unit as developed for *L-R* circuits. This quantity is now 0.7*RC* and represents the time interval for a 50% change in current, and also in the voltage across the capacitor. An illustration will again show the application of the practical time constant.

Example 19-9

A capacitor of 1.0 μF and a resistor of 500 000 Ω are connected in series across a 1000-V supply. Tabulate data for the curve of charging current and capacitor voltage versus time.

Solution

1. Practical time constant $= 0.7RC = 0.7 \times 0.5 \times 1 = 0.35$ s

2. Initial current $= I = \dfrac{E}{R} = \dfrac{1000}{500\ 000} = 2.0$ mA

Time (in 0.7 RC units)	0	1	2	3	4	5	6	7	8	9
Time (s)	0	0.35	0.70	1.05	1.40	1.75	2.10	2.45	2.80	3.15
i (mA)	2 mA	1.00	0.50	0.25	0.125	0.063	0.032	0.016	0.008	0.004
CEMF (V)	0	500	750	875	938	969	984	992	996	998

Notice that with each successive time interval the current is reduced by 50% of its previous value, and the counter EMF across the capacitor increases by 50% of the remaining differences between line voltage and capacitor voltage.

19-19 Capacitor discharge curves.   If a resistor is connected across a *charged* capacitor, electrons will flow from the negative plate through the resistor to the positive plate, equalizing the charge on the plates. At the instant the circuit is closed, the capacitor voltage is at its maximum value. The discharge current is also at its maximum value and equal to $I = E/R$. As the charge on the plates is becoming equalized, the difference in potential between plates decreases. Naturally, the discharge current also decreases, until, when the charge is completely equalized, the voltage across the capacitor is zero and current stops flowing. At the start, the voltage is high, the discharge current is high, and the capacitor discharges rapidly. Toward the end, the voltage is low and decreasing, the current is low and decreasing, and the capacitor is discharging more and more slowly.

The discharge time depends directly on R and C. The greater the resistance across the capacitor, the lower the discharge current and the slower the discharge of the capacitor. Also, the larger the capacitor, the more charge it holds and therefore more time will be required to discharge it. Obviously, the discharge current curve is exponential, and identical to the charge curve. (However, the direction of current flow is now reversed). Mathematically, then, the equation for the discharge current curve is

$$i(\text{discharge}) = -I_m \epsilon^{-t/RC} \qquad\qquad (19\text{-}13)$$

Since the capacitor voltage is also decaying from its maximum value, it follows that

$$e_C(\text{discharge}) = E(\epsilon^{-t/RC}) \qquad\qquad (19\text{-}14)$$

The instantaneous values of current and voltage on discharge can be evaluated using any of the three methods—exponential equation, universal time-constant curve, or practical time constant—as shown above.

Again, the time constant RC (seconds) corresponds to the time required for both the discharge current and the capacitor voltage to drop by 63% of their starting (or maximum) values. The practical time constant, $0.7RC$ (seconds) is the time required for both current and voltage to drop by 50% of their previous value.

Solution of the above equations requires the use of tables of exponential functions or natural logarithms, or calculators with these features. Quite often none of these may be available. On the other hand, common logarithms, base 10, are more readily obtainable. Fortunately, we can change a natural logarithm to a common logarithm by multiplying it by the constant 2.30. Using this technique, and solving for the time required to obtain a specified voltage across a capacitor—as a base 10 logarithm:

$$\text{On charge:} \qquad t = 2.3RC \log \frac{E_T}{E_T - E_C} \qquad \textbf{(19-15A)}$$

$$\text{On discharge:} \quad t = 2.3RC \log \frac{E_T}{E_C} \qquad \textbf{(19-15B)}$$

In either case, E_T is the supply voltage and E_C is the voltage across the capacitor.

19-20 Energy stored in a capacitive circuit.

While the capacitor in Fig. 19-7 is being charged, energy is being delivered by the supply source during the entire charging period. Part of this energy is dissipated as heat in the resistor, the rest is stored in the dielectric field of the capacitor. Meanwhile, the capacitor voltage builds up to the full supply value. The energy stored in the capacitor is equal to the work done to put the charge on the capacitor, in opposition to the built up counter voltage e. For an infinitesimal time period, the work done is

$$dw = e\, dq$$

But from the capacitance relation ($Q = CE$),

$$dq = C\, de$$

and

$$dw = e\, dq = Ce\, de$$

The total work done, and the total energy stored, is then the summation of all these variable quantities, dw, as the capacitor charges to the full supply voltage E. This is an integration process, or

$$W = \int dw = C \int_0^E e\, de$$

Solving this calculus equation, the total energy stored in the capacitor is

$$W = \tfrac{1}{2}CE^2 \qquad \textbf{(19-16)}$$

19-21 Measurement of capacitance. There are a number of methods for measuring capacity. Since most of these methods employ ac supply and principles, they will only be mentioned at this time.

1. *Ohmmeter-Type Circuit.* Since the current (ac) in the circuit increases with the capacitance value, a milliameter is calibrated to register capacitance directly.

2. *Wheatstone Bridge Type.* Alternating current is used for supply and the standard resistor is replaced by a standard capacitor. The bridge is balanced and the unknown capacitor is equal to the standard capacitance value multiplied by the ratio arm.

3. *Resonant Circuit Type.* This is used mainly for small capacitors at high frequencies. Since resonance is a phenomenon encountered only in ac circuits, explanation of this method cannot be covered here.

4. *Ballistic Galvanometer Method.* This is a dc measurement, but it is employed mainly in delicate laboratory measurements. The ballistic galvanometer is an instrument of the suspension type. Its deflection is proportional to the charge sent through its moving coil. A standard capacitor is charged and then discharged through the galvanometer. The deflection, D_s, is noted. The unknown capacitor is then charged from the same supply voltage and is also discharged through the galvanometer. The deflection, D_x, is also noted. The basis for this method is the formula $Q = CE$.

$$Q_x = C_x E$$
$$Q_s = C_s E$$

Dividing one equation by the other and substituting deflectons for charges, we get

$$\frac{D_x}{D_s} = \frac{C_x}{C_s}$$

or

$$C_x = \frac{D_x}{D_s} \times C_s$$

REVIEW QUESTIONS

1. With reference to Fig. 19-1:
   (a) What do the negative signs below plate A represent?
   (b) What do the lines in between plates A and B represent?
   (c) What is the name and symbol given to this quantity?
2. (a) What is meant by the *elastance* of a dielectric circuit?
   (b) What three factors determine the elastance, and how does elastance vary with each?
3. Show the similarity between Ohm's law for the electric circuit, magnetic circuit, and the dielectric circuit.

4. Show the similarity between resistance, reluctance, and elastance.

5. (a) What is meant by the *dielectric constant* of a material?
   (b) What is meant by *dielectric strength*?
   (c) Is there any relation between the quality of an insulator and either dielectric constant or strength? Explain.

6. (a) What is meant by *capacitance*?
   (b) What is the relation between capacitance and elastance?
   (c) How is capacitance formed?

7. A capacitor has a capacitance of 2.0 μF. What will be the effect on its capacitance if:
   (a) The area of the plates is tripled?
   (b) The thickness of dielectric is increased from 2.5 mm to 10 mm?
   (c) The dielectric is changed from paraffin paper ($K = 2.5$) to mica (best grade)?

8. (a) What is the significance of the term *working voltage* of a capacitor?
   (b) What affects this factor?

9. When capacitors are connected in parallel, how is the total capacitance obtained?

10. Three capacitors (6 μF, 10 μF, and 8 μF) are connected in series. Without calculation:
    (a) What is the maximum possible value for the "total" capacitance?
    (b) What is the minimum possible value?

11. Two capacitors (a 2-μF and a 10-μF) each rated at 450 V, are connected in series across a 600-V supply. Is this a safe combination? Explain.

12. Name and describe briefly three types of losses that can occur in a capacitor.

13. Name four types of fixed capacitor and give an application for each.

14. Why are the terminals of an electrolytic capacitor marked with polarity signs ($+$ and $-$)?

15. What is meant by an adjustable capacitor? When are they used?

16. Explain how the capacitance of a variable capacitor changes with shaft rotation.

17. With reference to Fig. 19-7:
    (a) Why is the current high at $t = 0$?
    (b) Why does the current decrease with time?
    (c) What does the dashed-line curve represent?
    (d) Why does this curve start at zero at $t = 0$?
    (e) What is the final value of e_c?

18. Why does the charging time of a capacitor increase with:
    (a) Increase in the series resistance value?
    (b) Increase in the capacitance value?

19. (a) How is the time constant of a series R-C circuit evaluated?
    (b) Give the letter symbol for this quantity.
    (c) In what units is it measured?
    (d) What does this quantity signify with respect to the capacitor voltage?
    (e) What does it signify with respect to the circuit current.

20. With reference to Fig. 19-8:
    (a) On the horizontal axis, do the numbers 1, 2, etc. refer to 1s, 2s, etc.? Explain.
    (b) What is the advantage of using this type of a time axis?

(c) For any specific problem, how is the maximum (100%) current value found?

(d) From the curve, what are the time intervals corresponding to 10%, 50% and 90% changes?

(e) What are the current and voltage values corresponding to $1\tau, 2\tau, 2.5\tau$, and 5τ?

21. (a) What is the value $0.7RC$ called?

(b) Of what significance is this time interval?

(c) If the voltage across the capacitor is 60% of maximum, to what value will it rise in another time period of $0.7RC$? Explain.

(d) Continuing from the answer in part (c), to what value will e_c rise in another $0.7RC$ time period? Explain.

22. (a) How does the universal time-constant curve for an R-L circuit compare with either of the curves of Fig. 19-8?

(b) Can either of these two curves be used for the *decay* of current in an R-L circuit?

(c) Are separate time-constant curves needed for inductive and capacitive circuits?

23. (a) A charged capacitor C is being discharged across a resistor R. How can we evaluate the *discharge* time constant?

(b) Will either of the curves in Fig. 19-8 accurately show the capacitor voltage-time relation on discharge? Explain.

(c) Will either curve show the current-time relation? Explain.

24. (a) What happens to the energy delivered by the supply source while a capacitor is charging?

(b) Give the equation for evaluating the energy in a charged capacitor.

PROBLEMS

1. A capacitor of 0.1 μF is connected to a 200-V supply. How much charge will it hold?

2. What voltage must be applied to a 2-μF capacitor to store a charge of 0.1 C?

3. What size capacitor is required to store a charge of 0.5 C from a 500-V supply?

4. (a) Find the capacitance of a 0.1-, 0.25-, and a 0.5-μF capacitor in parallel.

(b) Repeat for these same units in series.

(c) By "observation" what are the minimum and maximum limits for this value?

5. A capacitor of 600 pF is connected in parallel with a 0.0004-μF capacitor across a 100-V supply.

(a) What is the total capacitance?

(b) What is the charge on each?

(c) What is the total charge?

6. The capacitors of Problem 5 are connected in series.

(a) What is the capacitance of the combination?

(b) What is the total charge?

(c) What is the charge on each?

(d) What is the voltage across each?

7. A capacitor of what capacity must be connected in series with a 0.001 μF to make the total capacitance 400 pF?

8. What is the time constant of a circuit having a capacitance of 200 pF and a resistance of 250 000 Ω?

9. In a high-fidelity receiver a time constant of 0.1 s is required in the AVC circuit. If the decoupling resistor in this circuit is 150 000 Ω, what value capacitor must be used?

10. A capacitor of 4 μF is connected in series with a 25 000-Ω resistor across a 200-V supply.
    (a) What is the time constant of this circuit?
    (b) Find the current through the circuit and the voltage across the capacitor after a time interval of τ seconds.
    (c) Find the current and voltage value after two such time intervals.
    (d) How long will it take for full charge?

11. (a) Using the circuit values of Problem 10, and the equations for i and e_C, find current and voltage values at the end of 0.2 s. How does this compare with the answer in Problem 10(c)? Explain.
    (b) Find i and e_C values using the universal time-constant curve. Compare.

12. (a) Using the circuit values of Problem 10, and the equations for i and e_C, find how long it will take for the capacitor voltage to reach 150 V.
    (b) How long will it take for the current to drop to 3.0 mA?
    (c) Repeat part (a) using the universal curve.
    (d) Repeat part (b) using the universal curve.

13. A capacitor of 12 μF is charged to 80 V, and then connected across a 300-V supply, in series with a 200 000-Ω resistor.
    (a) What is the initial charging current?
    (b) How long will it take to charge to the full 300 V?
    (c) Using the equation, find the voltage after 6 s, and how long it will take to charge to 200 V.
    (d) Repeat part (c) using the universal curve.

14. Using the circuit values of Problem 10 ($C = 4$ μF, $R = 25$ kΩ, $E = 200$ V), find the capacitor voltage after a time interval of 0.28 s:
    (a) Using the universal curve.
    (b) Using the equation.
    (c) Using the practical time constant.

15. (a) Using the practical time-constant technique, tabulate data and plot curves of E and I versus time for charging the capacitor in Fig. 19-9.

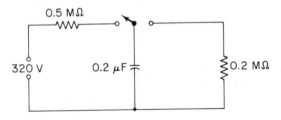

Figure 19-9

(b) What are the voltage across the capacitor, charging current, and charge in the capacitor after 0.3 s?

(c) How long will it take for full charge?

16. The capacitor in Fig. 19-9 is fully charged. Then it is discharged across a 0.2-MΩ resistor.

(a) How long will it take for full discharge?

(b) Using the practical time constant, how long will it take to discharge by 75%?

(c) What value of R would reduce the 75% discharge time to 0.050 s?

17. The capacitor in Fig. 19-9 is fully charged. Using the universal curve:

(a) How long will it take for 25% discharge?

(b) What are the current and voltage now?

18. The capacitor in Fig. 19-9 is fully charged again. Using the universal curve, find the current and voltage after:

(a) 0.1s of discharge.

(b) 0.064 s.

19. The capacitor of Fig. 19-9 is fully charged. Using the equations, find how long it will take to discharge the capacitor by 75%. [Compare your answer with that obtained in Problem 16(b).]

20. Repeat Problem 17(a) using the equations.

21. Repeat Problem 18(b) using the equations.

22. Find the energy stored in the capacitor of Problem 10 at full charge. ($C = 4 \, \mu F$, $R = 25 \, k\Omega$, $E = 200$ V.)

23. Find the energy stored in the capacitor of Problem 13 when fully charged. ($C = 12 \, \mu F$, $E = 300$ V, $R = 200 \, k\Omega$.)

20

DIRECT-CURRENT GENERATORS

20-1 Variation of speed of cutting. In an earlier chapter we saw how an induced voltage was developed by moving a conductor up and down through a magnetic field, or by moving a magnet up and down through a coil. We also noticed that the magnitude of the induced voltage depended on the speed of motion. To get high speed with reciprocating (up-and-down) motion is very difficult. The moving coil (or magnet) must be stopped dead at one end of its travel, then made to move up or down to the opposite end and stopped dead again. This abrupt reversal causes terrific strains on the moving parts. Also, the speed of motion must be slow at the ends and high at the center of its travel. Flux cutting can also be obtained by rotating the coil (or magnet) through a circular path at uniform speed. During half revolution, the coil (or magnet) is moving up. In the next half revolution it is moving down. To increase the speed, merely increase the speed of rotation. This method is used for all ac and dc machinery.

Figure 20-1 shows a conductor being rotated in a uniform magnetic field. Even though the speed of rotation is constant the speed of cutting will vary with the position of the conductor. For example, in sawing a block of wood, we cut through quickest if the blade of the saw is perpendicular to the surface of the wood. If we place the blade at an angle, more wood has to be cut, and the operation takes longer. Finally, if we place the blade parallel to the surface of the wood, we would never cut through.

The speed of cutting of our conductor varies similarly. In Fig. 20-1 we show the coil in various positions as it rotates. Let us see how the speed of cutting varies.

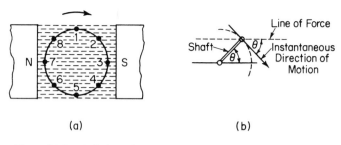

Figure 20-1 Variation of speed of cutting with conductor position.

Position 1: The conductor is moving parallel to the field; the speed of cutting is zero.

Position 2: The conductor is moving at an angle to the field; the speed of cutting is intermediate and increasing.

Position 3: The conductor is moving perpendicular to the field; speed of cutting is maximum.

Position 4: The conductor is again moving at an angle to the field; the speed of cutting is intermediate but decreasing, since the angle is decreasing.

Position 5: The conductor is now moving parallel to the field (the angle is zero); the speed of cutting is zero.

Positions 6, 7, and 8 are repetitions of 2, 3, and 4.

The speed of cutting is zero when the angle of cutting is zero (angle between direction of motion and line of force). The speed of cutting is a maximum when this angle is 90°. But the sine of the angle varies in the same manner! Therefore the speed of cutting varies as the sine of the angle of cutting.

20-2 Generation of sine-wave voltage. Instead of having one conductor rotating in the magnetic field, as above, let us use two conductors directly opposite each other and tie the back ends together to form a coil of one turn (see Fig. 20-2).

1. *Coil in Exact Vertical Position; Angle of Rotation 0°.* For a brief instant conductor A is moving parallel to the lines of force. The speed of cutting is zero; the induced voltage is zero; the potential of brush 1 is the same as brush 2.

2. *Coil in Vertical Position; Angle of Rotation 1°.* Conductor A is starting to move down through the magnetic field. By Lenz's law, or Fleming's rule, current (conventional) will begin to flow from D to C. However, remembering that electrons actually flow in the opposite direction, this makes point C positive because it loses electrons, and point D negative because it gains electrons. In the meantime, conductor B is starting to move up. The electron movement is from point E to F, making point E positive compared to point F. These two con-

ductors with their induced voltages can be considered similar to two cells. But we have connected the negative end of conductor *A* to the positive end of conductor *B*! This is similar to connecting cells in series. The total voltage for the coil is the sum of the induced voltages across conductors *A* and *B*. Point *C* is the positive terminal, point *F* the negative terminal; or the potential of point *C* is positive compared to point *F*. This induced voltage is applied to an external circuit by means of **slip rings** and **brushes**. (A slip ring is a continuous copper ring mounted on the same shaft as the coil and turning with the coil. Each end of the coil is connected to its own slip ring. Stationary carbon brushes maintain contact with the slip rings. The voltage induced in the coil will appear across these brushes.) Brush 1 becomes positive compared to brush 2.

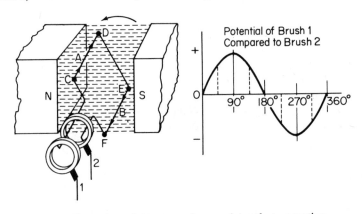

Figure 20-2   Generation of sine wave. Arrows show *electron* motion.

3. Coil in Horizontal Position; Angle of Rotation 90°. Conductor *A* is still moving down. The motion of its electrons is still toward point *D*. Brush 1 is still positive. Conductor *B* is still moving up. Its electrons are still moving toward point *F*. Brush 2 is still negative. But the speed of cutting is now at a maximum. The potential difference between brushes is at a maximum, and brush 1 is positive compared to brush 2. While the coil rotated from 0 to 90° this voltage built up from zero to a maximum. The speed of cutting varied as the sine of the angle of cutting. The rise in voltage follows a sine curve.

4. Coil in Vertical Position; Angle of Rotation 180°. Conductor *B* is now at the top and conductor *A* at the bottom. The conductors are moving parallel to the field. The speed of cutting is zero. The difference in potential between brushes is zero. As the coil rotated from 90 to 180°, the angle of cutting dropped from 90 to 0°. The speed of cutting dropped as the sine of this angle. The fall in voltage follows a sine curve.

5. Coil in Horizontal Position; Angle of Rotation 270°. The speed of cutting is now a maximum. The difference of potential between brushes is again a maximum. But conductor *B* is now moving down. The electrons in this wire will now move from *F* to *E*. This makes brush 2 positive. Conductor *A* is now moving up. The electron motion in the wire reverses. The electrons move toward point *C*. Brush 1 acquires a negative potential. The voltage across the brushes is a maximum, but brush 1 is now negative compared to brush 2. As before, the rise in voltage follows a sine curve.

6. *Coil in Vertical Position; Angle of Rotation* 360°. Conductor *A* is back on top, conductor *B* at the bottom. The speed of cutting is again zero. The induced voltage is zero. The angle of cutting has decreased from 90 to 0°. The difference in potential between brushes has dropped from maximum to zero. This drop also followed a sine curve.

We have completed one series of events, or a *cycle*. The graph representing the voltage changes is one cycle of a sine curve.

In the above simple generator we saw that the potential of brush 1 reversed from positive to negative after one half-cycle. This reversal occurred because conductor *A*, to which this brush is permanently connected (by the slip ring), reversed its direction of motion from downward to upward. Meawhile terminal *F* of conductor *B* has acquired a positive potential. At this moment (180° rotation), let us reverse our coil to slip ring connections. Terminal *F* will now be connected to brush 1. Brush 1 remains positive in potential! Terminal *C* of conductor *A* will be connected to brush 2; brush 2 will remain negative.

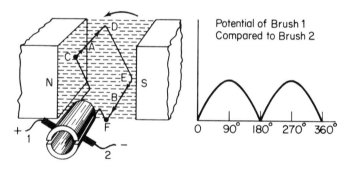

Figure 20-3   Rectification by commutator.

20-3 Commutator.   This reversal of coil terminal-to-brush connection is accomplished automatically by a *commutator*. Only one slip ring is used, instead of two separate rings. The ring is cut in half, and each half is insulated from the other. Each end of the coil is connected to one of the two segments so produced (see Fig. 20-3).

From Fig. 20-3 it is obvious that as conductor *A* is moving down, brush 1 acquires a positive potential. Conductor *B* is moving up; brush 2 is negative. This action is the same as explained previously with Fig. 20-2. For the first half revolution (180°) the voltage curve will be a sine wave as shown. As conductor *A*, which is now at the bottom, starts to move up, the induced voltage reverses. But at the same time, brush 1 makes contact with the commutator segment attached to point *F* of conductor *B*. Point *F* is now positive. Brush 1 remains positive! At the same time, brush 2 makes contact with the commutator section attached to point *C* of conductor *A*. Point *C* is now negative. Brush 2 remains negative! The voltage curve for a full rotation will be as shown in Fig. 20-3. The current flowing when this voltage is applied across a circuit will always be in the same direction. We have produced a direct current (dc). However, this direct current is not steady in value.

20-4 Multiple-coil generators.   In the above generator, let us suppose that two coils had been used. The coils are connected in series and the free ends are attached to the commutator segments. At any instant the total voltage will be the sum of the instantaneous voltages across each coil. Let us place the two coils at right angles. When one coil is moving parallel to the field, its induced voltage is zero. But at the same instant the other coil is moving perpendicular to the field, its induced voltage is a maximum. The voltage between brushes is never zero. The resultant voltage at any instant can be found by adding the voltage curves produced by each coil (Fig. 20-4). The resulting voltage is not steady, but notice that the variation (ripple) is much smaller.

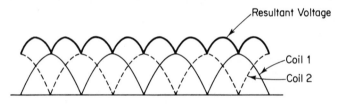

Figure 20-4   Voltage output from two coils at right angles.

In a practical generator many coils and commutator segments are used. The coils are evenly distributed so that at any instant one coil is generating maximum voltage. The *ripple* of the resultant wave is so small that it cannot be detected except by amplification.

20-5 Construction of dc generators.   To reduce the ripple and to get higher voltages in a practical machine, not only are many coils used but also more than two poles. Let us consider a four-pole generator (Fig. 20-5). In order to get strong magnetic fields, electromagnets with iron cores are used. The reluctance of the magnetic field is kept low by using generous iron paths and as small an air gap as possible.

The coils of a generator are placed in slots on the surface of a steel drum, called the **armature**. The conductors of the coil are so spaced that when one side of the coil is under a north pole, the opposite side is under a south pole. The commutator is mounted on the same shaft, at one end of this steel drum. The ends of each coil are connected to the commutator bars (Fig. 20-6). The commutator bars are insulated from each other by mica.

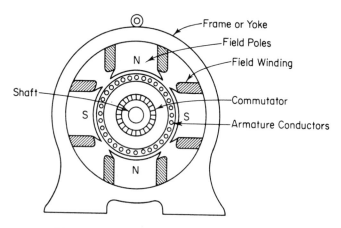

Figure 20-5 Sectional view of four-pole dynamo.

It is obvious from the foregoing that a generator must have two windings: a *field winding*, placed around the poles, to produce the magnetic field, and an *armature winding* in which the voltage is generated. In the four-pole machine shown, the four field coils are wired in series with each other to form the field winding. To generate a voltage, the armature conductors must cut the magnetic field produced by the field coils. The armature must be rotated by some outside means. The prime movers for this purpose are usually steam engines, steam turbines, water turbines, diesel engines, or gasoline engines.

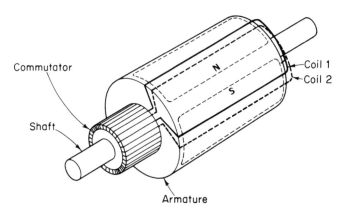

Figure 20-6 Two coils mounted on a four-pole armature.

In Chapter 17, Induced Voltage, we saw that the magnitude of the induced voltage depended upon the number of turns in the generating coil, the flux being cut, and the speed of cutting. In a complete generator, the number of turns is constant. Combining the number of turns with other constant values, we can say that

$$E = K\phi S \qquad (20\text{-}1)$$

where S is the speed of rotation. Since the generator should be driven at its rated speed, the only variable is ϕ. This flux depends on the ampere-turns (NI) of the field circuit. Since the number of turns in the field winding is constant, the flux, and therefore the voltage, depends on the field current.

20-6 Generator ratings.

The nameplate data on a dc generator specify the rated or full-load values for that machine. They include the speed at which it should be driven, the terminal voltage it will produce, and the normal maximum current that should be drawn from the armature. The rated or full-load current is limited by the allowable temperature rise. Every armature has some resistance. When current is taken from the armature, an I^2R power loss occurs in the winding. This power loss appears as heat, which must be dissipated. The temperature of the armature will rise. In addition, the I^2R loss in the field winding, the friction loss in the bearings, and the brush contact on the commutator all produce additional heat. These multiple sources of heat make it difficult for the armature to dissipate the heat developed in its windings. If excessive current is drawn from the armature, the heat produced may raise the temperature to the point where damage may occur to the insulation of the machine. The rated current of a machine is specified so as to keep the temperature rise within safe values. For short periods of time, however, it is permissible to operate the machine at higher than rated current. When a machine is to be operated intermittently, it will have a better opportunity to cool off. In this case the machine can be operated with higher current in the armature. A machine may therefore have two current and power ratings, one for continuous duty, and higher values for intermittent duty. The power rating of a generator is the product of the rated voltage and rated current.

The efficiency of a generator is the ratio of the electrical power output divided by the mechanical power input. Both powers must be expressed in the same units (horsepower or watts). Because of the losses mentioned above, the power output is always less than the power input. In general, the efficiency of larger machines is better than that of smaller units. In well-designed machines it may be higher than 90%. The maximum efficiency of any machine is obtained at approximately three-fourths of rated load. At higher load values, the I^2R loss in the armature increases faster than the output, $E \times I$. The efficiency drops. At lower values of load, the armature loss decreases, but the field power loss and the friction losses are approximately constant. The power output decreases directly with the load current. The output decreases faster than the losses, and the efficiency drops. Machines should be run as close to three-fourths load as possible for best efficiency.

20-7 Shunt generator—self-excited.

In a *shunt* generator the field winding is connected across the armature output. To control the generated voltage, a rheostat is used in series with the field winding. This type of field winding is called a **shunt field winding**.

The voltage of a self-excited generator builds up as follows:

1. The iron in the field circuit must have been previously magnetized. Because of its retentivity, it has some residual magnetism left.
2. As the armature rotates it cuts the residual flux. A small voltage (approximately 5% of rated) is generated and appears across the brushes.
3. The field is connected across the brushes. Field current flows. An additional flux, created by the field ampere-turns, adds to the residual flux.
4. The armature conductors cut a stronger field, and a higher voltage is generated.
5. This higher voltage in turn produces a stronger field current, and so on.
6. The process continues, but as the iron in the field circuit approaches saturation, the generated voltage rises more and more slowly. The voltage will stop rising at some point beyond the "knee" of the saturation curve of the iron. The exact voltage will depend on the resistance of the field circuit.

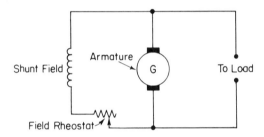

Figure 20-7   Connections for a shunt generator.

20-8   Failure to build up.   It may happen that the machine is connected as shown in Fig. 20-7 but will not build up to rated voltage. There are several possibilities for remedying this situation:

1. If the generated voltage is zero, disconnect the shunt field. If the voltage remains zero, obviously there is no residual magnetism left in the iron. To remedy this condition, connect the field winding across a battery for a few moments and remagnetize the iron. Then reconnect the machine properly. This process is called *flashing* the field.

2. Let us suppose that the generated voltage is zero, but that on disconnecting the field winding it rises to 5% of rated. Obviously, the iron still has residual flux. The current flowing in the field winding must be producing a flux that opposes or *bucks* the residual flux. The remedy is to reverse the field connections to the armature. This will reverse the direction of current flow in the field winding. The field magnetism will now aid the residual magnetism and the flux will build up.

3. The voltage generated is approximately 5% of rated but does not increase. Obviously, residual flux is present but no additional flux is being produced by the field winding. Either there is too much resistance in the field rheostat or an "open" in the field circuit. Check the rheostat setting. If resetting does not produce any change, look for the break in the circuit.

20-9 External characteristic. In a shunt generator, the armature current divides, part goes to the field winding, the remainder going to the load circuit. Since the rated current of any machine is limited, we want most of this current to be available for the load. The field current should be as small as possible. To get a strong field flux with low current, we need many turns. The field winding therefore consists of many turns of fine wire. The resistance of the winding is high; therefore, the field current will be low. Field current in a shunt machine is usually less than 5% of the rated armature current.

In order to obtain high and steady voltages, a generator armature consists of many coils in series. Each coil has many turns. Of necessity, the armature winding will offer resistance to the flow of current. In larger machines, the armature is wound with heavier wire; the armature resistance may be as low as 0.02 Ω.

When current is taken from the armature, an IR drop will result in the winding. The terminal voltage of the armature will be less than the induced voltage by an amount equal to the IR drop, or

$$V_T = E - I_a R_a$$

where I_a is the armature current, R_a the armature resistance, E the induced voltage, and V_T the terminal voltage.

As the load current is increased, the $I_a R_a$ drop increases. The terminal voltage drops as the load is increased. This effect is similar to the fall in terminal voltage of a dry cell. In addition, the drop in terminal voltage causes a slight reduction in field current, which in turn reduces the flux. In turn, this reduction in flux reduces the terminal voltage further. If the magnetic circuit of the machine is well saturated, this secondary effect is very small.

20-10 Voltage regulation. The ability of a generator to maintain constant terminal voltage is expressed in percent *regulation*:

percent regulation

$$= \frac{\text{voltage at no load} - \text{voltage at rated load}}{\text{voltage at rated load}} \times 100 \qquad \textbf{(20-2)}$$

The smaller the percent regulation, the more nearly constant the terminal voltage of the generator will remain as the load varies.

The voltage of a shunt generator can be maintained constant by adjusting the rheostat in the field circuit. This adjustment may be made manually by an operator, or automatically by a voltage-regulator relay mechanism. As the terminal voltage of the generator drops because of increased load current, the field rheostat is changed so as to increase the field current, thus increasing the flux and therefore the induced voltage. When the increase in generated voltage is equal to the $I_a R_a$ drop, the terminal voltage will remain constant.

20-11   Compound generator.   In a compound generator this increase in flux is obtained from a second field winding connected in series with the armature as shown in Fig. 20-8. This auxiliary winding is therefore called a

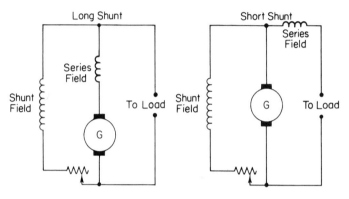

Figure 20-8   Connections for a compound generator.

series field. The current in the series field will be the same as the armature current. It is much larger than the shunt field current. Also, only a small increase in flux is desired. The series field coils are wired with a few turns of heavy wire. The coils are then placed on the field poles directly over the shunt field winding. Electrically, they are wired in series with the armature. The operation is as follows:

1. As the armature current increases, the $I_a R_a$ plus the $I_a R_s$ increases, tending to reduce the terminal voltage.
2. At the same time, this armature current flows through the series field, producing additional flux.
3. Since the flux increases, the generated voltage increases, offsetting the $I_a(R_a + R_s)$ drop.

20-12   Types of compounding.   The series field must be connected *cumulatively* so that it aids the shunt field flux. If the connections are reversed, the voltage will drop rapidly. If the series field winding has too many turns, the increase in flux and in generated voltage is greater than the voltage drops. The terminal voltage will rise as the load increases. This machine is *overcompounded*. When the series field is so designed that the increase in generated voltage is just equal to the voltage drops (*at rated load*), the terminal voltage remains substantially constant. This machine is called *flat compounded*. An **undercompounded** machine results if the effect of the series field is not enough to overcome the voltage drops. The voltage of this machine will drop as the load increases. The drop, however, will not be as great as in a shunt generator.

REVIEW QUESTIONS

1. Why is rotary motion of a generating coil preferable to reciprocating motion?
2. (a) Why does the speed of cutting vary as a conductor is rotated in a uniform magnetic field?
   (b) In Fig. 20-1, when is this speed a maximum?
   (c) When is this speed a minimum?
3. (a) In Fig. 20-2, as conductor *A* is moving down, why is brush 1 positive compared to brush 2?
   (b) Is this polarity contrary to the direction of electron flow from negative to positive? Explain.
4. Explain how a sine-wave voltage is generated in one revolution.
5. (a) What kind of current flows in the armature of a dc generator?
   (b) What is the purpose of a commutator?
   (c) Explain the action of the commutator in Fig. 20-3.
6. (a) Sketch a curve showing the voltage output of a dc generator with one coil.
   (b) Explain how the "resultant voltage" in Fig. 20-4 is obtained.
7. (a) What factors will be found on the nameplate of a generator?
   (b) What limits the speed at which the generator should be driven?
   (c) What limits the rating of a machine?
   (d) What difference would you expect to find between intermittent- and continuous-duty ratings?
8. (a) Draw a schematic diagram of a shunt generator with voltage control.
   (b) Explain the current distribution in this type of generator.
   (c) Explain how control of voltage is obtained in Fig. 20-7.
9. (a) Explain how a self-excited shunt generator builds up.
   (b) What limits the final voltage?
10. (a) Give three reasons for failure of a generator to build up.
    (b) How can the type of trouble be distinguished?
    (c) How is each corrected?
11. (a) What is meant by flashing the field?
    (b) When is this done?
12. (a) What is meant by the external characteristic of a generator?
    (b) Why does the voltage of a shunt generator drop with load?
13. (a) What is meant by percent regulation?
    (b) Should this be high or low? Explain.
    (c) Under what conditions would you get negative percent regulation?
14. (a) Why is a series field used?
    (b) Explain its action.

15. (a) What difference will be caused by short- or long-shunt connection in a compound generator?
    (b) Draw a diagram for each of these connections.

16. (a) Where are the series field windings located in a machine?
    (b) A generator has six leads coming out of its frame work. How can you distinguish between windings?

17. (a) What is the effect of reversing the series field connections of a compound generator?
    (b) What is meant by overcompounding?
    (c) What is meant by flat compounding?
    (d) What is meant by undercompounding?

21

DIRECT-CURRENT MOTORS

21-1 Principle of motor action. In Chapter 16, Direct-Current Measuring Instruments, we saw that when current was sent through the moving coil, the magnetic field produced by the coil reacted on the field of the permanent magnet, causing the coil to rotate. Without the opposition of the springs, the coil would rotate until its north and south poles were adjacent to the opposite poles of the permanent magnet. The torque developed by the magnetic reactions depended upon the flux of the permanent magnet (ϕ) and the current in the coil (I), or $T = K\phi I$, where K is a constant that includes such constant factors as reluctance of the circuit and number of turns in the coil.

The construction of a motor is identical with the construction of generators.

Let us consider the action that takes place in a two-pole machine with an armature of one coil. The shunt field is connected to a source of direct current. It will act similarly to a permanent magnet.

Now if a voltage is applied to the armature, making brush 1 negative and brush 2 positive (see Fig. 21-1), current will flow in the coil as shown. This direction of current is such that it produces a north pole at the face of the coil nearest the south pole of the permanent magnet. In the meantime a south pole is produced on the opposite face of the coil, near the north pole of the field magnet. The resulting torque will cause the coil to rotate clockwise into a vertical position. But as the coil comes up to this position, the commutator segments make contact to the opposite brushes, reversing the direction of the current flow in the coil. In turn, the magnetic polarity of our coil reverses. The south pole of the coil is now adjacent to the south pole of the permanent magnet. The two north poles are also adjacent. A torque will again be produced that will make the coil turn another 180°. But now the commutator connections are again reversed. The coil will continue to rotate.

The torque developed with one coil is not steady. It will vary from zero to a maximum, twice in each revolution. To obtain uniform torque in a practical machine, many coils at different angles, and more than two poles are used.

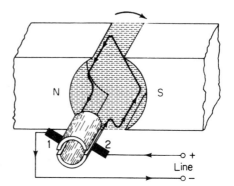

Figure 21-1 Principle of motor action. Arrows show conventional current flow.

As in the Weston movement, the torque developed by a motor depends on the strength of the magnetic field and the current in the armature:

$$T = K\phi I_a \qquad (21\text{-}1)$$

21-2 Counter electromotive force.   The armature of a motor rotates due to the torque produced by the interaction of the flux of the field and flux due to the armature current. The conductors in the armature will cut the flux produced by the field winding. But this is a condition that produces an induced voltage! Therefore a voltage is generated in the armature. Applying Lenz's law, we can see that this induced voltage is in opposition to the line voltage. It is therefore called a *counter electromotive force* (CEMF). The current that flows in the armature is therefore limited by the armature resistance and also by the CEMF. The "net" voltage in the circuit is the difference between the line voltage (V_L) and the counter EMF, or

$$I_a = \frac{V_L - \text{CEMF}}{R_a} \qquad (21\text{-}2)$$

In studying the dc generator, we saw that the generated voltage was given as $E = K\phi S$. The CEMF is also a generated voltage, or

$$\text{CEMF} = K\phi S$$

In other words, the stronger the field flux or the faster the motor turns, the greater the counter EMF.

21-3 Torque–current–speed relation.   Suppose that a motor has a high current in the armature winding. The torque developed will be strong. If the mechanical load on the motor shaft (opposing torque) is small, the motor is developing more torque than necessary, and it will speed up. This action is similar to an automobile going up a grade. When it reaches level ground, the opposing torque is reduced; if the gasoline feed is not reduced, the car speeds up. Similarly, our motor speeds up. As the speed starts to increase, the counter EMF increases—but this reduces the armature current. The torque developed decreases. The process continues until the torque developed is equal to the opposing torque.

If the mechanical load is increased, the motor tends to stall. But as the speed drops, the counter EMF drops. This allows more current to flow, increasing the developed torque.

The equations involved are

$$\text{CEMF} = K\phi S \tag{1}$$

$$I_a = \frac{V_L - \text{CEMF}}{R_a} \tag{2}$$

Solving equation (2) for CEMF, we get

$$\text{CEMF} = V_L - I_a R_a \tag{3}$$

Combining equations (1) and (3) and solving for speed, we get

$$S = \frac{V_L - I_a R_a}{K\phi} \tag{21-3}$$

21-4 Starting boxes.   At the instant the power is applied to a motor, the speed is zero and the counter EMF is zero. The only factor limiting the amount of current that flows in the armature is its own resistance. In fractional horsepower motors this resistance is sufficiently high to keep the armature current within reasonable value. But in the larger motors this resistance is extremely low. The current that would flow is enormous, sufficient to "blow" the line fuses, or burn out the feeders, or the armature winding itself. As a preventive measure, resistance must be inserted in series with the armature to limit the current to safe value while starting. This resistance is enclosed in a *starting box*. Starting boxes are designed for a particular size of machine, to limit the starting current to 150% of the rated value. The starting resistance is made variable in several steps. At the instant of starting, maximum resistance is needed. The motor gains speed and the counter EMF starts to build up; the armature current started at 150% rated and begins to drop as the counter EMF builds up. When the speed of the machine levels off, one step of the starting resistance is cut out. The motor starts to speed up again, the counter EMF builds up higher, and the current starts to drop again. When the speed of the machine levels off again, another step of the starting resistance is cut out. The process is continued until the entire starting resistance is eliminated. Commercially, starting boxes are made in automatic or manual types. In the automatic types the steps are cut out by relays, as the counter EMF builds up. In the manual types the operator moves the starting-box handle to cut out the steps as needed. Manual starting boxes used

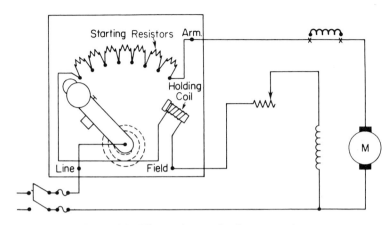

Figure 21-2   Three-point starting-box connections.

with shunt or compound motors are made in two types: 3-point box and 4-point box. In either type a holding magnet is included to keep the starting-box arm in the running position while the power is on. A spring is attached to the arm to return the arm to the OFF position when the power is turned off. Figure 21-2 shows the connections for a 3-point box; Fig. 21-3 shows the 4-point box.

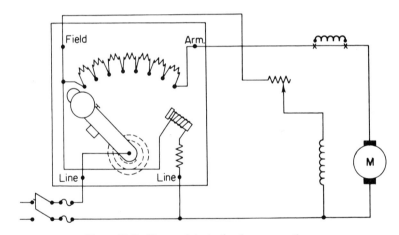

Figure 21-3   Four-point starting-box connections.

21-5 Shunt motor.   In a shunt motor the field circuit is connected across the line, in parallel with the armature. A rheostat is usually included in series with the field winding to control the flux. A schematic diagram is shown in Fig. 21-4.

The torque of a motor is given by $T = K\phi I_a$. Since the field is connected directly across the line, the flux is constant. The torque increases directly with the armature current.

The speed equation developed earlier in this chapter was

$$S = \frac{V_L - I_a R_a}{K\phi}$$

As the load increases, the armature current increases and the $I_a R_a$ drop increases. This reduces the numerator of our equation and the speed drops. In a good machine the armature resistance drop is small. The change in speed with load is also small.

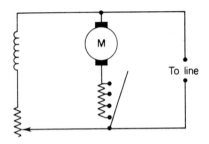

To line

Figure 21-4 Schematic diagram of shunt motor connections.

Speed regulation is a measure of the ability of a machine to maintain constant speed:

$$\text{percent regulation} = \frac{\text{speed at no load} - \text{speed at rated load}}{\text{speed at rated load}} \times 100 \qquad \textbf{(21-4)}$$

The smaller the percent regulation, the more nearly constant the speed will remain as the load varies. In a good shunt motor the speed regulation is approximately 5%.

21-6 Series motor. In a series motor, the field winding consists of relatively few turns of heavy wire. It is connected in series with the armature winding (Fig. 21-5).

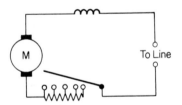

To Line

Figure 21-5 Schematic diagram of series motor connections.

Since the field is in series with the armature winding, the flux in this machine depends on the armature current. The torque therefore increases as the square of the armature current. At rated current this machine develops very strong torque.

Looking back at the speed equation, we notice that as the flux increases the speed will drop. In the series motor the flux increases directly with the armature current. The speed of this machine will drop rapidly with load. On the other hand, if the mechanical load is removed completely, the armature current, and therefore the flux, drops so low that the motor will reach dangerously high speeds. For this reason, series motors must be coupled to the load securely, Direct, gear, or chain drive should be used. Belt drive should be avoided.

21-7 Compound motor.   The compound motor has both a shunt and a series field. The schematic connection is shown in Fig. 21-6.

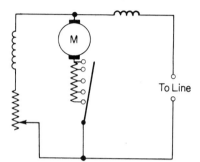

Figure 21-6 Schematic diagram of compound motor connections.

The machine could also be connected long-shunt. The characteristics of this type of motor depend on the magnetic polarity of the series field. The series field may be connected to aid the shunt field, *cumulative compound*, or to oppose the shunt field. *differential compound.*

21-8 Cumulative compound.   With the cumulative connection, the flux due to the series field increases with load. The total flux also increases with load. The torque of this motor will increase with load because of increase in armature current and also because of increase in flux. The torque increases faster than in the shunt motor. This motor has a better torque characteristic than a shunt machine. However, due to the increase in flux, the speed of this machine will drop faster than in the shunt motor. The regulation of the cumulative compound motor is poorer than the shunt motor.

21-9 Differential compound.   With the differential connection, the flux due to the series field increases with load, but it bucks the shunt field flux. The total flux will decrease as the load current increases. The torque of this motor will increase with current but at a slower rate than in the shunt motor, because of the decrease in flux. In the shunt machine, the speed tends to drop with load because of the increased $I_a R_a$ drop. In the differential motor, this tendency is offset by the decrease in flux. If the series field flux is correctly proportioned, the speed of this machine remains substantially constant as the load varies (flat compounded). If the series field flux is too strong, the speed will actually rise with load (overcompounded). On the other hand, if the series field flux is too weak, the speed will drop as the load increases, but the regulation will be better than with a shunt motor.

21-10 Methods of speed control. The formula for the speed of a motor, developed earlier in this chapter, was

$$S = \frac{V_L - I_a R_a}{K\phi}$$

The armature current adjusts itself automatically to meet the torque requirements of the mechanical load. We can, however, control the speed by varying the other three factors: ϕ, R_a and V_a.

1. *Field Resistance Control.* By decreasing the resistance of the field rheostat, the field current, and therefore the flux, increases. Examination of the speed equation shows that this decrease in resistance will decrease the speed. When the field rheostat is set at zero, the maximum flux is obtained. This flux corresponds to a speed of approximately 10% below rated. Lower speeds cannot be obtained by this method. Theoretically, by adding resistance in series with the field winding it is possible to increase the speed indefinitely. Practically, however, unless the machine is specially designed for this purpose, speeds higher than 50% above rated should not be attempted. Mechanical trouble and vicious sparking at the brush contacts will result. Speed control by this method has a normal range of from 10% below to 50% above rated speed. Since the field current is low, very little power is wasted in the field rheostat. This method is very efficient.

2. *Armature Resistance Control.* By adding resistance in series with the armature, the factor $I_a R_a$ increases. The numerator in the speed equation will decrease and the speed will drop. With this method it is impossible to obtain speeds above rated. By increasing the series resistance, the speed can be dropped down to standstill. This method, however, is very inefficient. Since the armature current is high, the power lost in the series resistance ($I^2 R$) at low speeds is tremendous. The power output of the motor decreases greatly as lower speeds are attempted.

3. *Armature Voltage Control—Ward–Leonard System.* In this method of control, the field circuit is connected to the constant voltage line as usual. The armature of the motor is connected to a variable voltage line. A separate motor-generator set is usually used for this purpose. The voltage output of the motor-generator set is varied by the field control of its generator. As the voltage across the armature is increased, the speed of the motor will rise. A decrease in the armature voltage will reduce the speed of the motor. By this method, smooth speed control from 0 to 150% of rated value can be obtained. Since there is no resistance in series with the armature, efficiency and power output of the system are good. The disadvantage of this method is that it requires more space and more initial expense for the separate motor-generator set.*

---

*In recent years, electronic power supplies are being used to supply this variable voltage.

21-11 Dynamotor.   A dynamotor is used when it is desired to transform from a fixed high-voltage direct current to a fixed lower value, or vice versa. This machine is provided with two armature windings on a common core that is arranged to rotate in a single field structure. The two armature windings are sandwiched between each other. Each winding has its own commutator and a separate set of brushes. The commutators are located one at each end of the armature.

When the field winding and the brushes of one commutator are connected to a source of supply, the armature winding connected to this commutator will operate as a motor. At the same time, the second armature is rotating in the magnetic field, and a voltage is generated in this winding. This winding will supply a voltage to the load connected to its commutator through the second set of brushes.

Both armatures must rotate at the same speed and must cut the same flux. Applying the equation for induced voltages, $E \propto N\phi S$, to both armature windings, we get:

Generated voltage in winding 2:

$$E_2 \propto N_2 \phi S$$

Counter EMF in winding 1:

$$E_1 \propto N_1 \phi S$$

Neglecting the small $I_a R_a$ drop in each winding, E_2 is the same as the output voltage, and E_1 is the same as the input voltage. The ratio of these two voltages will then be

$$\frac{\text{output voltage}}{\text{input voltage}} = \frac{N_2}{N_1}$$

When it is desired to step up the supply voltage, the output armature winding of the dynamotor should have more turns than the input armature winding, and vice versa. The output voltage cannot be controlled by the common field-winding current. An increase in flux will result in decreased speed. The product ϕS remains constant.

These machines have good efficiency because there is only one field power loss and only two bearings to cause losses. A motor generator set would have two fields and four bearings to contribute to the power loss.

REVIEW QUESTIONS

1. Explain why a commutator is used on a dc motor.
2. Using Fig. 21-1 explain the principle of motor action.
3. (a) What is meant by the torque of a motor?
   (b) On what does the torque developed depend?

4. (a) Explain the production of counter EMF.

    (b) Does the CEMF ever equal the applied voltage? Explain.

    (c) When is the CEMF closest in value to the line voltage?

5. Explain why the armature current of a motor increases as the mechanical load is increased.

6. (a) Why are starting boxes used on large motors?

    (b) Why are they not used on very small motors?

    (c) Explain the principle involved in the design of a starting box.

7. With reference to Fig. 21-2, explain the purpose of each part.

8. How does an automatic starting box differ from the manual type?

9. Draw schematic diagrams of the following motors:

    (a) Shunt.

    (b) Series.

    (c) Differential compound.

    (d) Cumulative compound.

10. (a) What is meant by speed regulation?

    (b) Should this be high or low? Explain.

11. Explain the speed and torque characteristics of a shunt motor.

12. (a) Repeat Question 11 for a series motor.

    (b) What precaution must be observed when using a series motor?

13. Repeat Question 11 for a cumulative compound motor.

14. Repeat Question 11 for a differential compound motor.

15. What type of motor is best suited for service where the required torque varies over wide limits? Explain.

16. What type of motor is best suited to constant-speed service?

17. What are the advantages and disadvantages of shunt field control of speed?

18. Repeat Question 17 for armature resistance control.

19. Repeat Question 17 for the Ward–Leonard system.

22

KIRCHHOFF'S LAWS

Earlier in this book (Chapters 7 and 8) Kirchhoff's voltage and current laws were stated, and although no further mention was made of them, you applied these principles in the solution of all the series, parallel, and series-parallel circuits that followed. Every time you added voltages in a series circuit to find the total voltage—or subtracted known voltages from a total voltage to find the missing component voltage—you were unconsciously applying Kirchhoff's voltage law. Similarly, every time you added branch currents to find the total current in a parallel network—or solved for a branch current, when knowing the total current and the other branch currents—you were applying Kirchhoff's current law. So, Kirchhoff's laws are not new to you. In fact, these laws are merely a generalization or expansion of the basic Ohm's law. Heretofore, in the solution of a series–parallel circuit we applied Ohm's law, one step at a time, to the entire circuit, or to any one portion of a circuit wherever any two values (I, E, or R) were known and solved for the missing quantity. Using this simple step-by-step procedure we were able to solve a wide variety of circuit problems. The majority of problems that you will encounter will probably yield to this direct-attack method. But occasionally this system leads to a blank wall, either because a network of resistors has two or more supply voltages feeding it *from different points*, or because of the interconnections the network cannot be analyzed in terms of simple series or parallel components. Sufficient data is given, but there is no starting point where two factors are known. Such problems, containing more than one unknown at a time, can still be solved. In general, the technique uses Kirchhoff's laws to set up algebraic equations that can be solved simultaneously. Let us examine these laws in more detail.

22-1 Kirchhoff's voltage law.   In Chapter 7, on series circuits, this law was stated as: *In any one path of a complete electrical circuit, the sum of the EMFs and of each voltage drop* **taken with proper signs,** *is equal to zero.* This statement is in keeping with our earlier concept of series circuit voltages—the total voltage is equal to sum of the individual component voltages. But now we have added "*taken with proper signs.*" To explain this, let us go back to basic energy considerations.

In any source of EMF, energy in one form or another is converted into electrical energy. A battery, for example, produces its EMF from the chemical energy of its ingredients. Electrons are "piled" on one terminal of the battery, making that terminal negative, and the other terminal positive. Any source of EMF may be considered as an electrical hill, with—using conventional current—positive current carriers at the top of this hill having acquired potential energy. If a network of resistors is connected across this EMF, current will flow through any of the paths so formed from the positive terminal of the supply source to the negative terminal. The current carriers are moving "down the electrical hill" and losing potential energy. When they reach the negative terminal—the bottom of the electrical hill—their potential energy is zero. As the source of EMF—acting as an elevator—raises the current carriers back to the top of the electrical hill, the carriers again have potential energy, and the process is repeated.

Now let us get back to the original question of what is meant when speaking of EMFs and voltage drops taken with proper signs. When tracing through any path of an electric circuit, if we go "uphill" we are gaining potential energy, and the proper sign for this voltage is positive ($+$). But if we go "downhill" we are losing potential energy, and this voltage sign should be negative ($-$). This in turn means:

1. When tracing through a source of EMF from negative to positive terminal (uphill) this potential should be given a positive sign.
2. When tracing through a source of EMF from positive to negative terminal (downhill) the sign should be negative. [Notice that the sign of the EMF in (1) and (2) corresponds to the polarity of the second, or output terminal when tracing through the circuit.]
3. When tracing through a resistor *in the direction of current flow* (downhill) there is a voltage drop. The sign should be negative.
4. When tracing through a resistor in opposition to the direction of current flow (uphill) there is a rise in potential energy and the sign should be positive.

We can check these polarities by applying the above rules to the circuit of Fig. 22-1. Let us trace through the path *abcdea*. First we will assume what we consider the correct direction of current flow. In this simple problem, this is easy. Since E_A is bigger than E_B, E_A will determine the direction of flow—from *a* to *b* to *c*, and so on, back to *a* as shown on the diagram.

1. From a to b we are going through a source of EMF from $(-)$ to $(+)$ or uphill. This EMF should be given a positive sign.

2. Now, from b to c we encounter resistor R_1. We are moving in the direction of current flow, or downhill. This is a voltage drop and should be given a negative sign.

3. From c to d we again have a voltage drop across R_2 and again a negative sign should be given.

4. From d to e we encounter a second source of EMF. But this time we are tracing from $(+)$ to $(-)$—opposite to the direction of the "elevator" (battery action), or downhill. This EMF should be given a negative sign.

5. From e to a we are again moving through a resistor (R_3) in the direction of current flow, or downhill, and again this is a voltage drop or negative.

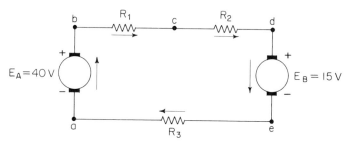

Figure 22-1

These five steps can be summarized in one equation as

$$E_A - IR_1 - IR_2 - E_B - IR_3 = 0$$

This is how we apply Kirchhoff's voltage law. No matter how complicated a circuit may be, such an equation can be written for *any* and *each* path of the network.

22-2 Kirchhoff's current law.   This law was stated in Chapter 8, on parallel circuits. Here it is again. *In any electrical network, the algebraic sum of the currents flowing to a junction must equal the algebraic sum of the currents flowing away from that junction.* We have already used this law in the analysis of current distribution in parallel and in series–parallel circuits (Chapters 8 and 9). These previous problems had only one source of supply. Now let us apply this law to a network containing two sources of EMF as shown in Fig. 22-2. The first step is to assume the most likely direction for the current distribution. Unfortunately, when the two supply sources are not of equal voltage, the *actual* direction of currents I_1 and I_2 is not definitely known. For example, if E_B is of lower value than E_A, then depending on the resistance values, E_B may act as a source and deliver current, or it may act as a load and *take* current. Current I_2 would then flow opposite to the direction shown in the diagram. If the latter condition is the true one, it will not affect the accuracy of the solution of this problem, but I_2 will come out as a negative value, the negative sign indicating, that the actual current flow is opposite to the assumed direction.

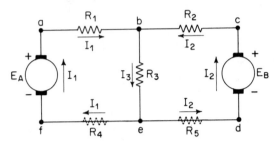

Figure 22-2

Once the direction of current flow has been assumed, we can apply Kirchhoff's current law to any junction. At point b, we have

$$I_3 = I_1 + I_2$$

or

$$I_2 = I_3 - I_1$$

or

$$I_1 = I_3 - I_2$$

From these relations, we can see that although the junction has three paths and three currents are involved, only two of these currents are unknowns, (independent values). The third current value is fixed (dependent) by the other two paths. Similarly if a junction had four paths, three current values are unknowns or independent, and the fourth current is a dependent value.

22-3   Problem-solving technique.   When applying Kirchhoff's laws in the solution of a problem, a systematic orderly technique is essential in reducing the labor involved and the possibility of errors. The following procedure will be found helpful.

1. Draw a neat diagram of the circuit.
2. Label all junction points (with letters) so that the various paths to be traced may be identified.
3. Indicate polarity and voltage for all known sources of EMF.
4. Indicate all known resistance values, in proper places.
5. Assume the most likely direction of current distribution, show these directions by arrows, and indicate all known values.
6. Identify all unknown circuit constants with suitable letters and subscripts (E_2, R_5, I_1, etc.) Keep the number of unknowns to a minimum by applying Kirchhoff's current law to the junction points.
7. Using Kirchhoff's voltage law, trace through any one complete path and

write the equation for that path. Repeat this step for as many paths as there are unknowns.

8. Solve the above equations simultaneously (or by determinants*).

Now let us apply this technique to the solution of a problem.

Example 22-1

Using the circuit of Fig. 22-2, find the voltage across R_3, the current through R_3, and how much of this current is delivered by each source if the circuit values are:

$$E_A = 100 \text{ V}, \quad R_1 = 10 \, \Omega, \quad R_4 = 0$$
$$E_B = 80 \text{ V}, \quad R_2 = 5 \, \Omega, \quad R_5 = 0$$
$$\phantom{E_B = 80 \text{ V}, \quad} R_3 = 2 \, \Omega$$

Solution

1. Taking path *fabef*, let us apply Kirchhoff's voltage law.

$$100 - 10I_1 - 2(I_1 + I_2) = 0$$

(Notice that we have eliminated one unknown, I_3, by using Kirchhoff's current law: $I_3 = I_1 + I_2$.)

2. By collecting terms and simplifying, we get

$$6I_1 + I_2 = 50 \tag{1}$$

3. Applying Kirchhoff's voltage law to path *fabcdef* yields

$$100 - 10I_1 + 5I_2 - 80 = 0$$
$$2I_1 - I_2 = 4 \tag{2}$$

Now we have two equations (1) and (2), with two unknowns, I_1 and I_2. We can solve these simultaneously by adding the two equations:

4.
$$\begin{array}{r} 6I_1 + I_2 = 50 \\ 2I_1 - I_2 = 4 \\ \hline 8I_1 = 54 \end{array}$$

$$I_1 = 6.75 \text{A (current delivered by source } E_A)$$

Substituting this value of I_1 in equation (2) gives us

5.
$$13.5 - I_2 = 4$$

$$I_2 = 9.5 \text{ A (current delivered by source } E_B)$$

6.
$$I_3 = I_1 + I_2 = 6.75 + 9.5 = 16.25 \text{ A}$$

7.
$$E_{be} = I_3 R_3 = 16.25 \times 2 = 32.5 \text{ V}$$

As a check we can solve for E_{be} by subtracting the voltage drop in R_1 from E_A.

*Use of determinants is reviewed in Appendix 5.

8.
$$E_{ab} = I_1 R_1 = 6.75 \times 10 = 67.5 \text{ V}$$
$$E_{be} = E_A - E_{ab} = 100 - 67.5 = 32.5 \text{ V}$$

Using this same network, let us see what happens if the resistance values are changed, and R_4 and R_5 are added.

Example 22-2

In Fig. 22-2, find the voltage across R_3, the current through R_3, and how much of this current is delivered by each source if the circuit values are

$$E_A = 100 \text{ V}, \qquad R_1 = R_2 = R_4 = R_5 = 2 \,\Omega$$
$$E_B = \quad 80 \text{ V}, \qquad R_3 = 28 \,\Omega$$

Solution

1. From path *fabef*,
$$100 - 2I_1 - 28(I_1 + I_2) - 2I_1 = 0$$
$$32I_1 + 28I_2 = 100 \qquad (1)$$

2. From path *facdf*,
$$100 - 2I_1 + 2I_2 - 80 + 2I_2 - 2I_1 = 0$$
$$4I_1 - 4I_2 = 20 \qquad (2)$$

Multiplying equation (2) by 7, and adding to equation (1), we get
$$32I_1 + 28I_2 = 100$$
$$\underline{28I_1 - 28I_2 = 140}$$
$$60I_1 + 0 \quad = 240$$
$$I_1 = 4 \text{ A (delivered by source } E_A)$$

Substituting this value of I_1 in equation (2) yields

3.
$$16 - 4I_2 = 20$$
$$I_2 = -1 \text{ A (delivered by source } E_B)$$

This means that the assumed direction of current I_2 is wrong and the "source" E_B is actually **taking** *current and acting as a load* (a battery being charged or a dynamo running as a motor).

4.
$$I_3 = I_1 + I_2 = 4 - 1 = 3 \text{ A}$$
5.
$$E_{be} = I_3 R_3 = 3 \times 28 = 84 \text{ V}$$

In the above problems, the circuits used contained two sources of EMF, and for this reason the basic Ohm's law solution was inadequate. This may give the impression that we must resort to Kirchhoff's laws only if the circuit contains more than one source of supply. Such an assumption is incorrect. There are networks with only a single source of supply that will require advanced circuit theory such as Kirchhoff's laws for their solution. An example of such a circuit is the unbalanced Wheatstone bridge.

Example 22-3

Find the current (magnitude and direction) flowing through the galvanometer in the bridge circuit of Fig. 22-3. The galvanometer resistance is 2000 Ω.

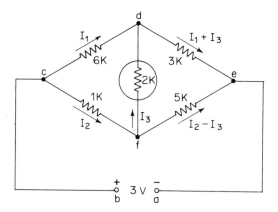

Figure 22-3

Solution   Let us *assume* that current I_3 flows up through the 2-kΩ galvanometer, as shown in Fig. 22-3. Then:

1. From path *abcdea*,

$$3 - 6000I_1 - 3000(I_1 + I_3) = 0$$

$$9000I_1 + 3000I_3 = 3 \tag{1}$$

2. From path *abcfea*,

$$3 - 1000I_2 - 5000(I_2 - I_3) = 0$$

$$6000I_2 - 5000I_3 = 3 \tag{2}$$

3. From path *fdef*,

$$-2000I_3 - 3000(I_1 + I_3) + 5000(I_2 - I_3) = 0$$

$$5000I_2 = 3000I_1 + 10\,000I_3$$

$$I_2 = 0.6I_1 + 2I_3 \tag{3}$$

4. Substituting this value of I_2 in equation (2), we get

$$6000(0.6I_1 + 2I_3) - 5000I_3 = 3$$

$$3600I_1 + 7000I_3 = 3 \tag{4}$$

5. Multiplying equation (1) by 2, equation (4) by 5, and subtracting, we have

$$
\begin{aligned}
18\,000I_1 + 6\,000I_3 &= 6 \\
\overline{\oplus 18\,000I_1 \; \oplus \; 35\,000I_3} &= \overline{\oplus 15} \\
\hline
- \, 29\,000I_3 &= -9 \\
I_3 &= 310 \; \mu A
\end{aligned}
$$

The galvanometer current is 310 μA flowing in the direction indicated on the diagram—from f toward d. Since the mathematics is laborious, it is quite possible to make errors in problems of this type. A check of the answer is desirable. In this case, it means solving for I_1 and I_2 and then applying Kirchhoff's law to an unused path.

Check

1. Substituting the value of I_3 in equation (1) gives us

$$9000I_1 + 3000(310 \times 10^{-6}) = 3$$

$$I_1 = \frac{3 - 0.93}{9000} = 230 \ \mu A$$

2. Repeating for equation (2) yields

$$6000I_2 - 5000(310 \times 10^{-6}) = 3$$

$$I_2 = \frac{3 + 1.55}{6000} = 758 \ \mu A$$

3. Taking path $cdfc$ and substituting current values, we have

$$-6000(230 \times 10^{-6}) + 2000(310 \times 10^{-6}) + 1000(758 \times 10^{-6}) = 0$$

$$-1.380 + 0.620 + 0.758 = 0$$

$$1.380 = 1.378$$

This check is close enough.

REVIEW QUESTIONS

1. In applying Kirchhoff's voltage law to a network path:
   (a) When is an EMF labeled positive?
   (b) When is the IR drop across a resistor labeled positive?
   (c) When is an EMF labeled negative?
   (d) When is an IR drop labeled negative?
2. In a complex network:
   (a) Can we predict in advance the correct direction of current flow through each path or resistor? Explain.
   (b) How are current directions assigned?
   (c) If the assigned direction is wrong, how will we ever know this?
   (d) How will this affect the *numerical* accuracy of the answers?
3. In the solution to Example 22-1:
   (a) Explain how each of the terms in the equation of step 1 is obtained.
   (b) How is the equation in step 2 obtained?
   (c) Could it be simplified further?
   (d) Explain how each term in step 3 is obtained.
4. In the solution to Example 22-2:
   (a) Current I_2 in step 3 is found to be -1 A. What does this mean?

   (b) In step 4 I_3 is equated to $I_1 + I_2$. Since I_1 is found to be 4.0 A, shouldn't I_3 be more than 4.0? Explain.
5. With reference to Fig. 22-3:
   (a) Is R_{c-d} in parallel or in series with R_{c-f}? Explain.
   (b) Is the 2-kΩ unit in series with R_{c-d} or R_{c-f}? Explain.
   (c) Using earlier techniques, how would you find the total resistance (R_T) between points c and e?
   (d) Can this circuit be solved using the simple Ohm's law methods of the earlier chapters?

PROBLEMS

1. Using the circuit of Fig. 22-2, find the magnitude and direction of current in resistors R_1, R_2, and R_3 for the following circuit constants: $E_A = 230$ V, $E_B = 250$ V, $R_1 = R_2 = 20\ \Omega$, $R_4 = R_5 = 10\ \Omega$, $R_3 = 5\ \Omega$.
2. Repeat Problem 1 for the following circuit values: $E_A = 220$ V, $E_B = 166$ V, $R_1 = R_4 = 10\ \Omega$, $R_2 = R_5 = 2\ \Omega$, $R_3 = 120\ \Omega$.
3. A section of a railway transit system between two substations is 10 km long. Substation A is supplying 600 V between third rail and track. Substation B has an output of 580 V. The track resistance is 0.03 Ω/km, and the third-rail resistance is 0.40 Ω/km. Generators at stations A and B have internal resistances of 0.05 and 0.04 Ω, respectively. A train, located 4 km from substation A draws a current of 200 A.
   (a) How much of this load does each substation supply?
   (b) What is the voltage between third rail and track at the train location?
4. (a) In Problem 3, at what location of the train will substation A supply the entire 200-A load?
   (b) What is the voltage across the train now?
   (c) What is the effect on the generator at substation B if the train is even closer to station A?
5. Two three-cell storage batteries are connected in parallel across a load. Due to difference in state of charge, battery A has an EMF of 6.6 V and an internal resistance of 0.02 Ω. Battery B has an EMF of 6.0 V and an internal resistance of 0.04 Ω. For each of the following load conditions find: the load current, the voltage across the load, and how much of the load is supplied by each battery.
   (a) A load resistance of 0.70 Ω
   (b) A load resistance of 0.20 Ω
   (c) A load resistance of 0.06 Ω
6. (a) The two batteris of Problem 5 are connected in parallel across a load of 0.08 Ω. Find load current, voltage across the load, and how much of the load is supplied by each battery.
   (b) Battery B is recharged so that its EMF and internal resistance are 6.44 V and 0.03 Ω, respectively. Solve for the above data.
   (c) Battery B is recharged again. It now has an EMF of 6.5 V and an internal resistance of 0.02 Ω .Solve as above.

7. Figure 22-4 shows a three-wire distribution system supplying three loads at some
distance from the generators. Each generator has an EMF of 120 V and internal
resistance of 0.05 Ω. Loads A and B are lighting loads of 85 and 50 A, respectively.
Load C is an industrial load of 160 A. The resistance of the outside feeders from
generators to lighting loads is 0.032 Ω each. The resistance of the neutral (center
feeder) is 0.084 Ω. The resistance of each feeder wire from lighting load to industrial load is 0.025 Ω. Find:

(a) The voltage across each lighting load when the industrial load is off.

(b) The voltage across each load when the industrial load is applied.

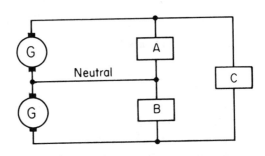

Figure 22-4

8. In the three-wire distribution system of Problem 7, load A has a resistance of
1.6 Ω, and load B has a resistance of 1.2 Ω. There is no other load. Find:

(a) The current in each feeder.

(b) The voltage across each load.

9. With loads as in Problem 8, what value of EMF is needed at each generator to
maintain a balanced voltage of 118 V at each load?

10. In the Wheatstone bridge circuit of Fig. 22-5, the galvanometer key was held down
during the balancing procedure and the movement was burned out. The galvanometer had a 200-μA movement with a resistance of 100 Ω. The bridge settings
were $R_1 = 100$ Ω, $R_2 = 40$ Ω, $R_s = 30$ Ω, and R_x was later found to be 25 Ω.
Find the current through the galvanometer just before it was damaged.

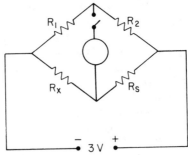

Figure 22-5

11. In Problem 10, with the ratio arm (R_1 and R_2) resistances as given, find two values
of R_s that would reduce the galvanometer current to 1.00 mA.

23

NETWORK THEORY

Any electrical circuit, regardless of the number of voltage sources, or of the complexity of the interconnections, can be solved using the Kirchhoff's laws technique shown in Chapter 22. However, the solutions can become very laborious. In an attempt to reduce this tediousness a variety of network theorems have been developed. Among these are: the *loop* (or *mesh*) method; the *nodal* method; the *superposition* theorem; the *reciprocity* theorem; *Thévenin's* theorem; *Norton's* theorem; the *delta–wye* transformation; and the *maximum power transfer* theorem. Depending on the problem, use of one of these methods may make the solution somewhat less laborious.

23-1 Loop or mesh method. This technique is almost identical to the more general Kirchhoff's law method shown in the previous chapter. There is a slight difference in the way the currents are labeled. This difference is readily seen from a comparison of Fig. 23-1(a) and (b). In diagram (a) the general Kirchhoff's law technique is used to label each individual current. Then one current is replaced by its dependent value, using the junction relation $I_3 = I_1 + I_2$. In the mesh method, diagram (b), only one current value is used for any one complete path or *loop*. Notice, however, that resistor R_3 is common to the two loops, so that currents I_1 and I_2 both must be considered when analyzing for voltage drops (applying Kirchhoff's voltage law). Therefore whether tracing path *abef* or loop 1, the identical equation will result:

$$E_A - I_1R_1 - R_3(I_1 + I_2) = 0$$

The rest of the procedure is as described in Chapter 22:

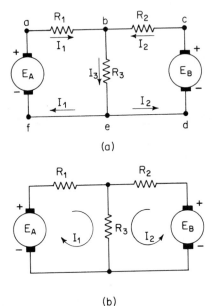

Figure 23-1   Circuit analysis by general
Kirchhoff's law and by loop method.

(a)

(b)

1. Apply Kirchhoff's voltage law to as many paths (loops) as there are unknowns.
2. Solve the resulting equations, simultaneously, or by determinants.

23-2   Nodal method.   In the loop method, Kirchhoff's voltage law is used to write an equation for the voltages in any loop—in terms of unknown current values. Now, instead, we will write current equations at one or more junctions—in terms of unknown voltages. This technique can be shown with the aid of Fig. 23-2(a). (This diagram is a repeat of Fig. 23-1.) In this diagram, points b and e are junctions of three (or more) elements, and are called **nodes**.* Applying this definition to Fig. 23-2(b), the nodes are junctions b, c, and g. Notice that all the junctions in the lower path are called g. These points are all at the same potential. In fact, they are really one and the same point—separated for convenience (drawing neatness). In applying the nodal method of analysis, one node is considered as a reference and the other node potentials are figured with respect to this reference node. In diagram (a), node e would be a suitable reference point; in diagram (b), the common point g would make the best reference.

In diagram (a), assuming the source voltages are given, there is only one unknown node voltage. Therefore, only one equation will be needed to solve for this voltage. Current directions have been assumed and are shown as I_1, I_2, and I_3. (As before, if an incorrect direction is assumed, that current will end up

*Some texts also call points a and c nodes (but to distinguish from b and e, they are classified as *dependent nodes*).

334

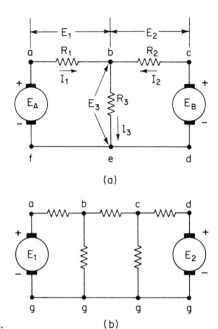

(a)

(b)

Figure 23-2 Nodal analyses.

as a negative quantity.) At node b, applying Kirchhoff's current law, we have

$$I_1 + I_2 = I_3 \tag{1}$$

By Ohm's law $I_1 = E_1/R_1$. But $E_1 = E_A - E_3$, so that

$$I_1 = \frac{E_A - E_3}{R_1}$$

In similar fashion, it can be seen that

$$I_2 = \frac{E_B - E_3}{R_2}$$

and obviously

$$I_3 = \frac{E_3}{R_3}$$

Replacing each current value in equation (1) by its equivalent Ohm's law value, we get

$$\frac{E_A - E_3}{R_1} + \frac{E_B - E_3}{R_2} = \frac{E_3}{R_3} \tag{2}$$

This can be rewritten as

$$\frac{E_A}{R_1} - \frac{E_3}{R_1} + \frac{E_B}{R_2} - \frac{E_3}{R_2} = \frac{E_3}{R_3}$$

Rearranging terms yields

$$-\frac{E_3}{R_1} - \frac{E_3}{R_2} - \frac{E_3}{R_3} + \frac{E_a}{R_1} + \frac{E_b}{R_2} = 0$$

335

Factoring (and changing signs), we have

$$E_3\left(\frac{1}{R_1} + \frac{1}{R_2} + \frac{1}{R_3}\right) - \frac{E_A}{R_1} - \frac{E_B}{R_2} = 0 \qquad (23\text{-}1)$$

Notice, in this final nodal equation, that it is of the form:

node voltage × (reciprocal of each adjacent branch R)

— (each adjacent voltage divided by the connecting element resistance)

Let us apply this technique to Example 22-1 of the previous chapter. (The diagram is redrawn for convenience.)

Example 23-1

In the circuit of Fig. 23-3, find the voltage across R_3, the current through R_3, and how much of this current is supplied by each source, if the circuit values are;

$$E_A = 100 \text{ V}, \qquad R_1 = 10 \, \Omega, \qquad R_3 = 2 \, \Omega$$
$$E_B = 80 \text{ V}, \qquad R_2 = 5 \, \Omega$$

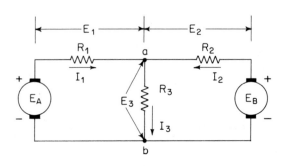

Figure 23-3

Solution Applying equation (23-1) to node a (node b is the reference node), we get

1.
$$E_3\left(\frac{1}{R_1} + \frac{1}{R_2} + \frac{1}{R_3}\right) - \frac{E_A}{R_1} - \frac{E_B}{R_2} = 0$$

$$E_3\left(\frac{1}{10} + \frac{1}{5} + \frac{1}{2}\right) - \frac{100}{10} - \frac{80}{5} = 0$$

$$0.8E_3 = 26$$

$$E_3 = 32.5 \text{ V}$$

2.
$$I_3 = \frac{E_3}{R_3} = \frac{32.5}{2} = 16.25 \text{ A}$$

3.
$$I_1 = \frac{E_A - E_3}{R_1} = \frac{100 - 32.5}{10}$$

$$= 6.75 \text{ A (supplied by source } E_A)$$

4.
$$I_2 = I_3 - I_1 = 16.25 - 6.75$$

$$= 9.5 \text{ A (supplied by source } E_B)$$

Notice that this solution is appreciably simpler than the Kirchhoff's law (or mesh) technique of the previous chapter.

The above T network resulted in only one unknown. Now let us try the nodal technique on a ladder network with two unknowns.

Example 23-2

In the circuit of Fig. 23-4, find the voltage across R_3 and R_4, if the circuit values are:

$$E_A = 100 \text{ V}, \qquad R_1 = R_2 = R_5 = 10 \, \Omega$$
$$E_B = 40 \text{ V}, \qquad R_3 = R_4 = 20 \, \Omega$$

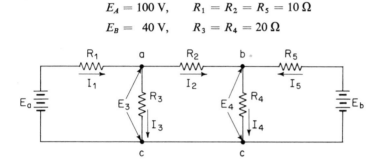

Figure 23-4

Solution

1. Node c will be used as reference.

2. Apply equation (23-1) at node a:

(a) $$E_3\left(\frac{1}{R_1} + \frac{1}{R_2} + \frac{1}{R_3}\right) - \frac{E_A}{R_1} - \frac{E_4}{R_2} = 0$$

(b) $$E_3\left(\frac{1}{10} + \frac{1}{10} + \frac{1}{20}\right) - \frac{100}{10} - \frac{E_4}{10} = 0$$

Multiply by the least common denominator (20), and collect terms:

(c) $$5E_3 - 2E_4 - 200 = 0 \qquad\qquad (1)$$

3. Apply the nodal equation at node b:

(a) $$E_4\left(\frac{1}{R_2} + \frac{1}{R_4} + \frac{1}{R_5}\right) - \frac{E_3}{R_2} - \frac{E_B}{R_5} = 0$$

(b) $$E_4\left(\frac{1}{10} + \frac{1}{20} + \frac{1}{10}\right) - \frac{E_3}{10} - \frac{40}{10} = 0$$

(c) $$5E_4 - 2E_3 - 80 = 0 \qquad\qquad (2)$$

4. Multiply equation (1) by 5, equation (2) by 2, and add:

(a) $\quad 25E_3 - 10E_4 - 1000 = 0$
(b) $\quad -\ 4E_3 + 10E_4 - 160 = 0$
(c) $\quad \overline{21E_3 - 1160 = 0}$
$$E_3 = 55.2 \text{ V}$$

5. Substitute this value of E_3 in equation (2):

$$5E_4 - 2(55.2) - 80 = 0$$

$$E_4 = 38.1 \text{ V}$$

6. As a check we can substitute these values for E_3 and E_4 in equation (1):

$$5(55.2) - 2(38.1) - 200 = 0$$

$$276 - 276.2 = 0$$

Any other voltage, or any current value can now be found by simple application of Ohm's law to that portion of the circuit.

In the above examples, equation (23-1) was used to find the node voltages. It should be stressed however, that this *special* form of the nodal equation applies only to the T or ladder-type networks shown in Figs. 23-2(a) and (b). The nodal method is ideally suited for these networks. For any other configuration, it will be necessary to fall back on the basic nodal technique: Write current equations at one or more junctions, in terms of the known and unknown voltages [as was done in developing equation (23-1).]

23-3 Superposition theorem. Any network containing more than one supply source can be solved using the *superposition* theorem—without the need for simultaneous equations. The technique is quite simple. All but one of the supply sources are replaced by their own internal resistance.* The circuit is then solved using simple Ohm's law relations. The process is repeated using each of the sources as the sole supply source. The actual current through any component (or portion) of the original circuit is the algebraic sum of the currents obtained when each source acted individually. To enable comparison with the loop and nodal methods we will illustrate this technique using the circuit of Example 23-1, Fig. 23-3.

Example 23-3
Using the superposition theorem and Fig. 23-3, find the voltage across R_3, the current through R_3, and how much of this current is delivered by each source. The circuit values are:

$$E_A = 100 \text{ V}, \qquad R_1 = 10 \, \Omega, \qquad R_3 = 2 \, \Omega$$
$$E_B = 80 \text{ V}, \qquad R_2 = 5 \, \Omega$$

Solution

1. Replacing E_B by a short circuit† produces Fig. 23-5(a). Obviously, R_2 and R_3 are in parallel, and the combination is in series with R_1. Therefore,

(a) $$R_T' = R_1 + \frac{R_2 R_3}{R_2 + R_3} = 10 + \frac{5 \times 2}{7} = 11.43 \, \Omega$$

*If the internal resistance is negligible compared to the circuit values, the source is replaced by a short circuit.

†This assumes the internal resistance of the source is negligible.

(b) $$I'_T = I'_1 = \frac{E_T}{R'_T} = \frac{100}{11.43} = 8.75 \text{ A}$$

At the junction of the three resistors this current divides, and by the current ratio method [equation (8-3)] the distribution is

(c) $$I'_3 = \frac{R_2}{R_2 + R_3}(I'_T) = \frac{5}{7} \times 8.75 = 6.25 \text{ A}$$

(d) $$I'_2 = \frac{R_3}{R_2 + R_3}(I'_T) = \frac{2}{7} \times 8.75 = 2.50 \text{ A}$$

Note the direction of these currents on the diagram.

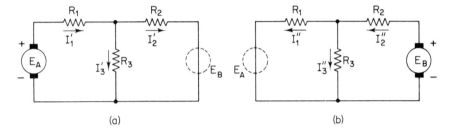

(a)   (b)

Figure 23-5 Equivalent circuits with alternate sources shorted.

2. Replacing E_A by a short circuit produces Fig. 23-5(b), from which

(a) $$R''_T = R_2 + \frac{R_1 R_3}{R_1 + R_3} = 5 + \frac{10 \times 2}{12} = 6.67 \ \Omega$$

(b) $$I''_T = I''_2 = \frac{E_T}{R''_T} = \frac{80}{6.67} = 12.0 \text{ A}$$

(c) $$I''_3 = \frac{R_1}{R_1 + R_3}(I''_T) = \frac{10}{12} \times 12 = 10.0 \text{ A}$$

(d) $$I''_1 = \frac{R_2}{R_1 + R_2}(I''_T) = \frac{2}{12} \times 12 = 2.0 \text{ A}$$

Note these current directions in Fig. 23-5(b).

3. Since both I_3 currents are in the same direction, the actual I_3 current is

$$I_3 = I'_3 + I''_3 = 6.25 + 10.0 = 16.25 \text{ A}$$

4. The I_1 currents are in opposite directions. Therefore,

$$I_1 = I'_1 - I''_1 = 8.75 - 2.0 = 6.75 \text{ A}$$

and the direction of I_1 is the same as I'_1, or source E_A *delivers* current (6.75 A) to R_3.

5. The I_2 currents are in opposite directions, with I''_2 the larger. Therefore, I_2 is in the direction of I''_2 and

$$I_2 = I''_2 - I'_2 = 12.0 - 2.50 = 9.50 \text{ A}$$

This is the current *delivered* by source E_B.

6. The voltage across R_3 is

$$E_3 = I_3 \times R_3 = 16.25 \times 2 = 32.5 \text{ V}$$

These answers agree very closely with those obtained by the nodal method (Example 23-1) and by the Kirchhoff's law (mesh) method (Section 22-3).

23-4 Thévenin's theorem.   In many instances where one or more supply sources deliver power to a load through some complex network, we are not interested in the details of voltage drops or current distribution within the network itself. All we want to know is the current through the load, voltage across the load, and power absorbed by the load. Solving such cases by the methods shown above can become very laborious. A theorem brought forth by M. L. Thévenin* can be used to reduce the work load. This theorem—*Thévenin's theorem*—is especially valuable when we wish to analyze a circuit for many different load values. The basic principle is quite simple. Any active network, no matter how complex, can be replaced by an *equivalent circuit* consisting of a constant-voltage source in series with a resistance, where

1. The voltage (E_{oc}) is the *open-circuit* voltage of the original network, measured across the load terminals, but with the load removed.
2. The resistance (R_i) for the equivalent circuit is the *internal resistance* of the original network as seen *looking in* from the load terminals, but with the voltage sources replaced by their internal resistances. (The sources can be replaced by short circuits if their internal resistances are negligible.)

For comparison purposes, let us apply this technique to the circuit of Example 23-1.

Example 23-4

Using Thévenin's theorem and the circuit of Fig. 23-6, find the current through R_3 and the voltage across R_3.

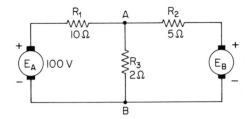

Figure 23-6

Solution

1. R_3 is considered as the load.
2. To find the voltage (E_{oc}) for the Thévenin equivalent circuit, we must remove the load and calculate the voltage at the load terminals (across A–B). With R_3 removed, the circuit becomes a simple series circuit with E_B opposing E_A. The net voltage is therefore $100 - 80$ or 20 V, and the

*M. Leon Thévenin, a French physicist, presented this theorem in 1883.

direction of current flow is determined by the higher-voltage source (E_A).
This is shown in Fig. 23-7(a). The voltage across $A-B$ is obviously $E_A - IR_1$
(or $E_B + IR_2$).

(a) $$I = \frac{E_{net}}{R_1 + R_2} = \frac{20}{15} = 1.33 \text{ A}$$

(b) $$E_{AB} = E_{oc} = E_A - (1.33 \times 10) = 86.7 \text{ V}$$

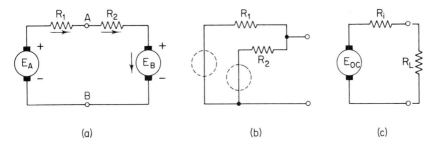

(a)   (b)   (c)

Figure 23-7   Analysis of Fig. 23-6 by Thévenin's theorem.

3. To find the Thévenin internal resistance, we short circuit both supply
   sources (internal resistance of each source assumed to be negligible).
   Looking in at the load end, we now see Fig. 23-7(b). Notice that R_2 and
   R_1 are now in parallel, so that

$$R_i = \frac{R_1 R_2}{R_1 + R_2} = \frac{10 \times 5}{15} = 3.33 \ \Omega$$

4. The Thévenin equivalent circuit can now be drawn [see Fig. 23-7(c)]. The
   current through the load R_L is therefore

(a) $$I_L = \frac{E_{oc}}{R_i + R_L} = \frac{86.7}{5.33} = 16.25 \text{ A}$$

(b) $$E_L = I_L R_L = 16.25 \times 2 = 32.5 \text{ V}$$

These answers check with the values obtained by each of the previous methods.
As far as the work done to obtain the answers, there is not much choice among
the various methods. However, the great advantage of the Thévenin method is
that if we now want load current and load voltage for several other values of
R_3, we need only use the simple step 4. With any of the previous methods, the
full procedure would have to be repeated.

23-5   Norton's theorem.   A variation of Thévenin's technique was
proposed by E. L. Norton.* It is based on the fact that any series circuit can be
replaced by an equivalent parallel circuit. Norton's theorem states that any
complex active network can be replaced by an equivalent circuit consisting of a

---

*Edward Lawry Norton is an American physicist, formerly with the Bell Telephone
Laboratories since 1922.

constant-current source *in parallel* with a resistance, where:

1. The constant-current value (I_{sc}) is the current that would flow through a short-circuited load, in the original network.
2. The shunt resistance (R_{sh}) for the equivalent circuit is the *internal resistance* of the original network as seen looking in from the load terminals, but with all voltage sources replaced by their own internal resistances. (This is the same value as is used in the Thévenin method.)

Let us apply this theorem to a problem. Again, for comparison, we will use the circuit of Example 23-1.

Example 23-5

Using Norton's theorem and the circuit of Fig. 23-6, find the current through R_3 and the voltage across R_3.

Solution

1. R_3 is considered as the load.
2. To find the current value (I_{sc}) for the Norton equivalent circuit, the load (R_3 in Fig. 23-6) must be shorted. The circuit can now be drawn as in Fig. 23-8(a). This circuit is readily analyzed by the superposition theorem.

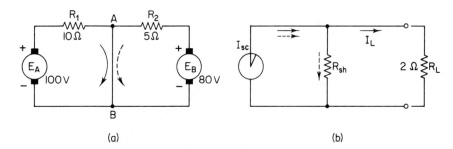

(a)   (b)

Figure 23-8 Analysis of Fig. 23-6 by Norton's theorem.

(a) When E_B is replaced by its internal resistance (shorted), R_2 is shorted, and the circuit consists only of E_A and R_1. The current through A–B is

$$I'_{AB} = \frac{E_A}{R_1} = \frac{100}{10} = 10 \text{ A}$$

(b) When E_A is replaced by its internal resistance,

$$I''_{AB} = \frac{E_B}{R_2} = \frac{80}{5} = 16 \text{ A}$$

(c) Therefore, the Norton constant-current source is

$$I_{sc} = I'_{AB} + I''_{AB} = 10 + 16 = 26 \text{ A}$$

3. To find the Norton shunt resistance (R_{sh}), the procedure is the same as discussed in step 3 of Example 23-4, and

$$R_{\text{sh}} = \frac{R_1 R_2}{R_1 + R_2} = \frac{10 \times 5}{15} = 3.33\ \Omega$$

4. The Norton equivalent circuit can now be drawn [see Fig. 23-8(b)]. By the current ratio method, it should be obvious that the load current (I_L) is

(a)
$$I_L = I_{\text{sc}} \frac{R_{\text{sh}}}{R_{\text{sh}} + R_L} = 26 \times \frac{3.33}{5.33} = 16.25\ \text{A}$$

(b) The load voltage is

$$E_L = I_L R_L = 16.25 \times 2 = 32.5\ \text{V}$$

23-6 Delta–wye transformation.   Figure 23-9(a) shows a three-terminal network. Because of the similarity of this configuration with the Greek letter delta (Δ), these resistors are said to be connected in *delta* (Δ). Three resistors can also be connected to form another three-terminal network as shown in Fig. 23-9(b). Here again, because of similarity with the letter Y (inverted) this configuration is called a *wye* (Y) connection.

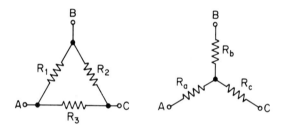

Figure 23-9 Delta and wye networks.

It is possible to replace any delta network with an equivalent wye network, or vice versa. To make this transformation all that is necessary is to maintain an equal resistance between equivalent terminals of the delta and the wye. Referring to Fig. 23-9, the resistance between terminals A and B is

1. For the wye circuit:
$$R_{A,B} = R_a + R_b$$

2. For the delta circuit:
$$R_{A,B} = \frac{R_1(R_2 + R_3)}{R_1 + R_2 + R_3}$$

Equating these two expressions for $R_{A,B}$ yields

$$R_a + R_b = \frac{R_1(R_2 + R_3)}{R_1 + R_2 + R_3} \tag{1}$$

Similarly, for $R_{B,C}$ and $R_{C,A}$ we get

$$R_b + R_c = \frac{R_2(R_1 + R_3)}{R_1 + R_2 + R_3} \tag{2}$$

and

$$R_c + R_a = \frac{R_3(R_1 + R_2)}{R_1 + R_2 + R_3} \tag{3}$$

Solving equations (1), (2), and (3) simultaneously

$$R_c \text{ (wye value)} = \frac{R_2 R_3}{R_1 + R_2 + R_3} \tag{23-2A}$$

$$R_a \text{ (wye value)} = \frac{R_1 R_3}{R_1 + R_2 + R_3} \tag{23-2B}$$

$$R_b \text{ (wye value)} = \frac{R_1 R_2}{R_1 + R_2 + R_3} \tag{23-2C}$$

These equations will enable us to replace any delta-connected resistors by an equivalent wye network. The equations are easy to remember if you notice that the denominators are all alike—the sum of the delta resistors—and the numerator is the product of the two delta resistors on either side of the terminal to which the wye resistor is connected.

Example 23-6

The delta network of Fig. 23-9 has the following circuit constants: $R_1 = 80\ \Omega$, $R_2 = 30\ \Omega$, and $R_3 = 70\ \Omega$. Find the resistance values for an equivalent wye network.

Solution

1. $R_a \text{ (terminal } A) = \dfrac{R_1 R_3}{R_1 + R_2 + R_3} = \dfrac{80 \times 70}{180} = 31.1\ \Omega$

2. $R_b \text{ (terminal } B) = \dfrac{R_1 R_2}{R_1 + R_2 + R_3} = \dfrac{80 \times 30}{180} = 13.3\ \Omega$

3. $R_c \text{ (terminal } C) = \dfrac{R_2 R_3}{R_1 + R_2 + R_3} = \dfrac{30 \times 70}{180} = 11.67\ \Omega$

Of the two networks, the delta versus the wye, the wye circuit is the easier to analyze. And so, conversion for circuit simplification is generally from delta to wye, as shown in Example 23-6. Should the opposite conversion—from wye to delta—be necessary, the equations for such conversion are*

$$R_1 \text{ (delta value)} = \frac{R_a R_b + R_b R_c + R_c R_a}{R_c} \tag{23-3A}$$

$$R_2 \text{ (delta value)} = \frac{R_a R_b + R_b R_c + R_c R_a}{R_a} \tag{23-3B}$$

$$R_3 \text{ (delta value)} = \frac{R_a R_b + R_b R_c + R_c R_a}{R_b} \tag{23-3C}$$

*These equations can be derived from the original basic principle that the resistance (or conductance) between similar terminals of the wye and its equivalent delta network must remain the same.

Notice in these equations for the equivalent delta values, that the numerator is the sum of the products of the three wye values—taken two at a time—while the denominator is the wye leg perpendicular (or opposite) to the delta leg for which we are solving.

Wye and delta networks are very commonly encountered in the power field, particularly in three-phase circuits. Similar interconnections of resistors are also used in electronic circuits. However, they are drawn in a slightly different manner, and they are called *pi* (π) and *tee* (T) networks. The comparison between the delta (or pi) network, and the wye (or tee) network can be readily seen in Fig. 23-10.

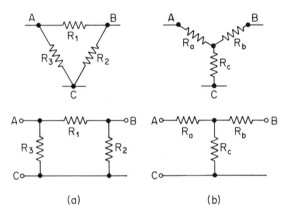

(a)   (b)

Figure 23-10   Delta/pi and wye/tee comparison.

Now to show the usefulness of these transformations in the solution of circuit problems, let us solve the Wheatstone bridge circuit, Example 22-3, and compare methods.

Example 23-7

Find the current flowing through the galvanometer in the bridge circuit of Fig. 23-11. The galvanometer resistance is 2000 Ω.

Solution

1. Let us replace the delta network between points *cdf* with its equivalent wye, as shown in Fig. 23-11(b). The equivalent wye-network values are:

(a) $\qquad R_a = \dfrac{R_1 R_2}{\sum R} = \dfrac{(6 \times 10^3)(1 \times 10^3)}{9 \times 10^3} = \dfrac{6 \times 10^6}{9 \times 10^3} = 666.6 \ \Omega$

(b) $\qquad R_b = \dfrac{R_3 R_1}{\sum R} = \dfrac{(2 \times 10^3)(6 \times 10^3)}{9 \times 10^3} = \dfrac{12 \times 10^6}{9 \times 10^3} = 1333 \ \Omega$

(c) $\qquad R_c = \dfrac{R_2 R_3}{\sum R} = \dfrac{(1 \times 10^3)(2 \times 10^3)}{9 \times 10^3} = \dfrac{2 \times 10^6}{9 \times 10^3} = 222.2 \ \Omega$

(Where $\sum R$ = sum of all R's = $R_1 + R_2 + R_3$).

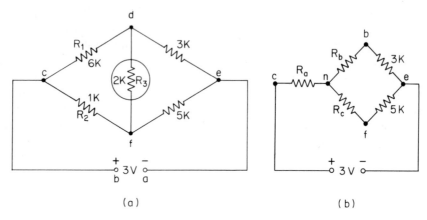

Figure 23-11

2. From Fig. 23-11(b):
   (a) R_b in series with 3 kΩ = 1333 + 3000 = 4333 Ω
   (b) R_c in series with 5 kΩ = 222 + 5000 = 5222 Ω
   (c) The 4333 Ω of step (a) in parallel with the 5222 Ω of step (b) is equivalent to

$$\frac{4333 \times 5222}{4333 + 5222} = 2368 \ \Omega$$

   (d) The 2368 Ω of step (c) is in series with R_a for a total resistance of

$$R_T = 2368 + 667 = 3035 \ \Omega$$

3. $$I_T = \frac{E_T}{R_T} = \frac{3}{3035} = 0.989 \text{ mA}$$

4. $$E_{cn} = I_T R_a = 0.989 \times 10^{-3} \times 666.6 = 0.659 \text{ V}$$

5. $$E_{ne} = 3.0 - 0.659 = 2.341 \text{ V}$$

6. $$I_{nde} = \frac{E_{ne}}{R_b + 3K} = \frac{2.341}{4333} \text{ A}$$

7. $$E_{de} = I_{nde} \times 3K = \frac{2.341}{4333} \times 3000 = 1.62 \text{ V}$$

8. $$I_{nfe} = \frac{E_{ne}}{R_b + 5K} = \frac{2.341}{5222} \text{ A}$$

9. $$E_{fe} = I_{nfe} + 5K = \frac{2.341}{5222} \times 5000 = 2.24 \text{ V}$$

10. $E_{df} = E_{fe} - E_{de} = 2.24 - 1.62 = 0.62 \text{ V}$ (with point d positive)

11. $$I_{gal} = \frac{E_{df}}{R_{gal}} = \frac{0.62}{2000} = 310 \ \mu\text{A}$$

Notice that this is the same answer as was obtained by the Kirchhoff's laws method of Example 23-3.

23-7 Maximum power transfer theorem.   At first thought, it might seem that the lower the resistance of a load, the higher the current it would draw from the source, and more power is transferred from source to load $(P = EI)$. This would be true, if the source had no internal resistance, and its terminal voltage remained constant. In large power distribution systems, this is "generally" true. Long before the fallacy of the above assumption is "exposed," the excessive current drawn by too low a load resistance would have burnt out the generator, the lines, or the battery supply.

In electronics applications however, the supply source quite often has appreciable resistance. Let us see, in such a case, how a change in load resistance affects the power delivered to the load.

Example 23-8

A dc power source has an EMF of 600 V and an internal resistance of 1000 Ω. Find the power dissipated by the load for various values of load resistance from zero to 20 000 Ω. Plot a curve of power output versus load resistance.

Solution

1. For each value of load resistance (R_L) the total resistance (R_T) is equal to $R_L + 1000$ (the internal resistance).
2. The current (I) in the circuit will be $E \div R_T$ (where $E = 600$ V).
3. The power output (P_o) will be $I^2 R_L$.
4. Using these relations, let us tabulate answers for convenient values of R_L (Table 23-1).
5. The curve of power output versus load is shown in Fig. 23-12.

TABLE 23-1   **Effect of Load Resistance on Power Output**
(R_{int} = 1000 Ω)

R_L (Ω)	R_T (Ω)	I (mA)	P_o (W)
0	1 000	600	0
200	1 200	500	50
500	1 500	400	80
1 000	2 000	300	90
2 000	3 000	200	80
5 000	6 000	100	50
11 000	12 000	50	27.5
19 000	20 000	30	17.1

Notice that a maximum value (90 W) is reached at $R_L = 1000$ Ω. Below or beyond this load value the power output decreases. But notice that 1000 Ω is also the value of the internal resistance of the supply source. This is not a coincidence. *Maximum power will be delivered to the load when the load resistance equals the internal resistance of the supply source.* This can be proven

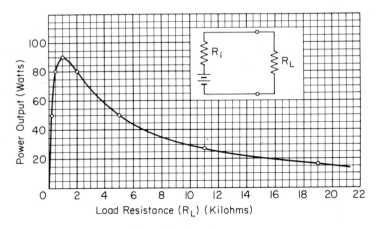

Figure 23-12 Effect of load resistance on power output.

mathematically as follows:

$$I = \frac{E}{R_i + R_L} \tag{1}$$

$$P_o = I^2 R_L = \frac{E^2 R_L}{(R_i + R_L)^2} \tag{2}$$

To solve for maximum power (while varying R_L) it is necessary to differentiate for power with respect to R_L, set the result equal to zero, and solve for R_L in terms of R_i. By this technique:

$$\frac{dP}{dR_L} = \frac{d\left[\dfrac{E^2 R_L}{(R_i + R_L)^2}\right]}{dR_L} = E^2 \left[\frac{(R_i + R_L)^2 \dfrac{dR_L}{dR_L} - R_L \dfrac{d(R_i + R_L)^2}{dR_L}}{[(R_i + R_L)^2]^2}\right] \tag{3}$$

$$= E^2 \left[\frac{(R_i + R_L)^2 - R_L[2(R_i + R_L)]}{(R_i + R_L)^4}\right] \tag{4}$$

$$= E^2 \left[\frac{(R_i + R_L) - 2R_L}{(R_i + R_L)^3}\right] \tag{5}$$

Setting equal to zero, and solving, we get

$$R_i + R_L - 2R_L = 0$$
$$R_L = R_i \tag{6}$$

So far, we have considered the effect of a source with internal resistance. However, this same principle also applies when a source feeds power to a load through any complex network. In such cases, if maximum power is to be delivered to the load, the resistance of the load should *match* (equal) the resistance looking back into the original circuit, with all sources replaced by their internal resistance. [Notice that this is the internal resistance (R_i) of the Thévenin equivalent circuit.]

REVIEW QUESTIONS

1. With reference to Fig. 23-1:
   (a) Which of these diagrams is more appropriate for solution by the *mesh* technique?
   (b) In diagram (a), what does current I_3 consist of?
   (c) In analyzing loop 1, what current value must be considered when evaluating the voltage drop across R_3?
   (d) Starting at point d, write the general Kirchhoff's equation for path *dcbed*.
   (e) Write the mesh equation for loop 2.
   (f) Compare the Kirchhoff's law method of Chapter 22 with the loop method.
2. With reference to Fig. 23-2(a):
   (a) Name the nodal points in this diagram.
   (b) With respect to the supply and nodal voltages, write an equation for current I_1.
   (c) Repeat part (b) for current I_2; for current I_3.
   (d) How are these three currents related?
   (e) How many *nodal equations* are necessary to solve this circuit? Explain.
3. With reference to Fig. 23-3:
   (a) Name the nodal points in this diagram.
   (b) Using supply and/or nodal voltages, write an equation for current I_5.
   (c) Repeat part (b) for current I_2.
   (d) How many nodal equations will be necessary to solve this circuit? Explain.
4. (a) What theorem is illustrated in Fig. 23-5?
   (b) In diagram (a), under what condition can source E_B be replaced by a short circuit?
   (c) Comparing diagrams (a) and (b), why are the current directions through R_1 reversed?
5. With reference to Fig. 23-7:
   (a) What does diagram (a) represent?
   (b) Why is the current in this circuit shown as flowing clockwise?
   (c) What is the "net" voltage in this circuit?
   (d) What does diagram (b) represent?
   (e) Is it always correct to show the supply sources as short circuits? Explain.
   (f) What does diagram (c) represent?
   (g) How is the value for E_{oc} obtained?
   (h) How is the value for R_i obtained?
6. Redraw Fig. 23-6 so as to be suitable for analysis by Thévenin's theorem, considering R_2 as the variable load.
7. With reference to Fig. 23-8:
   (a) What does diagram (a) represent?
   (b) What is this diagram used for?
   (c) What does diagram (b) represent?
   (d) How is the value for I_{sc} obtained?
   (e) How is the value for R_{sh} obtained?
   (f) In diagram (b), if the R_L value is reduced, what happens to the value of I_{sc}? Explain.

8. (a) In Fig. 23-9, under what condition can the two networks be considered equivalent?

(b) It is desired to replace the delta network by its equivalent wye network. What combination of delta values would be used to find the value for R_b? Explain.

(c) Repeat part (b) for R_b in Fig. 23-11(b).

9. (a) Give another name for a wye network.

(b) Where is this nomenclature used?

(c) What is a T network?

10. It is desired to replace a wye network by its equivalent delta. What combination of wye values would be needed to find the value of:

(a) R_2 in Fig. 23-9(a)?

(b) R_2 in Fig. 23-10?

(c) In general, how is the equivalent delta value found?

11. With reference to Table 23-1:

(a) In line 1, since the current is 600 mA, why is the power output zero?

(b) In line 3, the current has decreased to 400 mA, yet the power output has increased. Why?

(c) In line 6, the load resistance is ten times greater than in line 3. Why is the power output less?

(d) At what load value is maximum power output obtained?

PROBLEMS

1. In Fig. 23-1 the circuit values are:

$$E_A = 40 \text{ V}, \qquad R_1 = 10 \text{ }\Omega, \qquad R_3 = 5 \text{ }\Omega$$
$$E_B = 30 \text{ V}, \qquad R_2 = 20 \text{ }\Omega$$

Using the loop system, solve for the voltage across R_3, the current through R_3, and how much of this current is supplied by each source.

2. In Fig. 23-4, the circuit values are:

$$E_A = 20 \text{ V}, \qquad R_1 = R_2 = R_5 = 50 \text{ }\Omega$$
$$E_B = 10 \text{ V}, \qquad R_3 = R_4 = 100 \text{ }\Omega$$

Solve for the voltages across R_3 and R_4 using the loop method.

3. Using the loop method and the circuit shown in Fig. 23-13, solve for the current in R_L and the voltage across R_L.

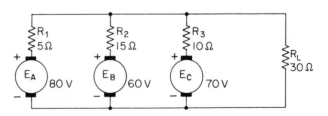

Figure 23-13

4. Using the loop method and the *lattice network* shown in Fig. 23-14, solve for the current through R_L and the voltage across R_L.

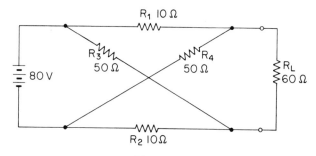

Figure 23-14

5. Using the loop method and the *bridged-T* network of Fig. 23-15, solve for the current through R_L and the voltage across R_L.

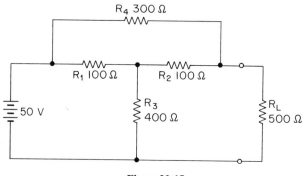

Figure 23-15

6. Repeat Problem 1 using the nodal method.
7. Repeat Problem 2 using the nodal method.
8. Repeat Problem 3 using the nodal method.
9. Repeat Problem 1 using the superposition method.
10. Repeat Problem 2 using the superposition method.
11. Repeat Problem 3 using the superposition method.
12. Repeat Problem 1 using Thévenin's theorem.
13. Repeat Problem 3 using Thévenin's theorem.
14. Repeat Problem 4 using Thévenin's theorem.
15. Repeat Problem 5 using Thévenin's theorem.
16. Repeat Problem 1 using Norton's theorem.
17. Repeat Problem 3 using Norton's theorem.
18. Repeat Problem 4 using Norton's theorem.
19. Repeat Problem 5 using Norton's theorem.

20. A transistor has an output of 26 V, and an internal resistance of 3000 Ω. Find the power it delivers to loads of:
   (a) 1000 Ω.
   (b) 9000 Ω.
   (c) Matched load.

21. (a) Using the circuit of Fig. 23-14, find the power delivered to the load.
   (b) What value of load would take maximum power from this circuit? What is the power output now?

22. Repeat Problem 21 for Fig. 23-15.

part two

Alternating Currents

24

INTRODUCTION TO
ALTERNATING CURRENTS

When electric utilities first started commercial distribution, the power supplied was primarily for lighting, and the distribution system was a two-wire direct-current (dc) system. The first problem they encountered was the choice of a suitable line voltage. They were faced with two conflicting considerations.

1. The higher the line voltage, the lower the current required to meet any given load demand. For example, assuming a load of 100 kW, if a 10-V distribution system were used, the current required to meet this load would be

$$I = \frac{P}{E} = \frac{100\,000}{10} = 10\,000 \text{ A}$$

On the other hand, if the line voltage were raised to 1000 V, the current required for the same load would drop to 100 A. Since the generator itself has resistance, and the distribution lines between the generating station and the consumers have resistance, when current is delivered to the load, there will be voltage drops (IR) and power losses (I^2R). The higher the line voltage, the lower the current and therefore the lower the line drop and line losses, for any given load. High currents present another problem in that larger size feeders must be used and the installation and maintenance costs of the distribution system would be increased. From an economic point of view, the power company would prefer to use as high a line voltage as possible.

2. The second consideration in selecting the line voltage for the distribution system is safety. High voltages can endanger the lives of the consumers. From a safety point of view, the lower the line voltage the better.

Obviously, some compromise was necessary between these two conflicting factors, so — 120 V was selected. It was not too dangerous, and since the lighting load in those early years was quite small, the feeder currents and line losses were within tolerable limits.

24-1 Three-wire distribution system.

The demand for electric power grew, and with the adoption of electric motors for industrial applications, it was no longer practical to meet the added power demand with a 120-V distribution system. This led to the development of a three-wire distribution system (Fig. 24-1). Heavy industrial loads were supplied with 240 V dc to reduce the current requirements, while lighting loads and residential loads were maintained at 120 V. In Fig. 24-1, *A* and *B* represent lighting loads and *C* represents industrial machinery loads. It is interesting to note that if the lighting loads represented at *A* and *B* were perfectly balanced, no current would flow in the neutral wire and it could be removed. However, in practice, perfect balance is impossible and the neutral wire does carry some current.

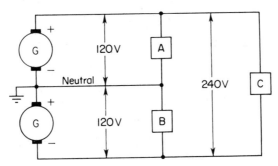

Figure 24-1   Three-wire distribution system.

24-2 AC distribution systems.

The above system proved to be only a temporary solution. More and more homes were electrified and in addition to lighting, electrical appliances came into use. Meanwhile industry, quick to realize the advantage of electric motors in stepping up production, also joined the bandwagon. The demand for electric power zoomed. How could this increased demand be satisfied and line currents still be kept within reason? Higher distribution voltages would solve the problem—but then what about safety? If high *distribution* voltage could be coupled with lower voltage at the consumer area, both aspects of the problem would be solved. This very desirable feature could be achieved by using an alternating-current (ac) distribution system. With an ac supply, a device called a *transformer** can be used at the generating station to step up (increase) the generator voltage before feeding the

---

*Transformer action is not possible in dc circuits, since it depends on the voltage induced by varying magnetic fields (see Chapter 17). Transformers are covered in Chapter 36.

transmission power lines. Then at the local consumer area another transformer can be used to step down (decrease) the voltage to safe values.

There was one fly in the ointment that prevented early adoption of the ac distribution system. Efficient motors with suitable characteristics were not available for operation on ac. So conversion to ac distribution system had to wait till the induction motor was developed. From then on no new dc installations were made, and gradually most of the earlier dc consumers were converted to ac operation. In addition, with improvement in equipment design the actual generated voltage has been steadily increased and distribution voltages are still climbing.

In 1966, 345-kV transmission systems were in use; by the mid-1970s, 765-kV systems were being adopted, and a 1200-kV test line was established in Oregon by the Bonneville Power Administration. Meanwhile, research programs by the Edison Electric Institute were investigating techniques and economic considerations for raising the transmission voltage to 1.5-MV. It is interesting to note that a 345-kV line is capable of carrying about six times the electric power of a 138-kV line. In turn, a 765-kV line can carry approximately five times the power of a 345-kV line, at an investment of only double the cost of the lower voltage line.*

It is fantastic to realize the growth of the electric power industries since Thomas Edison first transmitted power for his electric light bulb. The total capacity of the electric utilities in the United States in 1955 was 116 328 000 kW. By 1968, this capacity rose to over 270 million kW, and to over 600 million kW in 1982. The Consolidated Edison Company of New York alone had a total nameplate capacity of 10 054 MW (as of 1982). This consisted of 6724 MW of conventional units, 1013 MW of nuclear units, and 2317 MW of gas turbines. Their long range plans include construction of additional plants, and increased use of atomic energy as the primary source of power in place of coal or water power.

The distribution system in use is almost entirely ac 60 "cycles," three-phase, four wire.† The amount of dc distribution, as of 1968, was less than 5%. In this connection it is interesting to note that all power generated by the power companies is ac, and where they are obligated to maintain dc service they convert the ac to dc by electronic or rotating rectifiers.

Since 95% of the power delivered to consumers is ac it should be obvious why study of ac fundamentals and the behavior of circuit elements (resistor,

---

*Advances in solid-state technology have led to the use of high-voltage dc transmission systems with: ac generation at the power plant; step up of this voltage to very high transmission levels; rectification of this supply to dc; and transmission of this extra-high-voltage dc. Then, at the consumer area, the supply is inverted back to ac, and stepped down to a safe local distribution value. (For a given power level, a dc transmission system has a lower line-voltage drop, and a lower power loss then the equivalent ac line.)

†See Section 37-10.

inductors, and capacitors) on ac is necessary. This may set off another train of thought. Study of ac may be advisable for those interested in power, but how does it fit in if one's major interest is electronics? To explain this let us discuss, in a general way, a field of electronics with which you are all familiar, and show how ac principles fit in with radio.

In a broad sense radio communications can be considered as a means of transmitting intelligent sounds to a large group of listeners over a wide area without interference from other sounds, intelligent or otherwise. A brief discussion of sound and its characteristics will follow. However, if the student has not already covered this subject in a physics course, any good book on physics is recommended for details.

24-3 Sound. Any sound that is created, whether it is speech, music, ringing of a bell, or just noise, creates *pressure changes* in the air around the source of the sound. Normal air pressure is approximately 100 kilopascals.* We will use this value as our reference pressure (zero). Sounds will cause the air pressure alternately to rise above normal, drop back to normal, drop below normal and again return to normal. This *cycle* or series of variation of air pressure changes is repeated over and over again as long as the sound continues. We can represent this by the pressure versus time curve shown in Fig. 24-2.

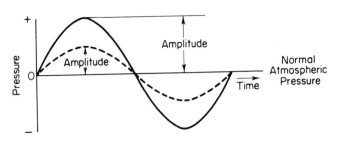

Figure 24-2 Sound waves.

The louder the sound, the greater the change in pressure or the greater the *amplitude* of the sound wave. As the pressure of the air surrounding the sound source changes, it reacts on the air just beyond—pushing it, or allowing it to expand. So, the air pressure just beyond also rises above and falls below normal in accordance with the vibrations of the sound source. This effect is transmitted to air farther and farther away. In this manner, pressure variations travel outward from the sound source in all directions. Sound is transmitted through the air. When these varying air pressure waves strike a listener's ear—through the mechanism of the ear, nervous system, and brain—these pressure changes register as sounds.

*This is the SI unit for newtons per square meter. In English units this value corresponds to approximately 15 lb/in².

Sound waves travel through the air at a speed of approximately 335 m/s.* (This speed is affected by the temperature and barometric pressure of the air.) If the listener is located some distance from the sound source it will take time for the sound to reach him. The truth of this statement should be obvious. If you have ever watched a baseball game from the bleachers, you see the batter hit the ball, and a short time later you hear the crack of the bat. Another example is lightning followed later by thunder.

If a sound source creates 100 "cycles" of air pressure variations in 1 second, the sound is said to have a *frequency* of 100 "cycles per second" (cps).† Such a sound would be recognized as a note of low pitch. A high-pitched sound, such as a whistle, might produce 10 000 cycles of pressure variations in one second. Its frequency would be 10 000 Hz. The *audio frequency* range is from 20 to 20 000 Hz. The term audio frequency is used to designate frequencies that can be heard (audible) directly by the human ear. Sounds below 20 Hz are also audible, but they are heard as separate sounds (such as the staccato tappings of a woodpecker) rather than a continuous tone. Sound waves above 15 000 Hz may be audible to some people. Probably very few, if any, human beings can hear 20 000 Hz.

Musical instruments produce frequencies as low as 20 Hz, and up to as high as 15 000 Hz. Yet if the frequencies below 50 and above 8000 Hz were lost, the sound would lose little of its quality. In fact, if only the frequencies between 100 and 5000 Hz were heard, musical instruments could be recognized and the quality, although not true, would still be acceptable. Real high-pitched sounds would of course be lost. For speech, the most important frequencies lie between 200 and 3000 Hz. However, speech does contain frequencies as low as 100 and as high as 10 000 Hz.

Obviously, the simplest way to transmit intelligent sounds (music or speech) to a group of listeners is to gather them into an auditorium or open space. Equally obvious is the fact that only a few people would hear these sounds. Special acoustically designed auditoriums, larger orchestras, or careful selection of speakers with loud, carrying voices would help, but the coverage would still be relatively low. For speech, a megaphone would help. In this way sounds that would normally be lost in back of the speaker are directed out to the audience, increasing the speech energy in the forward direction.

24-4 Audio amplifier systems.   If we could amplify the sounds before transmitting them to the audience, we could reach larger groups of people. This can be done electronically by use of microphones, amplifiers, and loud-speakers. Such a system is shown in Fig. 24-3.

The *microphone* picks up the pressure-variation sound waves from the sound source and converts them into a varying electrical potential (or current).

*In English units this is approximately 1100 ft/s.
†The SI unit for frequency, in cycles per second, is the hertz (Hz).

These potentials have the same frequencies as the frequencies produced by the original sound source, but they are very weak. The *amplifier*, by means of transistors, takes energy from the *dc power supply* and builds up the energy of these "signals." The amplified electrical signals are then fed to a *loudspeaker* that converts the variations in the potential of the electrical signals back into air

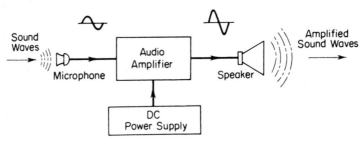

Figure 24-3   Audio amplifier system.

pressure variations or sounds. The sounds coming from the loudspeaker will have the same characteristics as the original sounds. But now they are very much louder. In order to reproduce all sounds faithfully, each component of the amplifier system must be capable of passing all audio frequencies with equal efficiency. An ideal high-fidelity system should amplify all frequencies from 20 to 20 000 Hz without discrimination.

By use of such amplifier systems, the number of listeners that can be reached is greatly increased. Even so, the size of audience and the distance covered by amplifier systems is still quite restricted. You have heard public address systems, so you know their range limitations. To increase range by increasing the power of the amplifier beyond certain limits would make the sound levels, near the speaker, too loud to be comfortable.

24-5 Principle of radio communication. Let us suppose we could convert these sound waves into some type of inaudible waves that could be transmitted through the air, and then, at many remote listening locations, reconvert these new waves back into sound waves. This would be a means of transmitting sound intelligence to a widely scattered audience. In this way the sound could not be heard except at the special listening points, and we would avoid the extremely loud sound levels near the original sound source. The problem is: What type of wave can travel through the air and be made to transmit sound?

In Chapter 14 we saw that:

1. A wire carrying current produced a magnetic field.

2. Magnetic fields reached out into space.

3. Magnetic fields were energy fields.

4. If a conductor were located in a varying magnetic field, a voltage was induced in the conductor.

5. The greater the rate of change of flux (i.e., rate of cutting), the greater the induced voltage.

There it is, five simple steps—can you see what is coming? A microphone will convert sound waves into electrical potentials or currents. An audio amplifier will increase the energy of these electrical signals. With sufficient amplification, we can build these signals up into very large currents, varying in magnitude at the same frequency as the original sound source. Now these currents can be made to flow through a wire (transmitting antenna). The magnetic field around this wire will reach far out into space. The magnetic field strength will vary exactly with the variation in signal current and sound pressure. We have transmitted intelligence into space.

In this discussion, the electromagnetic waves transmitted were assumed to be varying directly within the audio-frequency range. This would result in a serious disadvantage.

If several different programs were being transmitted at the same time, the air would be filled with electromagnetic waves from different sound sources, all in the same frequency range of 20 to 20 000 Hz. The listener would hear all of these programs at the same time—bedlam!

24-6 Simple radio transmitter. The solution to the above problem is to transmit the electromagnetic waves at frequencies much higher than the audio frequencies themselves. This higher frequency is called the *carrier frequency* of the transmitting station. By assigning a different carrier frequency to each station, each program would then have its own private "channel." The listener can then select the channel he wishes to hear. This eliminates the bedlam of sounds produced when transmitting the audio-frequency waves directly.

Figure 24-4 shows, in block form, the basic equipment necessary for a radio transmitter.

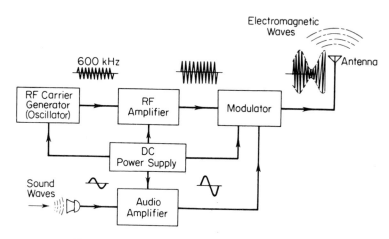

Figure 24-4 Block diagram of a radio transmitter.

Let us review the purpose of each unit.

1. *Oscillator.* This unit generates the much needed radio-frequency carrier wave. The oscillator circuit must have some means for "tuning" it to the desired carrier frequency. (You will recall that each transmitting station should use a different channel or carrier frequency.)

2. *RF Amplifier.* The output of any practical oscillator is rather weak. We must build up the energy in the carrier wave if we wish the electromagnetic field of the transmitter to reach far out into space. The purpose of the RF amplifier is to increase the energy or power level of the RF carrier to the required value.

3. *Microphone:* Converts sound (air pressure) waves into electrical waves of the same (audio) frequencies.

4. *Audio Amplifier.* Increases the energy or amplitude of the audio-frequency waves.

5. *Modulator.* Raises the audio intelligence to the desired high frequency. This process is known as *modulation.* Strong, modulated carrier-frequency currents are developed in the output of this stage.

6. *Transmitting Antenna.* Carries the strong modulated carrier-frequency currents and allows the electromagnetic field around itself to reach out or *radiate* into space.

7. *DC Power Supplies.* In the various units of the transmitter, signals are generated or amplified by solid-state or vacuum-tube circuits. The energy needed for these functions is supplied by the dc power supplies. In general, power supplies are operated from the ac house supply. They take energy from the ac line and convert it to ac or dc of proper voltage as required by the semiconductors or vacuum tubes. These, in turn, take energy from the power supply and convert this energy into increased signal energy.

24-7 Simple radio receiver. Now that we have produced this modulated carrier wave and have sent it out into space as an electromagnetic wave, let us examine the equipment required at the receiving location so that we can recover the original intelligence. For analysis, the receiver functions can be broken down into steps (see Fig. 24-5)

1. It must intercept these electromagnetic waves and convert them into varying electrical potentials. This is done by the *receiving antenna.* The antenna should be located in free space, where it can be cut by the electromagnetic waves of any transmitting stations in the area. The induced voltage in the antenna will have the same shape as the modulated carrier. Of which transmitting station? It will develop induced voltages for *all transmitting stations* within reach. But this sounds like bedlam again! No—each induced voltage will be of a different frequency, corresponding to the frequencies of the carriers from the various transmitters.

2. The receiver must be able to select one carrier frequency at a time and reject all other carrier frequencies. This is done by the *radio-frequency tuner and amplifier*. By means of capacitors and inductors, the tuner can be made to select only one frequency. If either the capacitance or the inductance is variable, the frequency selected can be changed. Amplification is necessary in the tuner because the induced voltage in the antenna is very low. The signal strength must be built up to higher amplitudes. Since the signals are radio frequencies, this section of the receiver is called the *radio-frequency amplifier*.

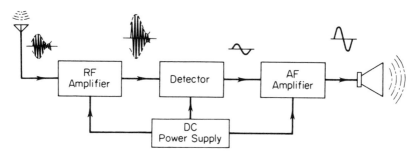

Figure 24-5   Receiver block diagram.

3. The next step is to get rid of the carrier frequency and retain only the audio components of the modulated wave. This is done by the *detector*, or *demodulator*.

4. Now we have audio frequencies again, but they are too weak to drive a loudspeaker and produce sufficient volume. An *audio-frequency amplifier* will build up these audio signals to the required amplitude and energy level.

5. Finally, a loudspeaker is used to convert these amplified audio-frequency electrical waves back into sound.

6. Again, a dc power supply is needed to supply the transistors in the above circuits with the energy that they will need to accomplish their specific purposes.

A block diagram of a receiver is shown in Fig. 24-5.

24-8 Radio-frequency bands.   In transmitting radio waves, we have shown that the carrier frequency used must be higher than audio frequencies. The transmitter used as an illustration had a carrier frequency of 600 kHz. What other frequencies are used for electronic work? The radio-frequency spectrum has been divided into five bands depending on the characteristics of these frequencies. These bands are shown in Table 24-1.

Radio frequencies extend beyond the limit of 3000 MHz shown in the table. Electronic equipment is already operating at frequencies around 30 000 MHz,* and development work is being done at even higher frequencies. The

*The K-band police speed radar frequency is on 24.15 GHz (24 150 MHz).

TABLE 24-1   Radio-Frequency Spectrum

Band	Frequency	Range Day	Range Night	Power Required
Low frequency (LF)	30–300 kHz	Long	Long	Very high
Medium frequency (MF)	300–3000 kHz	Medium	Long	High to medium
High frequency (HF)	3–30 MHz	Short	Medium to long	Medium
Very high frequency (VHF)	30–300 MHz	Short	Short	Low
Ultra high frequency (UHF)	300–3000 MHz	Short	Short	Low

radio-frequency spectrum extends as high as 500 000 MHz, although little is known about the characteristics of these "super-high" frequencies.

We have discussed two types of waves and have given their frequency range (air pressure or sound waves—20 to 20 000 Hz, and electromagnetic or radio waves—30 kHz to 500 GHz). Notice that the radio waves start just about where the sound waves leave off. The frequencies in between the audible sound waves and the radio waves are known as *supersonic waves*. These are "sound waves" just beyond the human audibility range. Supersonic waves are used in under water detection and depth measurement devices, and in a number of industrial applications such as cleaning and drilling.

From the preceding, it is not surprising to learn that there are other types of waves occupying the region above radio waves. To make the wave-spectrum picture complete, these waves will be listed with their approximate frequencies:

1. *Heat Waves:* from 5×10^{13} Hz to 1.5×10^{14} Hz.
2. *Light Waves* (from infrared to ultraviolet): from 1.5×10^{14} Hz to 1×10^{15} Hz.
3. *X-Rays:* from 5×10^{16} Hz to 1.5×10^{20} Hz.

You may have noticed that there still are "holes" in this frequency spectrum. At present little is known about waves in these regions.

REVIEW QUESTIONS

1. (a) What is the advantage of high voltage in a distribution system?
   (b) What is the advantage of low voltage?
2. Why was the three-wire dc distribution system preferable to the simple two-wire system?
3. (a) Explain briefly the advantage of an ac distribution system over dc.
   (b) What device makes this advantage possible?

4. **(a)** What type of distribution system is in common use today?
   (b) What maximum distribution voltage is being planned?
5. **(a)** What is a sound wave?
   (b) Describe how sound is transmitted through the air.
6. **(a)** What is the difference between a low-frequency and high-frequency sound? Give an example of each.
   (b) What determines the amplitude of a sound wave?
7. **(a)** What is the frequency range of sound waves?
   (b) What is the minimum frequency band for acceptable musical reproduction?
   (c) What is the frequency band for high-fidelity reproduction?
8. Name the components of an audio amplifier system, and give the function of each component.
9. What are the advantages of radio communication over direct sound communication?
10. **(a)** What is meant by *carrier frequency*?
    (b) Why is it necessary to use a carrier frequency for radio communication?
11. Name the components of a simple radio transmitter, and describe briefly the purpose of each component.
12. Name the components of a simple receiver, and state briefly the function of each component.
13. What frequency bands are normally used for radio-frequency transmission?

25

CHARACTERISTICS
OF SINE WAVES

In the preceding chapter we saw that the basic source of electric power in the home and in industry is alternating current (ac). We also saw that the *signals* in electronic equipment vary in amplitude. They too are some form of alternating current. So, whether our interest is in power or electronics, it becomes necessary to study the nature of alternating currents, and to analyze the behavior of circuit components (*R-L-C*) on such a supply source. This will be the aim of the chapters that follow. Since the simplest type of ac wave is the *sine wave*, our study will include an analysis of this wave. But first, let us review some basic trigonometry so that we can define the term "sine."

25-1 Trigonometry for ac circuits. Any three-sided figure is called a *triangle*—because it includes three angles. All the angles of the triangle may be acute angles (less than 90°); or one of the angles may be exactly 90°; or one of the angles may be an obtuse angle (greater than 90°). In all cases, however, the sum of the angles in any triangle is equal to 180°. Three such typical triangles are shown in Fig. 25-1.

Solution of problems in technical fields often requires the addition or subtraction of vector quantities—for example, forces or velocities in mechanics; voltages, currents, or impedances in electricity and electronics. These problems can be solved mathematically by the use of trigonometry. Although there are trigonometric formulas and methods for handling acute and obtuse triangles [such as Fig. 25-1(a) and (c)], it is distinctly preferable in electrical problems to use *right-triangle methods* throughout. Any other triangle can be resolved into

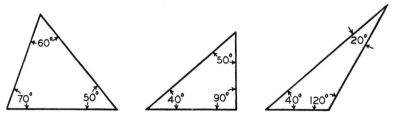

Figure 25-1 Typical triangles.

equivalent right triangles. The discussion and the formulas that follow are based on right-triangle principles.

Lines *A–B–C* and *A–D–E* in Fig. 25-2 form an angle at their junction. If you measure this angle with a protractor, you will find that it is 25°. *Regardless of the lengths of these lines (A–B–C and A–D–E), this angle does not change.*

From points *B* and *C*, let us drop two perpendicular lines down to the base line to form right angles at *D* and *E*, respectively. Now we have formed two right triangles *B–A–D* and *C–A–E*. Remembering that the sum of the angles of any triangle is 180°, and since these triangles have a common angle of 25° and a right angle of 90° (at *D* and *E*), the remaining angle (at *B* and *C*) must in each case be 65°. These two triangles are said to be *similar triangles*. (Any two triangles whose angles are equal are called similar triangles.) This is true regardless of the overall size of the triangles.

Figure 25-2 Formation of right triangle.

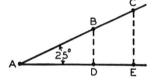

Whenever triangles are similar, the ratio of any two sides of one triangle is equal to the ratio of the corresponding two sides of the other triangle. For example in Fig. 25-2:

$$\frac{BD}{AD} = \frac{CE}{AE} = \text{a constant}$$

also

$$\frac{BD}{AB} = \frac{CE}{AC} = \text{another constant}$$

and

$$\frac{AD}{AB} = \frac{AE}{AC} = \text{another constant}$$

The values of these constants depend only on the size of the angle at *A*. (Of course, it must be realized that if this angle is increased, the angles at *B* and C must decrease, because the sum of all three angles must remain $180°$). In other words, we could also have said that the value of the ratio constant also depends on the size of the angles at B or C.

25-2 Functions of a right triangle.   For any given right triangle we therefore have three different constants depending on which two sides we used for our ratio. Also, the value of these constants varies with the size of the angles—or is a function of the angle. In order to identify which of the three ratios we are referring to at any one time, the ratios are specified by naming the three functions of the angle: the **sine**, the **cosine**, and the **tangent** of the angle.

Figure 25-3 shows a typical right triangle. The Greek lowercase letter theta (θ) is used to identify the angle under discussion in this triangle. Let us evaluate the constants for this angle. Remember that the actual size of the triangle does not matter as long as angle θ is of fixed value. First let us give appropriate names to the sides of this triangle:

1. The side opposite the $90°$ angle ($A–B$) is always called the **hypotenuse**.

2. Side $A–C$ is called the side *adjacent*. Of course, the hypotenuse is also an adjacent side—but it already has its own name. There should be no confusion as to which side we mean by adjacent.

3. Side $B–C$, since it is opposite to the angle under discussion, is called the side *opposite*.

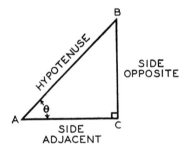

Figure **25-3** Evaluation of trigonometric functions of a right triangle.

If we were discussing the angle at B, it should be obvious that the hypotenuse would remain the same, but the side adjacent and side opposite would be reversed.

Now let us evaluate the functions of angle θ. By definition:

1. The *sine* of an angle is the ratio of the length of the side opposite to the length of the hypotenuse, or

$$\sin \theta = \frac{\text{opposite}}{\text{hypotenuse}} \qquad \textbf{(25-1A)}$$

2. The *cosine* of an angle is the ratio of the length of the side adjacent to the length of the hypotenuse, or

$$\cos \theta = \frac{\text{adjacent}}{\text{hypotenuse}} \qquad \text{(25-1B)}$$

3. The *tangent* of an angle is the ratio of the length of the side opposite to the length of the side adjacent, or

$$\tan \theta = \frac{\text{opposite}}{\text{adjacent}} \qquad \text{(25-1C)}$$

We know that the value of the above ratios depends *not* on the size of the triangle but only on the size of the angles. The question now is in what manner do these functions vary with size of angle? Let us check this relation for angles between zero and 90°. Figure 25-4 shows four triangles all with hypotensues of equal length. Let us make this length 10 units. (If you see only two triangles, just wait and you will see all four.)

1. *For $\theta = 0°$ (θ_1).* The triangle automatically collapses into a straight line. The hypotenuse is $O–C$; the adjacent side is also $O–C$; and side opposite is *zero*.

(a) $\qquad\qquad \sin \theta_1 = \dfrac{\text{opposite}}{\text{hypotenuse}} = \dfrac{0}{10} = 0.000$

(b) $\qquad\qquad \cos \theta_1 = \dfrac{\text{adjacent}}{\text{hypotenuse}} = \dfrac{10}{10} = 1.000$

(c) $\qquad\qquad \tan \theta_1 = \dfrac{\text{opposite}}{\text{adjacent}} = \dfrac{0}{10} = 0.000$

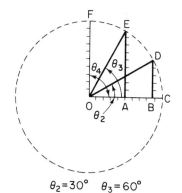

Figure 25-4 Variation of trigonometric functions from 0 to 90°.

$\theta_2 = 30°$ $\theta_3 = 60°$

2. *For $\theta = 30°$ (θ_2).* The hypotenuse is now $O–D$ (still 10 units); the side opposite has increased to $B–D$ (5.00 units); the side adjacent has decreased to $O–B$ (8.66 units).

(a) $\qquad\qquad \sin \theta_2 = \dfrac{\text{opposite}}{\text{hypotenuse}} = \dfrac{5.00}{10} = 0.500$

(b) $\qquad\qquad \cos \theta_2 = \dfrac{\text{adjacent}}{\text{hypotenuse}} = \dfrac{8.66}{10} = 0.866$

(c) $\qquad \tan \theta_2 = \dfrac{\text{opposite}}{\text{adjacent}} = \dfrac{5.00}{8.66} = 0.577$

3. *For $\theta = 60°$ (θ_3).* The hypotenuse $O-E$ is 10 units; the side opposite $A-E$ is 8.66 units; side adjacent $O-A$ is 5.00 units.

(a) $\qquad \sin \theta_3 = \dfrac{\text{opposite}}{\text{hypotenuse}} = \dfrac{8.66}{10} = 0.866$

(b) $\qquad \cos \theta_3 = \dfrac{\text{adjacent}}{\text{hypotenuse}} = \dfrac{5.00}{10} = 0.500$

(c) $\qquad \tan \theta_3 = \dfrac{\text{opposite}}{\text{adjacent}} = \dfrac{8.66}{5.00} = 1.732$

4. *For $\theta = 90°$ (θ_4).* Again the triangle collapses into a straight line ($O-F$). The hypotenuse $O-F$ is 10 units; the side opposite is also $O-F$ (10 units); the side adjacent is now zero.

(a) $\qquad \sin \theta_4 = \dfrac{\text{opposite}}{\text{hypotenuse}} = \dfrac{10}{10} = 1.000$

(b) $\qquad \cos \theta_4 = \dfrac{\text{adjacent}}{\text{hypotenuse}} = \dfrac{0}{10} = 0.000$

(c) $\qquad \tan \theta_4 = \dfrac{\text{opposite}}{\text{adjacent}} = \dfrac{10}{0} = \text{infinity} (\infty)$

To summarize, as the angle increases from zero to 90°:

1. The sine of the angle *increases* from zero to a maximum of 1.0.
2. The cosine of the angle *decreases* from 1.0 to zero.
3. The tangent of the angle *increases* from zero to infinity.

Notice in the above illustrations, that the sine of 30° is the same as the cosine of 60°. This same coincidence will be found for any other pair of angles whose sum is 90° (complementary angles). For example, the sine of 15° would equal the cosine of 75°.

To find the trigonometric functions or constants for angles of any value, we could, of course, draw the angle accurately, construct a right triangle, scale each side accurately, and get the ratios of the sides. This is laborious, and the graphical results are not too accurate. Luckily, this work has all been done, and these values can be found in published tables of trigonometric functions. These values are readily obtained using a calculator. First enter the angle value, and then press the appropriate key (SIN, COS, or TAN).

There are many other triangle relations covered in a complete course in trigonometry. However, only one other relation is of importance in electrical or electronic work. It is the relationship between the length of the three sides of a right triangle, or the *theorem of Pythagoras:*

The square of the hypotenuse is equal to the sum of the squares of the other two sides.

Expressing this theorem mathematically, if c is the hypotenuse, and a and b are the other two sides of a right triangle, then

$$c^2 = a^2 + b^2$$

This gives us two ways to solve for the hypotenuse of a right triangle:

1. If we know the other two sides,

$$\text{hyp} = \sqrt{(\text{adj})^2 + (\text{opp})^2} \qquad \textbf{(25-2A)}$$

and transposing this,

$$\text{adj} = \sqrt{(\text{hyp})^2 - (\text{opp})^2} \qquad \textbf{(25-2B)}$$

$$\text{opp} = \sqrt{(\text{hyp})^2 - (\text{adj})^2} \qquad \textbf{(25-2C)}$$

2. If we know one of the acute angles (θ), and one side, we can use the sine or cosine relations. From equation (25-1A),

$$\text{hyp} = \frac{\text{opp}}{\sin \theta}$$

or

$$\text{opp} = \text{hyp} \times \sin \theta$$

and from equation (25-1B),

$$\text{hyp} = \frac{\text{adj}}{\cos \theta}$$

or

$$\text{adj} = \text{hyp} \times \cos \theta$$

25-3 Generation of sine waves. If a conductor is rotated in a magnetic field, the conductor cuts the lines of force and a voltage is induced in the wire. But we know that the induced voltage varies with the rate of cutting, and that the polarity of the induced voltage depends on the direction of cutting. Even though the speed of rotation is constant, the rate of cutting and the direction of cutting *at any instant* will depend on the position of the conductor. Figure 25-5 shows eight positions of a conductor as it is rotated in a magnetic field.

Position 1: The conductor is starting to move up. At this instant it is moving parallel to the magnetic field. Rate of cutting is zero, and the induced voltage is zero.

Position 2: The conductor is moving up. At this instant it has rotated through 45°. It is now cutting the magnetic field, but it is cutting at a 45° angle. The rate of cutting has increased, but it has not reached a maximum. A voltage will be induced in the wire. This voltage is not at its maximum value; it is still rising. Since the conductor is moving up, the polarity of the induced electromotive force (EMF) is such as to make the far end of the conductor positive and the near end negative. Check this polarity by means of Lenz's law or Fleming's hand rule.

Position 3: The conductor has now rotated through 90° and is moving perpendicular to the field. The rate of cutting is maximum. The induced voltage is also at its maximum value. Since the conductor is still moving up, the polarity, as before, will make the far end of the conductor positive and the near end negative.

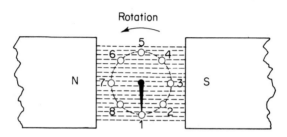

Figure 25-5   Variation of speed of cutting with conductor position.

Position 4: The conductor has rotated through 135°. It is again cutting at a 45° angle. The rate of cutting has dropped and is dropping further. The induced voltage will be the same as for position 2, but now it is decreasing. Since the conductor is still moving up, the polarity has not changed.

Position 5: The conductor has turned through 180°. It is moving parallel to the lines of force—rate of cutting is zero; induced voltage is zero. The conductor is now starting to move down through the mangetic field.

Position 6: The angle through which the conductor has rotated is 225°. Again the wire is cutting the field at an angle of 45°. This induced voltage will have the same magnitude as in position 2. But this time the wire is moving *down* through the magnetic field. The polarity has reversed—the far end of the wire is now negative.

Position 7: The conductor has rotated through 270°. It is cutting directly across the lines of force. The rate of cutting is maximum; the induced voltage is maximum. (Compare this with position 3.) Since the wire is moving down through the field, its far end is negative.

Position 8: The conductor has rotated through 315° and is cutting the field at a 45° angle. The value of induced voltage is the same as for position 2. The speed of cutting is dropping; the induced voltage is also dropping. Since the conductor is still moving down, the polarity of the induced voltage will make the far end of the conductor negative.

Position 1: The coil has completed one revolution. It is again moving parallel to the lines of force. The induced voltage will be zero.

With each succeeding revolution, the process just described in detail will take place: The induced voltage will vary in magnitude from zero to some maximum value; back to zero; reverse in polarity and rise to maximum value; back to zero; reverse in polarity; and repeat this sequence of events over and over again.

You should have noticed that:

1. Between 0 and 90° rotation, the voltage is rising to its maximum value.
2. At 90° the voltage is a maximum.
3. Between 90 and 180°, the voltage is dropping back to zero.
4. At 180° the voltage is again zero.
5. Between 180 and 270° rotation, the voltage is again rising to a maximum.
6. At 270° the voltage is again a maximum.
7. Between 270 and 360° the voltage is again dropping back to zero.
8. At 360° (zero) the voltage is once more zero.
9. Between 0 and 180°, the potential of the far end of the conductor is positive.
10. Between 180 and 360°, the potential of the far end of the conductor is negative.

But this variation in potential is exactly the same relation as the variation of the sine of an angle between 0 and 360°! In other words, the voltage induced in a conductor at any instant during its rotation is some portion of its maximum value (E_m), depending on the sine of the angle through which the coil has rotated. From this we can write the equation for the instantaneous voltage (e) induced in the conductor:

$$e = E_m \sin \theta \qquad\qquad (25\text{-}3)$$

where θ is the angle through which the coil has rotated. Since this is the equation of a sine curve, the potential developed in the conductor is referred to as a *sine wave*.

In commercial generators one conductor alone would not develop sufficiently high voltages. Since the magnitude of induced voltages depends on the number of turns that are being cut by the magnetic field, generators use coils of many turns instead of a single conductor. This method of generating alternating currents is used for power work (machinery, lighting, etc.) where the frequency is comparatively low (60 Hz). In electronic applications where the frequencies are much higher, sine waves are generated by electronic circuits called *oscillators*. Carrier frequencies used by transmitting stations are sine waves.

25-4 Plotting of sine waves.   There are two methods by which we can plot the curve for the sine wave of induced voltage. They are best illustrated by an example.

Example 25-1

A generator develops 100 V as its maximum value. Plot the curve for the instantaneous value of induced voltage versus degrees of rotation.

Solution A: trigonometric plot   Using the equation $e = E_m \sin \theta$, and 100 V for E_m, calculate the value of e for several values of the angle θ. The more points used, the more accurate the plot. Let us use 30° intervals, as a minimum: To evaluate the sine of the angles, we can use a calculator. Merely enter the angle value and $\boxed{\sin}$. Be sure that the calculator is set for degrees, and not for radians.

Values of e and the resulting curve are shown in Fig. 25-6.

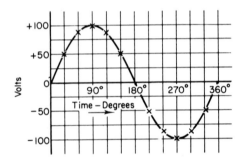

θ	$\sin \theta$	e	θ	$\sin \theta$	e
0°	0.000	0.00	180°	0.000	0.00
30°	0.500	50.0	210°	−0.500	−50.0
60°	0.866	86.6	240°	−0.866	−86.6
90°	1.000	100.0	270°	−1.000	−100.0
120°	0.866	86.6	300°	−0.866	−86.6
150°	0.500	50.0	330°	−0.500	−50.0
180°	0.000	0.00	360°	0.000	0.00

Figure 25-6 Sine wave—mathematical plot.

Solution B: radius vector method   This is a graphical solution. Pick a scale suitable to the size of paper you will use for the plot. (As in all graphical solutions, the larger the scale, the more accurate the plot.) Using the value of E_m as a radius, draw a circle. Since our plot is to be for 30° intervals, draw radii every 30°. The vertical projection of these radii is equal to the sine value for the respective angles. Since the radius itself is E_m, each vertical projection is equal to $E_m \sin \theta$ or e! The solution is shown in Fig. 25-7.

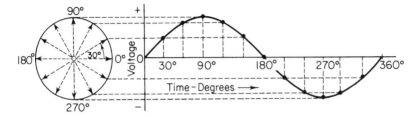

Figure 25-7   Sine wave—graphical plot.

25-5   Cycle.   With every revolution of a coil in a magnetic field, the induced-voltage wave goes through the series of variations shown in Figs. 25-6 and 25-7. Since it is a repetitive or periodic function, one complete series of events (the waveshape corresponding to one revolution) is called a *cycle*. Figure 25-8 shows three cycles of a sine wave.

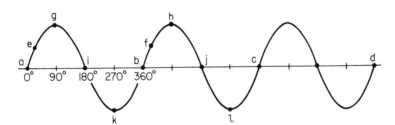

Figure 25-8   Sine wave—showing cycles.

It is obvious from our definition of a cycle that distance *a-b*, *b-c*, and *c-d* each represent one cycle; also that *a-c* is two cycles, and *a-d* is three cycles. A cycle does not have to start at zero degrees and end at 360°. A cycle can be measured from any point in one revolution of the coil to the corresponding point in the next revolution. For example: *g-h* is one cycle (90° point to 90° point); *i-j* is one cycle (180° point to 180° point); similarly, *k-l* is one cycle and *e-f* is one cycle!

How many cycles does *g-k* represent? Since it spans 90 to 270° of the same revolution, (or a total of 180°), it is only $\frac{1}{2}$ cycle. As for *a-h*, it spans one complete cycle from *a-b*, plus $\frac{1}{4}$ cycle from *b-h*, or a total of $1\frac{1}{4}$ cycles. Similarly, *a-l* would be $1\frac{3}{4}$ cycles, and *i-d* would be $2\frac{1}{2}$ cycles.

25-6 Frequency and angular velocity. The *frequency* of any wave is a measure of how many cycles occur in one second. For example, in the two pole machine of Fig. 25-5, if the generating coil made one revolution in one second, and since it produced one cycle in one revolution, the frequency of the sine wave would be one cycle per second or one hertz. If the speed of rotation were 100 revolutions per second, the frequency would be 100 Hz. What would be the frequency of the sine wave produced, if the speed of rotation were 1200 revolutions *per minute*? 1200 revolutions per minute would mean 20 revolutions *per second*, or 20 Hz.

It is obvious that frequency is directly related to speed of rotation, or the *angular velocity* of the rotating coil. There are a number of ways of expressing angular velocity. You have already seen one way: revolutions per second (rps) or revolutions per minute (rpm). In its basic form, angular velocity is a measure of the angle through which the coil has rotated per unit time. Angular velocity can therefore be expressed in degree per second, or preferably in **radians per second**. When measured in radians per second, angular velocity is denoted by the Greek letter omega (ω).

To explain the significance of angular measurements in radians, let us review some simple geometry. Figure 25-9(a) shows a circle with radius r and diameter D. If we use the diameter as a flexible meauring tape we would find that the circumference of the circle (C) is a little longer than three times the diameter, the exact value being 3.141 59 + times the diameter. This ratio (approximately 3.14) is true for any circle. For convenience, this ratio, since it is constant, is assigned the symbol π (pi). Expressing the above relationship by formula, we get

$$C = \pi D$$

or in terms of the radius

$$C = 2\pi r$$

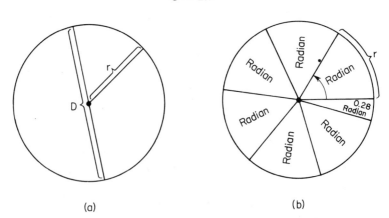

(a)   (b)

Figure 25-9 Radian measure.

Now let us divide the circumference into arcs equal in length to the radius, and join the ends of each arc to the center of the circle. The angle subtended at the center of the circle by an arc equal in length to the radius is called a *radian*. Since the circumference of the circle equals $2\pi r$, there must be 2π radians (or approximately 6.28 rad) in a circle. But we know that there are 360° in a circle. Therefore, 2π radians equals 360°; π radians equals 180°; and 1 rad equals $180/\pi$ (or approximately 57.3°). This relationship can be expressed as

$$\text{degrees} = \text{radians} \times \frac{180}{\pi} \qquad \textbf{(25-4)}$$

Example 25-2

Change 1.5 rad into degrees.

Solution

$$\text{Angle, in degrees} = \text{radians} \times \frac{180}{\pi} = \frac{1.5 \times 180}{\pi} = 86°$$

Example 25-3

A coil has rotated through 210°. What is this angle in radians?

Solution

1. The numerical answer in radians must be smaller.

2. $$\text{Radians} = \text{degrees} \times \frac{\pi}{180} = \frac{210\pi}{180} = 3.67 \text{ rad}$$

Let us get back to angular velocity. What is the relation between angular velocity and frequency? If a coil is generating a sine wave of 60 Hz, it must be making 60 rps. Since each revolution is 2π radians, the angular velocity is

$$\omega = 2\pi \times 60 = 377 \text{ rad/s}$$

Since frequency and revolutions per second are equal, angular velocity is directly proportional to frequency, or

$$\omega = 2\pi f \qquad \text{radians per second} \qquad \textbf{(25-5A)}$$

and since $2\pi = 360°$,

$$\omega = 360f \qquad \text{degrees per second} \qquad \textbf{(25-5B)}$$

Now we are ready to relate frequency and angular velocity to a sine wave. Previously, we developed the equation for the voltage at any instant as $e = E_m \sin \theta$, where θ was the angle through which the coil has turned at any instant. The angle θ must depend on the angular velocity of the coil and the time (t) that the coil has been rotating. This angle, *in radians*, is

$$\theta = \omega t \qquad \textbf{(25-6)}$$

An example will best tie this all together.

Example 25-4

An ac generator produces a sine wave having a frequency of 60 Hz and a maximum value of 100 V. What is the voltage developed at 0.005 s?

Solution

1. $\theta = \omega t = 360 \times f \times t = 360 \times 60 \times 0.005 = 108°$

2. $e = E_m \sin \theta = 100 \sin 108° = 100 \times 0.951 = 95.1$ V

Calculator entry:

$$1, 0, 0, \boxed{\times}, 1, 0, 8, \boxed{\sin}, \boxed{=}$$

From the above discussion, particularly since we have shown that $\theta = \omega t$, the equation of instantaneous voltage for a sine wave can now be written as

$$e = E_m \sin \omega t \qquad\qquad (25\text{-}7)$$

This form of the equation is more commonly used since it expresses frequency ($\omega = 2\pi f$) and time.

In Figs. 25-6 and 25-7 we plotted the equation for the instantaneous voltage of a sine wave ($e = E_m \sin \theta$). The plot was for voltage versus angle of rotation. Since time (in seconds) was not involved in the equation, such a plot would represent a sine wave of any frequency. Based on our new equation ($e = E_m \sin \omega t$), we can now plot a sine wave as voltage versus *time*. As an illustration, let us plot a sine wave having a frequency of 10 Hz. The time for one cycle will be $1/f$ or 0.1 s. A half-cycle (180°) would require 0.05 s; a quarter-cycle (90°) would take 0.025 s. A plot for this sine wave is shown in Fig. 25-10 (solid curve).

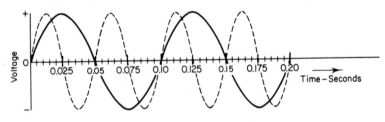

Figure 25-10   Sine waves of 10 and 20 "cycles per second" (hertz).

Is there any advantage in plotting sine waves against time instead of degrees? Plotted in degrees, one cycle would be 360°, regardless of frequency. It would be impossible to show the relation between two or more sine waves of different frequency! On the other hand, examine Fig. 25-10. The solid curve is the 10-Hz sine wave. Now notice the dashed curve. It completes one cycle in 0.05 s. Its frequency is $1/t$ or 20 Hz. Sine waves of any other frequency could be plotted on the same axis for comparison. Since such comparisons are necessary, sine waves are generally plotted with time as the horizontal axis.

25-7 Sine-wave values.   All ac calculations are based on sine waves. In our discussion so far we have only mentioned voltage; however, the resultant currents that flow, and the power that is expended, are also sinusoidal. There are four values of these sine waves that are of particular importance. We have already discussed two.

1. *Instantaneous Value.* The voltage, in an ac circuit is continuously changing. The value at any instant is represented by the lowercase letter (e), and is given by the equation of its curve (for example, $e = E_m \sin \omega t$). This value varies from zero to a maximum with time, depending on frequency (remember that $\omega = 2\pi f$). But since the voltage is continuously changing, the current flowing through the circuit and the power dissipated in the circuit must also vary correspondingly. The instantaneous values of current and power are also represented by lowercase letters (i, p) and are obtained from their equations:

$$i = I_m \sin \omega t \tag{25-8}$$

or

$$p = P_m \sin \omega t \tag{25-9}$$

2. *Maximum Value.* For two brief instants in each cycle the sine wave reaches a maximum value. This value is represented by a capital letter and the subscript m (E_m, I_m, P_m). It is often also referred to as the *peak* value. The two terms have identical meanings and are interchangeable.

3. *Average Value.* Since the instantaneous value of a sine wave is constantly changing, and since the maximum value occurs only twice in each cycle, it is often desirable to know what is the average value of a sine wave. As for any average, this value can be found by adding the individual values and dividing by the number of individual values. In the case of a sine wave, the individual values are the instantaneous values. But there are an infinite number of instantaneous values in one cycle of a sine wave! How many should be used? The more values averaged, the more accurate the final answer. The average of the instantaneous values taken every 10° should be sufficient (every 15 or 30° would require less work, but would not be as accurate).

If the average for a full cycle were taken, the answer would be zero. This is mathematically correct since the two half-cycles are identical, but the values for the second half-cycle are all negative, and the total sum of the instantaneous values would be zero. This answer seems to contradict a commonly seen relation, where the average value is given as 0.637 of the maximum value. Such an interpretation is really a special case *applicable only to pure sine waves.* (It was developed before complex waves were fully understood and does not consider the true meaning of "average value.") The origin of this *conventional* average value of a sine wave is as follows: An electric current will do work regardless of the direction of the current flow through the circuit. The negative half-cycle of a sine wave does just as much work as the positive half-cycle. For this reason, when speaking of the "average" value of a *sine wave*, it is generally understood to mean the average value *regardless of the sign or direction of the voltage or current*. Since each half-cycle is identical, the average value of the positive or negative half-cycle will be the same. So when finding this *conventional average value*, only half the cycle need be considered. *This special treatment can be used only when dealing with sine waves.* The importance of this statement will be seen when studying wave shapes other than sinusoidal, because the direction of current or the polarity of voltage cannot then be ignored. Let us apply this idea of conventional average value to a problem.

Example 25-5

Find the average value of a sine wave of voltage, whose maximum value is 100 V.

Solution

1. Tabulate the instantaneous values from the equation $e = 100 \sin \theta$, for every 10°, from 0 to 180°.
2. Find the sum of these instantaneous values.
3. Find the average of these instantaneous values. Since we have a total of 18 values of e, the average value is equal to $1143.2/18 \approx 63.5$ V.

Degrees	$\sin \theta$	e	e^2
10	0.1736	17.4	303
20	0.3420	34.2	1 179
30	0.5000	50.0	2 500
40	0.6428	64.3	4 135
50	0.7660	76.6	5 867
60	0.8660	86.6	7 500
70	0.9397	94.0	8 836
80	0.9848	98.5	9 722
90	1.0000	100.0	10 000
100	0.9848	98.5	9 722
110	0.9397	94.0	8 836
120	0.8660	86.6	7 500
130	0.7660	76.6	5 867
140	0.6428	64.3	4 135
150	0.5000	50.0	2 500
160	0.3420	34.2	1 179
170	0.1736	17.4	303
180	0.000	0.0	0
Sum		1143.2	90 084

Note: Disregard the last column (e^2); this will be used later.

In this problem, the average value is 63.5% of the maximum value.

Had we started with a sine wave having a maximum value of 10 V, the average value would have been 6.35 V, or again 63.5% of E_m. This is not a coincidence. This relation is true for all sine waves regardless of amplitude and regardless of frequency. However, our value 63.5 is not too accurate, because our sample points were too few. In the above solution we evaluated the instantaneous voltage values at intervals of 10°. For greater accuracy, let us use a much smaller angle, $d\theta$* radians. Again we can get the sum of all the instantaneous values over the range zero to π radians. But this summation is an integration process, or

$$\Sigma e = \int_0^\pi E_m \sin \theta \, d\theta$$

To get the average value we divide this sum by the number of such intervals between 0 and π, so that

$$E_{av} = \frac{1}{\pi} \int_0^\pi E_m \sin \theta \, d\theta$$

Solving this calculus equation, we get: $E_{av} = 0.6366 \, E_m$. We will use this more accurate equation, rounded out to three significant figures, or

$$\text{average value} = 0.637 \times \text{maximum value} \qquad \textbf{(25-10A)}$$

Average values are represented by capital letters with the subscript "av" (E_{av}, I_{av}, P_{av}).

Now that the relation between average and maximum values has been established, it is a simple matter to find the average value of any sine wave.

Example 25-6

Find the average value of a sine wave of current having a maximum value of 15 A.

Solution

$$I_{av} = 0.637 I_m = 0.637 \times 15 = 9.56 \, \text{A}$$

This same relation can be used to find the maximum value when the average value is known:

$$\text{maximum value} = \frac{\text{average value}}{0.637} \qquad \textbf{(25-10B)}$$

Inadvertently, you may sometimes divide by 0.637 instead of multiplying (or vice versa) in converting from one value to the other. The error should be immediately obvious, because you will end up with a maximum value that is *smaller* than the average value. This is impossible, and failure to correct such errors is inexcusable.

4. *Effective or rms Value.* In any given dc circuit, the voltage and current are constant. The power dissipated in the resistance of the circuit can be calculated from E^2/R or I^2R. What single value of E and I should be used for a similar ac circuit? Instantaneous and maximum values are definitely not the proper values. If we use the average value to find the power dissipated, the answer obtained is *less* than the actual power consumption. Looking at the power equations again, we notice that the power dissipated as heat depends on

*This d is a term used in calculus to represent an infinitesimally small value.

the *square* of the voltage or current. To get the same heating effect as with dc, the value of E and I chosen for an ac circuit should be obtained from the square of the instantaneous values. The procedure is similar to that used in Example 25-5. For convenience we will use the same data.

Example 25-7

Find the effective value of a sine wave of voltage whose maximum value is 100 V.

Solution

1. Find the instantaneous values of voltage for every 10° for a half-cycle (columns 1, 2, and 3, Example 25-5).
2. Square each of these instantaneous values (column 4).
3. Find the sum of these squared values (90 084).
4. Find the average or "mean" of these squared values. (Since there are 18 values in all, the average squared value is 90 084 ÷ 18, or 5005.)
5. Find the square root of this average (or mean squared value):

$$\sqrt{5005} = 70.7 \text{ V}$$

A sine-wave voltage of 100-V maximum would produce the same heating effect in a circuit as a steady voltage (dc) of 70.7 V. The 70.7 V is therefore called the *effective voltage*. From the method by which this value was found (square root of the mean of the squared values), this term is also called the **root mean square** (**rms**) **value**. Effective (rms) values are represented by capital letters *without subscript* (E, I).

In the above example, the effective value was found to be 70.7% of the maximum value. This relation is true for all sine waves regardless of maximum value and regardless of frequency:

$$\text{effective value} = 0.707 \times \text{maximum value} \qquad \textbf{(25-11A)}$$

On a calculus basis, since the rms value is dependent on a heating effect, which in turn is proportional to E^2, and since the evaluation can now be taken over a full cycle, the rms value of a sine wave of voltage is

$$E = \sqrt{\frac{1}{2\pi} \int_0^{2\pi} (E_m \sin \theta)^2 \, d\theta}$$

which results in

$$E = \frac{E_m}{\sqrt{2}} \qquad \textbf{(25-11B)}$$

Again the graphical and the mathematical analyses check.

Example 25-8

House voltage for lighting purposes has an rms value of 115 V. What is the peak value?

Solution

$$E_m = \frac{E}{0.707} = \frac{115}{0.707} = 163 \text{ V}$$

Sometimes the equation for maximum voltage is given as $E_m = 1.414E$. This is the same thing—the reciprocal of 0.707 is 1.414. Obviously, any of the above relationships ($E/0.707$; $E/\sqrt{2}$; or $1.414E$) can be used. However, to avoid numerical blunders, always remember that the maximum value must be larger than the effective value.

In general, ammeters and voltmeters for use in ac circuits are calibrated to read effective values. Instruments are also specially made to indicate peak values or sometimes average values. However, unless otherwise specified, current and voltage values for ac circuits should be understood to be effective (rms) values. The importance of understanding this relationship between the rms and maximum values of a sine wave cannot be overemphasized, because although equipment for use on ac is commonly rated in rms values, the equipment must be capable of withstanding the full peak value.

Since there is a fixed relation between the maximum and effective values, and also between the maximum and average values, it follows that there must also be a fixed relation between effective and average values. The effective value is higher than the average value (70.7% compared to 63.7%). The relation between them is the ratio of these constants:

$$70.7 \div 63.7 \quad \text{or} \quad 1.11$$

or

$$\text{effective value} = 1.11 \times \text{average value} \qquad \textbf{(25-12)}$$

By use of this last relation, we can change directly from average to rms values (or vice versa). However, it is not necessary to remember this constant. It can be evaluated at any time; or such a solution can always be made in two steps—from average to maximum, and from maximum back to effective (or vice versa).

The relation among all ac sine-wave values is shown clearly in Fig. 25-11.

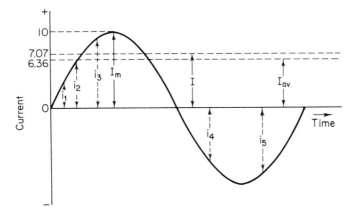

Figure 25-11 Sine-wave values of current.

REVIEW QUESTIONS

1. How does a right triangle differ from other triangles?
2. (a) Which side of a right triangle is called the hypotenuse?
   (b) Which side is called the side adjacent?
   (c) What is the third side called?
3. (a) Give the equation for evaluating the sine of an angle.
   (b) What is the largest value the sine can have?
   (c) For what angle value is the sine a maximum?
   (d) What is the sine value for an angle of 30°?
4. (a) Give the equation for the cosine of an angle.
   (b) What is the largest value the cosine can have?
   (c) For what angle is the cosine a maximum?
   (d) What is the cosine of 60°?
5. (a) Give the equation for the tangent of an angle.
   (b) What is the largest value the tangent can have?
   (c) For what angle is the tangent a maximum?
   (d) What is the tangent of 45°?
6. If the lengths of two sides of a right triangle are known, how can we find the length of the hypotenuse?
7. How does ac differ from dc energy?
8. What is the simplest type of ac wave called?
9. With reference to Fig. 25-5:
   (a) What do the blocks marked S and N represent?
   (b) What do the dashed lines in between represent?
   (c) What is the direction of the field?
   (d) What do the numbered circles represent?
   (e) Is the speed of rotation constant? Explain.
   (f) What condition must be satisfied in order for a voltage to be induced in a conductor?
   (g) At which conductor position is this voltage a maximum? Why?
   (h) Is the induced voltage ever zero? At which positions? Why?
   (i) At which positions will the induced voltage make the front end of the conductor positive? Why?
   (j) Will the polarity ever reverse? When? Why?
   (k) How often is this voltage sequence repeated?
10. Why is this induced voltage called a sine wave?
11. Give the equation for this voltage.
12. In electronics, how are sine waves produced?
13. With reference to Fig. 25-6, explain how this curve is obtained.
14. Explain how the curve in Fig. 25-7 is obtained.
15. With reference to Fig. 25-8:
    (a) What is the total number of cycles shown?

    (b) How many cycles are contained in each of the following spans: *a-i; a-c; a-j; a-g; a-k; a-h; i-b; i-h; g-h; g-l?*

16. (a) In electrical terminology, what is the basic unit used to measure angular velocity?

    (b) What symbol is used to represent angular velocity?

    (c) What does the term *radian* signify?

    (d) What is the relationship between radians and degrees?

    (e) What is the relationship between frequency and angular velocity? Give the equation.

    (f) What is the relationship between the angle of rotation and angular velocity?

    (g) Give the equation of a sine wave, in terms of angular velocity.

17. With reference to Fig. 25-10:

    (a) What are the axes of these plots?

    (b) How does this differ from the curve in Fig. 25-8?

    (c) What is the frequency of the solid curve in Fig. 25-10?

    (d) What is the frequency of the dashed-line curve in Fig. 25-10? Explain.

    (e) What is the frequency of the curve in Fig. 25-8?

    (f) What is the advantage of plotting sine waves on a *time* axis?

18. With reference to Fig. 25-8:

    (a) Give the first point in this wave where the voltage is a maximum value.

    (b) Is there any other such point in the first cycle?

    (c) What symbol is used to represent the maximum value of a current wave? Of a voltage wave? Of a power wave?

    (d) Give another name used to express maximum values.

19. (a) What does the term *instantaneous value* of a sine wave mean?

    (b) How may instantaneous values are there in one cycle of a sine wave?

    (c) In Fig. 25-8 would point *j* be an instantaneous value? What is the value at this instant?

    (d) Would *l* be an instantaneous value?

    (e) Name two other instantaneous values.

20. (a) In general, how is the *average* value of *any* wave obtained?

    (b) What is the true average value for a sine wave? Why is this so?

    (c) How is the "conventional" average value of a sine wave obtained?

    (d) Which is larger, this average value or the peak value? Give the equation.

    (e) Give the symbol used to represent the average values of current and voltage.

21. (a) What does the term *effective value* of an ac wave mean?

    (b) Do waveshapes other than sine waves have effective values? Explain.

    (c) Give another name for effective value.

    (d) How do we represent rms values by symbol?

    (e) Explain how the rms value of *any* ac wave can be obtained.

    (f) Which is larger, rms or maximum value? Give the equation for this relationship for a sine wave.

    (g) Does the relationship apply for a square wave? For any other wave?

    (h) Which is larger, rms or average value? Give this relation for a sine wave.

    (i) Does the equation apply to a pulse wave? Explain.

PROBLEMS

1. A right triangle has an adjacent side 6 units long, and an opposite-side length of 4 units. Find:
   (a) The tangent of the angle.
   (b) The angle value in degrees.
   (c) The sine of the angle.
   (d) The cosine of the angle.
   (e) The length of the hypotenuse.

2. A right triangle has a base angle of 40°, and an opposite side of 12 units in length. Find:
   (a) The sine of the angle.
   (b) The length of the hypotenuse.
   (c) The length of the adjacent side.

3. A right triangle has a base length of 8 units, and an opposite-side length of 5 units. Find:
   (a) The length of the hypotenuse.
   (b) The sine, cosine, and tangent values.
   (c) The angle value in degrees.

4. A sine wave of current reaches a maximum value of 8 A.
   (a) What is the equation of this curve?
   (b) Using 30° intervals, tabulate values and plot curve of current versus degrees of rotation (trigonometric method).

5. Plot the curve of Problem 4, using the rotating-radius vector method.

6. Make the following conversions:
   (a) 26° to radians.
   (b) 172° to radians.
   (e) 0.83 rad to degrees.
   (d) 2.3 rad to degrees.
   (e) π radians to degrees.

7. Sketch $1\frac{1}{2}$ cycles of a sine wave. Mark each zero and maximum points with successive letters (a, b, c, etc.). Using these letters, specify distances on the curve corresponding to:
   (a) Three one-cycle intervals.
   (b) Three half-cycle intervals.

8. The output of a generator is a 300-Hz sine wave. Find the angular velocity for this wave in:
   (a) Radians per second.
   (b) Degrees per second.

9. Repeat Problem 8 for a sine wave of 60 Hz.

10. What is the frequency of each of the following waves, if it completes:
    (a) One cycle in 0.1 s.
    (b) One half-cycle in 0.02 s.
    (c) Three cycles in 0.06 s.
    (d) One quarter-cycle in 0.0001 s.

11. A generator is producing a sine wave of voltage of 200 Hz and a maximum value of 40 V.
    (a) Write the equation for the instantaneous value of this voltage.
    (b) Find the instantaneous value of this voltage at 0.003 s.

12. A generator is producing a 60-Hz 120-V (rms) sine wave.
    (a) Write the equation for the instantaneous value of this voltage.
    (b) Find the instantaneous value of this voltage at 0.012 s.

13. In each of the following waves, find the maximum value and the frequency.
    (a) $e = 85 \sin 2512t$ V
    (b) $i = 250 \sin 31\,400t$ mA
    (c) $p = 377 \sin 377t$ W

14. The maximum value of a sine-wave voltage is 320 V. Find:
    (a) The rms value.
    (b) The average value.

15. The maximum value of a sine-wave current is 12.6 A. Find:
    (a) The average value.
    (b) The rms value.

16. The average value of a sine-wave voltage is 220 V. Find:
    (a) The rms value.
    (b) The maximum value.

17. The effective value of a sine-wave current is 60 mA. Find the average value and maximum value of this current.

18. A capacitor rated at 8 μF, 350 V is connected in an ac circuit. An ac voltmeter in the circuit indicates 300 V. Yet, in a short time, the capacitor breaks down. Explain.

26

RELATIONS BETWEEN SINE WAVES

In the previous chapter we discussed the characteristics of sine waves, considering only one sine wave at a time. But under actual operating conditions many sine waves may be present simultaneously in a piece of electronic equipment. For example, refer back to the simple transmitter of Fig. 24-4 (page 359). The modulator section is receiving RF and audio sine waves simultaneously. In addition, the sound waves entering the microphone, and the audio waves entering and leaving the audio amplifier may be several waves of different frequency. (A single speech sound or a single sound from a musical instrument is a combination of several frequencies, as will be shown in the next section.) It is therefore important that the relation between sine waves is clearly understood.

26-1 Harmonics. When middle C is sounded on a piano, that string produces a *fundamental*, or lowest frequency, sound wave of 256 Hz. In addition, the vibrating string also produces vibrations (overtones) of two, three, and four times the fundamental frequency. It is the presence of these overtones, in varying percentages, that causes a note on the piano to sound different from the same note on a violin or any other instrument.

In electronic work, multiples of the fundamental frequency are called *harmonics* instead of overtones. Oscillators designed to produce a certain fundamental frequency will often also produce harmonics of this fundamental. At times these harmonics are desirable. A typical case is when we wish to double or triple the frequency of the output voltage (as in frequency doublers or triplers used in FM transmitters); we merely "tune" the connecting circuit to the second or third harmonic of the fundamental frequency. At other times harmonics are undesirable and they must be eliminated or suppressed. For example, the audio amplifier shown in Fig. 24-3, if improperly operated, may generate second or third harmonics of the input signal. Since these harmonics were not in the original input, this is a form of distortion.

For purposes of analysis, harmonics are classified depending on their relation to the fundamental frequency. The *second* harmonic has a frequency *twice* the fundamental. The frequency of the third harmonic is three times the fundamental frequency. The fifth harmonic is five times the fundamental frequency, and so on. The relation between a fundamental, its second, and its third harmonics are shown in Fig. 26-1.

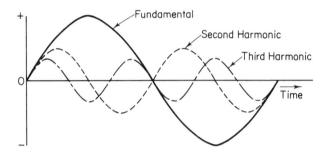

Figure 26-1   Relation of second and third harmonics to the fundamental.

The harmonics are shown at lower amplitudes than the fundamental to make the diagram easier to follow. However, it is true that harmonics are weaker than the fundamental for which the unit was designed. In general, the higher the order of the harmonic, the lower its amplitude. But this is not always true. For example, in many cases the third harmonic (odd) may be stronger than the second harmonic (even).

Since harmonics are multiples of the fundamental frequency, their equations must be of the same form as any sine wave. If the angular velocity of a sine wave is given by $\omega = 2\pi f$, then for its second harmonic the angular velocity is 2ω; for third harmonic, 3ω: and so on. The equations for such waves are:

Fundamental:	$e_1 = E_{1m} \sin \omega t$
Second harmonic:	$e_2 = E_{2m} \sin 2\omega t$
Third harmonic:	$e_3 = E_{3m} \sin 3\omega t$
Seventh harmonic:	$e_7 = E_{7m} \sin 7\omega t$

26-2   Phase and time relations.   Let us consider two coils mounted on the same shaft, in parallel planes and rotating in equal magnetic fields. Since they are on a common shaft, they will rotate at the same speed, and produce equal-frequency sine waves. Since the coils are parallel, the instantaneous voltage from each coil will be in step at all times. That is, they both have zero voltages and maximum voltages at the same instant, and they both reverse at the same instant. These waves are said to be *in phase* [Fig. 26-2(a)]. The equation for each of these waves is: $e = E_m \sin \omega t$.

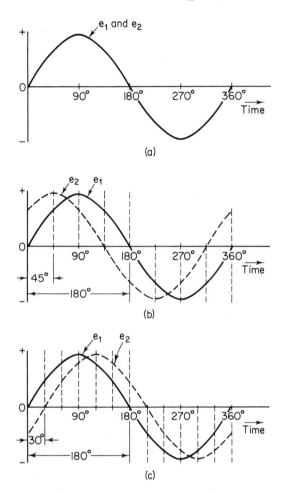

Figure 26-2 Sine-wave phase relations: (a) In phase. (b) e_2 leading by 45°. (c) e_2 lagging by 30°.

Now suppose the axis of coil 2 were shifted counterclockwise, 45° ahead of coil 1. (The shaft is rotating counterclockwise.) Coil 2 will go through its cyclic points 45° ($\frac{1}{8}$ cycle) ahead of coil 1. The voltage generated by coil 2 (e_2) *leads* the voltage from coil 1 (e_1) by 45°. This is shown in Fig. 26-2(b). Looking at the diagram, it might seem that e_1 is leading—it seems ahead of e_2. But remember these curves are plotted against time. The maximum point (and any other point) on e_1 curve occurs farther to the right—or later! Therefore, e_2 is leading. Notice that while e_1 is zero and starting positively, e_2 already is positive. This is the easiest way to recognize *a leading wave—it has a finite positive value when the other wave is just starting to go positive.* If the equation for coil 1 is expressed as $e_1 = E_m \sin \omega t$, the equation for e_2 must advance the angle of rotation by 45°, or

$$e_2 = E_m \sin(\omega t + 45°)$$

where ωt is expressed in degrees. If ωt is in radians, then since 360° equals 2π radians, 45° must equal $\pi/4$ radians, or

$$e_2 = E_m \sin\left(\omega t + \frac{\pi}{4}\right)$$

Now let us reverse the coil placement. Coil 2 is moved back, clockwise, till it is 30° behind coil 1. From our discussion above, it is obvious that the voltage from coil 2 now *lags* coil 1 by 30°. This is shown in Fig. 26-2(c). Again a sure determination of a lagging wave is that it is still negative while the other curve is at zero *and going positive.* As before, if the equation for coil 1 is $e_1 = E_m \sin \omega t$, the equation for coil 2 must retard the angle of rotation by 30°, or for angles in degrees,

$$e_2 = E_m \sin(\omega t - 30°)$$

for angles in radians,

$$e_2 = E_m \sin\left(\omega t - \frac{\pi}{6}\right)$$

In general, the equation for a sine wave that is leading or lagging another sine wave can be expressed by adding or subtracting some angle ϕ to the ωt value, or

$$e = E_m \sin(\omega t \pm \phi) \tag{26-1}$$

There are several special cases worthy or further analysis. These will be considered in turn.

1. *When the Angle of Lead or Lag (ϕ) Is Exactly* 90°. Notice in Fig. 26-3(a) that at $t = 0$, e_2 is positive while e_1 is zero. This indicates that e_2 leads e_1. Also notice that e_2 is at its maximum value when e_1 is at zero. This is a 90° phase difference, and confirms that e_2 leads e_1 by 90°. Applying the general equation (26-1) to e_2, we get

$$e_2 = E_m \sin(\omega t + 90°)$$

But from trigonometry we know that the sine of any angle $(\theta + 90°)$ is equal to the cosine of θ. So $\sin(\omega t + 90°) = \cos \omega t$, and the equation for e_2 can be rewritten as

$$e_2 = E_m \cos \omega t$$

Therefore, the equation for e_2 is a cosine curve. Notice in Fig. 26-3(a) how the maximum and minimum values for e_1 and e_2 compare. When e_1 is zero, e_2 is positive maximum, and that when e_1 is positive maximum, e_2 is zero. This is the same relation as exists between the sine and cosine of any angle—

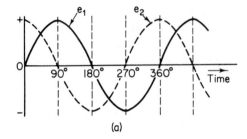

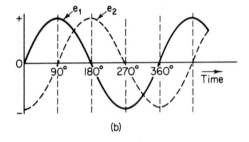

Figure 26-3 Cosine curves for e_2: (a) e_2 leading by 90° (positive cosine curve). (b) e_2 lagging by 90° (negative cosine curve).

sine is zero when cosine is maximum, and vice versa. This again shows that e_2 is a cosine curve.

In Fig. 26-3(b), the voltage e_2 is shown lagging by 90°. Obviously, the plot for e_2 is again a cosine curve. But notice this time that e_2 is at a *negative* maximum when e_1 is zero. Compare e_2 from this plot and e_2 of the previous plot [Fig. 26-3(a)]; e_2 lagging, is at all instants opposite in polarity to e_2 leading. This reversal of polarity can be expressed by formula. The equation of a curve *lagging* by 90° is

$$e_2 = -E_m \cos \omega t$$

2. When the Two Voltages Are 180° *Apart.* Such a curve is shown in Fig. 26-4. Several facts are apparent from the diagram.

(a) Both curves start at zero when time = 0, and reach zero and maximum values at same instant; they are both sine curves.

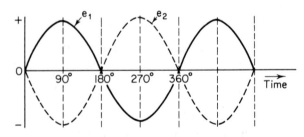

Figure 26-4 Sine waves, 180° out of phase.

(b) Voltage e_2 is at all instants opposite in polarity to e_1. Therefore,

$$e_1 = E_m \sin \omega t$$

$$e_2 = -E_m \sin \omega t$$

(c) The curve for e_2 leading or lagging by 180° would be the same curve! Therefore, it cannot be classified as leading or lagging, but merely 180° out of phase.

The equation for any curve 180° out of phase with the reference curve could be written as $E_m \sin (\omega t \pm 180°)$, but more specifically it is given by

$$e_2 = -E_m \sin \omega t$$

3. *When a Curve Leads by More Than* 180°. If e_2 leads e_1 by 250°, it can be considered as lagging by 110°. Any angle of lead (ϕ) greater than 180° can be considered as an angle of lag equal to $(360° - \phi)$. Similarly, any angle of lag (ϕ) greater than 180° is more logically considered as an angle of lead equal to $(360° - \phi)$.

This should be apparent from our previous discussion on two curves 180° apart. The 180° point marks the dividing line between lead and lag. It can also be explained from our basic generator principle. Figure 26-5 shows two conductors rotating in a magnetic field, on the same shaft but displaced by a fixed angle. Coil 2 may be considered as 300° behind coil 1, its voltage lagging e_1 by 300°; or coil 2 may be considered as 60° ahead of coil 1, and e_2 leading e_1 by 60°. Obviously, it is preferable to consider the smaller angle—e_2 leading e_1 by 60°.

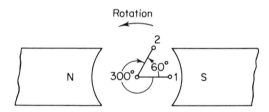

Figure 26-5   Angle of lead or lag.

26-3 Phase relations—different frequencies. So far we have been considering two coils rotating on the same shaft. The rotational speeds were the same, and the voltages developed in each coil had to be of the same frequency. As the coils were displaced on the shaft, we noticed a difference in phase between the two voltage waves. Whatever this phase angle was, it remained constant for cycle after cycle of the generators.

What would happen to phase relations if the coils were not on the same shaft, and one coil were rotating just the slightest amount faster? Suppose that coil 2 were rotating faster. Its frequency would be slightly higher. Assume that the coils start out parallel to each other—their induced voltages in phase. Coil 2 would finish its first cycle and would have started its second cycle before coil 1 completed one cycle. At this point coil 2 is leading by some angle. With each successive cycle this angle of lead will keep increasing. At some instant these waves will be 90° out of phase; later they will be 180° out; and so on. Finally, at some time they will be in phase again, when coil 2 has gained one full cycle on coil 1. These phase relations will be repeated over and over again.

Obviously, the phase relation between two waves of different frequency is continuously shifting. The greater the difference in frequency, the faster the phase relation between the waves will change. Due to this continuous change in phase angle, phase relations between waves of different frequency have little meaning *except when the waves are harmonically related.*

Earlier in this chapter we showed curves of a fundamental and its second and third harmonics (Fig. 26-1). Although the phase relation throughout one cycle of the fundamental is changing, notice that both harmonics start in phase with the fundamental and are back in phase at the end of the cycle. This applies to any harmonic. Whatever the phase relation is between fundamental and harmonic at the start of one cycle of the fundamental, that same phase relation will exist at the start of each successive fundamental cycle. It is therefore common practice to state the *starting* phase angle as the phase relation between harmonically related waves.

Check this for yourself. Sketch a few fundamental curves and some harmonic. You will notice, further, that all odd harmonics are also in phase with the fundamental at the start of the half-cycle of the fundamental. Also, at this half-cycle point all even harmonics are 180° out of phase with the fundamental. You will find later that these relations between harmonics and fundamentals are important when considering amplifier circuits and distortion.

In electronics, comparison of phase relations (input to output) is of great value when analyzing circuit operation. Any change in phasing between signals (phase shift) is a form of distortion. For example, in television, phase relations are of extreme importance. A change in phasing will cause blurring of picture detail. In order to study the causes and cures for these undesirable phase shifts, we must understand the meaning and representation of phase relations.

26-4 Addition of sine waves—resultant. In Fig. 24-4 (page 359) we see that two signals—the RF and the AF signal—are applied to the modulator section. The "effective signal" must be due to the sum of the applied signals. There are many such cases where more than one voltage or current are combined in electronic equipment. For example, as in dc, the voltages in series circuits, and the currents in parallel circuits are additive. But since voltages and currents in ac circuits are sine waves, it now becomes necessary to add sine waves. Let us see how these additions are performed in ac circuits. Graphically, the method is simple enough:

1. Plot each sine wave (amplitude versus time).
2. At several points along the time axis, measure the instantaneous value of each sine wave.
3. At each instant of time, add the individual instantaneous values to get the total instantaneous value. (The more points at which this is done, the more accurate the final result.)
4. Plot these total instantaneous values and join these points. This curve is the *resultant* waveshape.

This method is illustrated in Example 26-1.

Example 26-1
Draw the resultant of the two in-phase sine waves ($e_1 = 3 \sin \omega t$, and $e_2 = 2 \sin \omega t$) shown in Fig. 26-6.

Solution

1. Divide the time base into a number of equal intervals. (We will use 30° intervals.)
2. At each instant measure the instantaneous value of e_1 and e_2. The measured values are tabulated below.

Time Instant	e_1	e_2	e_R
0°	0	0	0
30°	1.5	1.0	2.5
60°	2.6	1.7	4.3
90°	3.0	2.0	5.0
120°	2.6	1.7	4.3
150°	1.5	1.0	2.5
180°	0	0	0

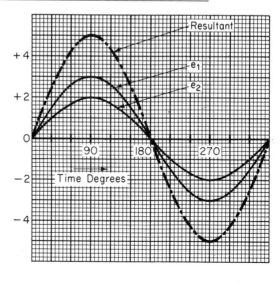

Figure 26-6 Addition of sine waves— in phase.

3. Add these instantaneous values to get the instantaneous total value. Tabulate (see column 4).
4. Locate these points on the graph. Draw the curve.

Obviously, the second half-cycle will be identical to the above tabulation, except that all values are now negative. The detailed steps are therefore not shown.

The technique can be used to find the resultant of any two, or more, waves—of the same frequency—regardless of their respective amplitudes, and regardless of their phase relations. Three illustrations are shown in Figs, 26-7(a), (b), and (c), as follows:

(a) Equal amplitudes—90° out of phase.
(b) Unequal amplitudes—90° out of phase.
(c) Unequal amplitudes—180° out of phase.

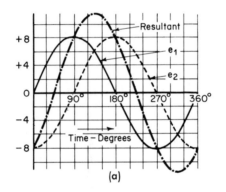

(a)

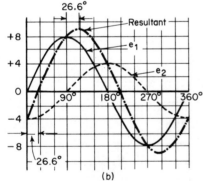

(b)

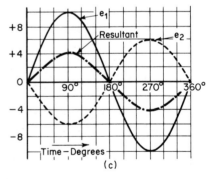

(c)

Figure 26-7 Further addition of sine waves—same frequency, but out of phase.

(a) Equal amplitudes—90° out of phase: $e_1 = 8 \sin \omega t$; $e_2 = 8 \sin (\omega t - 90°)$ $= -8 \cos \omega t$; $e_R = 11.2 \sin (\omega t - 45°)$.

(b) Unequal amplitudes—90° out of phase: $e_1 = 8 \sin \omega t$; $e_2 = 4 \sin (\omega t - 90°) = -4 \cos \omega t$; $e_R = 8.9 \sin (\omega t - 26.6°)$.

(c) Unequal amplitudes—180° out of phase; $e_1 = 10 \sin \omega t$; $e_2 = -6 \sin \omega t$; $e_R = 4 \sin \omega t$.

From Fig. 26-7, the following points should be noted.

1. The resultants are of the same frequency as the original waves.
2. When the sine waves are in phase, the resultant is also in phase.
3. When the sine waves are out of phase and of equal amplitude, the resultant has a phase shift half way between the original waves.
4. When the sine waves are out of phase but of unequal amplitude, the resultant phase angle will be in between the two original waves, but will be closer to the wave of larger amplitude.

This point-by-point addition of sine waves can be applied equally well to harmonically related waves. Figure 26-8 shows the addition of a fundamental and its second harmonic. The resultant wave is obviously not a sine wave. These waveshapes will be discussed further in the next chapter. However, the method for adding sine waves is still the same.

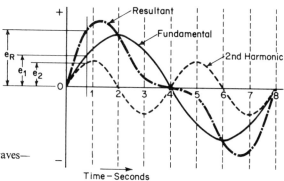

Figure 26-8 Addition of sine waves—
harmonic frequencies.

26-5 Phasor representation. From study of mathematics (or physics) you probably know that a vector quantity is a quantity having both *magnitude* and *direction*. For example, a force has a push or pull in newtons,* and a direction or angle at which it acts. Vector quantities are conveniently represented graphically by an arrow, whose length (to a convenient scale) represents the magnitude, and whose direction represents the direction or angle in which the force acts.

A similar technique can also be used in ac circuits to represent currents and voltages, and in the graphical solution of ac circuit problems. The length of the arrow is an indicator of the magnitude of the ac quantity, and the direction of the arrow is a measure of the phase angle. The horizontal direction—to the right—is the reference direction (phase angle equals zero). Phase angles are then measured *counterclockwise* from this reference line.

However, a phase angle is not a "direction" (in the sense that a force has a direction). It is really a time relation and not a space relation. Therefore currents and voltages are not true vectors. To avoid possible complica-

*This is the *SI* unit for force (equivalent to the pound in English units).

tions between vectors having directional meaning and ac quantities having *time-delay* information, these current and voltage arrows have been renamed *phasors.** Since they are of fixed length, phasors do not normally represent instantaneous values. They are generally used to represent effective values. However, phasors can just as readily be used with maximum or average values. (They can also be used with instantaneous values for *one specific instant of time*.)

Where a current and a voltage are being represented in the same diagram, different scales can be used for each. However, all voltages (or all currents) *in the same diagram* must be drawn to the same scale. This system of representation is shown in the following example.

Example 26-2

Show by phasor diagrams the following ac values:
(a) $E_1 = 60$ V, $E_2 = 30$ V, in phase.
(b) $E_1 = 25$ V, leading E_2 of 20 V by 30°.
(c) $E = 100$ V, $I = 4$ A, I leading E by 90°.
(d) $I_1 = 5$ A, lagging I_2 of 3 A by 45°.

Solution See phasor diagrams (Fig. 26-9).

This method of representing sine waves is not limited to just two at a time, but can be used for any number of sine waves simultaneously. There is one restriction. You will notice in Fig. 26-9 that there is a fixed phase angle between the two waves. This is true only if the waves are of the same frequency. Phasor representation cannot be used with waves of different frequency, because they would apply only at *one* instant—a moment later all phase relations would have changed.

26-6 Phasor addition—graphical. By use of phasors, addition of sine waves becomes a simple matter. However, remember that it can be used only when the sine waves are all of the same frequency. The procedure is as follows:

1. Pick a scale suitable to the magnitude of the given voltages or currents.
2. Draw the first phasor to scale and at the given phase angle.
3. Where the first phasor ends, start the second phasor, draw it to scale and at the given phase angle.
4. Where the second phasor ends, start the third phasor; and so on.
5. Join the starting point of the *first* phasor with the *head* of the *last* phasor (see Fig. 26-10).
6. This line is the *resultant*. Its length represents the magnitude of the resultant voltage or current, and its direction is the phase angle of the resultant.

*Prior to (about) 1960, this fine distinction was not observed, and the term "vector" was also used for current and voltage values.

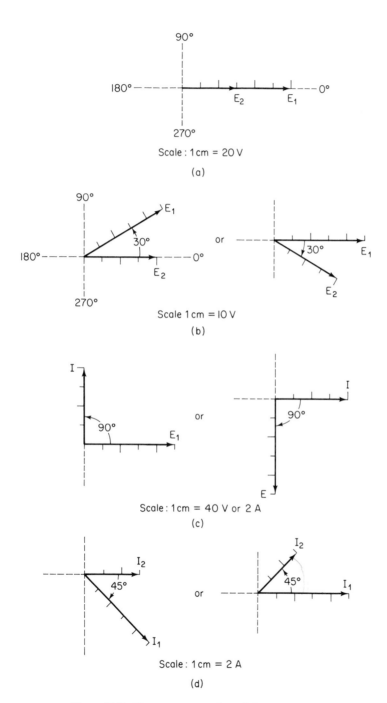

Figure 26-9 Phasor representation of sine waves.

If the given values are rms values, the resultant is also an rms value. If the given values were maximum values, the resultant would also be a maximum value.

Example 26-3

The branch currents in a parallel circuit are (rms values): $I_1 = 10$ A leading by $60°$; $I_2 = 7$ A at $0°$; $I_3 = 6$ A lagging by $30°$. Find:
(a) The resultant current.
(b) The phase angle of this current.
(c) The equation for the resultant current.

Solution   See Fig. 26-10.

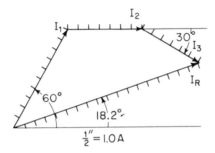

Figure 26-10 Graphical addition of phasors.

(a, b) $I_R = 18.1$ A leading by $18.2°$.

(c)   Maximum value of $I_R = \dfrac{\text{rms}}{0.707} = \dfrac{18.1}{0.707} = 25.6$ A

$$i_R = 25.6 \sin(\omega t + 18.2°)$$

This phasor method for addition of sine waves is much simpler and quicker than the previous method where the individual sine waves were plotted and then added. The phasor method will be used extensively in later chapters for the solution of ac problems involving only one frequency. In the preceding example, the problem was solved graphically. Careful plotting and large scales are necessary for accurate results. The phasor method can also be solved by trigonometry or complex algebra. The diagram can then be merely a freehand sketch of the circuit values. Mathematical solutions will be treated later.

REVIEW QUESTIONS

1. Do speech or musical sounds produce sine waves? Explain.
2. Why does a given musical note sound different when played on a trumpet as compared to a guitar?
3. (a) What is a *harmonic*?
   (b) Give an example of an undesirable harmonic situation.
   (c) Give an example showing where harmonics are desirable.

4. With reference to Fig. 26-1:
   (a) What are the coordinates of this plot?
   (b) Why is the short-dash curve a second harmonic?
   (c) Why is the long-and-short-dash curve a third harmonic?
5. Give the equation for the fifth harmonic of a fundamental whose angular velocity is ω.
6. (a) In the equation $e = E_m \sin \omega t$, what is the value of e at $t = 0$?
   (b) How is the above equation modified to indicate a curve that does not start at 0?
7. With reference to Fig. 26-2(b):
   (a) Compare the frequencies of curves e_1 and e_2. Why is this so?
   (b) In comparing phases, which of these two curves would be better as the reference? Why?
   (c) What is the phase of curve e_2 compared to e_1?
   (d) Why is the phase difference $45°$?
   (e) Why is e_2 leading?
   (f) Give the equation for curve e_2 using e_1 as reference.
   (g) What is the phase of e_1 *compared* to e_2?
8. With reference to Fig. 26-2(c):
   (a) Compare the frequencies of these two curves.
   (b) What is the phase of e_2 compared to e_1?
   (c) Why is this phase difference $30°$?
   (d) Why is e_2 lagging?
   (e) Give the equation of curve e_2 using e_1 as reference.
   (f) What is the phase of e_1 *compared to* e_2?
9. With reference to Fig. 26-3(a):
   (a) What is the phase of curve e_2 compared to e_1?
   (b) Give the equation of curve e_2 using e_1 as reference.
   (c) Give another equation for this curve.
   (d) Why is a lead of $90°$ the same as a cosine curve?
   (e) Why is it a positive cosine curve?
10. With reference to Fig. 26-3(b):
    (a) Give the equation for curve e_2 using e_1 as reference.
    (b) Give another equation for this curve.
    (c) Why is this a *negative* cosine curve?
11. With reference to Fig. 26-4:
    (a) Give the equation for e_1.
    (b) Give the equation for e_2.
    (c) Does e_2 lead or lag e_1? Explain.
12. When is it preferable to express a given angle of lead as a lagging angle?
13. Convert each of the following into more convenient expressions.
    (a) e_2 leads e_1 by $200°$ (b) e_2 leads e_1 by $270°$
    (c) e_2 leads e_1 by $310°$ (d) e_2 lags e_1 by $220°$
    (e) e_2 lags e_1 by $290°$
14. Wave e_1 leads e_2 by $30°$. The waves are *not* of the same frequency. Will this affect their phase relation? Explain.

15. With reference to Fig. 26-1:
    (a) What is the starting phase relation between the fundamental and second harmonic?
    (b) Between fundamental and third harmonic?
    (c) Do these phase relations remain the same throughout the cycle? Explain.
    (d) What are the phase relations at the start of the second cycle of the fundamental?
    (e) When would these phase relations be repeated again?
    (f) What does the term "phase" mean when applied to harmonically related waves?

16. (a) Give a step-by-step procedure for adding two (or more) sine waves.
    (b) What is the combined waveform called?
    (c) Can this technique be used to add nonsinusoidal waves?

17. With reference to Fig. 26-6:
    (a) What is the amplitude of e_1? of e_2?
    (b) What is the phase relation of e_1 to e_2?
    (c) What is the amplitude of the resultant?
    (d) How is this value obtained?
    (e) What is the phase of e_R with respect to e_1?
    (f) At 210°, what is the value of e_1? e_2? e_R?
    (g) At 300°, what is the value of e_1? e_2? e_R?

18. With reference to Fig. 26-7(a):
    (a) What is the amplitude of e_1? of e_2?
    (b) What is the phase of e_2 with respect to e_1?
    (c) What is the amplitude of the resultant?
    (d) Since e_1 and e_2 each have amplitudes of 10, why isn't e_R 20 V?
    (e) What is the phase of e_R with respect to e_1?
    (f) Give the equation for e_R using e_1 as reference.
    (g) At 60°, what is the value of e_1? of e_2? of e_R?

19. With reference to Fig. 26-7(b):
    (a) What is the phase of e_2 with respect to e_1?
    (b) What is the amplitude of e_2?
    (c) Give the equation for e_2 using e_1 as reference.
    (d) What is the phase of e_R with respect to e_1?
    (e) Give the equation for e_R.
    (f) At 150°, what is the value of e_1? e_2? e_R?

20. With reference to Fig. 26-7(c):
    (a) What is the phase of e_2 with respect to e_1?
    (b) Give the equations for e_2.
    (c) What is the amplitude of e_R? How do we get this value?
    (d) Give the equation for e_R using e_1 as reference.
    (e) At 30°, what is the value of e_1? of e_2? of e_R?

21. When sine waves of the same frequency are added:
    (a) Is the resultant a sine wave?
    (b) What is its frequency?
    (c) What is the phase of the resultant if the component waves are out of phase and of equal amplitudes?
    (d) If the amplitudes are also unequal, how does this affect the phase of the resultant?

22. With reference to Fig. 26-8:
    (a) Which of these waves are *not* sine waves?
    (b) What is the equation of the fundamental?
    (c) Give the equation for the second harmonic.
    (d) Give the equation for the resultant.
    (e) At what time instant does the resultant curve intersect the fundamental curve? Why?
    (f) During the time span 2–4, why is the resultant smaller than the fundamental?
    (g) When obtaining points for plotting the e_R curve, at what locations must you take sums—*as a minimum*?

23. (a) What two general properties does a vector quantity have?
    (b) How are these properties represented graphically?
    (c) How is this applied to ac quantities?
    (d) What name is given to "vectors" when used to represent currents or voltages?

24. In an ac series circuit, the following values exist: $E_1 = 20$ V, $E_2 = 3$ V, and $I = 80$ mA. A scale of 2 V per unit is selected to plot E_1.
    (a) Can you use a smaller scale to show E_2 more distinctly? Explain.
    (b) Can you use a larger scale to plot the current? Explain.

25. With reference to Fig. 26-9(a):
    (a) What is the value of E_2? of E_1?
    (b) What is the phase of E_2 with respect to E_1?

26. In Fig. 26-9(b)—*left-hand diagram:*
    (a) What is the value of E_2? of E_1?
    (b) What is the phase of E_1 with respect to E_2?
    (c) What is the distinction between the left- and right-hand diagrams?

27. With reference to Fig. 26-9(c):
    (a) Which phasor is used as reference in the left-hand diagram? In the right-hand diagram?
    (b) What is the value of E? of I?
    (c) Why are separate scales used to represent each quantity?
    (d) What is the E–I phase relation? Express two ways.

28. With reference to Fig. 26-9(d):
    (a) Which phasor is used as reference in these diagrams?
    (b) What is the value of I_1? of I_2?
    (c) Why aren't different scales used this time?
    (d) What is the phase relation? Express two ways.

29. Can phasors be used to represent two voltages of different frequencies? Explain.

30. In Fig. 26-10, why aren't I_1, I_2, and I_3 drawn starting from a common origin as shown in all diagrams of Fig. 26-9?

31. With reference to Fig. 26-10:
    (a) What is the magnitude and phase of I_1?
    (b) What is the magnitude and phase of I_2?
    (c) What is the magnitude and phase of I_3?
    (d) Since it does not start at the origin, how can we evaluate its phase?
    (e) What is the magnitude and phase of I_R?

(f) In drawing the I_R phasor, does it make any difference whether it is drawn from the top of I_3 to the origin or vice versa? Explain.

(g) State the general technique for adding more than two phasors.

PROBLEMS

1. (a) Accurately sketch two cycles of a sine wave.
   (b) Directly below, using the same time base, sketch a wave of twice the frequency.
   (c) Repeat for a wave of four times the frequency.
   (d) Repeat for a wave of one-half the frequency.
   (e) Using part (a) as the reference wave, what is the equation of each of the curves in parts (b) through (d)?

2. (a) Accurately sketch $1\frac{1}{2}$ cycles of a sine wave.
   (b) Directly below, using the same time axis, sketch a wave of the same frequency leading by 60°.
   (b) Repeat for a wave lagging by 45°.
   (d) Repeat for a wave lagging by 90°.
   (e) Using (a) as the reference wave, what is the equation for each of the curves?

3. In each of the following equations, specify amplitude, angular velocity, frequency, and angle of lead or lag:
   (a) $e = 25 \sin (377t + 40°)$ (b) $p = 4 \sin (628t - 80°)$
   (c) $i = 0.35 \cos 3150t$ (d) $e = -120 \sin 628t$

4. (a) Carefully sketch on graph paper one cycle of a sine wave having an amplitude of 10.
   (b) On the same axis, sketch a second wave of the same frequency and amplitude, leading by 60°.
   (c) Plot the resultant of parts (a) and (b).
   (d) What is the amplitude and phase angle of the resultant?

5. (a) Carefully sketch on graph paper one cycle of a sine wave having an amplitude of 10.
   (b) On the same axis sketch a third harmonic having an amplitude of 5 and starting in phase.
   (c) Plot the resultant of parts (a) and (b).

6. Represent the following sine wave values by phasors.
   (a) $E_1 = 20$ V lagging $E_2 = 15$ V by 170°
   (b) $E = 80$ V lagging $I = 10$ A by 80°
   (c) $I_1 = 40$ mA lagging $I_2 = 12$ mA by 40°
   (d) $I_1 = 10$ A leading $E_1 = 80$ V by 80°

7. Using the phasor method (graphical) find the resultant (magnitude and phase angle) of the following voltages.
   (a) Values for Problem 4.
   (b) Values from Problem 6(a).
   (c) Values from Problem 6(c).
   (d) $E_1 = 40$ V lagging by 40°.
       $E_2 = 60$ V leading by 80°.
       $E_3 = 30$ V at 0°.

27

NONSINUSOIDAL
OR COMPLEX WAVES

There is nothing complex about a complex wave—except the name. In fact, we have already seen an example of such a wave in the previous chapter (see Fig. 26-8). It was the resultant obtained when we added a fundamental sine wave and its second harmonic. Notice that this resultant waveshape is not a sine wave—and that, in a nutshell, is the whole story. *Any wave that is not a pure sine wave is called a complex wave. All complex waves are resultants of two or more sine waves of different frequencies.* One of the components of a complex wave may be of "zero frequency"—in other words, dc.

Of what value are complex waves in the study of electronics? The answer to that is quite simple. Complex waves occur in all phases of electronics. Music and speech (sound or audio waves) are complex waves because they are composed of fundamental and overtones or harmonics. Modulated carrier waves from transmitting stations, as shown in Fig. 24-4. are another example of such waves. Outputs from the rectifiers of a dc power supply, special waveshapes from electronic test equipment, radar, loran, navigational beacons, etc. are all complex waves. To analyze all complex waves would be impossible, but some waveshapes are of such common occurrence that special analysis is warranted.

27-1 Effect of odd harmonics on sine-wave shape. Sometimes electronic equipment, such as an amplifier, although supplied with a sine-wave input, will have a complex-wave output. This is a distortion of the original waveshape. Since all complex waves are the resultants from two or more sine waves of different frequencies, it means that the equipment has produced the new frequencies. Usually, these additional frequencies are harmonics of the input waves. When the output wave contains only odd harmonics of the input wave-shape, this can be easily recognized. The two alternations of the wave are identical, and the curve is said to have ***mirror-image symmetry.*** This effect can be shown graphically. Figure 27-1 shows the resultant of a fundamental and a third harmonic, in phase. The equation for such a wave is

$$e_R = E_{m1} \sin \omega t + E_{m3} \sin 3\omega t$$

If we cut this curve in half at the Y axis, and slide the lower half to the left, under the upper half, you will notice that each half is a mirror image of the other.

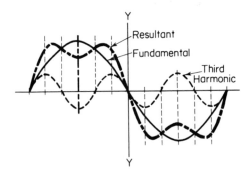

Figure 27-1 Fundamental and third harmonic—in phase.

A closer inspection will show that this waveshape has still another form of symmetry. Consider only one half-cycle—for example, the positive portion. Now imagine folding this portion of the curve in half along the quarter-cycle ordinate (shown darker in Fig. 27-1) as the centerline. Notice that the two quarter-cycle portions would coincide. This is ***left-right symmetry.***

When the third harmonic is not in phase, the output waveshape is entirely different, but the mirror image is maintained. Figure 27-2 shows the resultant wave when the third harmonic is 180° out of phase with its fundamental. The equation for this wave is

$$e_R = E_{m1} \sin \omega t - E_{m3} \sin 3\omega t$$

Notice again that the composite waveshape has left-right symmetry in addition to mirror-image symmetry. For ease of visual representation, the third

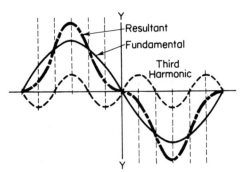

Figure 27-2 Fundamental and third harmonic—180° out of phase.

harmonic used in the above illustrations was either at zero or at 180° with respect to the fundamental, and the resultant wave had both left-right and mirror-image symmetry. For any other phase shift angle, left-right symmetry is lost. However, *mirror-image symmetry is maintained for any odd harmonic at any phase angle.* Let us see why.

Each half-cycle of the fundamental will contain $n/2$ half-cycles of the harmonic (two and one-half for the fifth; four and one-half for the ninth; etc.). So, whatever the phase relation between the fundamental and harmonic at the start of the fundamental cycle, the same phase relation must exist at the start of the second half of the fundamental cycle. This makes each alternation of the resultant wave identical and gives us mirror image.

27-2 Effect of even harmonics on sine-wave shape.   When the output waveshape contains even harmonics (with or without odd harmonics) the mirror image is destroyed. However, each alternation of the output wave may be similar in shape. This is illustrated in Fig. 27-3, which shows the effect of a second harmonic in phase (a), and 90° out of phase (b). The equations for these complex waves are

(a) $$e_R = E_{m1} \sin \omega t + E_{m2} \sin 2\omega t$$

(b) $$e_R = E_{m1} \sin \omega t + E_{m2} \cos 2\omega t$$

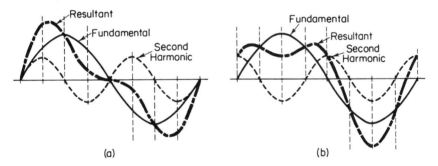

Figure 27-3 Effect of even harmonics on sine-wave shapes: (a) Fundamental and second harmonic in phase. (b) Second harmonic leading by 90°.

Notice that neither of the curves in Fig. 27-3 has mirror symmetry. Yet a certain symmetry is still present in Fig. 27-3(a). This type of symmetry may be called *Z-axis symmetry.* Hold a pencil at the midpoint of the fundamental wave, perpendicular to the plane of the paper. Your pencil will represent the Z axis. Now rotate the curves 180° around the Z axis. The resulting waveshape will be identical to the original curve. This is Z-axis symmetry. Such symmetry will occur whenever the even harmonics are either in phase or 180° out of phase. Now apply the Z-axis test to the curve in Fig. 27-3(b). Obviously, it does not have Z-axis symmetry. To illustrate the loss of Z-axis symmetry in Fig. 27-3(b), the even harmonic chosen was a second harmonic, and the phase angle between fundamental and harmonic was 90°. These specific values were chosen merely for ease of drawing and ease of visual interpretation by inspection. However, it must be emphasized that *any even harmonic at any phase angle other than 0 or 180° will result in complete lack of symmetry.*

In summarizing the effects of harmonics, the following conclusions can be made:

1. If a complex wave has mirror symmetry, it has no even harmonics. All its harmonics must be odd.
2. If a complex wave has Z-axis symmetry—but no mirror symmetry—it *must* have *even* harmonics. The even harmonics must be in phase or 180° out of phase. It may also have *odd* harmonics, but these also must be in phase or 180° out of phase.
3. If a complex wave has no symmetry, it must have even harmonics (not at 0 or 180°), and it may also have odd harmonics.

27-3 Effect of dc on waveshape.   Many waveshapes fed into or taken out of electronic equipment consist of dc plus some sine waves. For example, the input to the gate of a field-effect transistor is usually made up of a dc bias voltage and the ac signal voltage. The output from this same FET will be ac (the amplified signal voltage) and the dc power supply voltage. When such a signal is fed through a capacitor, the dc component is removed. In television it is necessary to include a special circuit, the "dc restorer," in order to add the correct level of dc to the signal. This is necessary for proper background illumination of the scenes. So, let us examine the effect of dc on a sine wave.

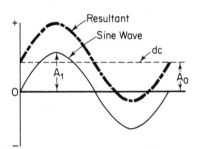

Figure 27-4   Effect of dc on sine-wave shapes.

Since dc can be considered a sine wave of zero frequency, addition of dc to a sine wave of any other frequency also results in a complex wave. The presence of dc in any complex wave can be easily recognized.

What was the average value of one full cycle of a sine wave? Zero—because the positive half-cycle is equal and opposite to the negative half-cycle. Check carefully the resultant wave in the earlier complex waves (Figs. 27-1 to 27-3). The average value in each case is still zero. This is obvious when you consider that the average value for each component is also zero! Now examine Fig. 27-4 showing the resultant of dc and a sine wave. The equation for this wave is

$$e_R = A_0 + A_1 \sin \omega t$$

Notice that the sine-wave shape is maintained, but it has been raised above the X axis by an amount equal to the dc component. The positive portion of the complex wave is much greater than the negative portion. The average value for one full cycle of the complex wave is positive. Had the dc component been negative, the average value would also be negative. The illustration shown is a simple one, but the principle applies to any complex wave. *If the average value of any complex wave, for one full cycle, is not zero, the wave has a dc component, equal to this average value.*

27-4   Fourier analysis.   So far, we have discussed complex waves in a qualitative manner, and we have seen that such waves consist of two, or more, sine waves of different frequencies. Although this gives us some idea of the harmonics (even or odd) present, and their phasing, we have no data as to the individual component amplitudes, or the specific frequency content. Such data can be obtained from a mathematical analysis using the Fourier* series method. The general equation for any periodic (repetitive) function can be expressed as an infinite power series of the form

$$f(t) = A_0 + A_1 \cos \omega t + A_2 \cos 2\omega t + \cdots + A_n \cos n\omega t \cdots$$
$$+ B_1 \sin \omega t + B_2 \sin 2\omega t + \cdots + B_n \sin n\omega t \cdots$$

This in turn, reduces to

$$A_0 + C_1 \sin(\omega t \pm \phi_1) + C_2 \sin(2\omega t \pm \phi_2) + \cdots + C_n \sin(n\omega t \pm \phi_n)$$

$$(27\text{-}1)$$

where $C = \sqrt{A^2 + B^2}$, and $\phi = \tan^{-1}(B/A)$. The maximum value of each component (A and B values) can be found by integration, or by graphical analysis.† We are not going to use Fourier analyses in this text. However, the *results* of such analyses when applied to commonly encountered waveshapes is of interest, and will be reviewed below.

27-5   Half-wave rectifier output.   Half-wave rectifiers are often used in electronic equipment. One typical use is in low-current dc power supplies as used for cathode-ray tubes, for example, in the high-voltage supply for oscilloscopes and television receivers.‡ If a sine wave is applied to such a rectifier, the output waveshape is shown in Fig. 27-5.

---

*Fourier, Jean Batiste, Baron, and French scientist (1768–1830).

†C. E. Skroder and M. S. Helm, *Circuit Analysis by Laboratory Methods*, Prentice-Hall, Englewood Cliffs, N.J., 1955.

‡J. J. DeFrance, *General Electronics Circuits* (2nd ed.), Holt, Rinehart and Winston, New York, 1976, Chap. 5.

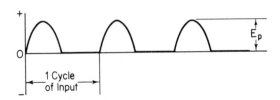

Figure 27-5 Output of half-wave rectifier.

This waveshape has often been called "pulsating dc." Such a description does not give a complete picture. It is a complex wave. Let us analyze it:

1. Since the average value is not zero, this wave has a dc component. The average value for the sine wave half-cycle is $(2/\pi)E_p$, or $0.637E_p$. The average for the next half-cycle is obviously zero. Therefore, the average for one full cycle is E_p/π, or $0.318E_p$. This is the value of the dc component.

2. By laboratory measurements, or by Fourier analysis, it can be shown that the next component is the same as the input frequency—the fundamental—and that its peak is equal to one-half the peak value of the complex wave.

3. Since the output wave does not have mirror image symmetry, it is certain to contain even harmonics. A wave analyzer would show that the next component is the second harmonic of magnitude equal to $0.212E_p$.

4. The fourth component is the fourth harmonic of the input frequency of magnitude $0.042E_p$.

5. Notice that the magnitude of the harmonics is getting quite small. The output wave has sixth and eighth, and so on, harmonics, but their magnitudes are so small as to be negligible.

Knowing the frequency and magnitude of each component, we can now write the equation for this wave:

$$e = 0.318E_p + 0.5E_p \sin \omega t - 0.212E_p \cos 2\omega t$$
$$- 0.042E_p \cos 4\omega t \ldots$$

(27-2)

This equation also shows the phase relation of the second and fourth harmonics to the fundamental. A minus cosine curve means a lag by 90°. The presence of a lead or lag (other than 180°) is obvious because the alternations of this waveshape are not similar. A verification of this analysis and equation is to plot the various components, add them, and compare the resultant with a typical half-wave rectifier output. This will be left as an exercise for the student (see Problem 6 at the end of this chapter).

27-6 Full-wave rectifier output. Another common complex waveshape is the output of a full-wave rectifier as shown in Fig. 27-6. Full-wave

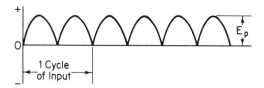

Figure 27-6   Output of full-wave rectifier.

rectifiers are probably the most common types of rectifier used in power supplies. To understand the choice between half-wave or full-wave rectification and the operation of the filtering circuits needed with these rectifiers, we must be familiar with the composition of the rectifier output waveshape.

The components of this complex wave are as follows:

1. dc equal to $0.367E_p$. The average value of one full cycle is $(2/\pi)$ E_p, or $0.637E_p$.
2. The next component is the second harmonic *of the input frequency*, of magnitude $0.425E_p$. There is no fundamental frequency in the output wave.
3. The third component is the fourth harmonic of amplitude $0.085E_p$.
4. The last component of any consequence is the sixth harmonic of amplitude $0.036E_p$.

The equation of this curve is

$$e = 0.637E_p - 0.425E_p \cos 2\omega t - 0.085E_p \cos 4\omega t$$
$$-0.036E_p \cos 6\omega t \dots$$

(27-3)

Comparing this equation with the equation for the output from a half-wave rectifier, we notice that the full-wave output has a higher dc component and lower ac components.

27-7 Square wave.   Square waves are often used in electronic work for test purposes because they are rich in harmonics. Any unit that can "pass" a square wave with little distortion is capable of handling a wide frequency range without discrimination. For example, let us assume that the audio amplifiers shown in block form in Chapter 24 must give true reproduction of frequencies up to 15 000 Hz. We can check this by feeding into the amplifier a square wave of 1500 Hz. If the output from the amplifier is still a good square wave, we know the amplifier is good to at least 15 000 Hz. Why this is so will be learned when studying the uses of the cathode-ray oscilloscope.*

A typical square wave is shown in Fig. 27-7.

Let us analyze this wave:

*J. J. DeFrance, *General Electronics Circuits*, App. B.

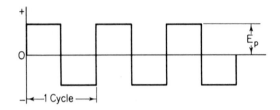

Figure 27-7 Typical square wave.

1. Since the average value for one cycle is zero, there is no dc component.
2. Since the curve has mirror image, there are no even harmonics.
3. The curve has steep sides, indicating a rapid change of instantaneous value with time. This means extremely high frequency, or high order of harmonics.

The equation for this curve is

$$e = E_m \sin \omega t + \tfrac{1}{3}E_m \sin 3\omega t + \tfrac{1}{5}E_m \sin 5\omega t + \tfrac{1}{7}E_m \sin 7\omega t \ldots \qquad (27\text{-}4)$$

where E_m is the maximum value *of the fundamental*, and is larger than E_p (the peak value of the square wave,) by the factor $4/\pi$, or

$$E_m = \frac{4}{\pi} E_p$$

The composition of the square wave is shown graphically in Fig. 27-8.
With the addition of the seventh harmonic, the resultant waveshape approaches a square wave. As higher harmonics are added, the sides of the wave will get steeper and the ripples at the top will be smoothed out. In order to duplicate the square wave exactly, an infinite number of harmonics must be added.

27-8 Triangular wave. A triangular wave is shown in Fig. 27-9.
From our earlier discussion, you should immediately summarize that this wave has:

1. No dc component (average value equals zero).
2. No even harmonics (mirror image symmetry). The harmonics must therefore all be odd harmonics. The equation of this wave is

$$e = E_m \sin \omega t - \tfrac{1}{9}E_m \sin 3\omega t - \tfrac{1}{25}E_m \sin 5\omega t$$
$$- \tfrac{1}{49}E_m \sin 7\omega t \ldots \qquad (27\text{-}5)$$

where E_m is the maximum value of the fundamental, E_p is the peak value of the triangular wave, and $E_m = (8/\pi^2)E_p$.

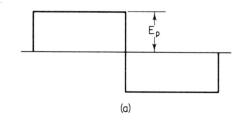

(a)

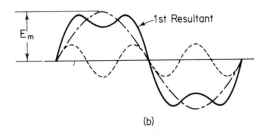

1st Resultant

(b)

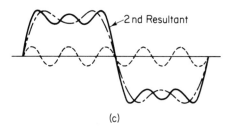

2nd Resultant

(c)

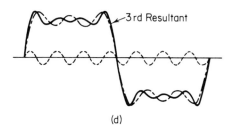

3rd Resultant

(d)

Figure 27-8 Composition of square wave: (a) Typical square wave. (b) Fundamental plus third harmonic. (c) First resultant plus fifth harmonic. (d) Second resultant plus seventh harmonic.

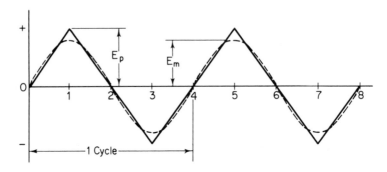

Figure 27-9 Triangular wave.

Since the amplitude of the harmonics is so low compared with the fundamental, it will be too difficult to show the composition of the triangular wave graphically.

27-9 Sawtooth wave. Sawtooth waves are used universally in sweep circuits of cathode-ray oscilloscopes and television receivers. They are also used quite frequently for trigger circuits. A typical sawtooth wave is shown in Fig. 27-10. It is obvious that the wave has no dc component since the area under each half-cycle is equal, and the average value is zero. In addition, the curve does not have mirror image. This means that it must contain even harmonics and may contain odd harmonics. However, Z-axis symmetry is present. Therefore, all harmonics present must either be in phase or 180 degrees out of phase with the fundamental. The actual equation for this curve is:

$$e = E_m \sin \omega t - \tfrac{1}{2}E_m \sin 2\omega t + \tfrac{1}{3}E_m \sin 3\omega t - \tfrac{1}{4}E_m \sin 4\omega t \ldots \qquad \textbf{(27-6)}$$

where E_m is the maximum value of the fundamental, E_p is the peak value of the sawtooth wave, and $E_m = (2/\pi)E_p$. The composition of this curve is also shown in Fig. 27-10. For analysis purposes, the sawtooth cycles shown in Fig. 27-10(b) to (e) are all for the time period 0–2. Notice that the odd harmonics are in phase with the fundamental and are positive sine curves, whereas the even harmonics are 180° out of phase, or negative sine curves. This checks with equation (27-6). On the other hand, a sawtooth, as more commonly seen, covers time period 1–3. This is a 180° shift in the fundamental component. Notice now—starting at time instant 1—that the fundamental and all harmonics are negative sine waves.

27-10 Pulse waves. A pulse wave is similar to the output of a half-wave rectifier, except that the duration of the pulse itself is only a very small fraction of the "cycle" time. The pulse may have any shape—peaked, rectangular, etc. A typical pulse wave is shown in Fig. 27-11.

It is obvious that this is a complex wave having a dc component and many even harmonics. A detailed analysis of this wave is too complicated.

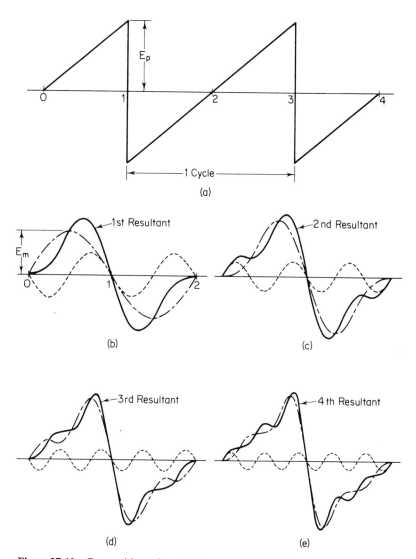

Figure 27-10 Composition of sawtooth wave: (a) Typical sawtooth wave. (b) Fundamental plus second harmonic. (c) First resultant plus third harmonic. (d) Second resultant plus fourth harmonic. (e) Third resultant plus fifth harmonic.

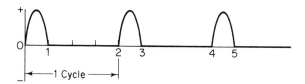

Figure 27-11 Typical pulse wave.

Furthermore, since there are many shapes of pulses used, no one equation would apply to all. In general, however, the steeper the sides of the pulse, the higher the harmonic content. Also, the flatter (more horizontal) the top of the pulse, the lower the frequency of its lowest frequency component. This generalization is true for any complex wave.

When speaking of pulse waves, the terms "frequency" and "cycle" are not used. Instead they are identified by new terms: *pulse duration* and *repetition rate*. Since the pulse duration is usually extremely short, time is measured in microseconds. For example, the pulse duration may be 1 μs or 20 μs, etc. The repetition rate is a measure of the number of pulses produced in 1 second. If pulses are produced at the rate of 500 pulses per second, the wave is said to have a pulse repetition rate (PRR) of 500.

Pulse waves have many applications in electronic circuits. Because of the exact timing between pulses, they are used for triggering and synchronizing purposes. (The horizontal synchronizing pulses used in television receivers are approximately 5 μs long and have a repetition rate of 15 750.) Pulses are also used in radar, loran, and navigational beacons. In this latter application, tremendous power can be sent out in the pulses (increasing the range of transmission), with relatively low power input. The reason for this can be explained readily. The power input is low, but it flows continuously for the entire "cycle." This energy is stored and then released in one brief instant at tremendous peak power. Such a waveshape has a high ratio of peak-to-average power.

27-11 Duty cycle. To explain the above more fully, let us consider another term common to pulse work—*the duty cycle*. This is the ratio of the duration of the pulse, to the time for one "cycle." For the pulse wave in Fig. 27-11, this can be evaluated as

$$\text{duty cycle} = \frac{\text{pulse duration}}{\text{time for one cycle}} = \frac{1 \text{ unit*}}{4 \text{ units}} = 0.25$$

Expressed in percent, this would be 0.25 × 100 or 25%.

More often, the time for one cycle is not given, but we do know the pulse repetition rate. But the time for one cycle is the inverse of the pulse repetition rate. Therefore, duty cycle is better expressed as

$$\text{duty cycle} = \text{pulse duration} \times \text{PRR} \qquad (27\text{-}7)$$

Since the repetition rate is in pulses per second, the pulse duration must also be expressed in seconds. Now, to illustrate the idea of tremendous peak power outputs for relatively low power inputs, let us use an example.

Example 27-1

A radar transmitter sends out a pulse of 2.5-μs duration with a peak power of 100 kW, at a rate of 200 pulses per second. Find:

*Both the pulse duration and the time for one cycle must be in the same time units.

(a) The duty cycle.
(b) The power input (average power) required for this transmission.

Solution

    (a)   Duty cycle = pulse duration $\times$ PRR = $2.5 \times 10^{-6} \times 200$

$$= 5 \times 10^{-4} \text{ or } 0.05\%$$

    (b) $P_{in} = P_{av} = P_{peak} \times$ duty cycle $= (100 \times 10^3)(5 \times 10^{-4})$

$$= 50 \text{ W}$$

[For an explanation of step (b), see Section 27-12.]

27-12 Average value of complex waves. When dealing with sine waves only, the average value (by convention) refers to the average for one half-cycle. This convention was adopted because each alternation is alike, and the average for a full cycle would be zero. However, when dealing with complex waves, the two alternations are not necessarily alike. The half-cycle average would have no meaning. This is especially true when the complex wave has a dc component. *The average value of complex waves should be taken for the full cycle.*

The method for obtaining an average is the same as explained for sine waves. A graphical method must be used unless the exact equation for the curve is known.

1. Plot or trace the wave (graph paper preferred).
2. Divide one *full* cycle of the wave into any number of equal parts (the more parts, the more accurate the final answer).
3. Measure the instantaneous values at each of these intervals—include all zero values.
4. Find the sum of all these instantaneous values.
5. Divide this sum by the number of values. This is the average value.

Another method of obtaining average values is sometimes easier:

1. Plot or trace the wave (graph paper preferred).
2. Calculate the net areas under the curve. If the curve has positive and negative values, the net area is the difference between the area under the positive and negative portions. For an irregular curve, a convenient way is to count the number of boxes (when using graph paper) under the curve.
3. Measure the length of the cycle. When boxes are used as a measure of area, then the length of the cycle can be measured in number of lines.
4. The average value is then: area divided by length of cycle (to the same scale as the curve is drawn).

Both methods are illustrated in the following problem:

Example 27-2

Find the average value of a half-wave rectifier output having a maximum value of 100 V.

Solution

1. *By instantaneous values:*

Degrees	e	Degrees	e
15	26	195	0
30	50	210	0
45	71	225	0
60	87	240	0
75	97	255	0
90	100	270	0
105	97	285	0
120	87	300	0
135	71	315	0
150	50	330	0
165	26	345	0
180	0	360	0

Sum of instantaneous values $= 762$

Number of instantaneous values $= 24$

Average value $= \frac{762}{24} = 31.8$ V

2. *By area method:* Plot the curve accurately (see Fig. 27-12). Then count:

number of squares under the curve $= 76\frac{1}{2}$

number of lines for one cycle $= 24$

average value $= \frac{76.5}{24} \times 10$ (volts per line) $= 31.9$ V

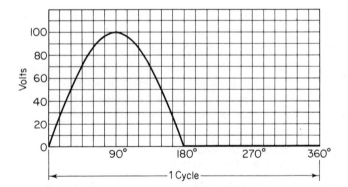

Figure 27-12   Half-wave rectifier output.

The exact answer to the above problem is 31.8 V. Notice that either method, if accurately done, gives reasonably accurate results.

A much simpler method can be used to evaluate average power if the pulses have high-peak values, are of short duration, and the interval between pulses is relatively long. We can assume that the pulse is a perfect rectangle. Then the product of (peak power) × (pulse time) equals the product of (average power) × (full-cycle time), and solving for average power, we get

$$P_{av} = \text{peak pulse power} \times \frac{\text{pulse time}}{\text{time for one cycle}}$$

and replacing with duty cycle, we get

$$P_{av} = \text{peak pulse power} \times \text{duty cycle} \qquad (27\text{-}8)$$

This is the equation that was used in Example 27-1, part (b).

27-13 Effective value of complex waves. The average value gives the dc component of a complex wave. But sometimes it is also desirable to know the combined effectiveness of all the components. We have already seen that the effective value of a sine wave is the root mean square of the simple sine function. Similarly, the effective, or rms value of any complex wave is the *square root of the mean of the squared rms values of the individual components.* In other words, the effective value of the current or voltage for any complex wave is given by

$$E \text{ or } I \text{ (complex wave)} = \sqrt{(\text{dc})^2 + \sum (\text{rms values})^2} \qquad (27\text{-}9\text{A})$$

If the maximum value of each sine-wave component is known, the rms values can be replaced by maximum values, and the equation becomes

$$E \text{ or } I \text{ (complex wave)} = \sqrt{(\text{dc})^2 + \sum \frac{(\text{peak values})^2}{2}} \qquad (27\text{-}9\text{B})$$

Example 27-3

Find the effective value of a sawtooth wave whose peak value is 133 V. (Use up to the fifth harmonic.) (No dc component.)

Solution

1. The maximum value of the fundamental is

$$E_m = \frac{2}{\pi} E_p = \frac{2}{\pi} \times 133 = 85 \text{ V}$$

2. The rms value of the fundamental is

$$E_1 = 0.707 E_m = 0.707 \times 85 = 60 \text{ V}$$

3. The rms values of the next four components are

$$E_2 = \tfrac{1}{2} E_1 = 30 \text{ V}$$
$$E_3 = \tfrac{1}{3} E_1 = 20 \text{ V}$$
$$E_4 = \tfrac{1}{4} E_1 = 15 \text{ V}$$
$$E_5 = \tfrac{1}{5} E_1 = 12 \text{ V}$$

4. The rms value of the sawtooth wave is

$$E = \sqrt{\sum (\text{rms})^2} = \sqrt{(60)^2 + (30)^2 + (20)^2 + (15)^2 + (12)^2}$$
$$= 72.6 \text{ V}$$

Although calculations in ac problems are based on sine waves, most waveshapes encountered in electronic work are complex waves. Calculations are therefore made on the individual components of the complex wave. Where the individual frequencies in any complex wave are widely separated, circuits handling such complex waves are analyzed in terms of the lowest-frequency component, medium-frequency component, and maximum-frequency component.

REVIEW QUESTIONS

1. (a) In its broadest sense, what is meant by the term *complex wave*?
   (b) What does a complex wave consist of?
   (c) A 200-Hz sine wave is added to another 200-Hz sine wave of $\frac{1}{2}$ amplitude, and leading by 40°. Is the resultant a complex wave? Explain.
   (d) Give two examples of where complex waves may be found.

2. With reference to Fig. 27-1:
   (a) Is the resultant a complex wave? Why?
   (b) Does this resultant have mirror-image symmetry?
   (c) Explain how this can be shown.
   (d) Give the equation for the resultant.

3. Still with reference to Fig. 27-1:
   (a) Does this wave have any other form of symmetry?
   (b) Explain how this can be shown.

4. With reference to Fig. 27-2:
   (a) Is the resultant a complex wave? Why?
   (b) Does the resultant have mirror-image symmetry?
   (c) Explain how this can be shown.
   (d) Give the equation for this resultant.
   (e) Does this wave have any other form of symmetry?

5. Comparing Figs. 27-1 and 27-2:
   (a) What are the component waves in Fig. 27-1?
   (b) What are the component waves in Fig. 27-2?
   (c) Why are the resultants so different?

6. For what phase condition of the odd harmonics will a resultant have:
   (a) Mirror-image symmetry?
   (b) Left-right symmetry?

7. A complex wave has mirror-image symmetry:
   (a) Does it contain any odd harmonics? Why?
   (b) Does it contain any even harmonics? Why?
   (c) Does it have a dc component?

8. With reference to Fig. 27-3:
   (a) Are the resultants complex waves? Why?
   (b) What are the components in diagram (a)?
   (c) What are the components in diagram (b)?
   (d) Why do the resultants differ?
   (e) What is the equation of the resultant in diagram (a)?
   (f) Give the equation for the resultant in diagram (b).
   (g) Which of these two resultants have mirror-image symmetry? Explain.
   (h) Which of these waves shows some form of symmetry?
   (i) What is this type of symmetry called?
   (j) Explain how to show this type of symmetry.

9. A complex wave has no mirror- and no Z-axis symmetry:
   (a) Does it have odd harmonics?
   (b) Does it have even harmonics?
   (c) What can be said of the phase relations between harmonics and fundamental?

10. A complex wave has no mirror symmetry, but it does have Z-axis symmetry:
    (a) Does it have odd harmonics?
    (b) Does it have even harmonics?
    (c) What is the phase relation between harmonics and fundamental?

11. With reference to Fig. 27-4:
    (a) What are the components of this resultant?
    (b) Is the resultant a complex wave? Explain.
    (c) Give the equation for this resultant wave.
    (d) If only the resultant were shown, how could you tell that it has a dc component?

12. With reference to Fig. 27-5:
    (a) Does this wave have a dc component? How can you tell?
    (b) What is the amplitude of this component? How is this value obtained?
    (c) Does this wave have mirror-image symmetry?
    (d) What does this mean with regard to its harmonic content?
    (e) Does it have Z-axis symmetry?
    (f) What does this mean with regard to its harmonic content?
    (g) In general, what happens to the amplitude of a harmonic as the order of the harmonic increases?

13. With reference to equation (27-2):
    (a) What is the magnitude and frequency of the first component?
    (b) What is the magnitude and frequency of the second component?
    (c) What is the frequency and phase of the third component?
    (d) What is the frequency and phase of the fourth component?
    (e) Does this wave have any other components? If so, which?
    (f) Why aren't these shown?

14. With reference to Fig. 27-6:
    (a) Does this wave have a dc component?
    (b) What is its value? How is this value obtained?
    (c) Does it have mirror-image symmetry?
    (d) Does it have Z-axis symmetry?
    (e) What does this mean with regard to its harmonic content?

15. With reference to equation (27-3):
  (a) What is the magnitude, phase, and frequency of the first component?
  (b) What is the magnitude, phase, and frequency of the second component?
  (c) What is the magnitude, phase, and frequency of the third component?
  (d) What is the magnitude, phase, and frequency of the fourth component?
  (e) Compared to equation (27-2), which has the higher dc component?
  (f) Which has the higher total ac component?

16. With reference to Fig. 27-7:
  (a) Is this a complex wave? Explain.
  (b) Does it have a dc component? Explain.
  (c) Does it have mirror-image symmetry?
  (d) What does this mean with regard to its harmonic content?
  (e) What does the steep rise and fall of the sides signify?

17. From the equation of a square wave:
  (a) What is the amplitude, frequency, and phase of the first component?
  (b) What is the amplitude, frequency, and phase of the second component?
  (c) What is the amplitude, frequency, and phase of the fourth component?
  (d) What is the amplitude of the ninth component?
  (e) Is this still a significant value?
  (f) How does the height of the square wave compare with the maximum value of the fundamental component?

18. With reference to Fig. 27-8:
  (a) What components are shown in diagram (b)?
  (b) What components are shown in diagram (c)?
  (c) What components are shown in diagram (d)?
  (d) Why is the resultant in (d) not a good square wave?

19. With reference to Fig. 27-9:
  (a) At what time marker does "one cycle" end?
  (b) Does this wave have a dc component? Explain.
  (c) Does this wave have mirror-image symmetry?
  (d) What does this mean with regard to its harmonic content?
  (e) From examination of the waveshape itself, how should the amplitude of these harmonics compare with the square-wave harmonics? Explain.

20. From equation (27-5) for a triangular wave:
  (a) What is the equation for the next component?
  (b) Is the eleventh harmonic of any significance? Explain.

21. With reference to Fig. 27-10(a):
  (a) Does the wave have a dc component? Explain.
  (b) Does it have mirror-image symmetry? Explain.
  (c) Does it have Z-axis symmetry? Explain.
  (d) What does this mean with regard to its harmonic content?
  (e) From the waveshape itself, would you expect the amplitude of the harmonics to be similar to the triangular or square wave?
  (f) What is the equation of the eighth harmonic component?

22. With reference to Fig. 27-10:
  (a) What components are added in diagram (b)?
  (b) What is the starting phase of the second harmonic?

(c) What additional component is shown in diagram (c)?

(d) What is the starting phase of this new component?

(e) What further component is shown in diagram (d)?

(f) How can you tell it is the fourth harmonic?

(g) Why isn't the resultant in diagram (e) a smooth sawtooth wave?

23. With reference to Fig. 27-11:

(a) Does this represent the output from a half-wave rectifier? Explain.

(b) What should this wave be called?

(c) What time interval is a measure of the pulse duration?

(d) What is a commonly used unit for measuring pulse duration?

(e) Why is the term *one cycle* put in quotes?

(f) If the time interval 0–2 is 500 μs, how many pulses are produced per second?

(g) What is the PRR for the above wave?

24. (a) Define duty cycle.

(b) Give the equation for duty cycle based on pulse repetition rate.

(c) How does duty cycle affect the relation between peak pulse power and average power?

(d) Of what advantage is a low duty cycle in radio transmissions?

25. (a) What does the *average value* of a pulse wave mean?

(b) What other term has the same value?

(c) Explain how the average of a complex wave can be obtained graphically.

(d) Explain another technique for getting this value.

26. Give the equation for the effective value of a complex wave:

(a) In terms of the rms value of each component.

(b) In terms of the maximum value of each component.

PROBLEMS

1. (a) Carefully sketch on graph paper a sine wave having an amplitude of 10.

(b) On the same axis, carefully sketch a fifth harmonic in phase, amplitude of 5.

(c) Plot the resultant of the two waves.

(d) What is the equation of this resultant?

2. Repeat Problem 1, using a third harmonic lagging by 90° as the second wave.

3. Repeat Problem 1 using a second harmonic 180° out of phase as the second wave.

4. Repeat Problem 1 using a fourth harmonic in phase as the second wave.

5. Repeat Problem 1 using dc (amplitude of 5) as the second wave.

6. (a) On the same axis plot the following waves.

1. Dc, amplitude 31.8 V

2. Sine wave, amplitude 50 V

3. Second harmonic lagging the fundamental by 90°, amplitude 21.2 V

4. Fourth harmonic lagging the fundamental by 90°, amplitude 4.2 V

(b) Plot the resultant of these four voltages.

(c) Write the equation of each component.

(d) Write the equation for the resultant.

(e) Compare this curve with the output of a half-wave rectifier. Explain any difference.

7. A full-wave rectifier has a dc output of 200 V. What is the amplitude and frequency of each ac component?

8. (a) A square wave has a frequency of 50 Hz. What is the frequency of the fundamental sine wave component?
   (b) What is the equation of the 15th harmonic component?

9. In a triangular wave, what is the equation of the 9th harmonic component?

10. In a sawtooth wave, what is the equation of the 8th harmonic component?

11. (a) Figure 27-13 shows a sawtooth wave. How does it differ from Fig. 27-10?
    (b) What is the equation of this sawtooth wave?

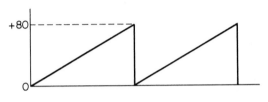

Figure 27-13

12. Draw a rectangular pulse wave having an amplitude of 10, pulse width of 0.02 s, and a pulse repetition rate of 10. Mark the "period" of this wave on the time base.

13. Calculate the duty cycle and the average value of the wave in Problem 12.

14. A radar transmitter sends out a rectangular pulse wave having a pulse duration of 5 μs. The pulse repetition rate is 500. The peak pulse power is 100 kW.
    (a) Draw the pulse wave, showing time intervals.
    (b) Calculate the duty cycle.
    (c) Calculate the average power per cycle.

15. The peak value of a rectified sine wave is 200 V. Calculate the rms value of the combined ac components:
    (a) For half-wave rectification.
    (b) For full-wave rectification.

28

RESISTANCE
AS A CIRCUIT
ELEMENT

In any electrical circuit you will find combinations of resistors, capacitors, and inductors used to control the circuit current in some desired manner. These components are called *circuit elements*. We have already discussed each of these elements from a dc standpoint. Their behavior on ac circuits will be considered in the following chapters.

28-1 Resistance concept for ac.   When studying direct currents, a resistor was considered as a definite piece of material (wire, carbon, or composition) having a definite length and cross-sectional area. When current passes through a resistor, electrical energy from the supply source is converted into heat energy in the resistor. The resistor dissipates electrical energy. This last statement is the basis for the concept of resistance in ac circuits. *Anything that dissipates electrical energy can be classified as a resistance.* It may be a real resistor such as was discussed in dc circuits, or it may be an *equivalent* resistance.

There are several ways by which these resistive effects are produced:

1. Eddy current and hysteresis losses in magnetic circuits dissipate electrical energy. These effects can be replaced by an equivalent resistance that will cause the same amount of power loss when inserted directly into the circuit. For example, a magnetic circuit carrying 100 mA has an iron loss (eddy currents and hysteresis) of 0.04 W. The equivalent resistance due to this iron loss is

$$R = \frac{P}{I^2} = \frac{0.04}{(0.1)^2} = 4 \ \Omega$$

2. Dielectric losses and leakage losses in insulation or dielectric materials will dissipate electrical energy. They can be replaced by an equivalent resistance which would result in the same power loss.

3. Antennas are used in transmitting stations to radiate electromagnetic waves into space. The electromagnetic energy came from the electrical energy supplied to the antenna circuit. As far as the circuit is concerned, power has been absorbed; therefore the circuit has an equivalent resistance. This radiation effect can be represented in the circuit by a resistance which will dissipate the same amount of power as was radiated into space. This resistance is known as the *radiation resistance* of the antenna. In an antenna, radiation is desirable. However, any wire or coil carrying current will also radiate electromagnetic energy to some degree. In this case, the power radiated because of the equivalent radiation resistance represents a power loss.

28-2 Effective resistance.   Any component (wire, coil, resistor, or plate) having a certain resistance in a dc circuit, will be found to have a higher resistance in an ac circuit. At low frequencies the increase in resistance may be small (in some cases almost negligible), but as the frequency of the supply voltage is increased the resistance of the unit will increase markedly. The resistance of any component in an ac circuit is called the *effective resistance. Unless otherwise stated, when a resistance value is specified in an ac circuit, this value is the effective resistance of the unit.* In view of the previous paragraph, it is not surprising to find that the resistance of circuit elements is higher on ac.

The effective resistance value of any unit is made up of the following components:

1. Dc or *ohmic* resistance, due to its length, cross-sectional area, and specific resistance of the material.
2. Eddy current losses in the unit itself, or in any metallic material within reach of the magnetic field produced when the unit carries current. This is particularly noticeable in RF coils with close fitting shields.
3. Hysteresis losses in the iron of the unit itself, or in any iron material within reach of its magnetic field.
4. Dielectric losses in the insulation of the unit, or in surrounding objects within reach of the dielectric field. This includes losses due to leakage currents through the insulating material, and also losses due to dielectric hysteresis effects in the insulation itself.
5. Radiation losses, due to magnetic field energy which reaches out into the space surrounding the unit while the current is building up, and does not return to the circuit when the current decays.
6. Skin effect (see next topic) which decreases the effective cross-sectional area of a wire.

Since all the preceding factors (except ohmic resistance) increase with frequency, effective resistance also increases with frequency. A resistance measurement made on dc or at low frequency may be valueless at radio frequencies. For accurate results at high frequencies, it is important that the effective resistance be known (or measured) at the frequency range desired.

28-3 Skin effect.   As pointed out above, skin effect reduces the effective area of a wire. This is due to magnetic effects. When a wire carries current, a magnetic field is produced which encircles the wire. These lines of force are present not only around the outside of the conductor, but also *within* the conductor itself. As the current rises and falls (ac) these lines cut the conductor. Since the lines originate at the center of the conductor, more lines cut the center of the conductor than the outer layers. The induced voltage produced by the cutting will also be greater at the center. But induced voltages oppose the flow of current—less current flows through the center than through the outer layers! At high frequencies (greater speed of cutting) the induced voltage at the center is so high that practically no current flows through this section. The conducting area of the wire is reduced, thereby increasing the resistance of the path. Larger-diameter wires would have to be used to compensate for this loss of effective area. Flat strip conductors are very poor. Current flows only along the outer edges, giving a very high ratio of effective to ohmic resistance.

In high-current, high-frequency circuits, where very low resistance is important, tubular conductors are often used in place of solid conductors. The center of the solid conductor would be ineffective, whereas all the material of the tube is effective. For the same weight of copper, the tube would have a much lower effective resistance.

Another way of reducing skin effect is by the use of *Litzendraht* (Litz) wire. This conductor consists of a large number of small enameled wires, insulated from each other except at ends, where they are carefully cleaned and paralleled. By proper interweaving of the strands, each single conductor will have the same average flux cuttings throughout its length. The current will divide equally among the strands. From another viewpoint, since current travels along the surface of the conductor (due to skin effect), using a larger surface area will reduce resistance. Many individual conductors of small diameter will have a larger surface area than a single conductor of the same total diameter.

28-4 Electron flow in an ac circuit.   In a dc circuit, one terminal is always positive, the other always negative. If the circuit is closed, electrons will flow from the negative terminal through whatever units are connected in the circuit, to the positive terminal; then, due to the action of the power supply, back to the negative terminal and round and round again if the circuit is closed for a long time. Such a circuit is shown in Fig. 28-1.

When the switch is closed, the ammeter indicates 2 A. An electron from *A*

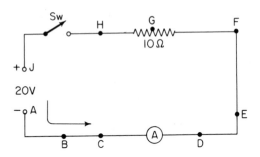

Figure 28-1 Electron flow in a dc circuit.

has started to move counterclockwise around the circuit. The switch is now opened. That particular electron may only have reached point *C*. Meanwhile, an electron from *B* may have reached point *D*; another electron originally at *C* may have reached point *E*; others may have progressed from *D* to *F*, from *E* to *G*, from *F* to *H*, from *G* to *I*, from *H* to *A*, and so on. No one electron went through the complete circuit, yet electrons were in motion *throughout* the circuit from *A* toward *J*. Current is flowing and power is being dissipated just as completely as if electrons had made complete "trips."

If the voltage were raised to 40 V, and the switch were closed and opened quickly, a current of 4 A would flow for a short time. Again no electrons may move through the complete circuit. Since the current is higher it means that more electrons are in motion from *A* to *J* throughout the circuit. Remember that current is a measure of the number of electrons *passing a given point* in the circuit in a given time.

Now let us suppose that in the above circuit the switch is closed and then opened sooner. Again, during the time the circuit is closed, electrons throughout the circuit are in motion from *A* toward *J*. This time, however, an electron from *A* may only reach *B*, another at *C* may only reach *D*, from *G* to *H*, and so on. Electrons will move a shorter distance—but these small electron motions take place everywhere in the circuit.

What is the nature of current flow in an ac circuit? The answer should now be obvious. Since the magnitude of the voltage is varying ($e = E_m \sin \theta$), the number of electrons in motion in any one direction is also varying. Since the polarity of the circuit reverses every half-cycle, electrons will move in one direction for one half-cycle and then in the opposite direction for the next half-cycle. Current flow for a complete cycle would be as follows:

1. At the beginning of a cycle, the current is zero—no electrons are in motion in one particular direction.
2. For the first quarter-cycle, current starts to flow, more and more electrons start to move *toward* one supply terminal until the current reaches a maximum.

3. For the next quarter-cycle, current still flows in the same direction, but the number of electrons in motion is decreasing, until finally the current is again zero.

4. For the third quarter-cycle, current now starts to flow in the *opposite* direction. The number of electrons in motion increases until the current reaches a maximum.

5. During the last quarter-cycle, current still is flowing in this opposite direction, but the number of electrons in motion is decreasing until the current is once more zero.

How far does any one electron travel around the circuit? This depends upon how long the supply voltage maintains any one polarity, or on the frequency of the supply voltage. If the frequency is very low, the electrons will have more time to travel in any one direction and may move quite a distance around the circuit. But consider instead a frequency of 1 MHz. The time for one cycle is 1 μs $(t = 1/f)$, and for one half-cycle only half a microsecond. How far does an electron move around the circuit? It just about starts to move in one direction when it stops and moves in the opposite direction! *But electrons are in motion throughout the circuit, in one direction and then the other. No matter how short the distance traveled in any one direction, it still constitutes current flow.*

In the above analysis we studied the electron motion, but this discussion could also apply to the assumed direction of conventional current flow. So current flow in an ac circuit may be considered as an oscillatory motion—like the motion of a pendulum—back and forth, with increasing and decreasing quantities of charges in motion.

28-5 Current and voltage relations in a resistive circuit.
Figure 28-2(a) shows a resistor connected to an $833\frac{1}{3}$-Hz supply. The time for one cycle is $1/f$ or 1.2 ms. Figure 28-2(b) shows how the potential of point B (with respect to A), varies with time.

The current (i) flowing in the circuit *at any instant* is determined by the resistance of the circuit and the instantaneous voltage at that time. This is the

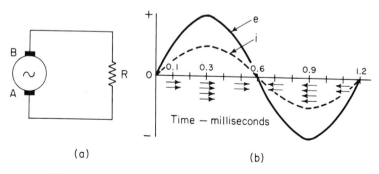

(a)   (b)

Figure 28-2 Current and voltage relations—resistive circuit.

same as Ohm's law in any dc circuit. However, since the voltage is varying, so must the current be.

1. At time equals zero, the voltage is zero and no current is flowing.
2. At time equals 0.1 ms (30° later), the potential of point B is positive, and the voltage has reached half its maximum value. A certain amount of current is now flowing. This is indicated by the double arrow in Fig. 28-2(b). Current is flowing from terminal B toward A.
3. At time equals 0.3 ms (90°), the potential of terminal B has reached a maximum. The current flowing from B toward A is also a maximum. In the diagram this is shown by four arrows.
4. Time equals 0.6 ms. There is no difference of potential between terminals A and B. No current flows.
5. Time equals 0.9 ms. Terminal B is now negative compared to terminal A, and the difference in potential is a maximum. The current flowing will be a maximum (four arrows) but current is flowing *from A toward B*.
6. Elapsed time is 1.2 ms. There is no difference in potential between the two terminals. No current flows.

The instantaneous value of the current flowing in the circuit is shown in Fig. 28-2(b) by the dashed curve. Notice that it is also a sine wave, and that it is in phase with the applied voltage. *The current flowing through a resistor is always in phase with the voltage across it.*

28-6 Ohm's law-ac resistive circuit.   Ohm's law can be applied to an ac resistive circuit in exactly the same way as we did for a dc circuit. The only point to remember is that since ac values may be instantaneous, average, rms, or peak values, the same ac value must be used for both current and voltage. For example, if the voltage given is an instantaneous value, the current will also be an instantaneous value. If the voltage is a peak value, the current will be the maximum value. As stated earlier (Chaper 25) effective values are most commonly used in ac work and, unless otherwise stated, current and voltages are effective (rms) values.

Example 28-1
A resistor of 20 Ω is connected across a 120-V 60-Hz supply. Find:
(a) The peak voltage across the resistor.
(b) The current in the circuit.
(c) The maximum current through the resistor.

Solution

(a) $$E_m = \frac{E}{0.707} = \frac{120}{0.707} = 170 \text{ V}$$

(b) $$I = \frac{E}{R} = \frac{120}{20} = 6.0 \text{ A}$$

(c)
$$I_m = \frac{I}{0.707} = \frac{6.0}{0.707} = 8.48 \text{ A}$$

or

$$I_m = \frac{E_m}{R} = \frac{170}{20} = 8.48 \text{ A}$$

28-7 Phasor representation—resistive circuits.   In Chapter 26 we showed how phasors could be used to represent effective (or maximum) values of currents and voltages, and the phase angles between them. Separate scales can be used for currents and voltages. However, all currents (and all voltages) in the same problem must be to the same scale. The previous problem (Example 28-1) can be represented by a phasor diagram as shown in Fig. 28-3. Notice that the current and voltage are drawn in phase.

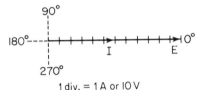

Figure 28-3 Phasor representation—
resistive circuit.
                                                          1 div. = 1 A or 10 V

28-8 Power in resistive circuits.   In any given dc circuit, currents and voltages are constant, and the power dissipated by the circuit is the product of the voltage and the current. On the other hand, currents and voltages in an ac circuit are continuously varying. But the power at any instant is equal to the product of the instantaneous values of the voltage and current at that instant, or

$$p = e \times i$$

A graphical plot of this relation is shown in Fig. 28-4. The ordinate of the power curve at any instant is the product of e and i.

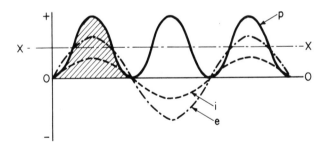

Figure 28-4 Power in a pure resistive circuit.

Examine the power curve carefully. There are several important facts brought out:

1. The power curve is always positive. During the second half-cycle, current and voltage values are negative, but the power (product of two negative

values) remains positive. *This means that a pure resistance takes power from the source during the entire cycle* regardless of the direction of current flow.

2. The instantaneous power varies from zero to a maximum value twice during each cycle of the current and voltage wave.

3. Since current and voltage are in phase, and their maximum values occur at the same time, the maximum power taken by the circuit is the product of the maximum value of current and voltage, or

$$P_m = E_m \times I_m \tag{28-1}$$

4. The power curve is also of sine-wave form, but at twice the frequency of the voltage (or current) wave. The zero axis of the curve is shifted upward from $O\text{–}O$ (the normal axis) to $X\text{–}X$.

This can be proven mathematically as follows:

$$p = ei = (E_m \sin \omega t)(I_m \sin \omega t)$$

$$= E_m I_m \sin^2 \omega t$$

replacing the $\sin^2 \theta$ function by its trigonometric equivalent, $\frac{1}{2}(1 - \cos 2\theta t)$,

$$p = \frac{E_m I_m}{2}(1 - \cos 2\omega t) = \frac{E_m I_m}{2} - \frac{E_m I_m}{2} \cos 2\omega t \tag{28-2}$$

This power equation is obviously a complex wave, containing two components—a steady value or dc ($E_m I_m/2$), and an ac component $[-(E_m I_m/2) \cos 2\omega t]$. This varying component must start at negative maximum at $t = 0$ (negative cosine curve); its frequency is twice the frequency of the current or voltage wave ($2\omega t$); and finally, since it is a cosine wave, its average value over a full cycle is zero. It therefore does not represent power dissipated in the resistor.

Now examine Fig. 28-4 again. Line $X\text{–}X$ portrays the steady value $E_m I_m/2$. This is the average value or dc component of the power curve, and represents the power (average) delivered to the load over a full cycle.

$$P = \frac{P_m}{2} = \frac{E_m I_m}{2} \tag{28-3A}$$

Replacing the denominator 2 by its equivalent $\sqrt{2} \times \sqrt{2}$, we get

$$P_{av} = \frac{E_m}{\sqrt{2}} \times \frac{I_m}{\sqrt{2}}$$

where $E_m/\sqrt{2}$ and $I_m/\sqrt{2}$ equal $0.707E_m$ and $0.707I_m$, respectively. But these are the rms values!

Therefore, the average power taken *by a resistive load* in an ac circuit is the product of the effective values of current and voltage, or

$$P = E \times I \tag{28-3B}$$

Notice in equation (28-3B), that the subscript "av" has been dropped. With regard to power, this is correct, because *average* power is the same as *effective* power, and by standard convention, no subscript is required to represent effective values.

Example 28-2

A resistor of 700 Ω is connected to an ac supply having a maximum value of 400 V. What is the power taken by the circuit?

Solution A

1. $$E = 0.707E_m = 0.707 \times 400 = 282.8 \text{ V}$$

2. $$I = \frac{E}{R} = \frac{282.8}{700} = 0.404 \text{ A}$$

3. $$P = E \times I = 282.8 \times 0.404 = 114.3 \text{ W, or}$$

3a. $$P = I^2R = (0.404)^2 \times 700 = 114.3 \text{ W}$$

Solution B

1. $$P = \frac{E^2}{R} = \frac{(0.707E_m)^2}{R} = \frac{0.5 \times 400 \times 400}{700} = \frac{800}{7} = 114.3 \text{ W}$$

28-9 Measurement of power. As you learned in your earlier studies, the power taken by a load can be measured directly by use of a wattmeter. Since use of wattmeters is quite important in ac applications, this topic is repeated here. The proper connections for a wattmeter are shown in Fig. 28-5.

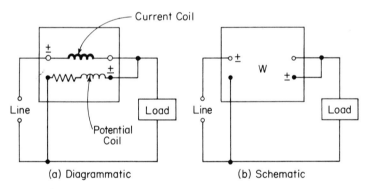

(a) Diagrammatic   (b) Schematic

(Note: ± Terminal of Potential Coil may be connected on the "Line" side of the Current Coil)

Figure 28-5 Wattmeter connections.

The wattmeter has four terminals—two circuits. The *current coil* terminals (usually larger in size) are connected in series with the circuit just as for an ammeter. The *potential coil* terminals are connected across the circuit like any voltmeter. Notice, however, that the ± terminal of the current coil is connected to the *line* side of the circuit and that the ± terminal of the potential coil is connected to the same line lead as the current coil. Reversal of either winding will result in a backward deflection.

REVIEW QUESTIONS

1. (a) What is meant by the term *circuit element*?
   (b) Give three examples of *passive* circuit elements.
2. In ac, what is the *general* meaning of the term resistance?
3. (a) Explain one way—other than by a real resistor—by which a resistive effect is produced in an ac circuit.
   (b) Repeat part (a) for another resistive effect.
   (c) Repeat part (a) for a third type of resistive effect.
4. (a) Does a component have a higher resistance in a dc or an ac circuit?
   (b) What is the ac resistance called?
   (c) State six factors that contribute to this ac resistance.
   (d) How are these factors affected by frequency?
   (e) Can a given value of effective resistance be used in any evaluation involving that same component? Explain.
5. (a) In what way does skin effect alter the characteristic of a conductor used in an ac circuit?
   (b) Explain why this is so.
   (c) At high frequencies, how will a hollow tube compare with a solid conductor of the same diameter, for carrying high currents?
6. (a) What is Litz wire?
   (b) Why is it used?
   (c) Explain why it is preferable to a single larger-diameter wire.
7. In Fig. 28-1, in order for current to flow in this circuit, is it necessary for electrons to complete the full trip around the circuit? Explain.
8. How does current flow in an ac circuit differ from flow in a dc circuit?
9. With reference to Fig. 28-2:
   (a) What does curve *e* represent?
   (b) What unit is used for the horizontal axis?
   (c) How is this specific time scale obtained?
   (d) Would this scale apply with a 60-Hz supply? Explain.
   (e) What do the two arrows at $t = 0.1$ ms signify?
   (f) Why are the arrows pointing to the left at time $= 0.7$ ms?
   (g) What is the phase of the current wave with respect to the voltage wave?
   (h) Why are the current and voltage waves in phase?
10. In an ac circuit, under what condition will current and voltage be in phase?
11. When applying Ohm's law to an ac circuit, can we use the *average* value of voltage with an *instantaneous* value of current, to find resistance? Explain.
12. (a) Can we use resistance with peak value of voltage, to find current?
    (b) What value of current would this be?
13. In Example 28-1, part (b), is this current an rms or a peak value? Explain.
14. With reference to Fig. 28-3:
    (a) What is the phase angle between E and I?
    (b) What is the value of E?

(c) Is this a peak, rms, average, or instantaneous value?

(d) What is the value of I?

(e) Is it a peak, rms, average, or instantaneous value?

15. (a) Is the power dissipated in an ac circuit a constant value? Explain.

(b) Since the current in a resistive circuit reverses, is the power negative for some portion of the cycle? Explain.

16. With reference to Fig. 28-4:

(a) What are the axes of these plots?

(b) When is the power dissipated a maximum?

(c) When is the power dissipated zero?

(d) Is the power curve a sinusoidal wave?

(e) What is the frequency of this curve?

(f) What does the line X–X represent?

17. (a) Give the equation for the maximum power dissipation.

(b) How does the average power per cycle compare with this maximum value?

(c) How does the effective power compare with the average power?

(d) Give an equation for average power in terms of rms values.

(e) To what type of load does this equation apply?

18. (a) What instrument can be used to measure power directly?

(b) Draw a circuit showing how to connect a wattmeter.

(c) What is the significance of the $\pm$ markings on one of the potential coil terminals?

(d) What may happen if this polarity is not observed?

(e) If the instrument deflects backward, how should this be corrected?

PROBLEMS

1. The resistance of an air-core inductor as measured with a Wheatstone bridge is 129 Ω. It is connected to a 400-Hz aircraft supply. The line current and circuit power as measured by an ammeter and wattmeter, respectively, are 1.46 A and 288 W.

(a) Find the resistance of the coil and account for the change.

(b) When a brass core is inserted through the center of the coil, the new instrument readings are 1.39 A and 290 W. Find the new resistance and account for the change.

(c) The brass core is replaced by an iron core. The instrument readings drop to 0.90 A and 145 W. Find the new resistance and account for the change.

2. A resistor of 4000 Ω is connected to a 500-Hz supply having a maximum value of 300 V.

(a) What value would an ammeter in the circuit indicate?

(b) Draw the phasor diagram.

3. In Problem 2:

(a) Write the equation for the power curve.

(b) Plot the curves for each of the two components.

(c) Plot the resultant power curve.

4. Find the average power dissipated in a resistive circuit, and the value of the missing quantity, E, I, or R.
   (a) $E_m = 200$ V, $I_m = 1.5$ A
   (b) $E_m = 500$ V, $R = 1800 \, \Omega$
   (c) $E = 300$ V, $I_{av} = 0.48$ A
   (d) $E = 40$ V, $I = 30$ mA
   (e) $I_m = 90$ mA, $R = 2500 \, \Omega$

5. A transmitter is rated at 200-W output. When the transmitter is fully loaded, the antenna current is 2.5 A. What is the radiation resistance of the antenna?

29

CAPACITANCE
AS A CIRCUIT
ELEMENT

Capacitors are used in electronic circuits for one of three basic purposes:

1. To "couple" an ac signal from one section of a circuit to another
2. To block out any dc potential from some component
3. To bypass or filter out the ac component of a complex wave

In the "power" field you will find capacitors used to supply starting torque, and to improve the running characteristics of single-phase motors. In a later chapter you will also see their use for power factor correction. In preparation for these applications, let us now study the action of capacitors as circuit elements.

From the action of capacitors in a dc circuit,* several important factors should be recalled. Let us review these briefly. When a capacitor is first connected to a steady voltage supply, the current flow is a maximum. The capacitor charges, building up voltage in opposition to the supply voltage. Due to this opposing voltage, the current drops to zero. The time required to charge the capacitor and for the current to drop to zero depends upon the time constant of the circuit (R-C). If the circuit contains no resistance, and the unit is a perfect capacitor (has no resistance), the time constant of the circuit is zero! This means that the capacitor will charge to the full supply voltage in zero time. Obviously, in a pure capacitive circuit with a dc supply, when the switch is first closed, the current is a maximum at the start, charges the capacitor to the full supply volt-

*See Chapter 19.

age, and drops to zero—all instantaneously! After that inrush *transient* condition, the current remains zero for as long as the supply voltage is constant.

In the same chapter, it was also shown that capacitance is the property of a circuit (or component) which allows a current to flow *when the voltage across it is changing*. The capacitance is one farad if one ampere flows through the circuit when the voltage changes at the rate of one volt per second. Now let us see how this applies to an ac circuit.

29-1 Current and voltage relation—pure capacitive circuit.

Figure 29-1 shows an ideal capacitor connected to an ac supply. The sine wave, solid curve, shows the variation of the applied voltage with time.

Let us assume that the circuit is closed when the supply voltage is at its maximum positive value (time A). At this time, the voltage, for a brief instant, is constant. For this brief instant the situation is similar to a dc circuit—there is an inrush of current, the capacitor charges to the E_m value, and the current drops to zero.

Neglecting the brief transient condition, we have the following conditions:

1. Line voltage: maximum positive
2. Capacitor countervoltage: maximum negative
3. Current: zero

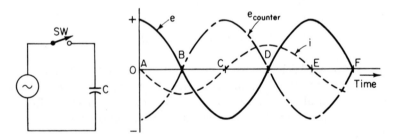

Figure 29-1   Current and voltage relations—pure capacitive circuit.

These conditions are shown in the Fig. 29-1 curves at time $= 0$. They are also shown in the schematic sketch, Fig. 29-2(a). But the supply voltage does not remain constant at this $t = 0$ value. During time interval A to B, the line voltage drops, so that the capacitor voltage tends to be slightly higher than the supply voltage. Therefore, the capacitor discharges into the line! This discharge current is in the "negative" direction compared to the supply-line polarity. At the start of this time interval (A to B), the supply voltage is dropping slowly, or the rate of change of voltage (de/dt) is low. The current will be low [see Fig. 29-2(b)]. As time approaches instant B in Fig. 29-1, the supply voltage is dropping faster and faster. The rate of change of voltage, de/dt, is increasing; the net circuit voltage is increasing; and the current will rise to a higher and higher value. At

B, the line voltage is dropping at its fastest rate; de/dt is a maximum; and the current is a maximum. By now it should be obvious that the instantaneous current value is dependent not on the actual value of the supply voltage, but on the rate of change of voltage, or

$$i \propto \frac{de}{dt} \qquad\qquad \textbf{(29-1)}$$

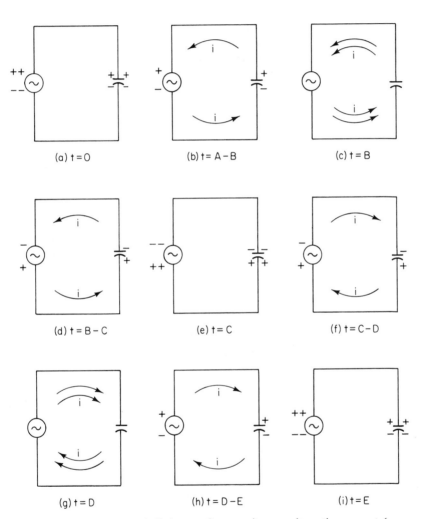

Figure 29-2 Charge and discharge of a capacitor on alternating current (see Fig. 29-1 for timing intervals). (a) Capacitor fully charged. (b) Capacitor discharging. (c) Capacitor fully discharged. (d) Capacitor charging with opposite polarity. (e) Capacitor fully charged again. (f) Capacitor discharging. (g) Capacitor fully discharged. (h) Capacitor charging with original polarity. (i) Capacitor fully charged. (Arrows show direction of conventional current flow.)

At time *B*, the new conditions are:

1. Line voltage: zero
2. Capacitor completely discharged and its countervoltage: zero
3. Current: maximum negative

This condition is shown in Fig. 29-2(c).

Between time *B* and *C* (Fig. 29-1), the supply voltage is rising toward its negative maximum. Since the capacitor is discharged, and the voltage is at first rising rapidly, the line now sends maximum current into the capacitor. The line voltage is negative, the current is negative, and the capacitor starts to charge with reverse polarity [see Fig. 29-2(d)]. As time approaches instant *C*, the supply voltage rises more slowly, the capacitor countervoltage builds up as it charges, and the current drops rapidly. Finally, at time *C*, we again reach a momentary static condition, as shown in Fig. 29-2(e):

1. Supply voltage: negative maximum
2. Capacitor countervoltage: positive maximum
3. Current: zero

During time interval *C* to *D* (Fig. 29-1), the supply voltage is dropping. Again the capacitor countervoltage tends to exceed the supply voltage. But the capacitor is now charged of opposite polarity. Current will flow from the capacitor into the line. This time the current will be positive. As before, the current will at first be low in value [see Fig. 29-2(f)], but as time approaches instant *D*, the supply voltage drops more rapidly and the current increases. At time *D*, the supply voltage is changing at its fastest rate—the current is a maximum. Meanwhile the flow of current out of the capacitor has discharged the capacitor. At time *D*, the conditions are [see Fig. 29-2(g)]:

1. Line voltage: zero
2. Capacitor countervoltage: zero
3. Current: positive maximum

From time *D* to *E* (Fig. 29-1), the supply voltage is rising toward its positive maximum value. Since the capacitor is discharged, current flows into the capacitor, charging it in the original positive direction. The current flowing is still positive, but as the capacitor charges, the line voltage rises more slowly, and the current is dropping. At time *E* the current is zero. Complete curves of supply voltage, capacitor countervoltage, and current are shown in Fig. 29-1. What are the phase relations?

1. Current leads the supply voltage by 90°.
2. Capacitor countervoltage and line voltage are 180° out of phase.

In any pure capacitive circuit, the line current leads the supply voltage by 90°.

29-2 Current flow in a capacitive circuit.   In dc theory, we learned that current will flow in a circuit only if the circuit is continuous and conducting. We also learned that a capacitor consists of two parallel conducting plates separated by the dielectric material, which is an insulator. Then, how does current flow in a capacitive circuit if the continuity of the circuit is broken by an insulating material? It does sound impossible—yet we know that it does!

The trouble is really in the terminology commonly used. We often speak of the current flow *through* a circuit, through a resistor, or through a capacitor, etc. In this text, we also have been using the same expressions. But notice that in starting this paragraph we have carefully used current flow *in* a circuit. You will recall from Section 28-4 that it was not necessary for electrons to flow completely around the circuit. Current flow was defined as *electrons in motion throughout the circuit in one general direction, regardless of how far they traveled in that direction.* Due to the alternating nature of the supply voltage, current flow in an ac circuit was found to be of an oscillatory nature—that is, the electrons move back and forth in the circuit.

A capacitor will allow such motion of electrons to take place. Therefore, current can flow in a capacitive circuit. In each cycle of the supply voltage, during the time that the supply voltage is rising, electrons will flow from the negative line terminal *toward* one plate of the capacitor. Electrons will be stored on this plate, giving it a negative charge. Electrons from the other plate will be repelled by this negative charge and also will be attracted by the positive line terminal. Electrons will flow from this plate *toward* the positive line terminal. Electrons were in motion *throughout the circuit from negative to positive line terminals.* Therefore, from our previous definition, current is flowing in the circuit.

Quite often you hear the expression "current flow through the capacitor." In the preceding explanation, electrons were stored on one plate of the capacitor; electrons were removed from the opposite plate of the capacitor; *but no electrons actually went through the capacitor!* No electrons can pass through the unit unless the dielectric breaks down, in which case the capacitor is ruined. Yet, since a capacitor does allow current to flow in a circuit, we often simplify the mental picture by saying that current is flowing through the circuit and through the capacitor. This "white lie" does not affect the accuracy of any calculations, and it does make reference to current flow in capacitive circuits much easier.

29-3 Capacitive reactance (X_C).   You will recall that as a capacitor is charged it builds up a countervoltage across its plates. The voltage developed depends on the capacitance of the unit, and on the charge (number of electrons) stored on the plates: $E = Q/C$. The larger the capacitance, the lower the voltage built up across the unit by any given number of electrons on its plates.

Now let us consider a capacitor of infinite capacitance connected across the ac supply. As the line voltage rises, the capacitor starts to charge. But due to the large capacitance of the unit, no opposing voltage is built up across its plates, and there is no limiting action on the flow of current in the circuit.

What would happen if the unit had a low capacitance? As the line voltage rises, the capacitor would charge, but now the back voltage builds up rapidly. The net voltage in the circuit is reduced, and the current flow is limited to a low value. This reaction of a capacitor is called *capacitive reactance* (X_C). Since the current-limiting action of a capacitor is similar in effect to resistance in a dc circuit, reactance is also measured in ohms. The larger the capacitance of a capacitor, the lower the countervoltage built up across the capacitor and the lower the reactance. In other words:

The reactance of a capacitor varies inversely with its capacitance.

Does frequency have any effect on the reactance of a capacitor? Definitely! The lower the frequency, the more time the capacitor has to charge, and the greater the limiting action. A more exact way of considering this relationship follows. Neglecting the initial transient charging current, no current flows in a capacitive circuit if the supply voltage is steady (dc circuit). If the voltage is increased, the capacitor charges further; current flows again. If the voltage decreases, current flows as the capacitor discharges. The amount of current flowing, in either case, will depend upon the *rate at which the voltage is changing* ($i \propto de/dt$). On low frequencies, the rate of change of voltage is slow; current is low. On high frequencies, the voltage changes rapidly with time; therefore, the current will be higher.

The reactance of a capacitor increases as the frequency of the applied voltage is decreased.

On dc, the reactance would be infinite. The rate of change of voltage in any ac circuit depends not only on frequency but more specifically on the angular velocity of the generating coil. In Chapter 25 we showed that the angular velocity (ω) was equal to $2\pi f$. Therefore, capacitive reactance varies inversely with angular velocity (ω) or $2\pi f$.

Since the reactance of a capacitor varies inversely with capacitance and with angular velocity ($2\pi f$ or ω), the equation for this effect is

$$X_C = \frac{1}{2\pi fC} = \frac{1}{\omega C} \tag{29-2}$$

If the frequency is expressed in hertz, and the capacitance in farads, then X_C will be in ohms. When the capacitor values are expressed in microfarads, we can convert the above equation by multiplying the numerator by 10^6. (For C in pF use 10^{12}.) Also, since the factor $1/2\pi$ occurs very often, the calculations are simplified if we replace the $1/2\pi$ by its equivalent 0.159. Combining these two points, we get

$$X_C = \frac{1}{2\pi fC} = \frac{0.159 \times 10^6}{fC} \qquad \text{(29-2A)}$$

where C is in microfarads.

Example 29-1

What is the reactance of a capacitor of 0.5 μF at 2000 Hz?

Solution

$$X_C = \frac{1}{2\pi fC} = \frac{0.159 \times 10^6}{(2000 \times 0.5)} = 159 \ \Omega$$

The parentheses in the denominator are added as a reminder for proper calculator entry, which in this case is:

$$\boxed{\cdot} \ , 1, 5, 9, \ \boxed{EE} \ , 6, \ \boxed{\div} \ , \ \boxed{(} \ , 2, 0, 0, 0, 0, \ \boxed{\times} \ , \ \boxed{\cdot} \ , 5, \boxed{)} \ , \ \boxed{=}$$

This relationship between capacitance, frequency, and reactance is very important. Let us check if you understand it. How would you find the reactance of the above capacitor if the frequency is tripled? Would you go through the complete calculation as for Example 29-1? If so, you would get the correct answer—but there is a much simpler way. Reactance, X_C, decreases with increase in frequency. Tripling the frequency will cut the reactance to one-third of the previous value or $159/3 = 53 \ \Omega$.

In general, if we know the reactance at any one frequency, we can find the reactance for any other frequency by multiplying the known reactance by the ratio of the two frequencies. If the new frequency is higher, the new reactance must be lower, or

$$\frac{X_{C2}}{X_{C1}} = \frac{f_1}{f_2} \quad \text{and} \quad X_{C2} = X_{C1} \frac{f_1}{f_2} \qquad \text{(29-3)}$$

Example 29-2

A capacitor has a reactance of 320 Ω at 5000 Hz. What is its reactance at 3500 Hz?

Solution

1. Since the new frequency is lower, the new reactance must be higher.

2.
$$X_{C2} = X_{C1} \frac{f_1}{f_2} = 320 \times \frac{50}{35} = 457 \ \Omega$$

Example 29-3

Find the capacitor needed in Example 29-2.

Solution

$$C = \frac{1}{2\pi fX_C} = \frac{0.159 \times 10^6}{5000 \times 320} = 0.1 \ \mu F$$

(As a reminder—the 0.159 is $1/2\pi$, and the 10^6 was used to get the answer in microfarads.)

29-4 Ohm's law—capacitive circuit. Ohm's law can be applied to a pure capacitive circuit in the same manner that it is used for resistive circuits. Since reactance is measured in ohms, we merely replace resistance by reactance:

$$I = \frac{E}{X_C}$$

In this equation no subscripts are shown for I and E. This means they are effective values or rms values. This equation can be used equally well for maximum or instantaneous values as long as both current and voltage values are maximum or instantaneous values.

Example 29-4

A capacitor of 0.02 μF is connected across a 500-V, 4000-Hz source. Find the current in the circuit.

Solution

1.
$$X_C = \frac{1}{2\pi fC} = \frac{0.159 \times 10^6}{4000 \times 0.02} = 1990 \ \Omega$$

2.
$$I = \frac{E}{X_C} = \frac{500}{1990} = 0.251 \ A$$

29-5 Phasor representation—capacitive circuits. Phasor representation is applied to capacitive circuits in exactly the same way as for resistive circuits. Again different scales can be used for current and voltage. The only point to remember is that current in a pure capacitive circuit leads the voltage by 90°. Figure 29-3 shows a phasor diagram for the values in Example 29-4.

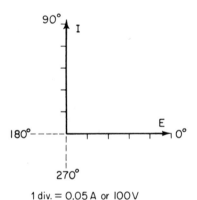

1 div. = 0.05 A or 100V

Figure 29-3 Phasor diagram—capacitive circuit.

29-6 Power in pure capacitive circuits. We have already seen that the instantaneous power in an ac circuit is equal to the product of the instantaneous voltage and the instantaneous current. In Fig. 29-4, the e and i curves represent the instantaneous voltage and current curves for a pure capacitive circuit. (Notice that the current curve leads the voltage curve by 90°.) Therefore, if at a number of time instants along the time axis, we multiply the instantaneous values of e and i, we can obtain sufficient points to plot the power curve p. Such a plot is shown as the solid-line curve in Fig. 29-4. Notice that, as in a resistive circuit, the power curve is a sine wave of twice the frequency of the current or voltage wave. However, the zero axis of this power curve is the same as the zero axis for the current and voltage waves. Let us analyze this curve more closely.

1. During the first quarter-cycle ($t = 0$–1), the current is positive. Current is flowing *into* the capacitor, charging it. The power is also positive, again showing that the capacitor is charging. It takes energy from the source and stores this energy in its dielectric field.

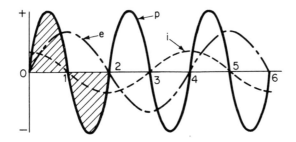

Figure 29-4   Power in a pure capacitive circuit.

2. During the second quarter-cycle ($t = 1$–2), the supply voltage is dropping, the capacitor is discharging, and the energy from the dielectric field is being *returned* to the line. The power is therefore negative.

3. During the third quarter-cycle ($t = 2$–3), the supply voltage is now rising to its negative maximum. The capacitor is charging again. It must take energy from the supply source and store it in its dielectric field. Therefore, the power is again positive.

4. During the fourth quarter-cycle ($t = 3$–4), line voltage is dropping. The capacitor must discharge and return its energy to the supply source. The power is negative.

During the first and third quarter-cycles, the capacitor takes power from the line, and the power is positive. In the other portions of the cycle, the capacitor returns this power to the line, and the power is negative. Therefore the total power dissipated in one full cycle is zero. *An ideal capacitor does not dissipate power.*

A mathematical analysis for the power curve follows the same technique as was used for the pure resistive circuit. The instantaneous power p is

$$p = ei$$

In a capacitive circuit, current leads voltage by 90°, and

$$p = (E_m \sin \omega t)(I_m \cos \omega t) = E_m I_m \sin \omega t \cos \omega t$$

From trigonometry, $\sin \theta \cos \theta = \frac{1}{2} \sin 2\theta$, and therefore

$$p = \frac{E_m I_m}{2} \sin 2\omega t \qquad\qquad (29\text{-}4)$$

Notice again, that there is no dc component, and that this is a sine curve of twice the frequency of the current or voltage curve. Also, since the power curve is a sine wave, the average value—over a full cycle—is zero, again showing that a pure capacitor does not dissipate any power.

29-7 Power factor.   If a capacitor is connected to an ac supply and a voltmeter, ammeter, and wattmeter are included in the circuit, we will find that:

1. The voltmeter will indicate the effective value of the supply voltage (E).
2. The ammeter will indicate the effective value of the circuit current (I).
3. The wattmeter will indicate zero! This is not surprising since we just explained by wave analysis that no power is dissipated by a pure capacitive circuit. Yet for resistive circuits we proved that the power dissipated is the product of the effective values of current and voltage. In a resistive circuit, the current and voltage are in phase, while in the capacitive circuit, current leads voltage by 90°. Obviously, the phase angle between current and voltage is responsible for this discrepancy. The power equation must be corrected by some term which includes phase angle. The term is called the **power factor** and is equal to the cosine of the phase angle. The general equation for power in an ac circuit is

$$P = EI \cos \theta \qquad\qquad (29\text{-}5)$$

where θ is the phase angle between line current and line voltage.

In a resistive circuit, current and voltage are in phase, the phase angle is zero, and the cosine of zero degrees is unity. Therefore the $\cos \theta$ term drops out, and the equation is the same as previously given for resistive circuits ($P = EI$). In a pure capacitive circuit, the phase angle is 90°. The cosine of 90° is zero. Therefore, the power is zero.

$$P = EI \cos 90° = EI \times 0 = 0$$

This is in agreement with our theoretical conclusions.

The product of voltage and current is sometimes referred to as the **apparent power**. Since it is not actually a true power measurement, it is more commonly referred to as **volt-amperes (VA)** or **kilovolt-amperes (kVA)**.

From the power factor of a circuit, we can immediately tell if there is a phase shift between current and voltage. The lower the power factor, the greater the phase shift (from 0 to 90°). However, we cannot tell from power factor, ammeter, voltmeter, or wattmeter reading whether the current is leading or lagging.

REVIEW QUESTIONS

1. (a) State three uses for capacitors in electronic circuits.
   (b) State two uses for capacitors in power applications.
2. (a) What does the time constant of a capacitive circuit mean?
   (b) What circuit parameters determine this value?
   (c) If a capacitive circuit has no resistance, what is its time constant?
   (d) How long will it take to charge the capacitor to the full supply voltage?
3. With reference to Fig. 29-1:
   (a) What does the solid curve represent?
   (b) What does the dot-dash curve represent?
   (c) How is this countervoltage developed?
   (d) What does the dash curve represent?
   (e) Why is the current zero when the supply voltage is at maximum?
4. In Fig. 29-1, at time instant *B*:
   (a) What is the supply voltage value?
   (b) What is the line current value?
   (c) How can current be a maximum, if the line voltage is zero?
5. (a) In a pure capacitive circuit, what is the phase relation between current and voltage?
   (b) In any circuit containing capacitance, what is the phase relation between the current through the capacitor and the voltage across the capacitor?
6. With reference to Fig. 29-2, what is the magnitude and direction of the current flow in each of the diagrams (a) through (i)?
7. Explain how current flows "through" a capacitor.
8. (a) Does a capacitor offer any opposition to current flow? Explain.
   (b) If the capacitance value is increased, will it offer more or less opposition? Why?
   (c) If the frequency of the supply voltage is increased, will a given capacitor offer more or less opposition? Why?
   (d) What term is used to denote the opposition of a capacitor?
   (e) What letter symbol represents this quantity?
   (f) Give the equation for capacitive reactance.
9. With reference to the solution of Example 29-1:
   (a) Where does the "0.159" come from?
   (b) Where does the "10^6" come from?
10. In each of the following, what happens to X_C:
    (a) If the capacitance value is quadrupled?
    (b) If the frequency is doubled?
    (c) If the capacitance is reduced to one-third its value?
    (d) If the frequency is reduced to one-fifth its value?
11. In the solution of Example 29-3, where does this equation come from?
12. (a) In a pure capacitive circuit, what two factors determine the current in the circuit?
    (b) Give the equation.

13. With reference to Example 29-4:
    (a) Why not start the solution directly with step 2?
    (b) In step 2, is this the peak current?
14. With reference to Fig. 29-3:
    (a) What is the value of voltage?
    (b) What is the value of current?
    (c) What is their phase relation?
    (d) What type of circuit must this represent? Why?
15. With reference to Fig. 29-4:
    (a) To what type of circuit do these e-i curves apply?
    (b) How can you tell it is a pure capacitive circuit?
    (c) What does the solid curve represent?
    (d) How is this curve obtained?
    (e) Why is the power curve positive during time 0–1?
    (f) What happens to this power (time 0–1)?
    (g) Why is the power curve negative during time 1–2?
    (h) What does this negative power mean?
    (i) Why is the power curve positive during time 2–3?
    (j) Why is the power curve negative during time 3–4?
    (k) What is the total power taken by the capacitor over one full cycle of input? Explain.
    (l) Is this answer reasonable? Explain.
16. If an ammeter, voltmeter, and wattmeter are properly connected to an ac pure capacitive circuit:
    (a) What will the voltmeter indicate?
    (b) What will the ammeter indicate?
    (c) What does the E/I ratio represent?
    (d) What will the wattmeter indicate? Explain.
    (e) Does the product $E \times I = 0$?
    (f) Account for the discrepancy between true power and the $E \times I$ product.
    (g) How is this discrepancy resolved?
    (h) What is this correction factor called?
    (i) How is this factor evaluated?
    (j) Give an equation for power that will apply to *any* ac circuit.
17. (a) What is the $E \times I$ product in an ac circuit called, to distinguish it from true power?
    (b) What is the power factor in a pure resistive circuit? Why?

PROBLEMS

1. (a) What is the reactance of a 5-μF capacitor at 100 Hz?
   (b) At 4000 Hz?
2. What size capacitor should be used to obtain 2000-Ω reactance at 1000 Hz?
3. At what frequency will a 100-pF capacitor have a reactance of 10 000 Ω?
4. In audio work, it is often stated that a bypass capacitor should have a reactance of

not more than one-tenth the value of the resistor it is shunting at the lowest frequency. If the resistance value is 1500 Ω, what capacitance value should be used for:

(a) Speech frequency band of 200 to 3000 Hz.

(b) High-fidelity band of 20 to 15 000 Hz.

5. A capacitor of 8 μF is connected to a 400-V 120-Hz supply.

(a) Find the current in the circuit.

(b) Draw the phasor diagram.

6. When a capacitor is connected to a 120-V 60-Hz line, it draws a current of 0.54 A. Find the capacitance of the unit.

7. In a pure capacitive circuit, the current and voltage are 8 mA and 350 V. The frequency is 15 kHz. What is the capacitance of the circuit?

8. Calculate the power dissipated in each of the following.

(a) Pure capacitive circuit, $I = 5$ mA, $E = 300$ V.

(b) Pure resistive circuit, $I = 80$ mA, $E = 200$ V.

(c) $I = 50$ mA, $E = 200$ V, current leading the voltage by 80°.

(d) $I_m = 100$ mA, $E_m = 500$ V, current lagging by 40°.

(e) What is the power factor in each of the above?

30

INDUCTANCE
AS A CIRCUIT
ELEMENT

Have you ever been curious enough to look up the circuit diagram of your radio receiver at home? Maybe you have seen the diagram of some other radio receiver. If so, you must have noticed inductors (coils) use in: the circuits that make it possible for you to tune in the desired station; or in the audio or power transformers; or as chokes in the power supply filter. These circuits and other applications of inductors will be covered in subsequent texts.* Meanwhile, before we can analyze such circuits we must understand the role of inductance as a circuit element.

When dc is applied to an inductive circuit, the current at first is zero. A short period of time is required before the current reaches its maximum, steady state, or Ohm's law value.† The time required for the current to reach its Ohm's law value depends on the time constant of the circuit (L/R). The lower the resistance of the circuit, the longer this time interval will be. This slow rise of current is due to the counter electromotive force induced in a coil whenever the current *tends* to change. Also, the magnitude of this back EMF depends on the inductance value (L), and the rate at which the flux produced by the coil cuts its own turns—or the rate of change of current (di/dt).

30-1 Current and voltage relation—pure inductive circuit. Let us consider an ideal inductor (resistance of the coil is zero) connected to an

---

*J. J. DeFrance, *General Electronic Circuits* (2nd ed.), Holt, Rinehart and Winston, New York, 1976; J. J. DeFrance, *Communication Electronic Circuits*, Holt, Rinehart and Winston, New York, 1972.

†See Chapter 18.

ac supply source, and that the circuit is closed at the instant the supply voltage reaches its positive maximum. The circuit and the sine waves of supply voltage, current, and counter EMF are shown in Fig. 30-1.

At the instant the switch is closed, the situation is identical to a dc circuit. The voltage is at its maximum positive, and at this point it is constant for a short duration. Since the circuit has no resistance, the current would tend to rise to a very high value. The rate of rise of current would be a maximum and a high induced voltage (CEMF) is developed. This induced voltage is equal and opposite to the line voltage. Therefore, the current is actually zero. Another way of explaining this is that the time constant (L/R) for this circuit is infinite. (Remember—the resistance of the circuit is zero.) So the current starts at zero and begins to rise to its Ohm's law value very gradually. As the current is rising, the line voltage starts decreasing. But the current still continues to rise! Why?

1. On dc, the current rises slower and slower as it approaches its maximum value. (It is an exponential curve.)
2. Due to the fall in supply voltage on ac, the current will rise even slower than in a dc circuit.
3. Since the rate of rise of current is decreasing, the induced voltage must also decrease.
4. The fall in line voltage is offset by the decrease in the opposing induced voltage—the current continues to rise until the line voltage reaches zero.

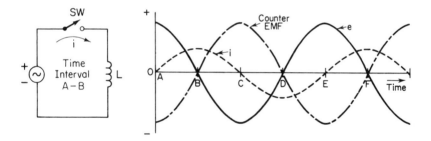

Figure 30-1 Current and voltage relations—pure inductive circuit.

How can a high value of current flow when the supply voltage is zero? Let's examine this more slowly. Just before the voltage reaches zero, it has some low value, let us say 0.001 V. Since the circuit resistance is zero (*pure inductive circuit*), the current, by Ohm's law, would be infinite. Even if the voltage were only one-tenth, one-hundredth, and so on, of this value, the current would still be infinite. However, since the circuit is inductive, as long as any line voltage is still applied and the current tends to rise just a trifle more—a small amount of induced voltage is also present to limit the flow of current, and so the current rises to some finite value right up to the point of zero line voltage. We have completed the conditions for the first quarter-cycle of current flow (*A* to *B* in

Fig. 30-1). The current in the circuit diagram has been flowing clockwise, starting at zero value and increasing to a maximum value.

During this first quarter-cycle, the magnetic field strength was increasing, and energy was stored in the magnetic field. Now, at time B, the supply voltage reverses and starts to build up in the opposite direction. In a resistive circuit, the current would also reverse. The current in this circuit is now at its maximum value. It cannot continue to rise. It should drop to zero and then reverse. But this is an inductive circuit! Any tendency for the current to decrease will give rise to an induced voltage in a *positive* direction, that will try to maintain the current flowing in the original direction. Therefore, the current drops slowly from its maximum value toward zero. The magnetic field is giving up its energy.

As the energy in the magnetic field decreases, the current drops faster and faster. Since the current is changing more rapidly, the counter EMF increases. Finally at time C, the supply voltage is a maximum negative; the current is zero, but it is changing at its fastest rate; the induced voltage is at maximum positive. This completes the first half-cycle.

Notice the similarity with the capacitive circuit. The current value depends not on the absolute value of the supply voltage, but on the "net" voltage, which in turn, is a function of the *rate of change* of voltage; or $i \propto de/dt$.

During the next half-cycle, the conditions are similar to those discussed above—except that all values are reversed in polarity. The current now starts to flow in a counterclockwise direction. The explanation for the phase relations is the same as before. Now examine the entire curve shown in Fig. 30-1. The applied voltage leads the current (or vice versa, the current lags the line voltage) by 90°. Also, the induced voltage and supply voltage are 180° apart. Comparing current and induced voltage, the induced voltage lags the current by 90°. All three curves are sine waves of the same frequency. *In a pure inductive circuit, the current lags the supply voltage by 90 degrees.*

30-2 Inductive reactance (X_L). We have just seen that, due to the induced voltage in an inductive circuit, the current never reaches its dc value. Obviously, the inductance produces an opposition to current flow in an ac circuit. *This effect is called inductive reactance (X_L).* As in a capacitive circuit, this opposition to current flow is also measured in ohms.

What factors affect the inductive reactance of any coil? We know it depends upon the induced voltage. But the induced voltage increases with the value of the inductance. It also increases with the rate of change of current, $e = -L(di/dt)$. But the rate of change of current is a function of the frequency of the applied voltage, or the angular velocity of the generating coil. Therefore, inductive reactance is equal to

$$X_L = 2\pi fL = \omega L \tag{30-1}$$

Example 30-1
Find the reactance of a 20-H coil at 120 Hz.

Solution

$$X_L = 2\pi f L = 2 \times \pi \times 120 \times 20 = 15\,080\ \Omega$$

From our analysis of inductive reactance, and also from the equation, you can see that reactance varies directly with frequency and with inductance. When it is necessary to find the reactance of a given coil at several frequencies, it is often more convenient to use the ratio method.

Example 30-2

Find the reactance of the inductance in Problem 30-1 at 50 Hz.

Solution

1. Since the frequency is lower, the reactance should be lower, by the same proportion.

2. $$X_{L_2} = X_{L_1}\frac{f_2}{f_1} = 15\,080 \times \frac{5}{12} = 6280\ \Omega$$

3. $$X_{L_2} = 2\pi f_2 L = 2 \times \pi \times 50 \times 20 = 6280\ \Omega \text{ (check)}$$

Example 30-3

An RF circuit requires a choke coil (inductor) having a reactance of not less than 50 000 Ω at 500 kHz. What is the minimum value of inductance that can be used?

Solution

$$L = \frac{X_L}{2\pi f} = \frac{0.159 \times 50\,000}{500 \times 10^3} = 15.9\ \text{mH}$$

30-3   Ohm's law—inductive circuit.   As in a capacitive circuit, Ohms' law can be used to show the relation between current, voltage, and reactance. In a *pure* inductive circuit,

$$I = \frac{E}{X_L}$$

The equation is stated in terms of effective values. Remember, it applies equally well to maximum, average, and instantaneous values, as long as both current and voltage are the same type of values.

Example 30-4

A coil of 50 mH is connected to a 20-V 12 000-Hz supply. Find the current in the circuit.

Solution

1. $$X_L = 2\pi f L = 2 \times \pi \times 12\,000 \times 50 \times 10^{-3} = 3770\ \Omega$$

2. $$I = \frac{E}{X_L} = \frac{20}{3770} = 0.005\,31\ \text{A} = 5.31\ \text{mA}$$

30-4   Phasor representation—inductive circuits.   The use of phasors to represent phase relations in an inductive circuit should be obvious.

The method is the same as for capacitive and resistive circuits. The only point to remember is that the current lags the supply voltage by 90°. As an illustration, Fig. 30-2 shows the results of Example 30-4.

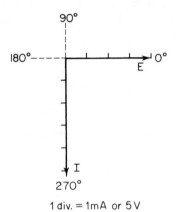

1 div. = 1mA or 5 V

Figure 30-2 Phasor representation— inductive circuit.

30-5 Power in pure inductive circuits.   Since the instantaneous power depends on the product of the instantaneous values of current and voltage, let us plot such curves (see Fig. 30-3). Then for several points along the time axis, the product of *e* and *i* will give the points for the power curve.

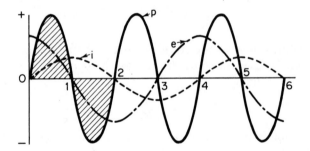

Figure 30-3 Power in a pure inductive circuit.

As in the case of a pure capacitive circuit, notice that:

1. The power curve is a sine wave of double frequency.
2. The average power, for one full cycle of current, is zero.

During the first quarter-cycle (time 0–1), as the current is rising, a magnetic field builds up around the inductance. Energy is stored in the magnetic field. The line supplies power to the circuit. As the current drops to zero, in the second quarter-cycle (time interval 1–2), the magnetic field around the coil is collapsing. Energy is returned to the line—the power is negative. In the third quarter-cycle, the current builds up to a negative maximum. The magnetic field again grows stronger, but with reversed magnetic polarity. The line is now supplying power to the circuit. Energy is stored in the magnetic field. The fourth quarter-cycle is a repetition of the second quarter-cycle—current drops to zero, magnetic field collapses, energy of the field is returned to the line, power is negative. This analysis again shows that *a pure inductive circuit dissipates no power.*

In terms of power factor, what does this mean? The power factor in a pure inductive circuit must be zero. This should have been obvious—the phase angle between current and voltage is 90°; therefore, cos θ, or the power factor, is zero.

30-6 Losses in inductors.   In the preceding discussions, we have been considering a *pure* inductive circuit. This assumes that the inductor and wiring have no losses. Practically, such a circuit does not exist except at cryogenic temperatures. A coil can be made to *approach* a perfect inductance. The better the quality of the coil, the closer we get to a pure inductance, but some losses still exist. These losses can be classified as follows:

1. *Ohmic (dc) Resistance of the Winding.* Since a coil is wound with wire, it still has a dc resistance, no matter what size wire is used. Naturally, the larger the diameter of the wire, the lower this loss.

2. *Effective Resistance.* As you learned in Chapter 28, resistance in an ac circuit is higher than the ohmic value because of *skin effect, core losses,* and *radiation losses.* Since these were discussed before, nothing further need be added. With careful design, the effective resistance of a coil can be reduced to a minimum.

3. *Effect of Coil Shields.* For use at radio frequencies, coils are often enclosed in a copper or aluminum shield. These shields are needed to prevent interaction with stray fields from other units. But the shields are of conducting material! Therefore, if they are cut by the field of the coil, eddy-current losses result. Coils for use at low frequencies (power or audio) are often encased in heavy cast-iron shields. This shields the unit against magnetic fields. However, again we have eddy-current losses and, in addition, hysteresis losses. Shields will therefore increase the effective resistance or losses of a coil. Good practice recommends that a shield be at least twice the diameter of the coil in order to minimize these losses.

Shielding will also affect the inductance of air-core or open-core coils. A magnetic shield will act as a part of the magnetic path decreasing the reluctance. This increases the flux and, in turn, increases the inductance of the coil. Non-magnetic shields act oppositely. They are good conductors, and the resulting eddy currents set up their own flux, which is opposite to the coil flux. This reduces the net flux, lowering the inductance of the coil.

4. *Figure of Merit* (*Q*). In order to compare the quality of coils, a comparison is made between the inductive reactance and the effective resistance of any coil. This ratio is called the *figure of merit or Q* of a coil.

$$Q = \frac{X_L}{R} = \frac{\omega L}{R} \tag{30-2}$$

The *Q* of a coil varies with frequency. Both the inductive reactance and effective resistance increase with frequency but not at the same ratio. It is, therefore, important when specifying the figure of merit of any coil that it be calculated for the band of frequencies on which the coil is to be used.

Example 30-5

A coil of 80 μH is to be used at a frequency of 500 kHz. Its resistance at that frequency is 3.0 Ω. What is the figure of merit of the coil?

Solution

1. $X_L = 2\pi f L = 2 \times \pi \times 500 \times 10^3 \times 80 \times 10^{-6} = 251 \ \Omega$

2. $Q = \frac{X_L}{R} = \frac{251}{3.0} = 83.7$

30-7 High-frequency effects. We have already seen that skin effect becomes more prominent at higher frequencies. This is one factor that would increase the effective resistance and reduce the *Q* of a coil. To reduce this loss, Litz wire or tubing is used to minimize the skin effect. With iron-core coils, core losses would become terrific. Laminating the iron is sufficient at low frequencies. At medium frequencies, powdered iron and ceramic cores have proven satisfactory. With such cores, the inductance values are obtained with smaller windings. The reduction of effective resistance of the winding more than offsets the core losses. The *Q* of the coil is increased. At higher frequencies, the core losses, even with powdered iron, are prohibitive.

At high frequencies, distributed capacitance of coils becomes important. Each turn of wire is a conductor. The insulation between wires is a dielectric. A small capacitor is formed between each turn. At very high frequencies, the inductive reactance may be higher than the capacitive reactance. More current will flow through the distributed capacity of the coil than through the winding itself! Unless special precautions are taken, an inductor may actually act as a capacitor.

To reduce distributed capacitance, the winding of a radio-frequency choke coil is often sectionalized. In addition, use of fine wire, space winding, and crisscross winding in place of parallel winding also minimize the capacitance between turns. Figure 30-4 shows the use of crisscross winding and sectionalized winding.

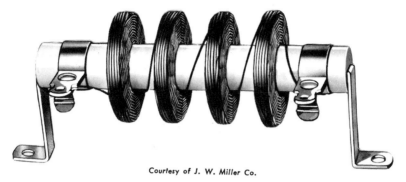

Figure 30-4   Coil construction to reduce distributed capacity.

REVIEW QUESTIONS

1. (a) What does the term *time constant* mean when applied to an inductive circuit?
   (b) What two factors determine its value?
   (c) Give the equation.
   (d) What is the value of time constant for a pure inductive circuit?

2. With reference to Fig. 30-1:
   (a) What does the solid curve represent?
   (b) What does the dashed curve represent?
   (c) Why is the current value zero at time instant *A*?
   (d) Why is the CEMF value a maximum at time instant *A*?
   (e) Why does the amplitude of the solid curve decrease during time interval *A–B*?
   (f) Why does the current increase?
   (g) Why does the CEMF decrease?
   (h) Why is the CEMF zero at time instant *B*?
   (i)   What is the phase relation between *E* and *I* in a pure inductive circuit?

3. (a) Does an inductance cause any opposition to current flow in an ac circuit?
   (b) What is the opposition called? Give its letter symbol.
   (c) What is the underlying cause for this opposition?

4. (a) State one factor that affects the value of inductive reactance.
   (b) Is this a direct or inverse relation? Explain.
   (c) Give another factor, and state how it affects reactance.
   (d) Give the equation for inductive reactance.
   (e) What is the unit used to evaluate inductive reactance?

5. What happens to the inductive reactance if:
   (a) The inductance is doubled?
   (b) The frequency is tripled?
   (c) The inductance is reduced to one-quarter?
   (d) The frequency is reduced to one-sixth?

6. In Example 30-2, compare the use of steps 2 and 3.

7. In the solution to Example 30-3:
   (a) Where does this equation come from?

    (b) Where does the 0.159 come from?

    (c) Why is the answer in millihenries?

8. (a) What two factors determine the current value in a pure inductive circuit?

    (b) How does the current vary with each of these factors?

    (c) Give the equation.

9. In the solution to Example 30-4, why don't we start directly with step 2?

10. With reference to Fig. 30-2:

    (a) What is the magnitude of E?

    (b) What is the magnitude of I?

    (c) What is their phase relation?

    (d) Is this correct? Explain.

11. With reference to Fig. 30-3:

    (a) What is the phase relation between the e and i curves?

    (b) What type of circuit must this represent?

    (c) How is the p curve obtained?

    (d) What is the total power per cycle of input?

    (e) Is this correct? Explain.

    (f) What does the positive power area during time interval 0–1 signify?

    (g) What happens to this power during time interval 1–2?

    (h) Give the equation for power in an inductive circuit?

12. What is the power factor in a *pure* inductive circuit?

13. Is a pure inductive circuit possible? Explain.

14. (a) State two factors that contribute to the effective resistance of a coil?

    (b) Is the effective resistance of a given coil a fixed value? Explain.

15. (a) Why are coils often enclosed in a metallic container?

    (b) Name three materials commonly used as shields.

    (c) When is each used?

    (d) What is the effect of a shield on a coil's losses?

    (e) What is the effect on a coil's effective resistance?

    (f) Which material produces the greatest losses? Why?

    (g) What is the effect of each shielding material on a coil's inductance? Explain.

16. (a) What does the *figure of merit* of a coil signify?

    (b) Give the letter symbol and the equation for this quantity.

    (c) Is this value constant for a given coil? Explain.

17. In coils used for high frequencies:

    (a) Can a change of wire reduce losses? Explain.

    (b) Can a change of core reduce losses? Explain.

    (c) What is meant by the *distributed capacitance* of a coil?

    (d) Does this exist at low frequencies? Why is it generally ignored?

    (e) How can distributed capacitance be minimized?

PROBLEMS

1. What is the reactance of a 20-mH coil at 1500 kHz?

2. What value inductance would be needed to obtain a reactance of 250 000 Ω at 4000 kHz?

3. At what frequency will a coil of 300 μH have a reactance of 3000 Ω?

4. A coil of 80 mH is connected across a supply of 30 V, 800 Hz.
   (a) What is the current in the circuit?
   (b) Draw the phasor diagram.

5. In a pure inductive circuit, the current and voltage are 120 mA and 40 V, respectively. If the frequency is 1200 Hz, what is the inductance value?

6. An inductance of 250 μH has a resistance of 45 Ω at 2000 kHz. What is the figure of merit of the coil?

7. What is the resistance of a 75-μH coil if its Q is 90 at 1.5 MHz?

8. A 320-μH coil, for use at 500 kHz, has a Q of 120. Find its resistance.

31

VECTOR ALGEBRA FOR AC CIRCUITS

Having discussed the action of resistance, capacitance, and inductance—taken individually—the next logical step would be to study the action of these components in combinations (in series, in parallel, and in series–parallel). We know from our studies of dc circuits, that voltages are additive in series circuits, and currents are additive in parallel circuits. Also, since there are no phase angles involved in dc circuits, we were able to solve all problems using plain algebra and straight numerical solutions. However, in ac circuits, phase angles enter the picture to complicate the solutions. In Section 26-6, we handled the situation by resorting to graphical methods. In general, such methods are not too accurate. Two other methods are available: *trigonometric*, and *vector-algebra* solutions.

Trigonometric treatment is quite satisfactory when the values involved are either in phase (0°), or at a 90° angle. But, this method is somewhat awkward when the phasors are at some angle other than 0 or 90°. Addition of such phasors require breaking down each quantity into its in-phase and quadrature components, adding the components, and then solving by the square root of the sum of the squares of the in-phase and quadrature values. What makes this method awkward is not so much the procedure, but rather, the terminology or "bookkeeping" system.

Since phasors have all the attributes of vectors, they can be represented graphically by vector methods, and they can be handled mathematically by vector-algebra methods. By use of vector algebra the terminology is simplified, and the solution of such problems becomes routine. Furthermore, by trigonometric means, we are unable to multiply or divide two vector quantities; nor could we square or take the square root of a vector quantity. All these things are made possible by use of vector algebra. In addition, even if you never solve a problem by vector algebra, it is still important that you learn the notations used in this system and their meanings. Many articles and texts in the electronic field use these methods of notation. Understanding of this new "language" will bring treasure chests of additional technical data within your reach.

31-1 Vector-algebra systems.

There are two systems of vector algebra in common use. Each method has its advantages. Each method has its limitations. The two systems are often used interchangeably. Very often we may start a problem in one system, continue it in the other, and later return to the first—using whichever system is simpler for the particular section of the problem on hand. One system uses *rectangular coordinates*; the other uses *polar coordinates*.

In the rectangular coordinate system, a vector (or a phasor) is expressed in terms of its in-phase and quadrature (right-angle) components. This method makes addition and subtraction of vector quantities simple. Multiplication, division, and squaring of vectors can also be done, but somewhat more laboriously. However, it is impossible to find the square root of a vector quantity in rectangular form.

On the other hand, the polar coordinate system makes multiplication, division, squaring, and taking the square root of vectors mere child's play. What are the limitations of this system? We cannot add or subtract vectors! It is necessary first to change the values from polar form to rectangular form; then to add (or subtract) the values while in rectangular coordinates; and finally convert the answer back to polar form. Obviously, it is necessary to understand both forms of vector algebra notation.

31-2 Representation by rectangular coordinates.

From previous studies in mathematics, you are probably familiar with X and Y axes, and how these axes divide an area into four quadrants. In the first quadrant, both X and Y values are positive; in the second quadrant, X values are negative, Y values are positive; in the third quadrant, both X and Y values are negative; in the fourth quadrant, X values are positive, but Y values are negative. This system of rectangular coordinates is shown in Fig. 31-1(a).

Using this system of coordinates, if we wish to represent a voltage at $0°$ phase angle, it would be drawn from the origin (O) horizontally *to the right*. For example, 50 V at $0°$ is shown in Fig. 31-1(b) as E_1. A second voltage (E_2) displaced by $180°$, would be the opposite of E_1, and would be drawn from the origin *to the left*. Since the first voltage is positive, this second voltage, rotated by $180°$, must be negative or -50 V. Vectorially, E_2 is similar to E_1 but *rotated through 180°*. Mathematically, E_2 is the same as E_1 but *multiplied by* (-1). Therefore a rotation of $180°$ corresponds to multiplying by an *operator* of (-1). Now suppose we wish to represent a voltage leading by $90°$. It would be drawn from the origin, *upward*. Since the operator for $180°$ rotation is (-1), the operator

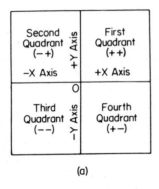

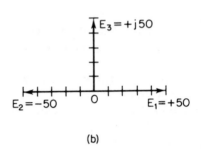

(a)   (b)

Figure 31-1 Rectangular coordinate system.

to denote a counterclockwise (leading) rotation of 90 degrees is taken as $\sqrt{-1}$. *This operator to indicate a 90° lead, ($\sqrt{-1}$). is designated by the symbol $+j$.* Therefore E_3, 50 V leading by 90°, is expressed as $+j50$. Using this system, a rotation of 90° clockwise (lagging) would be denoted by the operator $-j$, ($-\sqrt{-1}$). The negative sign preceding the operator denotes a lagging angle. It must be clearly understood that the operator j, whether positive or negative, denotes a 90° rotation only. *There is no notation for any other angles of rotation.*

Well then, how do we represent a voltage leading or lagging by some angle other than 90 or 180°? Simply resolve the voltage into two components—the in-phase and quadrature components—and specify this voltage in terms of these components. (That is exactly what we do in the trigonometric method, only now we have a simple way of expressing these components.) The form used is

$$E = a \pm jb$$

where a is the *in-phase* component of the voltage, and b is the *quadrature* component. One such voltage is shown in Fig. 31-2, at an angle of 30°. To evaluate these components, we again resort to trigonometry. Since

$$\cos\theta = \frac{\text{adjacent}}{\text{hypotenuse}} \quad \text{and} \quad \sin\theta = \frac{\text{opposite}}{\text{hypotenuse}}$$

then

$$\text{in-phase component} = E\cos\theta = a \tag{31-1}$$

$$\text{quadrature component} = E\sin\theta = b \tag{31-2}$$

If the voltage is leading, the quadrature component is prefixed with $+j$. If the voltage is lagging, we use $-j$ for the prefix. When the voltage is in phase (or at 180°), the quadrature component is zero. Therefore, the j term becomes zero. Let us try a few illustrations.

Example 31-1

Draw a phasor diagram, and express each of the following voltages in vector algebra notation. See Fig. 31-2.

(a) $E_1 = 100$ V at 0°
(b) $E_2 = 80$ V leading by 30°
(c) $E_3 = 60$ V leading by 90°
(d) $E_4 = 50$ V at 180°
(e) $E_5 = 75$ V lagging by 50°

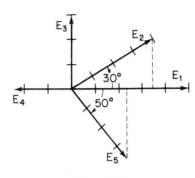

Figure 31-2   1 div. = 20 V

Solution

(a) E_1 being at 0° has no quadrature component; therefore,
$$E_1 = 100 + j0 \text{ V}$$

(b) E_2 at 30° has both in-phase and quadrature components, as follows:
$$\text{in-phase} = E \cos \theta = 80 \times 0.866 = 69.3 \text{ V}$$
$$\text{quadrature} = E \sin \theta = 80 \times 0.500 = 40 \text{ V}$$
Therefore,
$$E_2 = 69.3 + j40 \text{ V}$$

(c) E_3 is at 90°. It has no in-phase component. Therefore,
$$E_3 = 0 + j60 \text{ V}$$

(d) E_4 at 180° has no quadrature component. In addition, the voltage is opposite or negative compared to E_1. Therefore,
$$E_4 = -50 + j0 \text{ V}$$

(e) E_5 is lagging by 50°. Since the angle is not 0, 90, or 180°, this voltage has both in-phase and quadrature components:
$$\text{in-phase} = E \cos \theta = 75 \times 0.643 = 48.3 \text{ V}$$
$$\text{quadrature} = E \sin \theta = 75 \times 0.766 = 57.4 \text{ V}$$
Remembering that the voltage is lagging, we have
$$E_5 = 48.3 - j57.4 \text{ V}$$

So far we have been discussing how to represent voltages at various angles in the vocabulary of vector algebra. The same procedure applies equally well to currents at any angle of lead or lag. Merely express them in terms of their horizontal (in-phase) and vertical (quadrature) components, exactly as we did for the voltages in the above illustrations.

Can we express resistances and reactances by vector algebra? Neither of these quantities is a vector or a phasor. Yet these circuit constants do affect the phase angles of currents or voltages in the circuit. Therefore, they must be treated as phasors. In Chapter 28 we saw that current and voltage in a pure resistive circuit are in phase. Consequently, resistances are treated as in-phase components. On the other hand, in pure capacitive or pure inductive circuits (Chapters 29 and 30), current and voltage are 90° out of phase. Therefore, reactances must be treated as quadrature components—with capacitive reactances as a $-j$ quantity, and inductive reactance as a $+j$ quantity.

In a dc series circuit, the total opposition (resistance) to current flow is the sum of the individual component oppositions (resistances). This also applies to ac series circuits, (see Chapter 32). However, the oppositions are not just resistances, but may also be capacitive and/or inductive reactances. These additions must be made vectorially—taking the phase angles into consideration. Since such a sum is neither just resistance, nor just reactance, a new name—*impedance* (Z)*—is given to the sum of resistance plus reactance $(R + jX)$. Let us see how these quantities (resistance, reactance, and impedance) are expressed in vector algebra, rectangular form.

Example 31-2

Express each of the following oppositions in vector-algebra notation.
(a) Resistor 1200 Ω
(b) Capacitor of 750-Ω reactance
(c) Coil of 6240-Ω reactance
(d) (a) and (b) in series
(e) (a) and (c) in series
(f) A coil of 5 H and 500 Ω at 100 Hz
(g) A coil of 12 000-Ω impedance and a Q of 8

Solution

(a) $\qquad R = 1200$

(b) $\qquad X_C = -j750 \ \Omega$

(c) $\qquad X_L = j6240 \ \Omega$

(d) $\qquad Z = 1200 - j750 \ \Omega$

(e) $\qquad Z = 1200 + j6240 \ \Omega$

(f) $\qquad X_L = 2\pi f L = 2 \times \pi \times 100 \times 5 = 3140 \ \Omega$

$\qquad\qquad Z = 500 + j3140 \ \Omega$

(g) $\qquad Q = \dfrac{X_L}{R} = \text{arc} \tan \theta_L = 8$

$\qquad\qquad \theta = 82.9°$

$\qquad\qquad R = Z \cos \theta = 12\,000 \times 0.124 = 1490 \ \Omega$

*Impedance is covered in more detail in the next chapter on series circuits.

$$X_L = Z \sin \theta = 12\,000 \times 0.992 = 11\,900 \ \Omega$$
$$Z = 1490 + j11\,900 \ \Omega$$

31-3 Representation by polar coordinates. Any vector or phasor (or impedance) is expressed in polar coordinates by its magnitude and phase angle. For example, 50 mA leading by 20° is written as $50\underline{/20°}$; or 150 V in phase would be $150\underline{/0°}$. Could anything be simpler? When the vector is lagging, we merely put a minus sign in front of the degree value. For example, a capacitor having a reactance of 240 Ω would be $240\underline{/-90°}$; or 80 V lagging by 70° would be $80\underline{/-70°}$. In some texts or articles, when the vector is lagging, instead of using the minus sign notation, the angle sign is reversed. For example, a current of 1.35 A lagging by 38° would be $1.35\overline{\diagdown 38°}$.

31-4 Conversion between systems. Many times it is necessary, or convenient, to change from one system of notation to the other before continuing the solution of a problem. It is therefore important that we understand how to convert from polar to rectangular form, and vice versa. The method has already been discussed. In polar form we know the magnitude and phase angle of the vector (or phasor) quantity. To change to rectangular form we solve for the in-phase component (magnitude $\times \cos \theta$) and the quadrature component (magnitude $\times \sin \theta$). If the phase angle is positive, the quadrature or j component is also positive. When the phase angle is negative, the j component is also negative.

Example 31-3

Convert the following polar quantities into rectangular form.

(a) $\qquad\qquad\qquad\qquad E = 83\underline{/72°} \text{ V}$

(b) $\qquad\qquad\qquad\qquad I = 125\underline{/-43°} \text{ mA}$

(c) $\qquad\qquad\qquad\qquad Z = 2160\underline{/23°} \ \Omega$

Solution

(a) $E = 83 \cos \theta + j83 \sin \theta = 83 \cos 72° + j83 \sin 72° = 25.7 + j78.9 \text{ V}$

Calculator entries:

$8, 3, \boxed{\times}, 7, 2, \boxed{\cos}, \boxed{=}$ and $8, 3, \boxed{\times}, 7, 2, \boxed{\sin}, \boxed{=}$

(b) $\qquad I = 125 \cos 43° - j125 \sin 43° = 91.7 - j85.2 \text{ mA}$

(c) $\qquad Z = 2160 \cos 23° + j2160 \sin 23° = 1985 + j845 \ \Omega$

If you have a more powerful calculator*, these conversions from polar to rectangular form are performed more simply using calculator entries as follows:

*One such calculator (used here for illustration) is the National Semiconductor NS 108.

(a) For $E = 83/72°$ V

   Enter: 8, 3, $\boxed{\text{2nd}}$, $\boxed{R \leftarrow P}$, 7, 2, $\boxed{=}$; the display is 25.6. This is the in-phase component.

   Now enter: $\boxed{x - y}$ *; the display is 78.9, which is the quadrature component.

   The complete answer is: $25.6 + j78.9$ V.

(b) For $I = 125/{-43°}$ mA

   Enter: 1, 2, 5, $\boxed{\text{2nd}}$, $\boxed{R \leftarrow P}$, 4, 3, $\boxed{=}$; the display is 91.4. This is the in-phase component.

   Now enter: $\boxed{x - y}$; the display is 85.2—the quadrature component. Then, since the angle was a negative quantity, the answer is: $91.4 - j85.2$ mA.

(c) For $Z = 2160/23°$ Ω

   Enter: 2, 1, 6, 0, $\boxed{\text{2nd}}$, $\boxed{R \leftarrow P}$, 2, 3, $\boxed{=}$; the display is 1988.

   Enter: $\boxed{x - y}$; the display is 843.97.

   Answer: $1988 + j844$.

To convert from rectangular to polar coordinates takes two steps:

1. Find the magnitude (from the square root of the sum of the squares of the components):

$$\text{magnitude} = \sqrt{a^2 + b^2}$$

2. Find the phase angle (from the ratio of the component values):

$$\theta = \text{arc tan} \frac{b}{a}$$

The angle is leading if the j term is positive, and lagging if the j term is negative.

Before we try any examples of this type, a word of caution is appropriate with regard to the first step—finding the magnitude. Although calculators do not make mistakes, it is quite possible for us to do so. Press a wrong key, or whatever,—and we can commit boners. But we can catch these readily, if we examine the answers. Obviously, a correct answer must be larger than either of the original quantities (a or b), yet smaller than the numerical sum of $a + b$. A quick check to see that an answer lies between these two limits can save you from possible boners. While we are "estimating" answers for magnitude, consider also a time saver. If either of the quantities, a or b, is more than 10 times greater than the other, neglect the smaller one completely, and consider the magnitude as equal to the larger value only. Now let us try some examples.

*This is the x, y interchange key. Other calculators may use a somewhat different symbol for this key.

Example 31-4

Convert the following values from rectangular to polar form.

(a) $\qquad\qquad\qquad Z = 75 - j140 \ \Omega$

(b) $\qquad\qquad\qquad E = 24 - j86 \text{ V}$

(c) $\qquad\qquad\qquad I = 1.7 + j0.6 \text{ A}$

Solution

(a) $\qquad\qquad\qquad Z^* = \sqrt{(75)^2 + (140)^2} = 159 \ \Omega$

Calculator entry:

$$7, 5, \boxed{x^2}, \boxed{+}, 1, 4, 0, \boxed{x^2}, \ = , \boxed{\sqrt{x}}$$

$$\theta = \text{arc tan} \ \frac{140}{75} = 61.8°$$

Calculator entry:

$$1, 4, 0, \boxed{\div}, 7, 5, \boxed{=}, \boxed{\text{ARC}}, \boxed{\text{TAN}}$$

$$Z = 159\underline{/-61.8°} \ \Omega$$

(b) $\qquad\qquad\qquad E = \sqrt{(24)^2 + (86)^2} = 89.5 \text{ V}$

$$\theta = \text{arc tan} \ \tfrac{86}{24} = 74.4°$$

$$E = 89.5\underline{/-74.4°} \text{ V}$$

(c) $\qquad\qquad\qquad I = \sqrt{(1.7)^2 + (0.6)^2} = 1.8 \text{ A}$

$$\theta = \text{arc tan} \ \frac{0.6}{1.7} = 19.4°$$

$$I = 1.8\underline{/19.4°} \text{ A}$$

Now, let us see how the above conversions from rectangular to polar form would be made using a calculator such as the National NS 108.

(a) For $Z = 75 - j140$

Enter: 7, 5, $\boxed{\text{2nd}}$, $\boxed{R \to P}$, 1, 4, 0, $\boxed{=}$. The display is 158.8, which is the polar magnitude.

Enter: $\boxed{x - y}$; the display is 61.8, the polar angle.

Answer: $159\underline{/-61.8°} \ \Omega$

*An alternative method for finding the polar magnitude is to first solve for the angle θ, and then from the trigonometric relation that $\cos \theta = $ adjacent/hypotenuse, and since Z is the hypotenuse,

$$Z = \frac{\text{adj}}{\cos \theta} = \frac{75}{\cos 61.8°} = 159 \ \Omega$$

(b) For $E = 24 - j86$ V

Enter: 2, 4, $\boxed{\text{2nd}}$, $\boxed{R \rightarrow P}$, 8, 6, $\boxed{=}$; the display is 89.28.

Enter: $\boxed{x - y}$; the display is 74.4.

Answer: $89.3\underline{/-74.4°}$ V

(c) For $I = 1.7 + j0.6$ A

Enter: 1, $\boxed{\cdot}$, 7, $\boxed{\text{2nd}}$, $\boxed{R \rightarrow P}$, $\boxed{\cdot}$, 6, $\boxed{=}$. The display is 1.803.

Enter: $\boxed{x - y}$; the display is 19.4.

Answer: $1.80\underline{/19.4°}$ A

31-5 Addition and subtraction by vector algebra.

In series circuits, it is often necessary to add (or subtract) the voltages across the various units. Also, we can add the impedances of the various units. Similarly, in a parallel circuit, we have to add or subtract currents to find the total current or branch currents. Vector (and phasor) quantities expressed in rectangular form may be added or subtracted by treating them as ordinary binomials. Merely add all a terms, then all b terms *algebraically*. *Vector quantities in polar form cannot be added or subtracted directly.* First, you must convert the values to rectangular form. Then add them. Finally, convert the sum back to polar form.

Example 31-5

A resistor, capacitor, and commercial coil are connected in series. The voltage across each unit is

$$E_R = 50 + j0 \text{ V}$$
$$E_C = 0 - j120 \text{ V}$$
$$E_L = 27 + j36 \text{ V}$$

Find the supply voltage (magnitude and phase angle) and express in polar form.

Solution

(a) $$E_T = \dot{E}_R + \dot{E}_C + \dot{E}_L, \text{ or}$$

$$\begin{array}{r} 50 + j0 \\ 0 - j120 \\ 27 + j36 \\ \hline 77 - j84 \end{array}$$

(b) $$E_T = \sqrt{(77)^2 + (84)^2} = 114 \text{ V}$$

$$\theta = \text{arc tan } \tfrac{84}{77} = 47.5° \text{ lagging}$$

(c) $$E_T = 114\underline{/-47.5°} \text{ V}$$

Example 31-6

The line current in a parallel circuit is $I_T = 2.6\underline{/15°}$. The current in branch 1 is $I_1 = 1.8\underline{/-65°}$. Find the current in branch 2.

Solution

1. $\qquad I_T = 2.6 \cos 15° + j2.6 \sin 15° = 2.51 + j0.67$

2. $\qquad I_1 = 1.8 \cos 65° - j1.8 \sin 65° = 0.76 - j1.63$

3. Since $I_2 = I_T - I_1$:

$$\begin{array}{r} 2.51 + j0.67 \\ - + \\ \oplus\, 0.76 \ominus j1.63 \\ \hline I_2 = 1.75 + j2.30 = 2.89\underline{/52.7°} \end{array}$$

31-6 Multiplication: rectangular form. Vectors expressed in rectangular form were treated as ordinary binomials for the purpose of addition or subtraction. It is therefore not surprising to learn that they can also be multiplied or divided in exactly the same manner as any other binomials. Such a situation arises whenever you have to apply Ohm's law to the solution of an ac circuit. Since the values, current, voltage, and impedance can be treated as vector quantities, the solution by vector algebra is applicable.

There is one point to remember: since $j = \sqrt{-1}$, then $j^2 = -1$; and the j^2 term appearing in the product can be simplified in this manner.

Example 31-7

Multiply $9 + j5$ by $12 + j4$, and express the answer in polar form.

Solution

$$\begin{array}{r} 9 + j5 \\ 12 + j4 \\ \hline 108 + j60 \\ + j36 + j^2 20 \\ \hline 108 + j96 + j^2 20 \end{array}$$

Replacing j^2 by -1, we get

$$108 + j96 - 20 = 88 + j96$$

In polar form:

$$= 130\underline{/47.5°}$$

31-7 Division: rectangular form. Division of vectors expressed in rectangular form is accomplished by multiplying both the numerator and denominator by the *conjugate* of the denominator. This rationalizes the denominator: The j term is eliminated and we can simplify the answer.

Example 31-8

Divide $40 + j10$ by $6 + j4$, and express the answer in polar form.

Solution

$$\frac{40 + j10}{6 + j4} \times \frac{6 - j4}{6 - j4} = \frac{240 - j100 - j^2 40}{36 - j^2 16}$$

Simplifying the j^2 term, we have

$$\frac{280 - j100}{52} = 5.38 - j1.92$$

In polar form:

$$5.71\underline{/-19.6°}$$

31-8 Multiplication and division: polar form.

When vectors are expressed in rectangular form, multiplication or division of these vector quantities is somewhat laborious. The polar form is much easier to handle.

1. *To multiply in polar form, multiply the magnitudes, and add the angles algebraically.*
2. *To divide in polar form, divide the magnitudes, and subtract the angles algebraically.*

Even when the vector quantities are expressed in rectangular coordinates, it is often more convenient to change to polar form and then multiply or divide as required.

Example 31-9

Multiply $50\underline{/20°}$ by $32\underline{/-45°}$.

Solution

1. $50 \times 32 = 1600$

2. $20° + (-45°) = -25°$

3. $1600\underline{/-25°}$

Example 31-10

Divide $147\underline{/64°}$ by $840\underline{/35°}$.

Solution

1. $147 \div 840 = 0.175$

2. $64° - 35° = 29°$

3. $0.175\underline{/29°}$

31-9 Squares and square roots by vector algebra.

Vector quantities can be squared, using either the rectangular or polar systems. Since squaring means multiplying by itself, merely follow the methods shown above for the multiplication of vector quantities. In polar form this becomes very simple—square the magnitude and double the angle.

We pointed out earlier in this chapter that it was impossible to find the square root of a vector quantity, using the rectangular coordinate system. The polar system makes this operation very simple:

1. Take the square root of the magnitude.
2. Divide the angle by 2.

If a vector is expressed in rectangular form, we must first change to polar form before extracting the square root. As a check on the square-root process, let us start "backwards" by finding what the square of a given complex number would be.

Example 31-11

Square $75 + j50$.

Solution

$$
\begin{array}{r}
75 + j50 \\
75 + j50 \\
\hline
5625 + j3750 \\
+ j3750 + j^2 2500 \\
\hline
5625 + j7500 + j^2 2500 = 3125 + j7500
\end{array}
$$

Example 31-12

Find the square root of $3125 + j7500$.

Solution

1. $3125 + j7500 = 8125\underline{/67.38°}$
2. $\sqrt{8120\underline{/67.4°}} = 90.1\underline{/33.69°}$; converting back to rectangular form:
3. $75 + j50$; this checks with Example 31-11.

We have now covered all the elements of vector algebra that we will need. In the chapters that follow we use this mathematical tool in the solution of series circuits, parallel circuits, and series–parallel circuits. However, it should be noted that series circuits and *pure* parallel circuits are just as readily solved by trigonometric methods, without vector algebra.

REVIEW QUESTIONS

1. Give two reasons for using vector algebra in ac circuit problems.
2. Name two forms of vector-algebra notation.
3. In the rectangular coordinate system:
   (a) What does the term "j" signify?
   (b) What does the term "$-j$" signify?
   (c) What does the term "j^2" signify numerically? Vectorially?
   (d) What does the term "$-j^2$" signify numerically? Vectorially?
4. With reference to Fig. 31-2:
   (a) What is the magnitude and phase of E_1?
   (b) Express E_1 in vector-algebra notation.

   (c) What is the magnitude and phase of E_2?

   (d) Give an approximate value for E_2 in vector-algebra notation.

   (e) Express this value accurately in a trigonometric equation.

   (f) What is the magnitude and phase of E_3?

   (g) Express E_3 in vector-algebra notation.

   (h) What is the magnitude and phase of E_5?

   (i) Give an approximate value for E_5, in vector-algebra notation.

   (j) Express this value accurately by a trigonometric equation.

5. (a) Can impedances be expressed in vector-algebra notation?

   (b) How would a pure resistor be denoted?

   (c) How would a pure inductor be denoted?

   (d) How would a pure capacitor be denoted?

   (e) How would a series R–C value be denoted?

   (f) How would a series R–L value be denoted?

6. In the solution to Example 31-2:

   (a) What does Z in step (d) represent?

   (b) What does Z in step (e) represent?

   (c) What does Z in step (f) represent?

7. How is a vector (or phasor) quantity specified in polar form?

8. Express each of the following in polar form.

   (a) 20 V as the reference phasor

   (b) 60 V lagging by $30°$

   (c) 25 mA leading by $70°$

   (d) 30 mA leading by $90°$

   (e) 5 A lagging by $45°$

   (f) A coil, $Z_L = 200 \, \Omega$, and $Q = 1$

9. With reference to Example 31-3(a):

   (a) What does the value 83 represent?

   (b) What does the $72°$ represent?

   (c) Is this a leading or lagging phase angle? Explain.

   (d) How is this voltage converted to rectangular form?

   (e) Which will be larger, the in-phase or the quadrature component? Explain.

10. In Example 31-3(b), what is the significance of the $\underline{/-43°}$?

11. What does the Z in Example 31-3(c) represent physically?

12. With reference to Example 31-4(a), by inspection only:

   (a) What does this Z represent physically?

   (b) Between what two values will the polar magnitude lie?

   (c) Will the polar form have a positive or negative angle?

   (d) Will this angle be more or less than $45°$?

13. In Example 31-4(c), what type of circuit would produce such a current?

14. Explain the method for adding vectors while in polar form.

15. Explain briefly, vector addition in rectangular form.

16. (a) In Example 31-5, explain how E_L could have a value such as $27 + j36$?

   (b) What is the approximate Q of this coil?

17. Explain how subtraction of polar vectors must be done.

18. In the solution to Example 31-6, step 3, what is the significance of the circled $+$ and $-$ signs?

19. In the solution to Example 31-7, explain how the number 88, in $(88 + j96)$, is obtained?

20. Explain briefly, the technique for dividing vectors in the rectangular form.

21. In the solution to Example 31-8:
   (a) Where does the $(6 - j4)$ come from?
   (b) Will multiplying by this value change the problem? Explain.

22. **(a)** In adding by polar form, how are the angles treated?
   (b) In multiplying by polar form, how are the magnitudes treated?
   (c) In multiplying by polar form, how are the angles treated?
   (d) In dividing by polar form, how are the magnitudes treated?
   (e) In dividing by polar form, how are the angles treated?

23. **(a)** In the solution to Example 31-9, why is the final angle $-20°$?
   (b) In the solution to Example 31-10, why is the final angle $29°$?

24. **(a)** How do we extract the square root of a vector in rectangular form?
   (b) How do we take the square root of a vector in polar form?

25. In step 2 of the solution to Example 31-12:
   (a) Where does the number 90.1 come from?
   (b) Where does the angle $33.69°$ come from?

PROBLEMS

1. Express each of the following in rectangular form.
   (a) $E = 145$ V at $30°$ lead
   (b) $I = 37.6$ A at $180°$
   (c) $E = 25$ V at $48°$ lag
   (d) $I = 85$ mA at $90°$ lag
   (e) $I = 4.3$ A at $0°$
   (f) Capacitive reactance of 1200 Ω
   (g) Inductive reactance of 850 Ω
   (h) Resistance of 600 Ω
   (i) (f) and (g) in series
   (j) (f) and (h) in series
   (k) (g) and (h) in series
   (l) (f), (g), and (h) in series

2. Express each of the values in Problem 1(a) through (k) in polar form.

3. Convert each of the following from polar to rectangular form.
   (a) $E = 80\underline{/30°}$ V
   (b) $I = 65\underline{/-20°}$ mA
   (c) $Z = 380\underline{/65°}$ Ω
   (d) $I = 3.5\underline{/80°}$ A

4. Convert each of the following from rectangular to polar form.
   (a) $I = 50 + j30$ mA
   (b) $E = 10 - j45$ V

    (c) $Z = 120 + j200 \ \Omega$

    (d) $Z = 15 - j8 \ \Omega$

5. Add the following values and express the sum in polar form.

    (a) $I_1 = 15 + j7, \ I_2 = 8 - j5, \ I_3 = 4 + j2$

    (b) $E_1 = 140 - j60, \ E_2 = 30 + j80, \ E_3 = 20 - j50$

    (c) $Z_1 = 25 + j70, \ Z_2 = 18 - j30, \ Z_3 = 10 + j15$

    (d) $E_1 = 40\underline{/30°}, \ E_2 = 25\underline{/-48°}, \ E_3 = 60\underline{/0°}$

6. Find the missing values and express in polar form.

    (a) $E_T = 80\underline{/0°}, \ E_1 = 75\underline{/40°}, \ E_2 = \ ?$

    (b) $I_T = 30 + j25, \ I_1 = 15 - j20, \ I_3 = 4 + j30, \ I_2 = \ ?$

    (c) Series circuit $Z_T = 120 - j85, \ Z_1 = 70 - j100, \ Z_2 = 20 + j15, \ Z_3 = \ ?$

7. Find the product of $2.5 + j0$ and $15 - j20$ using the rectangular form.

8. Repeat Problem 7 using the polar form.

9. Find the product of $20 + j15$ and $5000 + j3500$:

    (a) Using the rectangular form.

    (b) Using the polar form.

10. Divide $40 + j0$ by $10 + j20$, using the rectangular form.

11. Repeat Problem 10 using the polar form.

12. A current of $3.0\underline{/30°}$ A flows through a 65-Ω resistor. Find the power dissipated, using the I^2R power equation.

13. Find the square root of each of the following values.

    (a) $86\underline{/48°}$

    (b) $420\underline{/-30°}$

    (c) $25 + j68$

    (d) $1500 - j900$

32

SERIES CIRCUITS—
BASIC PRINCIPLES

Alternating-current series circuits "obey" all the principles previously discussed for direct-current circuits.* The correlation is complete if the circuit contains resistors alone. Where more than one type of circuit element (resistance, capacitance, or inductance) is involved, phase angles must be considered. However, the principles are still the same:

1. The current is the same in all the components. (In phasor diagrams, current is therefore generally used as the reference.)
2. The current in the circuit depends on the total opposition in the circuit. This opposition may consist of resistance, inductive reactance, capacitive reactance, or any combination thereof.
3. The total opposition is the sum of the oppositions of each component.
   (a) If the circuit contains resistors alone, this total opposition is a resistance:
$$R_T = R_1 + R_2 + R_3 + \cdots$$
   as in any dc circuit.
   (b) If the circuit contains capacitors alone, the total opposition is a reactance:
$$X_{C_T} = X_{C_1} + X_{C_2} + \cdots$$
   (c) If the circuit contains inductors alone, the total opposition is a reactance:
$$X_{L_T} = X_{L_1} + X_{L_2} + \cdots$$

*See Chapter 7.

(d) If the circuit contains a combination of circuit elements, we cannot call the total opppsition a resistance nor can we call it a reactance. It is given a new name, *impedance* (*Z*).

Each of these cases is taken up in more detail in this chapter. It is recommended, particularly where the circuit contains more than one type of circuit element, that a phasor diagram be drawn, even though a mathematical solution is used.

32-1 Resistors in series. In a pure resistive circuit, whether the circuit contains one or a hundred resistors, the current is in phase with the line voltage. Since the current flows through each of the resistors, there is a voltage drop across each:

$$E_1 = IR_1, \qquad E_2 = IR_2 \ldots$$

Each of these voltage drops is in phase with the current. Since the phase angle is zero for the entire circuit, and also for any part of the circuit, this case is identical to a dc circuit and can be treated in the same manner. Therefore, no further discussion is necessary.

32-2 Capacitors in series. When capacitors are connected in series, the circuit is still a pure capacitive citcuit. In such a circuit, we know that the current leads the applied voltage by 90°. How much current flows in the circuit? That will depend upon the applied voltage and the *total* reactance. The total reactance depends on the total capacitance. In Chapter 19, we learned that when capacitors are connected in series the total capacitance is less than that of the smallest unit, or

$$\frac{1}{C_T} = \frac{1}{C_1} + \frac{1}{C_2} + \frac{1}{C_3} + \cdots$$

1. From this relation we can find the total capacitance, C_T.
2. Knowing C_T we can find the total reactance, X_{C_T}.*
3. From the applied voltage, we can find the line current, I.
4. The current through each component is the same. Therefore, we can find the voltage across each unit.

$$E_1 = IX_{C_1}, \quad E_2 = IX_{C_2}, \quad E_3 = IX_{C_3}, \quad \ldots$$

This is exactly the same as we would do with a resistive circuit to find voltage drops.

5. Since the total circuit is a pure capacitive circuit, and since each unit is a pure capacitor, *the line voltage, and also the voltage across each capacitor, will lag behind the line current by 90°.*

---

*Total reactance can also be obtained by finding the reactance of each capacitor, and adding (see step 2, Example 32-1).

Example 32-1

A 2-μF, 4-μF, and 8-μF capacitor are connected in series across a 1000-Hz 150-V supply. Find the line current and the voltage across each unit.

Solution

1.

$$\frac{1}{C_T} = \frac{1}{2} + \frac{1}{4} + \frac{1}{8}; \qquad C_T = 1.14 \; \mu F$$

2(a).

$$X_{C_T} = \frac{1}{2\pi f C} = \frac{0.159 \times 10^6}{1000 \times 1.14} = 139 \; \Omega$$

or

2(b).

$$X_{C_1} = \frac{1}{2\pi f C_1} = \frac{0.159 \times 10^6}{1000 \times 2} = \;\; 79.5 \; \Omega$$

$$X_{C_2} = \frac{1}{2\pi f C_2} = \frac{0.159 \times 10^6}{1000 \times 4} = \;\; 39.8 \; \Omega$$

$$X_{C_3} = \frac{1}{2\pi f C_3} = \frac{0.159 \times 10^6}{1000 \times 8} = \;\; 19.9 \; \Omega$$

$$X_{C_T} = X_{C_1} + X_{C_2} + X_{C_3} \qquad = \overline{139.2 \; \Omega}$$

3.

$$I = \frac{E_T}{X_{C_T}} = \frac{150}{139} = 1.08 \; A$$

4A.

$$E_1 = IX_{C_1} = 1.08 \times 79.5 = \;\; 85.8 \; V$$

$$E_2 = IX_{C_2} = 1.08 \times 39.8 = \;\; 43.0 \; V$$

$$E_3 = IX_{C_3} = 1.08 \times 19.9 = \underline{\;\; 21.5 \; V}$$

$$E_T = E_1 + E_2 + E_3 \qquad = 150.3 \; V \; \text{(check)}$$

Notice that all the voltages are in phase (with each other). Therefore, we can add them numerically. Their sum is equal to the applied voltage. A phasor diagram of this problem is shown in Fig. 32-1. Notice that the current is being used as the reference phasor (0°), and that each voltage lags the current by 90°. Phasor diagrams are best drawn on graph paper for convenient scales.

Since all the component voltages in Example 32-1 are in phase, a vector-

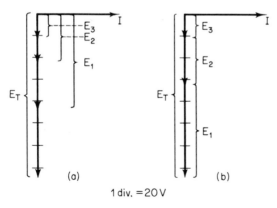

1 div. = 20 V

Figure 32-1 Phasor diagram—capacitive series circuit. (a) Individual voltages from zero reference. (b) Additive voltage reference.

algebra solution is not really necessary. However, let us repeat step 4 of the solution, using vector algebra "for experience." In any series circuit, since the current is the same in all components, we generally use current as the reference phasor. Therefore, the current value found in step 3 becomes

$$\dot{I}* = 1.08 + j0 \text{ A}$$

Also, remembering that all capacitive reactances are $-j$ quantities, then:

4B.
$$\dot{E}_1^* = \dot{I}\dot{X}_{C_1} = (1.08 + j0)(0 - j79.5) = -j85.8 \text{ V}$$
$$\dot{E}_2 = \dot{I}\dot{X}_{C_2} = (1.08 + j0)(0 - j39.8) = -j43.0 \text{ V}$$
$$\dot{E}_3 = \dot{I}\dot{X}_{C_3} = (1.08 + j0)(0 - j19.9) = \underline{-j21.5 \text{ V}}$$
$$\dot{E}_T = -j150.3 \text{ V}$$

This shows the voltages lagging the current by 90°.

32-3 Inductors in series.   We have already learned that when inductors are connected in series, the total inductance is the sum of the individual inductances.†

$$L_T = L_1 + L_2 + L_3 + \cdots$$

We have also seen that in a pure inductive circuit, the current lags the applied voltage by 90°. This, coupled with the treatment we just went through for capacitors in series, should make solution of this type of problem obvious:

1. Find the total inductance.
2. Find the total reactance and the reactance of each unit.
3. Find the current in the circuit by Ohm's law.
4. Find the voltage across each unit.

The phasor diagram will be similar to Fig. 32-1. However, since we use current as the reference phasor, and since current now *lags* the applied voltage by 90°, all voltages will be drawn *upward* instead of downward. Rather than using a similar type of problem to illustrate inductors in series, we will try a variation.

---

*The dots over the symbols indicate that these quantities have a phase angle as well as a magnitude. Consequently, addition and multiplication must be done using vector-algebra methods. Some texts use boldface type to represent phasor quantities. However, there are two reasons why the "dots" system is preferred in this text:

1. The distinction between boldface and standard type may escape the student—but the dot cannot be ignored.
2. When writing phasor equations on paper (or on the blackboard), one cannot use boldface—the dot system has no such limitation.

†See Chapter 18.

Example 32-2

Three inductances, 30, 50, and 60 mH are connected in series across an ac supply. The current in the circuit is 0.5 A. The voltage across the 30-mH coil is 40 V. Find the supply voltage, frequency, and voltage across each coil.

Solution A: *Trigonometric*

1.
$$X_{L_1} = \frac{E_1}{I} = \frac{40}{0.5} = 80\,\Omega$$

2.
$$f = \frac{X_{L_1}}{2\pi L_1} = \frac{80}{2 \times \pi \times 30 \times 10^{-3}} = 425\,\text{Hz}$$

3.
$$X_{L_2} = 2\pi f L_2 = 2 \times \pi \times 425 \times 50 \times 10^{-3} = 134\,\Omega$$

4.
$$X_{L_3} = 2\pi f L_3 = 2 \times \pi \times 425 \times 60 \times 10^{-3} = 160\,\Omega$$

5a.
$$X_{L_T} = X_{L_1} + X_{L_2} + X_{L_3} = 80 + 133 + 160 = 373\,\Omega$$

or

5b.
$$L_T = L_1 + L_2 + L_3 = 30 + 50 + 60 = 140\,\text{mH}$$
$$X_{L_T} = 2\pi f L_T = 2 \times \pi \times 425 \times 140 \times 10^{-3} = 374\,\Omega$$

6.
$$
\begin{aligned}
E_1 &= && Given && 40.0\,\text{V}\\
E_2 &= IX_{L_2} = 0.5 \times 134 = && 67.0\,\text{V}\\
E_3 &= IX_{L_3} = 0.5 \times 160 = && 80.0\,\text{V}\\
E_T &= E_1 + E_2 + E_3 && = 187.0\,\text{V}\\
E_T &= IX_{L_T} = 0.5 \times 374 = && 187.0\,\text{V (check)}
\end{aligned}
$$

(Since all the voltages are in phase with each other, we can add them numerically for a check.) A phasor diagram for this solution is shown in Fig. 32-2.

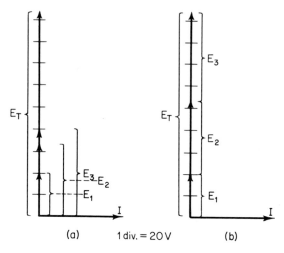

(a)   1 div. = 20 V   (b)

Figure 32-2 Phasor diagram—inductive series circuit. (a) Individual voltages from zero reference. (b) Additive voltage reference.

Solution B: *Using vector algebra* (again for experience)

1. Since this is a series circuit, we will use current as the reference phasor.

$$\dot{I} = 0.5 + j0$$

Also, since X_{L_1} is pure inductive, voltage leads current by 90°, and

$$\dot{E}_1 = 0 + j40$$

Then

$$X_{L_1} = \frac{\dot{E}_1}{\dot{I}} = \frac{+j40}{0.5} = +j80 \, \Omega$$

2. See solution A. (Phase angles do not apply here.)

3, 4, and 5. Since these reactances are all inductive, merely put a "$+j$" in front of each numerical value found in solution A.

6.
$$\dot{E}_1 = \text{(given)} \qquad\qquad\qquad +j40.0$$
$$\dot{E}_2 = \dot{I}\dot{X}_{L_2} = (0.5)(+j134) = \quad +j67.0$$
$$\dot{E}_3 = \dot{I}\dot{X}_{L_3} = (0.5)(+j160) = \quad +j80.0$$
$$\overline{\qquad\qquad\qquad\qquad E_T = \quad +j187.0}$$

Notice that the zero values have been omitted from the above phasor values. For example, current was shown as 0.5 instead of $0.5 + j0$; and X_{L_2} was $+j134$ rather than $0 + j134$. This simplification is perfectly permissible.

32-4   Resistance and capacitance in series.

So far we have been considering series circuits containing only one type of circuit element. These circuits were comparatively simple because phase angles were the same in all parts of the circuit, as well as for the total circuit (0° for pure resistive circuits and 90° for pure reactive circuits). We were therefore able to add oppositions and voltage drops numerically. Now, however, we begin to analyze circuits that contain more than one type of circuit element. Such circuits are very common in all types of electronic equipment.

If a resistor and a capacitor are connected in series, the current that flows in the resistor must be the same current that flows through the capacitor. This is true because it is a series circuit. We know that the voltage drop across a resistor (IR) is in phase with the current. But the voltage across a capacitor (IX_C) *lags* the current by 90°. As in any series circuit, the total voltage is equal to the sum of the voltages across each unit. Since the individual unit voltages are out of phase, this addition must be a *phasor* addition.

Example 32-3

An $R–C$ series circuit contains $R = 15 \, \Omega$ and $X_C = 20 \, \Omega$. Find the supply voltage needed to produce a circuit current of 2.0 A. Use:

A. Graphical solution for the line voltage.

B. Trigonometric solution.

C. Vector-algebra solution.

Solution

1. The voltage across the resistor, $E_R = IR = 30$ V, in phase with the current.

2. The voltage across the capacitor, $E_C = IX_C = 40$ V, lagging the current by 90°.

3A. *Graphical:* To get the total voltage, E_T, graphically, we draw E_R and E_C to scale, and at the proper angles using the phasor method explained in Section 26-6. This is shown in Fig. 32-3. Measuring the length of the resultant, E_T, we find that the total voltage is 50 V! Obviously, the component voltages IR and IX_C cannot be added numerically.

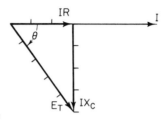

Figure 32-3 Phasor diagram—*R-C* series circuit.

1 div. = 10 V

3B. *Trigonometric:* Graphical solutions need special "tools" and are never too accurate unless drawn to a large scale and with great care. However, using Fig. 32-3 as a sketch, we can solve for E_T by trigonometry. Since we have a right triangle with E_T as the hypotenuse, then

$$E_T = \sqrt{E_R^2 + E_C^2} = \sqrt{(30)^2 + (40)^2} = 50 \text{ V}$$

It must be remembered that this mathematical solution applies only when the phasors are at 90°.

Whenever making phasor additions, it is good practice to first set minimum and maximum limits for the answer. Obviously, the resultant of two right-angled phasors must be greater than the larger value, but less than their numerical sum. Furthermore the increase (above the large value), is never more than 50% of the smaller value. (In the above example, the resultant should fall between limits of 40 and 55.)

3C. *Vector algebra:* To solve for E_T using vector algebra, we first note that $E_R = IR$ is in phase with the current, while $E_C = IX_C$ lags the current by 90°. Expressing these two component voltages in vector form, we have $E_R = 30$ and $E_C = -j40$. We are using the rectangular form, since polar quantities cannot be added. Then

$$\dot{E}_T = \dot{E}_R + \dot{E}_C = 30 - j40$$

Now, converting this to polar form, we get

$$E_T \text{ (magnitude)} = \sqrt{(30)^2 + (40)^2} = 50 \text{ V}$$

(The angle θ is discussed in Section 32-6.) Notice that the actual mathematics to find the magnitude of E_T is exactly the same as in the trigonometric method.

32-5 Impedance (Z).   In order to find the current in a series circuit, we must know the total opposition of the circuit. An *R-C* circuit has two types of oppositions, a resistance (due to R) and a reactance (due to C). Each of these oppositions is measured in ohms. We have already learned that the total opposition in this type of circuit is called *impedance* (Z), and that it is measured in ohms. Can we add resistance and reactance numerically to get impedance? No! The resistance and reactance effects are not in phase. Again, phasor addition must be used. This can be shown as follows:

1. Current flowing through the resistor produces a voltage drop, IR, in phase with the current.
2. Current flowing through the capacitor produces a second voltage, IX_C, lagging the current by 90°.
3. The total voltage must be equal to the product of this same current and the total opposition, IZ. But we also know that the total voltage is equal to the phasor *sum* of IR and IX_C. Therefore, IZ is equal to the phasor sum of IR and IX_C.

Notice, in Fig. 32-3, that IR, IX_C, and E_T form a right triangle, with E_T or IZ as the hypotenuse of the triangle. Also notice that the current I is the same for each of these voltages. Therefore, if we divide each value by I, we still have a right triangle with each side reduced in scale, by the value of I (in this case reduced by 2, since $I = 2$ A). This new triangle is called the *impedance triangle* (see Fig. 32-4). The sides are $R = 15\ \Omega$, $X_C = 20\ \Omega$, and $Z = 25\ \Omega$. This again shows that to find the impedance of a circuit, the resistance and reactance values cannot be added numerically. They must be added in vector fashion.

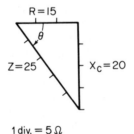

1 div. = 5 Ω

Figure 32-4 Impedance triangle—R-C series circuit.

In drawing the impedance triangle, notice that no arrows are shown. This is because resistance, reactance, and impedance by themselves are *not* phasor quantities. They act with the current, which is a phasor, to produce voltages which are also phasors. Since resistance and reactance "operate" on the current to produce voltages of different phase angles, they must be drawn at their "operating" angle. That is, resistance must be drawn horizontally (zero phase angle), and reactance must be drawn vertically (90° phase rotation). But the impedance triangle (Fig. 32-4) is a right triangle, with the impedance Z as the hypotenuse. We can, therefore, solve for Z trigonometrically, or

$$Z = \sqrt{R^2 + X_C^2} \qquad (32\text{-}1)$$

Using this method with our values of R and X_C above, we get

$$Z = \sqrt{R^2 + X_C^2} = \sqrt{(15)^2 + (20)^2} = 25\ \Omega$$

This checks with the graphical answer. *A valuable time-saving point is that if one quantity (R or X) is more than 10 times the other, the Z value can be considered equal to the larger of the two individual values* (see Problem 7, at the end of the chapter.

As we saw in Section 31-2, impedance can be expressed in vector form. In the above problem, the impedance (in rectangular form) would be $R - jX_C$, or $15 - j20$. This impedance can also be expressed in polar form. The magnitude, we already know from above, is $25\ \Omega$. How to find the phase angle is shown below.

32-6 Phase angle: R-C series circuit.   In the problem illustrated in Fig. 32-3, we have three phase angles to consider:

1. The angle between the current, and the voltage across the resistor, ($E_R = IR$). This angle is always zero degrees.
2. The angle between the current, and the voltage across the capacitor, ($E_C = IX_C$). This angle is always $90°$.
3. The *circuit phase angle* (θ). This is the angle between the line current and the total voltage, ($E_T = IZ$). Applying a protractor to Fig. 32-3, this angle is measured as $53°$. If we apply the protractor to the impedance triangle (Fig. 32-4) we get the same answer, $53°$.

This gives us another means for finding the circuit phase angle—by trigonometry. We can use sine, cosine, or tangent function:

1. The cosine of the circuit phase angle is the ratio of the adjacent side (R) to the hypotenuse (Z), or

$$\cos \theta = \frac{R}{Z} \qquad\qquad \text{(32-2A)}$$

and since arc cosine means "*the angle whose cosine is*"

$$\theta = \text{arc cosine*}\ \frac{R}{Z} = \text{arc cosine}\ \frac{15}{25} = 53.1°$$

2. The sine of the circuit phase angle is the ratio of the opposite side (X_C) to the hypotenuse (Z) or

$$\sin \theta = \frac{X}{Z} \qquad\qquad \text{(32-2B)}$$

and

$$\theta = \text{arc sin*}\ \frac{X}{Z} = \text{arc sin}\ \frac{20}{25} = 53.1°$$

*These functions are also shown as $\cos^{-1}$ and $\sin^{-1}$.

3. The tangent of the circuit phase angle is the ratio of the opposite side (X_c) to the adjacent side (R) or

$$\tan \theta = \frac{X}{R} \tag{32-2C}$$

and

$$\theta = \text{arc tan*} \frac{X}{R} = \frac{20}{15} = 53.1°$$

In evaluating phase angles, it is again good practice to estimate the approximate value. Obviously, if R and X are equal, the angle is 45°. If R is greater than X, θ is less than 45, and approaches zero. For R values greater than $50X$, the angle can be considered zero. Conversely, as X increases to over $50R$, the angle can be considered equal to 90°.

32-7 Ohm's law: R-C series circuit.

We have already covered all the theory pertinent to Ohm's law in an R-C series circuit. We have discussed the principles applicable to series circuits, and we have seen how to calculate impedance and phase angles. Now let us apply this theory to a problem.

Example 32-4

A resistor of 500 Ω is connected in series with an 0.05-μF capacitor across a 100-V 8000-Hz supply. Find:
(a) The current in the circuit.
(b) The voltage across each unit.
(c) The circuit phase angle.
(d) Draw a phasor diagram.

Solution A: *Trigonometric*

1.
$$X_C = \frac{1}{2\pi fC} = \frac{0.159 \times 10^6}{8000 \times 0.05} = 398 \ \Omega$$

2.
$$Z = \sqrt{R^2 + X_C^2} = \sqrt{(500)^2 + (398)^2} = 640 \ \Omega$$

(a)
$$I = \frac{E}{Z} = \frac{100}{640} = 0.156 \ \text{A}$$

(b)
$$E_R = IR = 0.156 \times 500 = 78.2 \ \text{V}$$
$$E_C = IX_C = 0.156 \times 398 = 62.2 \ \text{V}$$

(c)
$$\theta = \text{arc tan} \frac{X}{R} = \text{arc tan} \frac{398}{500} = 38.5°$$

(d)

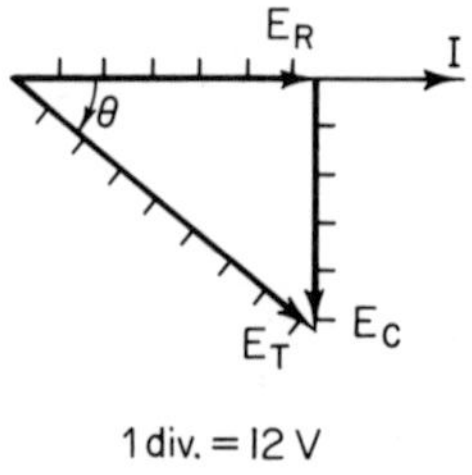

1 div. = 12 V

Figure 32-5 Phasor diagram—R-C series circuit.

*This function is also shown as $\tan^{-1}$.

The above solution was by formula throughout. In such cases the phasor diagram need not be drawn to scale, but can be a sketch to represent circuit conditions.

Solution B: *Using vector algebra*

1. From solution A, step 1, we found $X_C = 398\ \Omega$. Now, we write this as $-j398$, and
2. $Z = R - jX_C = 500 - j398$; and converting this to polar:
   (a) Calculator entry: 5, 0, 0, $\boxed{x^2}$, $\boxed{+}$, 3, 9, 8, $\boxed{x^2}$, $\boxed{=}$, $\boxed{\sqrt{x}}$. The display is 639. This is the magnitude.
   Then enter: 3, 9, 8, $\boxed{\div}$, 5, 0, 0, $\boxed{=}$, $\boxed{\text{TAN}^{-1}}$. The display is 38.5. This is the angle.
   Therefore,

$$Z = 639\underline{/-38.5°}$$

   (b) Alternative entries using a NS 108-type calculator:
   Enter: 5, 0, 0, $\boxed{\text{2nd}}$, $\boxed{R \rightarrow P}$, 3, 9, 8, $\boxed{=}$. The display is 639 (magnitude).
   Enter: $\boxed{x - y}$; the display is 38.5 (angle). Again, $Z = 639\underline{/-38.5°}$.
3. Using the supply voltage as the reference phasor, we get

$$\dot{I} = \frac{\dot{E}}{\dot{Z}} = \frac{100\underline{/0°}}{639\underline{/-38.5°}} = 0.156\underline{/38.5°}\ \text{A}$$

The current *leads* the voltage by 38.5°. (This is really the same as in Fig. 32-5, where voltage *lags* current by 38.5°.)

Let us try another problem, this time using only vector algebra.

Example 32-5

A current of 1.25 A flows in a series circuit containing 40-Ω resistance and 70-Ω capacitive reactance. Using vector algebra, solve for the voltage applied to the circuit.

Solution

1. Using current as the reference phasor, $I = 1.25 + j0$.
2. $Z = 40 - j70$
3. Multiplying step 1 by step 2 ($E = IZ$):

$$
\begin{array}{r}
1.25 + j0 \\
40 - j70 \\
\hline
50 + j0 \\
- j87.5 - j^2 0 \\
\hline
E = IZ = 50 - j87.5
\end{array}
$$

In polar form:

$$E = 100.7 \underline{/-60.3°} \text{ V}$$

The line voltage is 100.7 V and lags the current by 60.3°.

32-8 Power in an R-C circuit.   In a pure resistive circuit (Section 28-8) the power curve (versus time) was always positive. On the other hand, in a pure capacitive circuit (Section 29-6) the power curve was half positive and half negative—the average power for the cycle was zero. This was because the capacitor absorbed energy while charging, and returned energy to the line when it discharged. What happens to the power curve when the circuit contains capacitance and resistance in series? For simplicity, we will analyze such a circuit where the value of capacitance and resistance, are so selected that X_C is equal to R. In the impedance triangle, the tangent function (X_C/R) is equal to 1.0. The phase angle between current and line voltage is 45°, with the current leading. Now we can plot the curves for e, i, and p versus time. These curves are shown in Fig. 32-6.

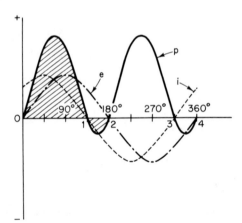

Figure 32-6 Power in an R-C series circuit.

Examine this curve (Fig. 32-6) carefully.

1. The power curve is partly negative, but mainly positive.
2. The total power (positive areas minus negative areas) for the cycle is positive. The energy is dissipated in the resistance of the circuit.

We know that the power in any circuit where current and voltage are out of phase is given by

$$P = E_T I_T \cos \theta$$

Let us rewrite this as

$$P = (E_T \cos \theta) I_T$$

Now examine the phasor diagram [Fig. 32-7(a)]; $E_T \cos \theta$ is equal to E_R! The power dissipated is therefore the product of current and *voltage across the resistor*. This should not be surprising. We know the capacitor dissipates no power, and therefore all the power dissipated in the circuit must be dissipated by the resistor.

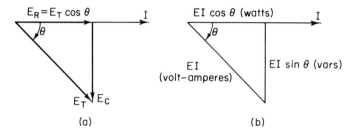

(a)   (b)

Figure 32-7   Phasor diagram—*R-C* series circuit.

32-9 Reactive volt-amperes (vars).   The phasor diagram Fig. 32-7(a), shows that $E_R = E_T \cos \theta$. But from our knowledge of trigonometry we can also see that $E_C = E_T \sin \theta$. If we multiplied the three voltage values by the *numerical* value of the current in the circuit, we would have another right triangle of the same relative proportions. This new triangle is shown in Fig. 32-7(b). The horizontal line (in phase with I) would now be $EI \cos \theta$, or would represent the circuit power. The hypotenuse would now be EI and would represent the circuit volt-amperes. The third side, the vertical component (in quadrature with I) would now be $EI \sin \theta$. Since it also is a product of voltage and current (amperes), it is called the *reactive volt-amperes* of the circuit. The basic unit for reactive volt-amperes is the *var*. This term is often used in the power field for circuits containing resistance and reactance, and is equal to the product of line voltage, line current, and the *sine* of the circuit phase angle, or

$$\text{reactive volt-amperes (vars)} = E_T I_T \sin \theta \qquad (32\text{-}3)$$

32-10 Measurement of power.   We saw in Chapter 5 how a wattmeter could be used in a pure resistive circuit to measure power directly. If a wattmeter were connected into a circuit containing resistance and reactance, would the meter indicate the true power ($EI \cos \theta$), or the volt-amperes (EI) taken by the circuit? We know that the torque or deflecting force in a wattmeter depends on the voltage across its potential coil and the current through its current coil. More specifically, the torque *at any instant* is proportional to the product of voltage and current *at that instant*. Now examine the wave diagrams of Fig. 32-6. The product of voltage and current at any instant is the instantaneous power curve p, and is the true circuit power. Since the wattmeter movement, due to its inertia, cannot follow the instantaneous variation in power, the meter indication will be proportional to the average power per cycle, or the true power taken by the load.

Commerical loads, particularly where motors are used, have both resistance and reactance. (Although the reactance is generally inductive, its effect on circuit power is similar to capacitance.) In such cases, since the circuit constants (R, L, C) or circuit power factor are seldom known, wattmeters are indispensable for finding the power taken by the load. In fact, when circuit power factor is desired, it is calculated from wattmeter, voltmeter, and ammeter readings.

Example 32-6

A $\frac{3}{4}$-ton air conditioner draws 8.0 A from a 120-V 60-Hz line. The power taken by this load, from a wattmeter reading, is 720 W. Find:

(a) The volt-amperes of the load.
(b) The power factor of the load.
(c) The phase angle between line current and voltage.
(d) The reactive volt-amperes.

Solution

(a) $\qquad$ Volt-amperes $= EI = 120 \times 8.0 = 960$ VA

(b) $\qquad$ From $P = EI \cos \theta$,

$$\cos \theta = \frac{P}{EI} = \frac{720}{960} = 0.75$$

(c) $\qquad$ $\theta = \text{arc cos } 0.75 = 41.4°$

(d) $\qquad$ vars $= EI \sin \theta = 960 \times 0.6613 = 635$ vars

32-11 Resistance and inductance in series. Here again we are faced with a circuit containing more than one type of circuit element. The analysis will be quite similar to an R-C series circuit. When a resistor and an inductor are connected in series, the current that flows through the resistor must be the same current that flows through the inductance. This current will produce a voltage drop (IR) across the resistance. We know that this voltage is in phase with the current. However, the voltage drop across the inductance (IX_L) will *lead* the current by 90°. As in any series circuit, the total voltage is the sum of the voltages across each unit—but due to the phase angles of these voltages, the addition must be a phasor addition.

Example 32-7

A current of 150 mA is flowing through a resistance of 1000 Ω and an inductance of 2 H. The frequency of the supply voltage is 50 Hz. Draw a phasor diagram and find the supply voltage and circuit phase angle.

Solution

1. $\qquad$ $E_R = IR = 0.15 \times 1000 = 150$ V

2. $\qquad$ $X_L = 2\pi fL = 2 \times \pi \times 50 \times 2 = 628 \ \Omega$

3. $\qquad$ $E_L = IX_L = 0.15 \times 628 = 94.2$ V

4A. *Graphical* (Fig. 32-8): $E_T = 175$ V, leading by $\theta = 32°$.

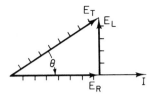

Figure 32-8 Phasor diagram—*R-L* series circuit.

1 div. = 20 V

4B. *Trigonometric:* Since E_L and E_R are at right angles, we can solve for line voltage and circuit phase angle by trigonometry:

(1) $$E_T = \sqrt{E_R^2 + E_L^2} = \sqrt{(150)^2 + (94.2)^2} = 177 \text{ V}$$

(2) $$\theta = \text{arc tan} \frac{E_L}{E_R} = \text{arc tan} \frac{94.2}{150} = 32.1°$$

4C. *Vector algebra:* Using current as the reference phasor,

$$E_R = 150 \text{ V} \qquad E_L = +j94.2 \text{ V}$$
$$\dot{E}_T = \dot{E}_R + \dot{E}_L = 150 + j94.2$$

Converting to polar form, we get

$$\text{Magnitude} = \sqrt{(150)^2 + (94.2)^2} = 177$$
$$\text{Angle } \theta = \text{arc tan} \frac{94.2}{150} = 32.1°$$
$$E_T = 177\underline{/32.1°} \text{ V}$$

32-12 Impedance: R-L circuit. Compare Fig. 32-8, for an inductive circuit, with Fig. 32-3, for the capacitive circuit. Notice that E_L is drawn upward (leading the current by 90°), whereas E_C was drawn downward (lagging the current by 90°). Since the voltage triangle, where

$$E_T = IZ, \quad E_L = IX_L, \quad E_R = IR$$

is the basis for the impedance triangle, the impedance triangle for an *L-R* circuit should be drawn upward.

Example 32-8

Find the impedance and phase angle for the circuit in Example 32-7.

Solution A: Graphical (Fig. 32-9) By direct measurement from the impedance triangle, $Z = 1175 \ \Omega$.

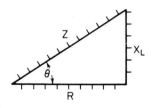

Figure 32-9 Impedance triangle—*R-L* series circuit.

1 div. = 100 Ω

487

Solution B: Trigonometric

$$Z = \sqrt{R^2 + X_L^2} = \sqrt{(1000)^2 + (628)^2} = 1181 \ \Omega$$

$$\theta = \text{arc tan} \frac{X_L}{R} = \text{arc tan} \frac{628}{1000} = 32.1°$$

Solution C: Ohm's law   (Using E_T and I from Example 32-7)

$$Z_T = \frac{E_T}{I} = \frac{177}{0.15} = 1180 \ \Omega$$

θ—no other method.

Solution D: Vector algebra

$$Z_T = R + jX_L = 1000 + j628$$

Converting to polar form, we have

$$\text{Angle } \theta = \text{arc tan} \frac{628}{1000} = 32.1°$$

$$\text{Magnitude (hypotenuse)} = \frac{\text{adj}}{\cos \theta} = \frac{1000}{\cos 32.1°} = 1181 \ \Omega$$

$$Z_T = 1181\underline{/32.1°} \ \Omega$$

Now, let us try another example using only vector algebra.

Example 32-9

A resistor of 60 Ω is connected in series with an inductor of 90-Ω reactance across a 50-V supply. Using vector algebra, solve for the magnitude and phase angle of the current in the circuit.

Solution

1. Using the supply voltage as the reference phasor, we get

$$E = 50 + j0$$

2. $$Z = 60 + j90$$

3. $$I = \frac{E}{Z} = \frac{50 + j0}{60 + j90} \times \frac{60 - j90}{60 - j90} = \frac{3000 - j4500}{3600 - j^2 8100}$$

$$= \frac{3000 - j4500}{11\ 700} = 0.256 - j0.384 \ \text{A}$$

In polar form

$$I = 0.463\underline{/-56.3°} \ \text{A}$$

The current is 0.463 A lagging the line voltage by 56.3°.

You have noticed that the impedance triangle was drawn *upward* for an inductive circuit (Fig. 32-9) and *downward* for the capacitive circuit (Fig. 32-4). This is strictly correct. Yet very often the impedance triangle for the capacitive circuit is also drawn upward. Since impedance, reactance, and resistance are not true phasors, this variation is permissible. This does not affect the value of impedance, nor the *value* of circuit phase angle. However, be careful to remember that this *R-C* circuit phase angle should be *negative*, that is, the line voltage in the *R-C* circuit lags the current.

32-13 R-L series circuit—commercial coil.
In the above *R-L* illustrations, we assumed the coil is a perfect inductance, and therefore the current through the coil lags the voltage drop across the coil by 90°. But a commercial coil has resistance as well as inductance! If the resistance of the coil is low, it can be neglected, and the above assumptions are correct. For example, a coil with a Q of 60 or better would introduce an error of less than 1°.* But when the Q of the coil is low, we cannot assume that the coil resistance is negligible. How do we handle such a case?

If we know the resistance and inductance of the coil, the problem is easily solved. Merely add the coil resistance in series with the coil inductance, and proceed as before.

Example 32-10

A coil of 5 H and 400-Ω resistance is connected in series with a 600-Ω resistor across a 120-V 60-Hz supply. Find:
(a) The line current.
(b) The circuit phase angle.
(c) The coil voltage.
(d) The coil phase angle.
(e) The coil Q.

Solution

1. $\qquad X_L = 2\pi fL = 2 \times \pi \times 60 \times 5 = 1885 \ \Omega$

2. $\qquad R_T = R + R_L = 600 + 400 = 1000 \ \Omega$

3. $\qquad Z_T = \sqrt{R_T^2 + X_L^2} = \sqrt{(1000)^2 + (1885)^2} = 2134 \ \Omega$

(a) $\qquad I = \dfrac{E_T}{Z_T} = \dfrac{120}{2134} = 0.0562 \ \text{A} = 56.2 \ \text{mA}$

(b) $\qquad$ Circuit phase angle $= \arctan \dfrac{X_L}{R_T} = \arctan \dfrac{1885}{1000} = 62.1°$

4. Impedance of coil $Z_L = \sqrt{R_L^2 + X_L^2} = \sqrt{(400)^2 + (1885)^2} = 1927 \ \Omega$

(c) Coil voltage $= IZ_L = 0.0562 \times 1927 = 108 \ \text{V}$

(d) Coil phase angle $= \arctan \dfrac{X_L}{R_L} = \arctan \dfrac{1885}{400} = 78.0°$

(e) Q (of the coil) $= \dfrac{X_L}{R_L} = \dfrac{1885}{400} = 4.71$

5. E_R (across the real resistor) $= IR = 0.0562 \times 600 = 33.7 \ \text{V}.$

6. E_{R_L} (across the "R" of the coil) $= IR_L = 0.0562 \times 400 = 22.5 \ \text{V}.$

7. E_{X_L} (across the "X" of the coil) $= IX_L = 0.0562 \times 1885 = 106 \ \text{V}.$

*($Q = X_L/R = 60$, but X_L/R is also the tangent of the phase angle. Arc tan 60 $= 89.05°$. This is almost the 90° shift of the perfect coil.)

The above problem was simple to solve, because we knew the component values R and L of the coil. When we know the inductance and Q *of the coil*, the solution is equally simple. Q is equal to X_L/R. Therefore, we first find X_L from L and the frequency of the applied voltage. Then, from the Q value, we can find the resistance of the coil. The problem is now the same as before.

Sometimes we are given an unknown commercial inductance. We would like to know its resistance and inductance. If we have a wattmeter and an ammeter of proper ranges, we can measure the resistance of the coil.* *An ohmmeter will not do.* It gives only the ohmic resistance; we want the effective resistance! Figure 32-10 shows us a method for getting R and L of the coil, using only a voltmeter.

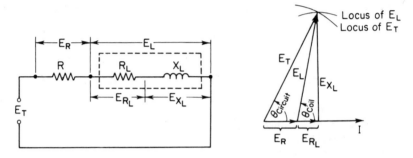

Figure 32-10   Phasor diagram—commercial-coil series circuit.

1. Connect the coil in series with a known resistor across an ac supply voltage.
2. Measure E_T, E_R, and E_L. We know that the phasor sum of E_R and E_L must equal E_T. Therefore:
3. Draw E_R to scale, horizontal direction. E_L starts at the end of E_R. We know its magnitude, and we know its direction is at some angle between zero and 90°, upward. Therefore, *from the end of E_R*, strike an arc of radius equal to E_L. We also know the magnitude of E_T and that it starts at the origin of E_R. With radius equal to E_T and with center at the start of E_R, strike a second arc to intersect the E_L arc. Draw E_T and E_L.
4. Measure the circuit phase angle and coil phase angle.
5. From the intersection point, drop a perpendicular to the horizontal axis. This resolves E_L into its two components E_{X_L} and E_{R_L}.
6. From the E_R and R values calculate the current in the circuit.
7. Using this value of current, and the E_{X_L} and E_{R_L} values from step 5, we can calculate R_L and X_L.

*From $P = I^2R$.

Example 32-11

An unknown coil is connected in series with a 500-Ω resistor across a 40-V 25-Hz supply. The voltage across the resistor is 20 V; across the coil 28 V. Find the inductance and resistance of the coil.

Solution

1. Construct the phasor diagram (Fig. 32-11).
2. By measurement from the diagram:*
   (a) $E_{R_L} = 10.4$ V
   (b) $E_{X_L} = 26$ V
   (c) Circuit phase angle $= 41°$
   (d) Coil phase angle $= 70°$

3. $$I = \frac{E_R}{R} = \frac{20}{500} = 0.04 \text{ A}$$

4. $$R_L = \frac{E_{R_L}}{I} = \frac{10.4}{0.04} = 250 \ \Omega$$

5. $$X_L = \frac{E_{X_L}}{I} = \frac{26}{0.04} = 650 \ \Omega$$

6. $$L = \frac{X_L}{2\pi f} = \frac{0.159 \times 650}{25} = 4.14 \text{ H}$$

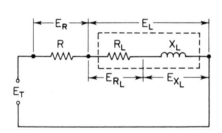
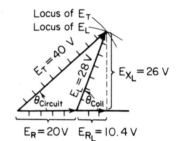

Figure 32-11   Graphical solution—commercial coil.

32-14   Power in R-L series circuit.   In a circuit containing resistance and inductance in series, the voltage will lead the current by some angle less than 90°. If we plot the sine waves of e, i, and p, we shall find that the power curve has positive and negative areas, but the positive portions are greater than the negative portions. The circuit takes power from the line. The inductance itself does not dissipate energy. We already know that it stores energy in its magnetic field and returns it to the line. The power taken by the circuit is the power dissipated in the series resistance (IE_R). If we draw a phasor diagram of such a circuit, we will see that $E_R = E_T \cos \theta$, where E_T is the line voltage. The

---

*A mathematical solution could be made, in place of this graphical evaluation, using the law of sines or cosines, as appropriate.

power dissipated is therefore $E_T I_T \cos \theta$. This is quite similar to the analysis of power in a R-C circuit. Again, no further discussion is necessary.

32-15 Resistance, inductance, and capacitance in series.   We now shall consider circuits containing all three types of circuit elements. Circuits containing resistance, inductance, and capacitance in series are very common in electronic work. Since we have already considered the effect of each of these circuit elements alone, and in combinations containing resistance and either inductance or capacitance, very little new theory is necessary. In fact, all we need to do is put previous knowledge together.

32-16 Impedance: L-C-R circuit.   Let us consider an R-C series circuit connected to a source of ac. The current flowing in the circuit will be limited to some value by the combined opposition ($\sqrt{R^2 + X_C^2}$) of the circuit elements. We shall use an ammeter to measure the current. Now we open the circuit and insert an inductance in series. The current, as indicated by the ammeter, *increases*. But why? We have added another circuit element. The total opposition should be greater—*No!* The total opposition is *less*.

Think back. In a pure capacitive circuit, the voltage across the capacitor lags the current by 90°. But the voltage across an inductance leads the current by 90°. These two voltages are 180° apart! *The net reactance voltage is decreased.* It is the difference between E_C and E_L. The net opposition in the circuit is decreased.

From another point of view, reexamine the impedance triangle for a capacitive, and for an inductive circuit. Notice that X_C is drawn downward, $(-j\,X_C)$ while X_L is drawn upward $(+j\,X_L)$. Again we see that these reactances counteract each other, and the *net reactance* (X_0) is less than either X_L or X_C alone.

In any L-C-R series circuit, the impedance is found as follows:

1. Calculate X_L and X_C.
2. Find the next reactance $X_0 = X_L - X_C$. If X_L is greater, the circuit is inductive; X_0 will be positive, meaning upward in direction when drawing the impedance triangle. If X_C is greater, the circuit is capacitive; X_0 will be negative, meaning downward in direction in the impedance triangle.
3. Calculate the impedance $Z = \sqrt{R^2 + X_0^2}$.

Steps 2 and 3 are sometimes combined as
$$Z = \sqrt{R^2 + (X_L - X_C)^2}$$
Either method is acceptable. Once we have found the impedance of the circuit, it is a simple matter to solve for the current and voltage relations in the circuit.

Example 32-12
An L–C–R series circuit has the following circuit constants: $L = 1$ H, $R = 20\,000\ \Omega$, and $C = 0.005\ \mu$F. The circuit is connected to a 100-V 1000-Hz supply. Find:

(a) The line current.
(b) The voltage across each component.
(c) The circuit phase angle.
(d) The power dissipated.
(e) Draw the phasor diagram.

Solution

1. $$X_L = 2\pi f L = 2 \times \pi \times 1000 \times 1 = 6280 \ \Omega$$

2. $$X_C = \frac{1}{2\pi f C} = \frac{0.159 \times 10^6}{1000 \times 0.005} = 31\ 800 \ \Omega$$

3. $$X_0 = X_L - X_C = 6280 - 31\ 800 = -25\ 520 \ \Omega$$

The circuit is capacitive.

4. $$Z_T = \sqrt{R^2 + X_0^2} = \sqrt{(20\ 000)^2 + (25\ 520)^2} = 32\ 400 \ \Omega$$

(a) $$I = \frac{E_T}{Z_T} = \frac{100}{32\ 400} = 0.003\ 09 \text{ A} = 3.09 \text{ mA}$$

(b) $$E_L = IX_L = 3.09 \times 10^{-3} \times 6280 = 19.4 \text{ V}$$

$$E_R = IR = 3.09 \times 10^{-3} \times 20\ 000 = 61.8 \text{ V}$$

$$E_C = IX_C = 3.09 \times 10^{-3} \times 31\ 800 = 98.3 \text{ V}$$

(c) $$\theta = \arctan \frac{X_0}{R} = \arctan \frac{25\ 520}{20\ 000} = 51.9°$$

Since the circuit is capacitive, the current leads the line voltage by 51.9°.

(d)

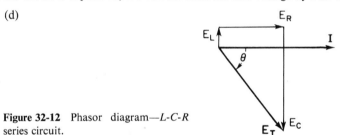

Figure 32-12 Phasor diagram—*L-C-R*
series circuit.

(e) $$P = E_T I_T \cos \theta = 100 \times 3.09 \times 10^{-3} \times 0.616 = 0.19 \text{ W}$$

or

$$P = I^2 R = (3.09 \times 10^{-3})^2 \times 20\ 000 = 0.19 \text{ W}$$

or

$$P = IE_R = (3.09 \times 10^{-3}) \times 61.8 = 0.19 \text{ W}$$

Example 32-13

A commercial coil ($L = 250 \ \mu$H, $R = 150 \ \Omega$) is in series with a resistor of
1350 Ω, and a capacitor of 150 pF. The supply voltage is 60 V, 600 kHz. Find:
(a) The line current.
(b) The circuit phase angle.
(c) The power dissipated.
(d) The magnitude and phase angle of each component voltage.
(e) Draw the phasor diagram.

Solution

1.
$$X_L = 2\pi f L = 2 \times \pi (0.6 \times 10^6)(250 \times 10^{-6}) = 942\ \Omega$$

2.
$$X_C = \frac{1}{2\pi f C} = \frac{0.159 \times 10^{12}}{(0.6 \times 10^6)(150)} = 1768\ \Omega$$

3.
$$X_0 = X_L - X_C = 942 - 1767 = -826\ \Omega$$

The circuit is capacitive.

4.
$$R_T = R_L + R = 150 + 1350 = 1500\ \Omega$$

5.
$$Z_T = \sqrt{R_T^2 + X_0^2} = \sqrt{(1500)^2 + (826)^2} = 1710\ \Omega$$

(a)
$$I = \frac{E}{Z_T} = \frac{60}{1710} = 35.1\ \text{mA}$$

(b)
$$\theta_C\ (\text{circuit}) = \text{arc tan}\ \frac{X_0}{R_T} = \text{arc tan}\ \frac{825}{1500} = 28.8°$$

(c) Power dissipated:
$$P = E_T I_T \cos \theta_C = 60(35.1 \times 10^{-3})(0.876) = 1.85\ \text{W}$$
$$= I^2 R_T = (35.1 \times 10^{-3})^2(1500) = 1.85\ \text{W}$$

(*Note:* We cannot use IE_R, since this does not include the power lost in the resistance of the coil.)

(d)
$$E_L = IZ_L \quad (\text{where } Z_L = \sqrt{R_L^2 + X_L^2});$$
$$Z_L = \sqrt{(150)^2 + (942)^2} = 954\ \Omega$$
$$E_L = (35.1 \times 10^{-3})(954) = 33.4\ \text{V}$$

To find the phase angle between coil voltage and coil current:

$$\theta_L = \text{arc tan}\ \frac{X_L}{R_L} = \text{arc tan}\ \frac{942}{150} = 80.9°\ (E_L \text{ leads } I)$$

$E_R = IR = (35.1 \times 10^{-3})(1350) = 47.4\ \text{V in phase with } I\ (\text{pure } R)$
$E_C = IX_C = (35.1 \times 10^{-3})(1768) = 62.0\ \text{V at } 90° \text{ lag (pure } C)$

(e)

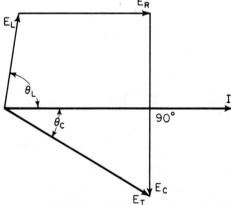

Figure 32-13   Phasor diagram, Example 32-13.

Now let us solve a similar problem using vector algebra.

Example 32-14

A capacitor of 0.02 μF, a resistor of 4000 Ω, and a commercial coil of 1.0 H and 1800-Ω resistance are connected in series across a 120-V 1000-Hz supply. Find the magnitude of the line current and its phase relation to the supply voltage.

Solution

1.
$$X_C = \frac{1}{2\pi f C} = \frac{0.159 \times 10^6}{1000 \times 0.02} = 7950 \ \Omega$$

$$X_C = 0 - j7950$$

2.
$$R = 4000 + j0$$

3.
$$X_L = 2\pi f L = 2 \times \pi \times 1000 \times 1 = 6280 \ \Omega$$

$$Z_L = 1800 + j6280$$

4.
$$Z_T = \dot{X}_C + \dot{R} + \dot{Z}_L = 5800 - j1670 = 6040\underline{/-16^\circ}$$

5.
$$E_T = 120\underline{/0^\circ} \text{ (reference phasor).}$$

6.
$$I = \frac{E_T}{Z_T} = \frac{120\underline{/0^\circ}}{6040\underline{/-16^\circ}} = 0.0199\underline{/16^\circ}$$

$$I = 19.9 \text{ mA leading by } 16^\circ$$

As in dc circuits, series circuits can assume many variations. It would be impossible to give illustrations for each variation. However, the same general rules apply to all series circuits:

1. The total opposition—impedance—is the *phasor* sum of each individual opposition.

2. The current in the circuit depends on the total impedance.

3. The same current flows through each component of the series circuit.

4. The applied voltage is the *phasor* sum of the voltage across each component.

5. Ohm's law can be applied to the entire circuit or to any portion of the circuit, as long as *all* the values used (E, R, X, or Z) apply to the entire circuit or to the same portion of the circuit.

We have now covered the basic principles that apply to any ac series circuit. The steps used in solving any specific case will vary depending on the data given. A thorough mastery of these principles is necessary in order for you to understand the analysis of electronic circuits that will follow in a later volume. To insure mastery of these fundamentals, numerous problems are included in the Problems section.

REVIEW QUESTIONS

1. In an ac series circuit:
   (a) How does the current in each component compare? Why?
   (b) What does the line current depend on?
   (c) How is the total opposition obtained?
   (d) Is this always a numerical sum? Explain.
   (e) Specify three "special" conditions when a numerical sum will give the correct value, of total opposition.
   (f) What term is used to signify the combined opposition of more than one type of circuit element?
   (g) Give the letter symbol for this term.

2. If several unequal capacitors are in series in an ac circuit:
   (a) Between what limits will the value of total capacitance fall?
   (b) Give the equation for total capacitance.
   (c) How does the current in the largest capacitor compare with the current in the smallest unit? Explain.
   (d) What determines the value of line current?
   (e) How does the relative magnitude of the voltage across each capacitor compare?
   (f) Give the equation for finding each component voltage.
   (g) What relation exists between the component voltages and the line voltage?
   (h) Is this a phasor sum or a numerical sum? Explain.
   (i) What is the circuit phase angle? Why?

3. With reference to Example 32-1:
   (a) What are we looking for?
   (b) Why then do we solve for C_T as the first step?
   (c) Why is it necessary to also find the X_C for each unit?
   (d) How does X_{C_T} compare with the individual X_C values?

4. With reference to Fig. 32-1(a):
   (a) What is the phase relation between E_1 and I? Why?
   (b) What is the phase relation between E_2 and I? Why?
   (c) What is the phase relation between E_T and I? Why?

5. In the solution to Example 32-1, step 4B, what is the significance of the dots over the E, I, and X_C symbols?

6. If several unequal inductors are connected in series:
   (a) How can we find the total inductance?
   (b) How does the total reactance compare with the individual X_L values?
   (c) Which inductor will have the higher current?
   (d) What determines the line current value?
   (e) Which component will have the highest voltage drop?
   (f) What is the phase relation between line current and line voltage? Why?

7. With reference to Example 32-2:
   (a) In which coil does the given current (0.5 A) flow?
   (b) Why do we start with X_{L_1}? Why not with L_T and X_{L_T}?
   (c) Why don't we solve for X_{L_2} and X_{L_3} immediately after X_{L_1}?
   (d) How does X_{L_T} compare with the sum of the individual X_L values?

(e) Why are the voltage phasors in Fig. 32-2 drawn upward?

8. In Solution B, of Example 32-2, step 1:
   (a) How is the value of $0.5 + j0$ obtained for the current?
   (b) Is this a good choice? Explain.

9. When a series circuit contains R and C:
   (a) What is the total opposition called?
   (b) Can the value of total opposition be found by numerical sum of the individual component oppositions? Explain.
   (c) How does the current value through each component compare?
   (d) How does the *phase* of the current through R compare with the current through C?
   (e) What is the phase of the voltage drop across a resistor, as compared to the line current?
   (f) What is the phase of the voltage drop across a capacitor with respect to the line current?
   (g) Is the total (line) voltage the numerical sum of the individual component voltages? Explain.

10. With reference to Fig. 32-3:
    (a) What is used as the reference phasor?
    (b) Why is this a good choice?
    (c) Why is IR drawn in phase with I?
    (d) What is the phase of IX_C? Explain.

11. With reference to Fig. 32-4:
    (a) What is this diagram called?
    (b) Why are R and X_C drawn at right angles?
    (c) How is Z evaluated?

12. (a) Give the mathematical equation for Z in terms of R and X_C.
    (b) Give an equation for X_C in terms of R and Z.

13. (a) What is meant by *circuit phase angle* in an R–C series circuit?
    (b) Is it shown in Fig. 32-3? Explain.
    (c) Is it shown in Fig. 32-4? Explain.
    (d) From Fig. 32-4, if the values of R, X_C, and Z are known, give three equations suitable for finding the circuit phase angle.

14. With reference to Example 32-4:
    (a) Since line current is the first required answer, why does the first step solve for X_C?
    (b) Between what two limits should the value of Z fall?
    (c) Can θ be greater than $45°$? Explain.
    (d) Should it be close to $45°$? Explain.

15. By inspection, what is the value of Z if:
    (a) $R = 100\ \Omega$ and $X_C = 5.0\ \Omega$?
    (b) $R = 100\ \Omega$ and $X_C = 1200\ \Omega$?

16. In the solution of Example 32-5:
    (a) How is the $40 - j70$ obtained for Z?
    (b) Since it is easier to multiply in polar form, why is the multiplication in step 3 made in rectangular form?

    (c) In step 3, polar form, how is the value 100.7 obtained? How is the angle value 60.3 obtained? Why is it a negative angle?

17. With reference to Fig. 32-6:
    (a) What type of circuit must this represent? Why?
    (b) Why is the power negative during time interval 1–2?
    (c) Why is the power positive during time interval 2–3?
    (d) Is the total power over one cycle of input, positive, negative, or zero? Explain.
    (e) Where does this power go?

18. With reference to Fig. 32-7(a):
    (a) What is the relation between E_R and E_T?
    (b) What is the relation between E_C and E_T?
    (c) How many components make up E_T?
    (d) With respect to the current direction, what can the E_R component be called?
    (e) What can the E_C component be called?

19. In any series ac circuit having resistive and reactive elements:
    (a) How can true power be determined?
    (b) Give a second method.
    (c) Give a third method.
    (d) What does *volt-amperes* mean?
    (e) What does the term *vars* stand for?
    (f) How is the vars for a circuit obtained?

20. In an ac circuit, will a wattmeter indicate volt-amperes, vars, or true power? Explain.

21. If you were asked to find the power factor of a load by test methods:
    (a) What instruments would you need?
    (b) How would you find power factor from the instrument readings?

22. In an ac series circuit containing resistance and inductance:
    (a) How does the current through the coil compare with the current through the resistor, in magnitude? In phase?
    (b) How does the phase of the voltage across the coil compare with the voltage across the resistor?
    (c) Can we add E_R and E_L directly to get the total voltage? Explain.

23. With reference to Example 32-7:
    (a) What is used as the reference phasor?
    (b) Is this a good choice? Explain.
    (c) In step 4A, how are the values of E_T and θ obtained?
    (d) What relationship is applied in step 4B(1)?

24. With reference to Example 32-8:
    (a) What are the values of R and X_L?
    (b) By inspection, within what limits must the value of Z fall?
    (c) Which method is generally preferable for finding Z, solution B, solution C, or solution D? Explain.
    (d) By inspection, what is the maximum value θ could have? Explain.

25. With reference to Example 32-9:
    (a) Why is the voltage taken as $50 + j0$?
    (b) Why is the Z value taken as $60 + j90$?

(c) In step 3, where does the $60 - j90$ come from?

26. Under what condition can the resistance of a coil be neglected:
    (a) When evaluating the impedance of the coil?
    (b) When evaluating the phase angle of the coil?

27. With reference to Example 32-10:
    (a) Can the resistance of this coil be neglected? Explain.
    (b) By inspection, between what limits must the Z_T value fall?
    (c) Should the value of θ_C be greater or smaller than $45°$? Why?
    (d) What is the distinction between Z_T and Z_L?
    (e) Which impedance should be greater Z_T or Z_L? Explain.
    (f) Which phase angle should be greater, θ_C or θ_L? Explain.

28. An inductor is to be used in an ac circuit. Its resistance and inductance are unknown.
    (a) Can we measure its resistance with an ohmmeter? Explain.
    (b) Can we calculate its resistance from current and voltage readings? Explain.
    (c) Can we obtain its resistance using ammeter and wattmeter readings? Explain.

29. With reference to Fig. 32-10:
    (a) How many actual components are there in this circuit?
    (b) What actual voltage readings can be made?
    (c) Since phase angles are *not* given, how is the direction for E_R, E_L, and E_T determined?
    (d) How is the value of E_{R_L} obtained? Explain.
    (e) How can this E_{R_L} value be used to evaluate R_L?

30. With reference to Example 32-11:
    (a) Specify the given quantities.
    (b) Can the phasor diagram be a *rough* sketch? Explain.
    (c) How is the value of θ_C obtained?
    (d) Can this value be obtained from right triangle relations? Explain.
    (e) Can this value be obtained mathematically? Explain.
    (f) Can X_L be evaluated directly from E_L? Explain.

31. A series circuit contains $X_L = 100 \, \Omega$, and $R = 100 \, \Omega$. It dissipates 200 W.
    (a) How much of this power is dissipated in the reactive portion?
    (b) How much in the resistive portion?
    (c) Does the inductive portion take any power from the line at any time?
    (d) What happens to this power?

32. A large capacitor is added to a series R–L circuit. How will this affect the total impedance? Explain.

33. In a series L–C–R circuit:
    (a) What does the term *net reactance* mean?
    (b) What symbol is used to represent this quantity?
    (c) How is its value obtained?
    (d) How is the total circuit impedance obtained?
    (e) How does the current through the inductance portion compare to the current through the capacitor in magnitude? In phase? Explain.

34. With reference to the solution of the L–C–R circuit, Example 32-12:
    (a) Why is this circuit called a capacitive circuit in step 3?

   (b) Is it a pure capacitive circuit? Explain.

   (c) Will the line current lead the line voltage by 90°?

   (d) Will the current lead the capacitor voltage by 90°? Explain.

   (e) In step 4(e), show the correlation between $EI \cos \theta$ and I^2R.

   (f) Show the correlation between I^2R and IE_R.

35. With reference to the phasor diagram, Fig. 32-12:

   (a) What is used as the reference phasor?

   (b) Is this a good choice? Explain.

   (c) Why is E_L drawn upward, yet E_C is drawn downward?

   (d) Why is E_R drawn horizontally, and to the right?

36. With reference to the solution, Example 32-13:

   (a) What does R_T (step 4) consist of?

   (b) Why is this step (finding R_T) necessary?

   (c) Why is it necessary to find the phase angle of the voltage E_L?

   (d) Why isn't this angle 90°?

37. In the solution to Example 32-14:

   (a) Why is X_C given as $0 - j7950$?

   (b) Why is Z_L given as $1800 + j6280$?

   (c) Why is Z_T given as $5800 - j1670$?

   (d) Why is the phase angle (16°) a minus quantity?

   (e) Why is the voltage E_T given as angle 0°?

   (f) How is an angle of $+16°$ obtained for I?

   (g) What does this $+$ angle mean?

   (h) Is this correct, considering the total circuit impedance?

PROBLEMS

1. Three resistors, 20, 40, and 50 Ω, are connected in series across a 120-V 60-Hz supply. Find the line current, circuit phase angle, and voltage across each unit.

2. Three capacitors, 0.1, 0.25, and 0.05 μF, are connected in series across a 20-V, 5000-Hz supply. Find the line current, voltage across each unit, circuit phase angle, and draw the phasor diagram.

3. Two capacitors, 50 pF and 200 pF, are connected in series to a 30-kHz supply. The voltage across the 200-pF capacitor is 40 V. Find:

   (a) Line current.

   (b) Voltage across the 50-pF capacitor.

   (c) Line voltage.

   (d) Total reactance of the circuit.

4. Three choke coils are connected in series across a 50-V 120-Hz supply. The inductances are 10 H, 15 H, and 3 H. Find:

   (a) Line current.

   (b) Voltage across each unit.

   (c) Circuit phase angle.

   (d) Draw the phasor diagram.

5. Two coils are connected in series across a 30-V 1000-Hz supply. The voltage across L_1 is 12 V. L_2 has an inductance of 20 H. Assuming that both coils are perfect inductances, find:
   (a) Voltage across L_2.
   (b) Line current.
   (c) Inductance of L_1.
   (d) Draw the phasor diagram.

6. A resistor of 600 Ω is connected in series with a capacitor of 800-Ω reactance, across a 120-V supply. Find:
   (a) Line current.
   (b) Voltage across each unit.
   (c) Circuit phase angle.

7. Find the impedance of the following R–C series circuits, if $R = 1000\ \Omega$ and:
   (a) $C = 0.1\ \mu F, f = 1000$ Hz.
   (b) $C = 500$ pF, $f = 200$ kHz.
   (c) $C = 500$ pF, $f = 50$ kHz.
   (d) $C = 500$ pF, $f = 30$ kHz.
   (e) $C = 500$ pF, $f = 1500$ kHz.
   (f) $C = 0.005\ \mu F, f = 200$ kHz.

8. A current of 2.5 A flows through an impedance of $15 - j20\ \Omega$. Using current as reference, solve for the voltage across the impedance by vector algebra, rectangular form.

9. Repeat Problem 8, using polar form.

10. A 100-Ω resistor is connected in series with a 200-pF capacitor, across a 500-V 1500-kHz supply. Find:
    (a) The circuit impedance.
    (b) The line current.
    (c) The voltage across each unit.
    (d) The circuit phase angle.
    (e) Draw the phasor diagram.

11. A capacitor of 10 μF is connected in series with a resistor of 120 Ω, across a 440-V, 60-Hz supply. Find:
    (a) The current in the circuit.
    (b) The voltage across each unit.
    (c) The circuit phase angle.
    (d) The power dissipated.

12. In a series R–C circuit, the current voltage and power dissipation are 1.5 A, 120 V, 60 Hz, and 50 W, respectively. Find the values of R and C.

13. A commercial load draws a current of 120 A from a 220-V line at a power factor of 0.8 lagging. Find:
    (a) The power taken by the load.
    (b) The reactive volt-amperes.

14. Data taken at full load on a repulsion-induction motor shows that it takes 1600 W from a 230-V line at a current of 9.6 A. Find:
    (a) The power factor of the load.
    (b) The reactive volt-amperes.

15. A series circuit contains a resistor of 150 Ω and an inductance of 0.5 H. The supply voltage is 220 V, 60 Hz. Find:
  (a) The circuit impedance.
  (b) The line current.
  (c) The voltage across each unit.
  (d) The circuit phase angle.
  (e) The power dissipated.

16. In an *L–R* series circuit, a current of 0.6 A flows when the supply voltage is 220 V, 120 Hz. The current lags the voltage by 70°. Find:
  (a) The power dissipated.
  (b) The impedance of the circuit.
  (c) The values of *R* and *L*.

17. A current of $20 + j15$ mA flows through an impedance of $5000 + j3500$. Find the voltage across the impedance, in rectangular and in polar form.

18. An impedance, $10 + j20$ Ω, is connected across a supply $40 + j0$ V. Solve for current using vector algebra, rectangular form.

19. The impedance of a commercial coil is 150 Ω at 200 Hz. Its resistance is 42 Ω. What is its inductance? What is the *Q* of the coil?

20. A commercial coil (having resistance and inductance) is connected in series with a resistor. Draw the phasor diagram, and find the circuit phase angle and coil phase angle (use graphical solution): when
  (a) $E = 120$ V, $E_R = 80$ V, $E_L = 70$ V.
  (b) $E_L = 200$ V, $E = 450$ V, $E_R = 300$ V.
  (c) $E = 5$ V, $E_L = 3.5$ V, $E_R = 2.5$ V.

21. In each of the parts of Problem 20, find *L* and R_L if:
  (a) $R = 80$ Ω and $f = 100$ Hz.
  (b) $R = 30\,000$ Ω and $f = 1000$ Hz.
  (c) $R = 5$ Ω and $f = 60$ Hz.

22. A coil having a reactance of 500 Ω is connected in series with a resistor of 200 Ω. If the current in the circuit is 0.8 A, find:
  (a) The voltage across each unit.
  (b) The line voltage.
  (c) The circuit phase angle.
  (d) The power dissipated.

23. In each of the following cases, a coil, capacitor, and resistor are connected in series. Find the impedance of the circuit, circuit phase angle, and state whether the circuit is capacitive or inductive.
  (a) $L = 3.5$ H, $C = 12$ μF, $R = 500$ Ω, $f = 60$ Hz
  (b) $L = 90$ μH, $C = 250$ pF, $R = 20$ Ω, $f = 1000$ kHz
  (c) $L = 1.5$ H, $C = 5$ μF, $R = 200$ Ω, $f = 60$ Hz

24. In each of the parts of Problem 23, if the applied voltages are 400 V, 1.5 V, and 230 V, respectively, find:
  (a) The line current.
  (b) The voltage across each component.
  (c) The power dissipated.

25. Find the applied voltage in a series circuit containing R, C, and L, when the voltage drops are:

$$E_C = 25 \text{ V, pure capacitance.}$$
$$E_L = 10 \text{ V, pure inductance.}$$
$$E_R = 15 \text{ V, pure resistance.}$$

26. Repeat Problem 25, for:

$$E_C = 100 \text{ V, lagging the current by } 90°.$$
$$E_R = 120 \text{ V, in phase.}$$
$$E_L = 80 \text{ V, coil } Q \text{ equal to 5.}$$

27. Repeat Problem 25, for:

$$E_L = 12 \; V \; (X_L = 100 \; \Omega, \; R_L = 50 \; \Omega).$$
$$E_R = 20 \text{ V, pure resistance.}$$
$$E_C = 10 \text{ V, capacitor has no losses.}$$

28. In each of the following cases, the circuit consists of a coil, capacitor, and resistor in series. The coil and capacitor may not necessarily be pure reactances. All phase angles given are with respect to the line current. In each case, use both a graphical and a mathematical solution.

(a) $E_T = 80$ V, leading by $30°$.
$E_R = 50$ V, in phase.
$E_C = 25$ V, lagging by $90°$.
Find E_L, both magnitude and phase angle.

(b) $E_T = 20$ V, lagging by $20°$.
$E_R = 15$ V, in phase.
$E_L = 8$ V, leading by $70°$.
Find E_C, both magnitude and phase angle.

(c) $E_T = 45$ V, lagging by $10°$.
$E_C = 20$ V, lagging by $85°$.
$E_L = 15$ V, angle unknown.
Find (1) E_R; (2) phase angle of E_L.

33

SERIES CIRCUITS—
ELECTRONIC ASPECTS

In the previous chapter we discussed the basic principles applicable to any series circuit. The topics covered and the treatment of each topic was general and applied equally well to the power or electronic fields. But in electronics, series circuits have special significance. So now we will analyze these circuits in more detail with regard to their electronic aspects.

33-1 Resonance.   The condition known as *resonance* occurs only in circuits containing both inductance and capacitance. The circuits may also have resistance—but although resistance greatly affects the current and voltage values when a circuit is in resonance, it does not determine *when* resonance will occur. Any circuit containing inductance and capacitance will be "resonant" at some frequency—or by proper choice of these circuit elements we can make resonance occur at any desired frequency. For example, the choice of L and C values in the RF carrier generator and RF amplifiers of the transmitter shown in Fig. 24-4 will determine the carrier frequency of that station. In Chapter 24 we also presented a salient feature of a receiver—the need for *selectivity*. Again resonance is the underlying principle that makes it possible for us to tune in any desired station, and that gives the circuits the necessary selectivity.

Notice that in both of the above illustrations, the circuits used were operated at radio frequencies (RF). It is in RF circuits, where only one frequency or a relatively narrow band of frequencies is desired at any one time, that resonance plays an important role. (Whereas in audio amplifiers, such as shown in Figs. 24-3 to 24-5, the frequencies these units must handle at any one time may range from 30 to 15 000 Hz. Video amplifiers used in television should also have uniform characteristics—from 20 Hz to 4.5 MHz. Resonance is undesirable in these units. In fact these amplifiers could well be called nonresonant amplifiers.)

Since resonance is mainly applicable to RF circuits, a complete discussion will be delayed to a later text on resonant circuit applications.* At this time we shall develop:

1. How to find the resonant frequency of a circuit.
2. How to make a circuit resonant at a given frequency.
3. What are the characteristics of a circuit when in resonance.

Any series circuit containing both inductance and capacitance will have a resonant frequency regardless of the values of these components. Resonance will occur whenever the reactances X_L and X_C are equal. Quite often in electronic circuits we need to know at what frequency this resonant condition will result. The solution is quite simple. Since X_L must equal X_C, then

$$2\pi f L = \frac{1}{2\pi f C}$$

Solving for f, we get

$$f_0 = \frac{1}{2\pi\sqrt{LC}} \qquad (33\text{-}1)$$

(The symbol f_0 is used to represent the resonant frequency.)
Let us apply this to a problem.

Example 33-1

A series circuit contains the following components: a 1250-Ω resistor, an inductance of 5 mH, and a capacitor of 500 pF. At what frequency will the circuit become resonant?

Solution

1. The resistance of the circuit is of no consequence in determining the resonant frequency.

2. $f_0 = \dfrac{1}{2\pi\sqrt{LC}} = \dfrac{1}{2 \times \pi \times \sqrt{5 \times 10^{-3} \times 500 \times 10^{-12}}} = 100.6 \text{ kHz}$

    Calculator entry:

    5, $\boxed{\text{EE}}$, $\boxed{+/-}$, 3, $\boxed{\times}$, 5, 0, 0, $\boxed{\text{EE}}$, $\boxed{+/-}$, 1, 2, $\boxed{=}$,

    $\boxed{\sqrt{x}}$, $\boxed{\times}$, 2, $\boxed{\times}$, $\boxed{\pi}$, $\boxed{=}$, $\boxed{1/x}$

Note that it is necessary to enter the square-root part first, otherwise the calculator (in some models) will take the square root of the entire denominator.

---

*J. J. DeFrance, *Communications Electronics Circuits*, Holt, Rinehart and Winston, New York, 1972.

This problem can also be set up for straightforward calculator sequencing, using the *parenthesis* key. The technique will be more readily evident, if we rewrite the equation as

$$f_0 = \frac{1}{2\pi(LC)^{1/2}} = \frac{1}{2 \times \pi \times (5 \times 10^{-3} \times 500 \times 10^{-12})^{1/2}} = 100.6 \text{ kHz}$$

The calculator entries are now:

2, $\boxed{\times}$, $\boxed{\pi}$, $\boxed{\times}$, $\boxed{(}$, 5, $\boxed{\text{EE}}$, $\boxed{+/-}$, 3, $\boxed{\times}$, 5, 0, 0, $\boxed{\text{EE}}$, $\boxed{+/-}$,

1, 2, $\boxed{)}$, $\boxed{\sqrt{x}}$, $\boxed{=}$, $\boxed{1/x}$

Now that we have a resonant circuit, let us investigate the characteristics of this circuit. This can best be seen by another problem.

Example 33-2

In Example 33-1, if the applied voltage is 100 V at the resonant frequency (100.6 kHz), find:
(a) The total circuit impedance.
(b) The line current.
(c) The voltage across each unit.
(d) The circuit phase angle.

Solution

1. It is always desirable to start by drawing the circuit diagram [see Fig. 33-1(a)].

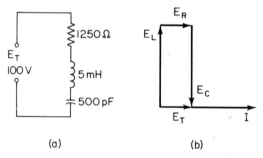

(a)   (b)

Figure 33-1   Series resonant circuit.

2. The total impedance is given by $Z_T = \sqrt{R^2 + X_0^2}$
   Therefore, let us first find the net reactance (X_0)

   (a) $X_L = 2\pi f L = 2 \times \pi \times 100.6 \times 10^3 \times 5 \times 10^{-3} = 3160 \, \Omega$
             $= +j3160$

   (b) $X_C = \frac{1}{2\pi f C} = \frac{0.159 \times 10^{12}}{100.6 \times 10^3 \times 500} = 3160 \, \Omega = -j3160$

   (c) $X_0 = X_L - X_C = 3160 - 3160 = 0 \, \Omega$

This result should have been obvious, since the resonant frequency found in Problem 33-1 was the frequency at which X_L would equal X_C. (A word of caution: Due to calculation inaccuracies in determining the resonant frequency, X_L or X_C, this cancellation may not at times seem perfect.)

(d) $$Z_T = \sqrt{R^2 + X_0^2} = \sqrt{(1250)^2 + (0)^2} = 1250 \ \Omega$$

Notice that the total impedance is equal to the resistance of the circuit! This is true for any *series-resonant* circuit.

3. $I_T = \dfrac{E_T}{Z_T} = \dfrac{100}{1250} = 80 \text{ mA}$

4. $E_R = IR = 80 \times 10^{-3} \times 1250 = 100 \text{ V (in phase with } I)$

$E_L = IX_L = 80 \times 10^{-3} \times j3160 = j253 \text{ V (leading the current by 90°)}$

$E_C = IX_C = 80 \times 10^{-3} \times -j3160 = -j253 \text{ V (lagging the current by 90°)}$

Notice that the voltage across each reactive component is higher than the line voltage. This can readily happen in an *R-L-C* circuit if close to resonance, and the circuit resistance is low compared with X_L or X_C. Remember that the reactive voltages are 180° out of phase and will cancel out. A voltmeter across both *L* and *C* will read zero!

5. To find the circuit phase angle, first draw the phasor diagram showing the relation between line current and component voltages. Use current as the reference phasor [see Fig. 33-1(b)]. Obviously, the line voltage is in phase with the line current; the circuit is operating at unity power factor; and the circuit phase angle is zero, just as if it were a purely resistive circuit.

Let us go back to Example 32-12 in the previous chapter (page 492), and examine the given data and calculated results. It was also a series *L-C-R* circuit:

Given: $L = 1 \text{ H}, C = 0.005 \ \mu\text{F}, R = 20\,000 \ \Omega$

$E = 100 \text{ V and } 1000 \text{ Hz}$

Calculated: $X_L = 6280 \ \Omega, X_C = 31\,800 \ \Omega$

$Z_T = 32\,400 \ \Omega, I_T = 3.08 \text{ mA}$

$\theta_c = 51.9° \text{ (current leading)}$

Is this circuit in resonance? Obviously not. We can see this from three aspects:

1. X_L does not equal X_C.
2. Z_T is higher than the resistance value.
3. θ_c is not zero—current is leading.

The next question is: How could we make this circuit resonant? Again there are several answers possible:

1. Any circuit containing L and C is resonant at some frequency. We could change the frequency. Which way—increase or decrease? Since X_L is too low, and X_C is too high, the frequency should be increased. As the frequency is increased, X_L will increase. At the same time X_C will decrease. At some higher frequency, X_L will be equal to X_C (See Example 33-3 below.)

But suppose that we wanted the circuit to be resonant at the given frequency of 1000 Hz. We still have two other methods for making $X_L = X_C$.

2. Change the value of the inductance. In Example 32-11, X_L is too low. Therefore, we must increase the inductance.
3. The circuit could also be brought into resonance by changing the capacitance value. Since X_C is too high, the value of capacitance must be increased.

Now let us find at what specific frequency, inductance value, or capacitance value the circuit of Example 32-11 could be made resonant.

Example 33-3

At what frequency will the above circuit ($L = 1$ H; $C \doteq 0.005 \, \mu$F) be in resonance?

Solution

$$f_0 = \frac{1}{2\pi\sqrt{LC}} = \frac{0.159}{\sqrt{1 \times 0.005 \times 10^{-6}}} = 2250 \text{ Hz}$$

Example 33-4

What value of inductance is needed to bring the circuit into resonance at 1000 Hz with the given capacitor of 0.005 μF?

Solution

1. X_C was found previously to be 31 800 Ω.
2. X_L must equal X_C or 31 800 Ω.
3. $L = \dfrac{X_L}{2\pi f} = \dfrac{0.159 \times 31\ 800}{1000} = 5.05$ H

Example 33-5

Find the capacitance value that would give a resonant condition at 1000 Hz with the original inductance of 1 H.

Solution

1. X_L at 1000 Hz $= 6280 \, \Omega$.
2. X_C must equal X_L or 6280 Ω.
3. $C = \dfrac{1}{2\pi f X_C} = \dfrac{0.159 \times 10^6}{1000 \times 6280} = 0.0253 \, \mu$F

In practical applications either method is used to "tune" a circuit into resonance. For example, in broadcast band receivers, tuning is done by varying capacitance, whereas in television tuners and in many FM tuners, inductance variation is more common.

Let us summarise the main points concerning resonance:

1. Resonance is obtained when $X_L = X_C$.
2. Resonant condition can be obtained by varying L, C, or frequency.
3. At resonance:
   (a) The impedance is a minimum and is equal to the resistance of the circuit.
   (b) The current is a maximum.
   (c) Current and line voltage are in phase.
   (d) The circuit acts as a pure resistive circuit.
   (e) The voltage across L, and across C may exceed the supply voltage.
4. Below resonance, X_C is greater than X_L, and the circuit acts as an R-C circuit.
5. Above resonance, X_L exceeds X_C, and the circuit acts as an R-L circuit.

If there is any possibility that a circuit might operate at or close to resonance, the voltage ratings of the coil and capacitor must be checked carefully. We have seen that the voltage across these units may exceed the supply voltage. This is particularly true if the circuit resistance is low compared with the reactance of the coil or capacitor. For instance, in Example 33-2, if the resistance were only 125 Ω, the current would have been 10 times higher (800 mA). But the voltages across the coil and the capacitor would have been 2530 V! It is, therefore, important that the units used have the proper voltage rating.

As mentioned earlier in this discussion, the details—such as effect of resistance and L-C ratio on the resonance curve—will be delayed till we are ready to apply resonance to practical circuitry.

33-2 Phase shift *R-C* circuits.   We know that in a pure capacitive circuit, the current leads the voltage by 90°. But we saw in the previous chapter that if we have resistance and capacitance in series, the current leads by some angle less than 90°. Obviously, the combination of resistance with capacitance reduced the angle of phase shift between current and voltage. This phase angle can be obtained from the impedance triangle for the circuit (see Fig. 32-4, page 480). In Fig. 33-2(a) is shown a group of impedance triangles with fixed resistance value and several values of capacitive reactance. Notice that the larger the X_C value, the greater the phase shift.

Now refer to Fig. 33-2(b). This time the reactance value is kept constant, while the resistance value is increased. What is the effect of resistance on the phase angle? The greater the resistance value, the smaller the phase shift. Obviously, the degree of phase shift depends not on resistance or reactance

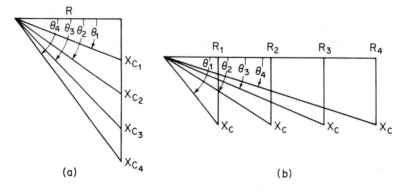

Figure 33-2 Effect of R or C on phase shift.

values alone, but on the ratio of reactance to resistance. This merely confirms what we had previously seen—the circuit phase angle, θ_c, is the angle whose tangent is equal to the ratio of reactance to resistance:

$$\theta_c = \text{arc tan}\,\frac{X_C}{R}$$

Very often in electronic work, R-C series circuits are used to feed the ac output of one circuit or "stage" to the input of the next stage. This is called *coupling*. R-C coupling is widely used in audio amplifiers, video amplifiers (television), in special circuits such as used in loran, radar, and in timing or trigger circuits. Sometimes phase shift is not too important (as in simple audio amplifiers). However, in video circuits, phase shift is detrimental, particularly at the lower frequencies. But we know that reactance (X_C) increases at lower frequencies! Therefore, unless capacitance and resistance values of the coupling circuit are carefully chosen, phase shift in television circuits would distort the picture. On the other hand, in some special circuits a definite phase shift is desired. Here the problem is to pick R and C values so as to produce the desired phase shift.

Figure 33-3 shows a typical R-C coupling circuit. The input voltage (E_i) is fed to the total circuit, while the output (E_o) is taken off across the resistor, and fed to the next stage. If this next stage has a high input impedance, as is the case with field-effect transistors (FETs), we can neglect the shunting effect

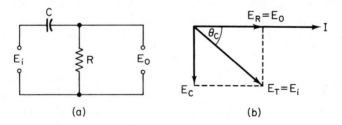

Figure 33-3 Typical R-C coupling circuit.

of such loads across the resistor R. Otherwise, the input impedance of the next stage must be considered in parallel with R.

Calculation of phase shift between the input and output voltage is simple. We know how to find the phase angle between the applied voltage (E_t) and the circuit current. The output voltage is the voltage across the resistor. But we know that this voltage is in phase with the current. Therefore, the phase angle between input and output voltage is the same as the circuit phase angle, and the output voltage *leads* the input voltage. This is shown in the phasor diagram, Fig. 33-3(b).

Example 33-6

In Fig. 33-3, find the phase relation between input and output voltage, if $R = 50\ 000\ \Omega$, $C = 0.05\ \mu\text{F}$, and the frequency of the applied voltage is 1000 Hz.

Solution

1. $$X_C = \frac{1}{2\pi f C} = \frac{0.159 \times 10^6}{1000 \times 0.05} = 3180\ \Omega$$

2. $$\theta_c = \arc\tan\frac{X_C}{R} = \arc\tan\frac{3180}{50\ 000} = 3.6°$$

If the frequency of the applied voltage were increased to 15 000 Hz, would the phase shift be greater or less? At the higher frequency, X_C, is less—15 times less; the phase shift would be less than 0.3°. But if the frequency were to be reduced to say 50 Hz, the reactance would be 20 times greater, and the phase shift would increase to 51.7°! Certainly, if phase shift is undesirable, and the circuit must operate on a frequency range that goes as low as 50 Hz, this circuit is inadequate.

How can we eliminate or reduce phase shift to a negligible value? Merely select a capacitor and resistor value, such that *at the lowest frequency* that the circuit will operate, the reactance is small compared with the resistance. How small? This depends on the specific circuit requirements. If X_C is less than $\frac{1}{60}$ of R, then the tangent of the phase angle is less than 0.0167, and the phase shift is less than 1.0°. At any higher frequency, the phase shift is even lower. Notice that this does not specify the value of the capacitor or resistor, but only the ratio of their respective oppositions. Quite often the value of either one or the other circuit component is fixed or limited by other circuit considerations.

Example 33-7

An R-C coupling circuit is to operate on a frequency range of 30 to 50 000 Hz. The maximum capacitor value is fixed by other considerations at 0.1 μF. What value resistor must be used in order to limit the phase shift to 3.0°?

Solution

1. $$X_C = \frac{1}{2\pi f C} = \frac{0.159 \times 10^6}{30 \times 0.1} = 53\ 000\ \Omega$$

2. For 3.0°, tan θ = 0.0524.

3. From tan θ = X_L/R,

$$R = \frac{X_C}{\tan \theta} = \frac{53\,000}{0.0524} = 1\,011\,000, \text{ or approximately } 1.0 \text{ M}\Omega$$

When a high degree of phase shift is desired, the capacitance value is made small and the resistance is reduced. This results in a high X_C to R ratio. The phase shift will be large.

We have seen how we can reduce undesired phase shift to negligible values, or how we can produce any given phase shift. In the latter case, the output voltage (taken across the resistor) leads the input by an amount equal to the circuit phase angle. Suppose, however, that we want a phase shift, but that the output voltage should *lag* the input. Can this be done with our simple *R-C* circuit? Yes—examine Fig. 33-3(b) again. Notice that the voltage E_C across the capacitor lags the input voltage. If we reverse C and R in Fig. 33-3(a), and take the output across the capacitor we can achieve the desired result. Now, to evaluate the phase-shift angle, refer to the phasor diagram [Fig. 33-3(b)]. Notice that the angle between E_C and E_T is $(90 - \theta_c)$, where θ_c is the circuit phase angle. An example will clarify the procedure.

Example 33-8

What value of capacitor is needed in an *R-C* phase-shift circuit to produce an output voltage that will lag the input voltage by 30° at 2000 Hz? The resistor value is 10 000 Ω.

Solution

1. The circuit used will be similar to Fig. 33-3(a) but with R and C reversed, and the output taken across the capacitor.

2. Since 30° phase shift is desired, the circuit phase angle θ_c must be $(90 - 30)$, or 60°.

3. $$\tan \theta = \frac{X_C}{R} = \tan 60° = 1.732$$

4. $$X_C = R \tan \theta = 10\,000 \times 1.732 = 17\,320 \ \Omega$$

5. $$C = \frac{1}{2\pi f X_C} = \frac{0.159 \times 10^6}{2000 \times 17\,320} = 0.004\,59 \ \mu\text{F}$$

33-3 Phase-shift *R-L* circuits.   Just as a capacitor causes a phase shift between line current and line voltage, so will an inductor. However, since the voltage across a pure inductance leads the current by 90°, the phase shift in any series *R-L* circuit will cause the line voltage to lead the line current—just the reverse of the conditions we saw for an *R-C* circuit. This angle of lead will depend on the amount of resistance and inductance in the circuit. The greater the resistance, the smaller the phase angle between the line current and the line voltage. Remember that if the circuit contains resistance only, the phase angle is zero. On the other hand, the greater the inductance, the greater will be the reactance (or the higher the frequency, the greater the reactance). This will cause a greater angle of shift. This can be verified by reference to the impedance triangle: The circuit phase angle is given by

$$\theta_c = \text{arc tan } \frac{X_L}{R}$$

Circuits containing resistance and inductance in series are used as coupling and filter circuits. Since these circuits produce phase shifts, the component values must be carefully selected if negligible phase shift (or a definite phase shift) is required. Phase-shifting circuits are shown in Fig. 33-4. If the output voltage is taken across the resistor, it will *lag* the input voltage by an angle equal to the circuit phase angle. When the output voltage is taken across the inductor, the output will *lead* the input voltage by an angle equal to 90 minus the circuit phase angle. This should be obvious from the phasor diagram for each condition.

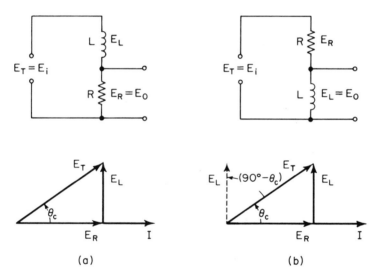

Figure 33-4   Phase shift in *L-R* circuits. (a) E_o lags E_i by θ_c°. (b) E_0 leads E_i by $(90 - \theta_c)^{\circ}$.

Example 33-9

It is desired to build an *R-L* circuit that will cause the output voltage to lead by 30° at 10 000 Hz. The resistance value is fixed at 50 000 Ω. What inductance value must be used?

Solution

1. Since the output voltage is leading, the circuit should be that of Fig. 10-4(b). The output voltage is taken across the inductor.
2. E_o will lead E_i by 30°, if the circuit phase angle is 60°, $(90 - \theta)$.

3. $$\theta_c = \text{arc tan} \frac{X_L}{R} = 60°$$

4. $$\tan 60° = 1.73$$

5. $$\frac{X_L}{R} = 1.73; \; X_L = 50\,000 \times 1.73 = 86\,500 \; \Omega$$

6. $$L = \frac{X_L}{2\pi f} = \frac{0.159 \times 86\,500}{10\,000} = 1.38 \; \text{H}$$

33-4 Frequency discrimination: *R-C* filter. In the study of dc circuits, we saw that voltage dividers could be made by using several resistors in series and taking the output voltages from taps.* Such a circuit is shown in Fig. 33-5. The output voltage is obviously less than the input voltage. Now if the output voltage is fed to a high-impedance circuit, the shunting effect of the "load" across R_2 is negligible. (A typical example is when the output voltage is fed to a FET.) Under such conditions, the output voltage is less than the input voltage by the same ratio that R_2 bears to the total resistance $(R_1 + R_2)$,† or

$$E_o = E_i \frac{R_2}{R_1 + R_2}$$

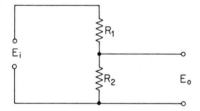

Figure 33-5   Simple voltage divider (attenuator).

This circuit works equally well with ac. Regardless of frequency, the above voltage relation will still hold true. Any change in resistance value or distributed capacitance or inductance in the resistors will affect R_2 and $(R_1 + R_2)$ in the same ratio. A resistive voltage divider will, therefore, show no frequency discrimination; that is, if the frequency of the input voltage is varied, the ratio of output to input voltage remains constant. What is the purpose of such a circuit? When the amplitude of the input voltage is too high for the circuit components that follow, this circuit will reduce the amplitude of all input signals. Any circuit that reduces the amplitude of the input signal voltage is called an ***attenuator***. (The volume control in a radio receiver is an example of a *variable* attenuator.) A resistive voltage divider attenuates all frequencies by the same proportion.

Sometimes the input voltage is too high at only certain frequencies. In this case we should attenuate only those frequencies. At other times, certain frequency

---

*See Chapter 10.

†If the load resistance is low compared to R_2 (such as when feeding a bipolar junction transistor), then an equivalent resistance equal to R_2 in parallel with R_{Load}, must be used in place of R_2.

components of the input voltage (either at the high- or low-frequency end) are undesirable. (For example, scratch produced by the phonograph needle is mainly in the high end of the audible frequency range.) Under these conditions, we would like to attenuate only the undesirable frequencies. An *R-C* voltage divider will do the trick.

Figure 33-6 shows two methods of connecting an *R-C* frequency discriminating circuit. Which of these circuits will attenuate the higher frequencies? Let us consider the circuit in Fig. 33-6(a). As the frequency increases, X_C will decrease, and more of the input voltage will appear across *R*. The output has increased with frequency. Whereas, at the lower frequencies, X_C increases, the drop across the capacitor increases, and the output voltage will decrease. Obviously, the series capacitor tends to block passage of low frequencies. This circuit, therefore, attenuates low frequencies. Since it *passes* the higher frequencies, it is also known as a **high-pass filter**. Now consider the circuit shown in Fig. 33-6(b). The output voltage is developed across X_C. As the input frequency increases, X_C decreases, and the voltage across the capacitor must decrease. The shunt capacitor tends to short-circuit the higher frequencies. Consequently, this circuit attenuates the high frequencies, but passes the low frequencies. A high-frequency attenuator is therefore also a **low-pass filter**. Notice that, depending on the relative positions of *R* and *C*, an *R-C* circuit can act to attenuate high or low frequencies. These circuits are often used as simple tone controls merely by making the resistance element variable.

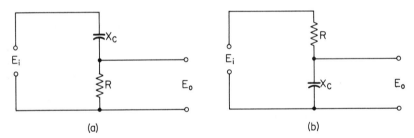

Figure 33-6 Frequency discriminating circuits. (a) Low-frequency discrimination. (b) High-frequency discrimination.

In Fig. 33-6(a), the output voltage (E_o) is equal to IR. In turn, $I = E_i/Z$; replacing current I by its equivalent we get
For low-frequency attenuation:

$$E_o = E_i \frac{R}{Z} \qquad \text{(33-2A)}$$

By similar reasoning:
For high-frequency attenuation:

$$E_o = E_i \frac{X_C}{Z} \qquad \text{(33-2B)}$$

From these relations, we can calculate the variation of output voltage with frequency for any given circuit, or the circuit values required in order to reduce the output voltage to some definite value at any desired frequency.

Example 33-10

In Fig. 33-6(b), $R = 100\ 000\ \Omega$, $C = 0.003\ \mu\text{F}$. The input is 10 V. Calculate the output voltage for a frequency range of 50 to 20 000 Hz. Plot. (Start at 50 Hz and double the frequency each time.)

Solution

1. At 50 Hz

$$X_C = \frac{1}{2\pi fC} = \frac{0.159 \times 10^6}{50 \times 0.003} = 1.06\ \text{M}\Omega$$

2.
$$Z = \sqrt{R^2 + X^2} = \sqrt{(0.1)^2 \times (1.06)^2 \times 10^6} = 1.06\ \text{M}\Omega$$

f (hertz)	X_C	Z	X_C/Z	E_o (volts)
50	1.06 MΩ	1.06 MΩ	1.00	10.0
100	0.53 MΩ	0.539 MΩ	0.985	9.85
200	0.265 MΩ	0.283 MΩ	0.936	9.36
400	0.132 MΩ	0.166 MΩ	0.795	7.95
800	66 300 Ω	120 000 Ω	0.55	5.50
1 600	33 100 Ω	106 000 Ω	0.31	3.10
3 200	16 600 Ω	101 500 Ω	0.163	1.63
6 400	8 300 Ω	100 000 Ω	0.083	0.83
12 800	4 140 Ω	100 000 Ω	0.0414	0.414
25 600	2 070 Ω	100 000 Ω	0.0207	0.207

3. See Fig. 33-7 for the plot.

(In checking the tabulated values above, they try some "shortcuts." Since the frequency is doubled in each successive step, X_C should be half of the previous value. Therefore, only one calculation need be made for X_C. Next, notice in the calculations that when X_C is greater than $10R$, $Z = X_C$, and R is negligible. Also, when R is greater than $10X_C$, $Z = R$, and X_C is negligible. These facts will often save you many unnecessary calculations.)

Examine Fig. 33-6(a) again. We discussed this circuit before. It is a typical coupling circuit such as is used in audio and video amplifiers. But we just learned that this circuit is a frequency discriminating circuit that attenuates the lower frequencies! What will happen to the fidelity of the output voltage? If the values of R and C are properly chosen, the frequency discrimination is negligible. From equation (33-2A), (and also from common-sense analysis), $E_o = E_i(R/Z)$, the larger the capacitance, and/or the larger the resistance, the closer the output voltage approaches the input value. Again, either the capacitance or resistance value may be fixed by other circuit considerations.

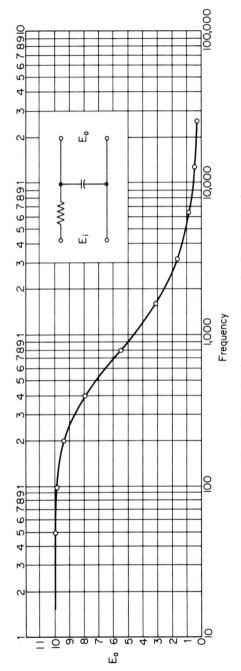

Figure 33-7 High-frequency attenuation by *R-C* circuit.

Example 33-11

In an R-C coupling circuit, the resistor value is 250 000 Ω. What value capacitor should be used in order to obtain an output voltage of at least 95% of the input voltage at 30 Hz?

Solution

1. For $E_o = 0.95E_i$, R/Z must equal 0.95.

2. $$Z = \frac{R}{0.95} = \frac{250\ 000}{0.95} = 263\ 000\ \Omega$$

3. $$X_C = \sqrt{Z^2 - R^2} = \sqrt{(263)^2 - (250)^2} \times 10^3 = 81\ 700\ \Omega$$

4. $$C = \frac{1}{2\pi f X_C} = \frac{0.159 \times 10^6}{30 \times 81\ 700} = 0.0647\ \mu\text{F}$$

The above problem illustrates how, by proper design, an R-C coupled amplifier can be made to give good low-frequency response.

33-5 Frequency discrimination: R-L circuit. The R-L series circuits shown in Fig. 33-4 also act as voltage dividers. The output voltage will be less than the input voltage due to the drop in the series circuit element. Since X_L varies with frequency, it is obvious that the ratio of output to input voltage will vary with frequency. These circuits inherently are, therefore, frequency discriminating circuits. The higher the frequency, the greater the X_L value and the higher the voltage developed across the inductance. When the output is taken across the resistor [Fig. 33-4(a)], the output decreases with increase in frequency. This circuit will attenuate high frequencies (low-pass filter). Using the connections shown in Fig. 33-4(b), the drop across the inductor increases with frequency, giving a greater output voltage. This circuit attenuates low frequencies (high-pass filter). If frequency discrimination is desired, one or the other of the two R-L circuits can be used, depending upon which end of the frequency range it is desired to attenuate. The choice of R and L values will depend on the amount of discrimination desired at any particular frequency. Since the action of these circuits is quite similar to the R-C circuits previously studied, no further discussion is necessary.

33-6 Differentiator and integrator circuits. In spite of their "sophisticated" names, these circuits are nothing more than the R-C series circuits we have been discussing, but now they are specifically designed to distort the output waveshape. In a differentiator circuit the output is taken across the resistor, and *for good differentiating action the time constant of the R-C circuit must be short*. If the time constant is too long, the distorting effect is lost and we revert to the simple coupling action of the circuit in Fig. 33-3. Conversely, if the output is taken across the capacitor and *the circuit has a long time constant, good integrator action will be achieved*.

This distorting effect of differentiating and integrating circuits is apparent only when the input is a complex wave, because each component frequency of the complex wave is shifted in phase by a different amount. With sine-wave input, there is no distortion. The output will be reduced in amplitude (voltage-divider action of R-C circuits), and shifted in phase (phase-shift action of R-C circuits) as compared with the input voltage, but that is all.

Those of you who are familiar with calculus will recall that the differential of the sine of an angle is the positive cosine. This would mean that if a sine wave is applied to a differentiating circuit, the output should be a positive cosine wave, or leading by 90°. That this is so can be readily seen from circuit analysis. The circuit for differentiating action is the same as shown in Fig. 33-3(a). Remembering that the time constant should be very short, it means that X_C should be very large (C is small), and that R should be small. The circuit impedance approaches pure capacitive reactance, and the current I will lead the input voltage E_i by 90°. The output voltage E_o taken across the resistor is in phase with I, and therefore also leads E_i by 90°. This is a positive cosine wave.

Similarly, by calculus, the integral of the sine of an angle is the negative cosine; by phasor diagram analysis (remember that an integrating circuit should have a long time constant—low X_C and large R), θ_c approaches 0 degrees and the output across the capacitor will lag the input voltage by $(90° - \theta)$, or 90°. The output lagging by 90° is a negative cosine wave.

Before we can discuss the action of these circuits, it will be necessary to review briefly the charging action of a capacitor in an R-C series circuit with dc supply—and particularly the concept of time constant.* Such a circuit is shown in Fig. 33-8.† At the instant that the power is turned on, there is no charge on the capacitor. Therefore the only opposition to current flow is the series resistor R. The initial value of current ($I = E/R$) will be a maximum. But just as soon as current starts to flow, the capacitor begins to charge and a countervoltage builds up across its plates. The net voltage, which causes the current to flow, is the difference between the line voltage and this countervoltage. Obviously, the current flow will decrease. Finally when the capacitor is fully charged and the countervoltage equals the supply voltage, the current drops to zero. These conditions are shown pictorially in Fig. 33-8(a) to (c). (The current arrows show conventional current flow.)

Will the capacitor charge increase linearly with time? No! At first the current flowing in the circuit is high, and the capacitor charges rapidly. But as the capacitor charges, the increase in countervoltage causes the current to drop. Due to the reduced current flow, the capacitor charges more slowly. This effect is cumulative—the current drops more and more slowly, and the capacitor charges more and more slowly. The curves for e and i are shown in Fig. 33-8(d). Curves of this shape are called *exponential* curves. The equation for the current and voltage *at any instant* are:

*For details, see Chapter 19.

†For a full-page graph refer to Fig. 19-8.

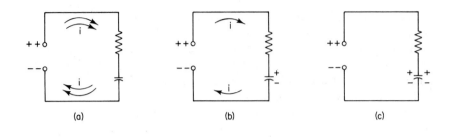

(a)   (b)   (c)

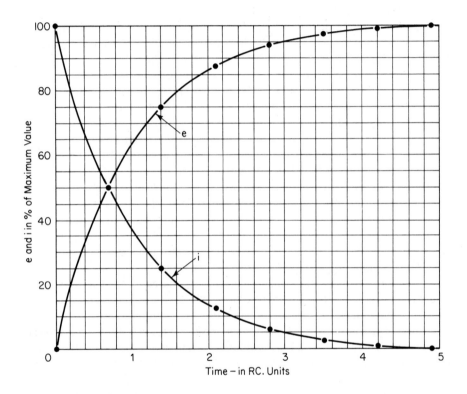

i(%)	Time	e(%)
100	0	0
50	0.7 RC	50
25	1.4 RC	75
12.5	2.1 RC	87.5
6.25	2.8 RC	93.75
3.13	3.5 RC	96.9
1.56	4.2 RC	98.4
0.78	4.9 RC	99.2

(d)

Figure 33-8 Charge of capacitor in a dc *R-C* circuit.

$$i = \frac{E}{R} \epsilon^{-t/RC}$$

$$e = E(1 - \epsilon^{-t/RC})$$

Obviously, the time required for the instantaneous current or voltage to reach some specific value will depend on the RC product. This RC product is known as the *time constant* (τ) of the circuit, and represents the time required for the capacitor to charge to 63% of its final value. (Simultaneously, the current will drop by 63% from its initial value.)

Four other important time intervals for R-C circuits can be derived from the equations above. These are:

1. $0.7RC$—the time for a 50% change in e or i. (This is known as the *practical time constant*.)
2. $0.1RC$—the time for 10% rise in e.
3. $2.3RC$—the time for e to rise to 90% of its final value.
4. $5.0RC$—the time for full charge.

Theoretically, the capacitor is never fully charged. Practically, however, in a time of approximately $5RC$, the capacitor is more than 99% charged and can be considered fully charged.

Example 33-12

How long will it take for a capacitor of 0.5 μF to charge completely if it is connected in series with a 200 000-Ω resistor across a 200-V supply?

Solution

1. Time constant (τ) $= RC = 0.2 \times 0.5 = 0.1$ s.
2. Full charge $=$ approximately $5RC = 0.5$ s.

The curves of Fig. 33-8(d) show how the charging current and voltage across the capacitor vary with time, time being expressed in RC units. There are two good reasons for using this time unit. First, in Fig. 33-8, no specific values were selected for R and C. Therefore, we cannot calculate the time constant in seconds. Second, had we chosen specific values for R and C, calculated the actual time in seconds for each step of the curve, and plotted the curve with "time in seconds" as the abscissa, this curve would now apply *only* to this specific situation. A new curve would have to be drawn for any other value of C and/or R, whereas the curve as shown is a *universal* time constant curve and can be applied for any value of C or R.

Example 33-13

A circuit contains a capacitor of 0.1 μF and a resistor of 220 000 Ω. The supply voltage is 230 V (dc).
(a) Find the time constant of the circuit.
(b) How long will it take for the capacitor voltage to reach 190 V?

Solution

(a) Time constant $(\tau) = RC = 0.22 \times 0.1 = 0.022$ s.

(b) 190 V is 190/230 or 82.6% of the maximum. From the universal curve [Fig. 33-8(d)], 82.6% corresponds to 1.8 RC units. Time $= 1.8 \times 0.022 = 0.04$ s.

Now we are ready to analyze the differentiating and integrating action of *R-C* circuits. The degree of distortion produced by these circuits depends on the time constant of the circuit as compared to the *period* (time for one cycle) of the input wave. For example, if the frequency of the complex wave input is 100 Hz, its period is 0.01 s. At this frequency, a time constant of 0.1 s or longer can be considered long, whereas a time constant of 0.001 s (or less) should be considered short. For pulse waves, the period of the wave has little meaning. These waves are specified in terms of pulse width (or duration) and pulse repetition rate. The degree of distortion produced by differentiating and integrating circuits again depends on the time constant of the circuit, but now this time constant must be compared to the time duration of the pulse. In general, *if the time constant exceeds 10 times the period of the input waveshape (or 10 times the pulse width) the circuit is said to have a long time constant; and when the circuit has a time constant less than one-tenth of the period of the input waveshape (or one-tenth pulse width), the time constant is short.*

Example 33-14

A pulse wave has a repetition rate of 500, and a pulse width of 80 μs. What value of time constant would be considered (a) long; (b) short?

Solution

(a) Long time constant $= 10 \times 80 = 800$ μs (or more).

(b) Short time constant $= \frac{80}{10} = 8$ μs (or less).

Example 33-15

At a frequency of 5000 Hz, what value of time constant will be considered (a) long; (b) short?

Solution

$$\text{Period} = \frac{1}{f} = \frac{1}{5000} = 0.0002 \text{ s} = 200 \text{ } \mu\text{s}$$

(a) Long time constant: $10 \times 200 = 2000$ μs (or more).

(b) Short time constant: $\frac{200}{10} = 20$ μs (or less).

Differentiating and integrating circuits are used quite commonly in pulse wave equipment such as loran, radar, racon, and also in timing, trigger, and synchronizing circuits, as in television. The best way to explain their action is to analyze the action of such a circuit, with inputs like square waves or sawtooth waves. Figure 33-9 shows an *R-C* series circuit with square-wave input. The values of *R* and *C* are so chosen, with respect to the period of the input wave, that the time constant is short, in order to give good differentiator action. A short time constant means that the capacitor will charge completely in less than one-half cycle. (Remember that a short time constant is less than one-tenth of the period of the wave, and full charge—5 *RC* units—will be less than five-tenths of the period, or less than one-half cycle.)

1. When the circuit is closed at time *A*, the supply voltage rises immediately from zero to + 100 V. Since the capacitor is not charged, the only opposition to current flow is the resistance *R*. The current will rise immediately to its Ohm's law value. The voltage drop across the resistor will equal *IR*, or 100 V.

2. During time period *A* to *B*, the action is similar to a dc circuit. The

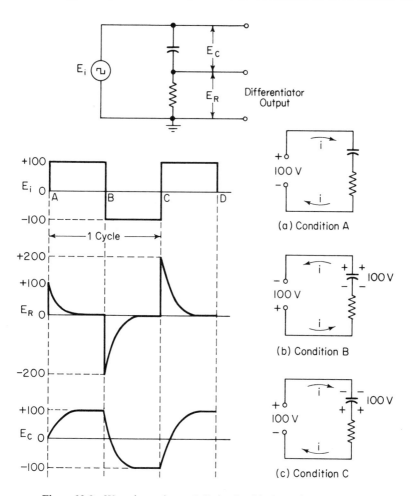

Figure 33-9   Waveshapes in an *R-C* circuit with short time constant.

capacitor starts to charge, and as the voltage across the unit builds up, the current and E_R will drop exponentially. Since the time constant of the circuit is short, the capacitor is fully charged (to $+100$ V) before the end of the half-cycle. The current and the voltage across the resistor will drop to zero for the remainder of the half-cycle.

    3. At time B, the potential of the circuit reverses. The circuit conditions at this instant are shown in Fig. 33-9(b). The applied voltage and the voltage across the capacitor are now additive. At this instant, the circuit voltage is 200 V. The current will reverse sharply and E_R will rise to a negative maximum of double the value at A. The voltage drop across the resistor (E_R) will be -200 V!

    4. During the time period B to C, the capacitor discharges exponentially. The net circuit voltage is decreasing; the current and E_R will decrease. When the capacitor is completely discharged, it begins to charge with reversed polarity and opposes the line voltage, causing the current and E_R to drop further. Again, since the time constant is short, the capacitor charges completely; current drops to zero; E_R becomes zero before the second half-cycle is completed.

    5. At time C, the applied voltage suddenly reverses. But, due to the reversed charge on the capacitor, supply voltage and capacitor voltage are again additive. The current will reverse sharply and rise to a positive maximum. The voltage across the resistor (E_R) will rise to $+200$ V.

    6. For time period C to D, the capacitor discharges and then recharges with opposite polarity, reducing the net circuit voltage. The current and E_R drop exponentially to zero.

    The values (E_R, I, E_C) for the first half-cycle differ in magnitude or slope from the curves shown for the remainder of the wave. This first half-cycle should be ignored. It is a *transient* condition that occurs only at the closing of the circuit. From then on, all the cycles will be identical. The output of the differentiator circuit, E_R, is of special interest. Notice that:

1. The waveshape is very peaked. Such a circuit is often called a *peaker or peaking circuit*.

2. The amplitude of the output voltage is twice the input amplitude.

3. Also notice that at any instant the sum of e_C (across the capacitor) and e_R (across the resistor) is equal to e_i (the input voltage). This is in accordance with Kirchhoff's voltage law for a series circuit.

    In the above circuit, the time constant was such that the capacitor was charged completely in less than one-half cycle of the square-wave input. This resulted in highly distorted output from the differentiator circuit. Meanwhile, the voltage across the capacitor (E_C) approaches the original square wave. Suppose now that the circuit constants are increased—either the capacitor or the resistor value or both are made larger. The time constant of the circuit is increased, and it will take longer for the capacitor to charge. Let us see how the waveshapes will look if the capacitor charges to only 20% of the supply voltage in one half-cycle. The curves for this condition are show in Fig. 33-10.

The circuit action follows the previous discussion, except that now the "added" capacitor voltage is only 20 V.

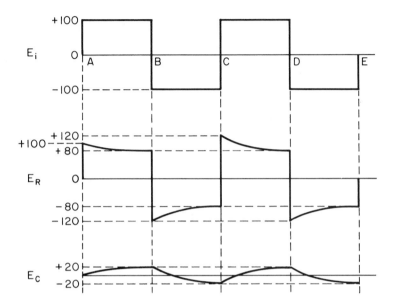

Figure 33-10   Effect of long time constant on *R-C* circuit waveshapes.

The waveshape of E_R and E_C should be obvious. Again the transient values obtained in the first half-cycle are slightly lower in value than the rest of the cycles and should be ignored in analyzing the action of the circuit. Notice that the voltage E_R across the resistor approaches the original square-wave input. Differentiator action is practically lost due to the long circuit time constant. On the other hand, good integrating action is obtained. Notice also that the voltage E_C across the capacitor looks like a triangular wave. Here the amplitude of the integrator output is low.

As the time constant of the circuit is made longer and longer, the capacitor will charge to a lesser and lesser degree. The current in the circuit will hardly drop. E_R will remain more nearly constant. At the points where the supply voltage reverses (B, C), the capacitor adds little to the net circuit voltage. The current will rise very little above its first maximum value. E_R will hardly exceed the supply voltage. The voltage E_R across the resistor is almost identical to the square-wave input. Meanwhile, since the capacitor charges very little, the integrator output is very low—approaching zero—and the waveshape approaches a perfect triangular wave. (This type of circuit—very long time constant, and the output taken across R—is desirable for amplifier coupling, since the input waveshape is not distorted.)

Effective integrator action is best realized when the charging time is appreciably longer than the discharge time (or the discharge *rate* is appreciably slower). Such a condition will occur if the input voltage is an unbalanced rectangular wave as shown in Fig. 33-11(b). Let us assume that the time constant of this circuit, Fig. 33-11(a), is such as to allow the capacitor to charge to 10% of the input voltage during time interval *A–B*. The capacitor voltage E_C (or output voltage E_o) is shown in Fig. 33-11(c). Since the rise is well below half-charge, the charging curve is a straight line. Now at time *B*, the supply voltage drops to zero. But the discharge time (*B–C*) is quite short. The capacitor discharges only slightly before the next pulse of energy arrives at time *C*. The capacitor must begin to charge again, and it charges to a higher value. The effect is cumulative (integrating). It should be obvious that after several input cycles, the capacitor will charge to the maximum value of the input voltage.

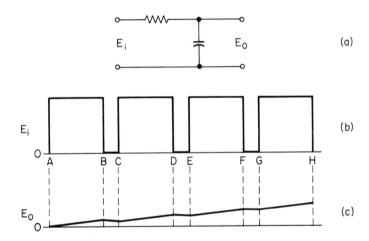

Figure 33-11 Cumulative action of a long-time-constant integrating circuit.

An *R-C* circuit may also be used with some other type of complex wave. Let us examine the effect when a sawtooth wave is applied as the input voltage, and the circuit has a very short time constant. Figure 33-12 shows the resulting waveshapes.

1. At *A* the supply voltage is zero and the current must be zero. As the voltage starts to rise, the current tends to rise to a high value. But since the time constant of the circuit is short, the capacitor charges quickly, and the capacitor voltage follows the input voltage closely. The countervoltage of the capacitor will prevent the current from rising to a high value. Since the supply voltage increases at a constant rate from *A* to *B*, the charging current in the circuit will be constant. E_R will rise quickly to some low value, and then remain constant for the remainder of the time *A* to *B*.

2. At *B* the supply voltage begins to drop to zero. But the capacitor is

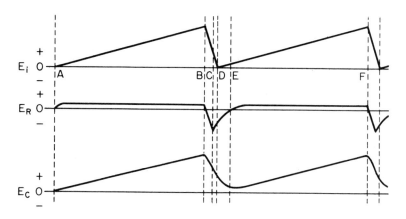

Figure 33-12 Waveshapes from a short-time-constant *R-C* circuit with sawtooth input wave.

charged to the full supply voltage. A high current, in the opposite direction, will flow as the capacitor starts to discharge rapidly. This produces a high negative voltage across the resistor. This condition prevails during the interval *B* to *C*.

3. At *C*, the capacitor has discharged sufficiently to reduce the current flow. The negative voltage across the resistor starts to decrease.

4. At *D*, the line voltage once again begins to rise in a positive direction, but the rate of rise in supply voltage is low. Meanwhile, the capacitor voltage is still higher than the supply voltage and the capacitor still continues to discharge, but at a slower rate. The voltage drop across the resistor is still negative.

5. Finally, at *E*, the line voltage is now higher than the capacitor voltage. The direction of current flow reverses, the capacitor once again begins to charge, and the voltage across the resistor is now of positive polarity. From there on, the operation repeats as above. The differentiator circuit output is of special interest. It produces sharp negative *pips* that are often used for timing, trigger, or synchronizing pulses in special electronic circuits.

REVIEW QUESTIONS

1. (a) What circuit components are necessary to produce resonance?
(b) In a series circuit, how does resistance affect the frequency for resonance?
(c) Does resistance have any other effect? Explain.
(d) Of what value is resonance in transmitter circuits?
(e) Of what value is resonance in receiver circuits?
(f) Of what value is resonance in audio amplifier circuits?

2. (a) What is the basic condition for resonance in a series circuit?
(b) Give the equation for the resonant frequency.
(c) What are the units for *L*, *C*, and *f* in this equation?

3. With reference to the solution for Example 33-2:
   (a) In step 2(c), why does $X_0 = 0$?
   (b) Was it necessary to solve for X_L and X_C to find Z_T? Explain.
   (c) Does Z_T always equal R in a series-resonant circuit?
   (d) In step 4, compare the E_R value to the line voltage. Is this reasonable, considering there are two other components in the circuit?

4. With reference to Example 32-12, page 492:
   (a) What components does this circuit contain?
   (b) Compare the calculated values of X_L and X_C. Is this circuit resonant? Explain.
   (c) Compare the calculated Z_T value against the given data. Is this circuit resonant? Explain.
   (d) From the calculated value, step 4(c), is this a resonant circuit? Explain.

5. Still with reference to Example 32-12, it is desired to produce a resonant condition. Can this be done by:
   (a) Changing the frequency? Which way? Explain.
   (b) Changing the L value? Which way? Explain.
   (c) Changing the C value? Which way? Explain.
   (d) Changing the R value? Which way? Explain.

6. With reference to the solution for Example 33-4, (page 508):
   (a) Why is X_L made equal to X_C?
   (b) How can this be done?

7. With reference to the solution for Example 33-5, (page 508):
   (a) In step 2, why is X_C made equal to X_L?
   (b) How can this be done?
   (c) In receivers, what is this technique called?

8. When a circuit is series-resonant, what distinguishing characteristic applies to:
   (a) The line current?
   (b) The total impedance?
   (c) The circuit phase angle?

9. In an ac L-C-R series circuit, the applied line voltage is 50 V. The voltage across the capacitor is 90 V. Is this possible? Explain.

10. (a) In a pure capacitive circuit, what is the phase angle between current and voltage?
    (b) If series resistance is added, what happens to this phase angle?

11. With reference to Fig. 33-2(a):
    (a) What type of diagram is this?
    (b) How many impedance diagrams does this represent?
    (c) To what type of circuits do these triangles apply?
    (d) Is the resistance value C changed?
    (e) Give a statement regarding the effect of changing the X_C values.
    (f) Repeat for changing C values.

12. With reference to Fig. 33-2(b):
    (a) Compare the component values for each of the four cases shown.
    (b) Draw a conclusion from these four cases.

13. (a) What two parameters determine the circuit phase angle in any ac series circuit?
    (b) Give the equation for evaluating this phase angle.

14. With reference to Fig. 33-3:
   (a) Why is this called a *coupling* circuit?
   (b) State two functions of the capacitor in this circuit.
   (c) Why is current used as the reference phasor?
   (d) Why is E_R also called E_0?
   (e) Why is E_R drawn in line with I?
   (f) Why is E_C drawn downward?
   (g) Why is E_T also called E_i?
   (h) What is the phase of the *output* compared to the input?
   (i) What two phasors determine the *circuit* phase angle?
   (j) How does the circuit phase angle compare with the *input–output* phase angle?
   (k) Is this relation always true for this type of circuit? Explain.

15. In the schematic of Fig. 33-3:
   (a) For what relative circuit values would the phase shift be 45°?
   (b) For what relative values would the phase shift be negligible?
   (c) If the input-signal frequency is increased, would this tend to increase, or decrease the phase shift? Explain.

16. A coupling circuit for TV must handle frequencies from 30 Hz to 4.0 MHz. Phase shift is undesirable. Which frequency value is more important when selecting the coupling component values? Explain.

17. **(a)** In the coupling circuit of Fig. 33-3(a), what value of R, or C will cause the output to *lag* the input voltage?
   (b) What must be done to produce a lagging output voltage?
   (c) Draw a circuit suitable for producing a lagging output voltage.
   (d) Will the change in circuit cause a change in the phasors shown?
   (e) Which phasor in Fig. 33-3(b), will now represent E_0?
   (f) Which phasor will now represent E_i?
   (g) What angle represents the phase shift between output and input?
   (h) How is this angle evaluated?

18. In the solution to Example 33-8, step 2, why is the angle 60° involved?

19. **(a)** Can phase shift between input and output voltages be obtained with other than R-C circuitry?
   (b) How will this change affect the phase shift?

20. **(a)** Draw the phasor diagram (current and voltages) for a series R-L circuit.
   (b) Across which component should the input voltage be applied so that E_0 will lead E_i? Explain.
   (c) Across which component should E_0 be taken so that E_0 will lead E_i? Explain.
   (d) Across which component should E_0 be taken so that E_0 will *lag* E_i? Explain.

21. **(a)** What value of phase shift is desired in Example 33-9?
   (b) In step 3 of this solution, why is a θ of 60° considered when the desired phase shift is 30°?

22. With reference to Fig. 33-5:
   (a) Will the output be more, less, or equal to the input voltage? Explain.
   (b) Give two names used to describe this circuit action.
   (c) By what ratio is the output voltage reduced?
   (d) At what frequency will the attenuation be greatest?

23. What is meant by the term *frequency discrimination*?

24. With reference to Fig. 33-6(a):

(a) By what ratio is the output voltage reduced?

(b) If the frequency is increased, what happens to R? to Z? to E_0? Explain.

(c) Does this circuit discriminate against high or low frequencies?

(d) Give another name for such a circuit.

(e) If $R = X_C$, by what ratio is the output reduced?

25. In Fig. 33-6(b):

(a) By what ratio is the output voltage reduced?

(b) If the frequency is *increased*, what happens to X_C? to E_0? Explain.

(c) Does this circuit discriminate against high or low frequencies?

(d) Give another name for this circuit.

26. In the solution to Example 33-10:

(a) What is the value of R?

(b) In step 2, why is Z the same value as X_C?

(c) In the table, how are the values for E_0 obtained?

(d) Does this circuit discriminate against high or low frequencies?

(e) By what other name can this circuit be called?

27. With reference to Fig. 33-7:

(a) What type of graph paper is this?

(b) Why is this type of graph paper used?

28. (a) What is the circuit of Fig. 33-6(a) called?

(b) What was it called earlier in this chapter?

(c) How can we prevent low-frequency distortion in this coupling circuit?

29. In Example 33-11, would a larger capacitor (such as 0.1 μF) be satisfactory? Explain.

30. (a) Can *R-L* circuits be used for frequency discrimination?

(b) Would the circuit in Fig. 33-4(a) discriminate against high or low frequencies? Explain.

(c) In Fig. 33-4(b), as the frequency is *increased*, will E_0 increase or decrease? Why?

(d) What is the ratio of E_0 to E_i in Fig. 33-4(b)?

31. (a) What is a *differentiating circuit*?

(b) What is an *integrating circuit*?

(c) What is the *general* effect of such circuits?

(d) Are these circuit actions obtained with sine-wave inputs? Explain.

32. In an *R-C* circuit with dc input:

(a) When the circuit is first energized, is the current high or low? Explain.

(b) Does the current remain constant at this value? Explain.

(c) Does the current decrease at a linear rate? Explain.

(d) What is this type of curve called?

33. With reference to Fig. 33-8(d):

(a) What are the coordinates of this curve?

(b) What does "time in *RC* units" mean?

(c) Why isn't the curve plotted directly in seconds?

(d) How long (in *RC* units) does it take for the capacitor to charge completely?

(e) What is the circuit current when the capacitor is fully charged?

(f) What does the time interval $0.7RC$ correspond to?

(g) Explain how the third point of the charted values is obtained?

(h) Explain how the fourth point is obtained.

34. In Example 33-12, explain how the time for full charge is obtained.

35. With reference to Example 33-13:

    (a) Does the curve of Fig. 33-8 apply to this problem? Explain.

    (b) In step 2, why do we bother to find "per cent of maximum voltage"?

    (c) In step 3, why do we multiply the time of 0.022 s by 1.8?

36. In connection with integrating and differentiating action:

    (a) Will any R-C circuit produce a noticeable effect? Explain.

    (b) What determines the amount of "distortion" produced?

    (c) When is a circuit considered to have a *long* time constant?

    (d) When is a circuit considered to have a *short* time constant?

37. In Example 33-14:

    (a) Why is 800 μs considered a long time constant?

    (b) Is less than 800 μs a short time constant? Explain.

38. With reference to Example 33-15:

    (a) What does the term *period* mean?

    (b) Why is 20 μs considered a short time constant?

    (c) Would 50 μs therefore be a long time constant? Explain.

39. Give two common uses for differentiating and integrating circuits.

40. With reference to Fig. 33-9:

    (a) What effect is desired from this R-C series circuit?

    (b) What two factors ensure this effect?

    (c) How much current will flow at time $t = 0$? Explain.

    (d) Why does $E_R = 100$ V at $t = 0$? What is the output voltage?

    (e) Why does $E_C = 0$?

    (f) At a time instant approximately halfway between A and B:

        1. Why does E_R drop to almost zero?

        2. Why does E_C rise to almost 100 V?

    (g) At any instant between A and B, what is the sum of E_R and E_C?

    (h) How much current will flow just before time instant B?

    (i) How much current will flow at time B, as compared to time instant A? Why?

    (j) What is the direction of the current at time B, as compared to the current at time A? Why?

    (k) What is the output voltage at time B?

    (l) What is the voltage across the capacitor (magnitude and polarity)?

    (m) What is the total voltage $(E_C + E_R)$ now?

    (n) Why does E_R drop to zero during time interval B–C?

    (o) Why does E_C rise to -100 V during time interval B–C?

    (p) What is the sum of E_R and E_C during time interval B–C?

    (q) What is the net voltage across the circuit at time C? Explain.

    (r) How much current will flow at C as compared to time A and B?

    (s) In what direction is this current flowing, as compared to the current at A and B?

    (t) What is the magnitude and polarity of the output voltage at time instant C?

41. Why is a differentiator circuit sometimes called a *peaking circuit*?

42. For integrator action:
- **(a)** Can an *R-C* circuit be used?
- **(b)** Across which component is the output taken?
- **(c)** What value of time constant is appropriate for this action?

43. With reference to Fig. 33-10:
- **(a)** At time *A*, why does E_R rise sharply to +100 V?
- **(b)** Why does E_R drop to 80 V during interval *A–B*?
- **(c)** Why does E_C rise to +20 V during interval *A–B*?
- **(d)** Why doesn't the capacitor charge to the full input voltage?
- **(e)** What is the sum of E_R and E_C at any instant during *A–B*?
- **(f)** What is the "net" circuit voltage at time *B*?
- **(g)** How is this net value obtained?
- **(h)** Why does the voltage drop across the resistor reverse at time *B*?
- **(i)** Why is it 120 V?
- **(j)** During time interval *B* to *C*:
  1. Why does the voltage across the resistor decrease?
  2. What happens to the voltage across the capacitor?

44. With reference to Fig. 33-11:
- **(a)** What type of circuit action is desired here?
- **(b)** To what value does the capacitor charge during time interval *A–B*? Why?
- **(c)** Why doesn't the capacitor discharge completely during time interval *B–C*?
- **(d)** What is the charge on the capacitor at time *D*? Explain.
- **(e)** For how long will this effect continue?

45. With reference to Fig. 33-12:
- **(a)** Why is the E_C waveshape almost identical to the E_i waveshape during time interval *A–B*?
- **(b)** What happens to the current during this period? Explain.
- **(c)** Why is E_R constant during this period?
- **(d)** Does $E_C = E_i$ during this period? Explain.
- **(e)** Why does E_R reverse during time *B–C*?
- **(f)** Why does the magnitude of E_R decrease during time *C–D*?
- **(g)** Since the input voltage is positive and rising during time *D–E*, why is E_R still negative?
- **(h)** When does the current flow reverse? Why?

PROBLEMS

1. A capacitor of 0.002 μF is connected in series with a coil of 0.25 H. Find:
- **(a)** The resonant frequency.
- **(b)** The total circuit impedance at resonance.

2. A 200-μH coil, a 320-pF capacitor, and a 200-Ω resistor are connected in series across a 50-V 1990-kHz supply source. Is this circuit in resonance? Explain the reason for your answer.

3. In Problem 2, the capacitor is reduced to one-tenth of its previous capacitance. Is the circuit now resonant? Explain.

4. Using the values of Problem 3, find:
   (a) The line current.
   (b) The voltage across each component.
   (c) The circuit phase angle.

5. In Problem 2, at what frequency would the circuit become resonant if the capacitance is not changed?

6. In Problem 2, what value of inductance would produce resonance if the frequency and capacitance are unchanged?

7. A capacitor of 1.0 μF is connected in series with a variable resistor. The input frequency is 100 Hz. Find the phase shift between E_{Line} and E_R for the following values of resistance.
   (a) 250 Ω (b) 2500 Ω (c) 5000 Ω (d) 15 000 Ω

8. Repeat Problem 7 for the phase angles between E_{Line} and E_C.

9. Draw the circuit diagram and determine the capacitance value for an R-C circuit that will cause the output voltage across a 600-Ω resistive load to lead the input voltage by 40° at 2000 Hz. Draw the phasor diagram.

10. Draw a circuit diagram suitable for coupling in an R-C coupled amplifier. Indicate the input and output voltage connections.

11. In the circuit of Problem 10, the coupling capacitor is 0.01 μF, and the shunt resistor is 0.25 MΩ. Find the phase shift between the input and output voltage:
    (a) At 100 Hz.
    (b) At 10 000 Hz.
    (c) If the resistor is not changed, how can the phase shift at 100 Hz be reduced to 5°?

12. In Problem 11 the input voltage is 10 V. Find the output voltage:
    (a) At 100 Hz.
    (b) At 10 000 Hz.
    (c) Repeat part (a), using a coupling capacitor of 0.1 μF.
    (d) Repeat part (b) for the 0.1-μF capacitor.

13. By means of circuit diagrams and supporting phasor diagrams, show two ways by which the output voltage can be made to lag the input voltage.

14. A frequency attenuating circuit consists of a 0.001-μF capacitor in series with a 200 000-Ω resistor. The applied voltage of 10 V is varied in frequency from 50 to 10 000 Hz. Plot the output voltage across the resistor for this range of frequency. (Start at 50 Hz and double. This plot should be made on semilog paper.)

15. (a) Draw a circuit diagram showing an R-C circuit connected for differentiating action.
    (b) Repeat for integrator action.

16. Find the time constant of the following R-C series circuits.
    (a) $R = 2$ MΩ, $C = 2$ μF   (b) $R = 0.5$ MΩ, $C = 0.05$ μF
    (c) $R = 0.1$ MΩ, $C = 200$ pF   (d) $R = 50\ 000$ Ω, $C = 40$ pF

17. For a frequency of 15 000 Hz, what values of time constants in Problem 16 would be considered:
    (a) Long?   (b) Short?

18. At what frequencies would the R-C circuit of Problem 16(c) become a long time constant?

19. A square wave having a frequency of 200 Hz is fed to an R-C series circuit having a resistance of 100 000 Ω. What value of capacitor would be required for:
    (a) A long time constant?
    (b) A short time constant?

20. A pulse wave has an amplitude of 10 V. The pulse duration is 50 μs and the pulse repetition rate is 2000. Find:
    (a) The maximum time for a short time constant.
    (b) The minimum time for a long time constant.
    (c) The maximum value of resistance that would still produce a short time constant if $C = 0.005$ μF.

21. The pulse wave of Problem 20 is fed to an integrating circuit consisting of $R = 250\,000$ Ω and $C = 0.001$ μF. To what value will the capacitor charge during the first pulse?

22. The pulse wave of Problem 20 is fed to a differentiating circuit consisting of $R = 50\,000$ Ω and $C = 250$ pF. To what value will the capacitor charge during the first pulse?

23. Figure 33-13(a) shows a typical pulse wave used for horizontal synchronization in a television receiver. Draw the output waveshape that would be obtained from:
    (a) A very short time constant differentiating circuit.
    (b) A long time constant integrator circuit where C charges to $\frac{1}{3}$ pulse amplitude.

24. Repeat Problem 23 for the vertical synchronizing signal shown in Fig. 33-13(b).

Figure 33-13   Typical TV synchronizing signals. (a) Horizontal sync pulses. (b) Vertical sync pulses.

34

PARALLEL
AND SERIES-PARALLEL
CIRCUITS

Parallel circuits are encountered quite frequently in all phases of electronic work. A very common example is the tuning circuits in a radio or TV receiver. Whenever you change the dial setting of your receiver, you are actually changing the circuit constants in several parallel circuits. Bypass capacitors, used in many portions of a receiver, also form parallel circuits. Filter networks and coupling circuits between stages of electronic units are further examples of parallel circuits. Sometimes the branches of these parallel circuits are pure components, that is, resistance alone, capacitance alone, or inductance alone. Such circuits are easily handled. But more often any branch of the parallel circuit may contain more than one type of circuit element. Actually, these circuits are no more difficult, but they are more laborious to solve. In order to understand the role of such circuits in electronic work, we must learn how to analyze and solve these circuits.

34-1 Rules for parallel circuits. All the principles we learned for handling dc parallel circuits apply equally well to ac circuits. However, now we must take phase angles into consideration. Let us summarize these principles with due respect to phase relations:

1. The voltage is the same across each branch of the parallel circuit. Since the voltage is common to all the branches, it is generally used as the reference value for phasor diagrams.
2. The current in any branch can be found from the voltage across the branch and the impedance of the branch.

(a) If the branch contains resistance only, the branch current is in phase with the applied voltage.

(b) If the branch contains capacitance only, the branch current will lead the applied voltage by 90°.

(c) If the branch contains pure inductance only, the branch current will lag the applied voltage by 90°.

(d) If the branch contains a combination of circuit elements, the current value will depend on the impedance (Z) of the branch, and will lead or lag the line voltage by some angle between zero and 90°. We can calculate this phase angle from the ratio of reactance (X_0) to resistance (R) for the branch.

3. Once the individual branch currents and their phase angles have been found, we can get the total line current by adding the branch currents. *This must be a phasor addition.* The addition may be made graphically, mathematically from trigonometric relations, or by vector algebra.

4. The total impedance can be found from the total current and the supply voltage:

$$Z_T = \frac{E_T}{I_T}$$

(When the voltage across the parallel circuit is not known, it is sometimes easier to assume a voltage and solve for the parallel circuit impedance, and then continue with the rest of the problem.)

In studying direct currents, we learned that

$$\frac{1}{R_T} = \frac{1}{R_1} + \frac{1}{R_2} + \frac{1}{R_3} + \cdots \qquad \text{and} \qquad R_T = \frac{R_1 R_2}{R_1 + R_2}$$

We used these relations to find the total resistance of a parallel circuit, *the second formula being used when only two branches were involved.* A similar statement can be made for ac parallel circuits:

$$\frac{1}{Z_T} = \frac{1}{\dot{Z}_1} + \frac{1}{\dot{Z}_2} + \frac{1}{\dot{Z}_3} + \cdots \tag{34-1}$$

and

$$Z_T = \frac{\dot{Z}_1 \dot{Z}_2}{\dot{Z}_1 + \dot{Z}_2} \tag{34-2}$$

However, since impedances have phase angles, the solution to the above equations must be done by vector algebra. This method is shown later in the chapter.

We also learned, in dc theory, that the total resistance was less than the smallest branch resistance. *Due to phase angles of impedance, such a general statement cannot be made for ac circuits:*

1. If the parallel circuit contains resistive and inductive branches only, the total circuit impedance will be less than the smallest branch impedance.

2. The above statement will also be true when the branches are resistive and capacitive.

3. When a parallel circuit contains resistive, inductive, and capacitive branches, the total impedance may be greater than if the capacitive or inductive branches were disconnected. (The reason for this will become apparent later in the chapter, under *Parallel Resonance*.)

34-2 Equivalent series circuit.   In analyzing complex dc circuits, we often replaced parallel networks by one resistance of equivalent value. Very often in electronics it is also easier to analyze a circuit by replacing parallel networks by their *equivalent series circuits*. This is not difficult. As far as the supply source is concerned, if the magnitude of the current being delivered, and the phase angle between that current and the supply voltage are unchanged—it makes no difference whether the load is a parallel or a series circuit. To match line current and phase angle, the total impedance (Z_T) and the circuit phase angle (θ_C) for the equivalent circuit must be identical to the original parallel circuit values. But in the series circuit impedance triangle, Z_T is the hypotenuse of the right triangle, with R as the horizontal (or in-phase) leg; and X as the vertical (or quadrature) leg of the triangle (see Fig. 34-1). Obviously, the equivalent series resistance is $Z_T \cos \theta_C$; and the equivalent series reactance is $Z_T \sin \theta_C$.

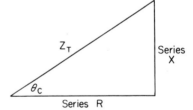

Figure 34-1 Equivalent series circuit values.

Example 34-1

A parallel circuit has a total impedance of 50 Ω. The supply source is 20 V, 400 Hz, and the line current leads the supply voltage by 30°. Find the constants for the equivalent series circuit.

Solution

1. Z_T of series circuit $= Z_T$ of parallel circuit $= 50\ \Omega$.
2. θ_c of series circuit $= \theta_c$ of parallel circuit $= 30°$.
3. $R_{\text{equiv}} = Z_T \cos \theta_c = 50 \times 0.866 = 43.3\ \Omega$.
4. $X_{\text{equiv}} = Z_T \sin \theta_c = 50 \times 0.500 = 25.0\ \Omega$.
   (Since current leads voltage, X is capacitive, i.e., X_C.)
5. $$C = \frac{1}{2\pi f X_C} = \frac{0.159 \times 10^6}{400 \times 25} = 15.9\ \mu\text{F}$$

That is just about all there is to parallel circuits. There are numerous variations for the application of these principles. It would be impossible to illustrate each. Yet a few problems now will "clinch" the foregoing principles.

34-3 Resistance and pure inductance in parallel. When a resistor and an inductor are connected in parallel, the line current consists of two components:

1. The current through the resistor—in phase with the line voltage.
2. The current through the inductor—lagging the voltage by 90°.

Example 34-2

A resistor of 100 Ω is connected in parallel with an inductance of 0.5 H across a 60-Hz, 120-V line. Find:
(a) The line current.
(b) The circuit phase angle.
(c) The total circuit impedance.
(d) The power dissipated.
(e) The equivalent series circuit.

Solution

1. Draw the circuit, and show the current relations [see Fig. 34-2(a)].

2. $$X_L = 2\pi fL = 6.28 \times 60 \times 0.5 = 188 \ \Omega$$

3. $$I_L = \frac{E_L}{X_L} = \frac{120}{188} = 0.638 \ \text{A, lagging by 90° (pure } L)$$

4. $$I_R = \frac{E_R}{R} = \frac{120}{100} = 1.20 \ \text{A, in phase (pure } R)$$

5. $$I_T = \dot{I}_L + \dot{I}_R \ \text{(the dots indicate a } phasor \text{ addition)}$$
    $$= 1.20 - j0.638$$

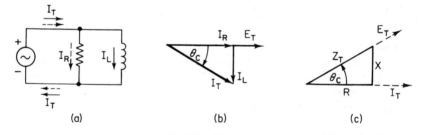

(a)   (b)   (c)

Figure 34-2 *R-L* parallel circuit.

Since the two currents are at right angles:
$$I_T = \sqrt{(I_R)^2 + (I_L)^2} = \sqrt{(1.20)^2 + (0.638)^2} = 1.36 \ \text{A}$$

6. Circuit phase angle

$$\theta_c = \text{arc tan} \frac{I_L}{I_R} = \text{arc tan} \frac{0.638}{1.20}$$

$$= 28.0°, \text{ current lagging (inductive circuit).}$$

or $I_T = 1.36\underline{/-28.0°}$ A.

7. Total circuit impedance

$$Z_T = \frac{\dot{E}_T}{\dot{I}_T} = \frac{120\underline{/0^\circ}}{1.36\underline{/-28.0^\circ}} = 88.2\underline{/28.0^\circ}\,\Omega$$

8A. Power dissipated equals

$$E_T I_T \cos\theta_c = 120 \times 1.36 \times 0.884 = 144\text{ W}$$

But we also know that a pure inductance does not dissipate power. In this circuit the power dissipated is in the resistive branch alone, or

8B. $$P = E_R I_R = 120 \times 1.20 = 144\text{ W}$$

9. The total circuit has an impedance of 88.2 Ω, and the current lags by 28.0°. *A series circuit would produce the same effect if it also had an imped-ance of 88.2 Ω—and the impedance triangle had a circuit phase angle of 28.0°.* Such an impedance is shown in Fig. 34-2(c). Therefore, the equiva-lent series circuit would be

$$R = Z_T \cos\theta = 88.2 \times 0.884 = 77.9\,\Omega$$

$$X_L = Z_T \sin\theta = 88.2 \times 0.468 = 41.4\,\Omega$$

$$L = \frac{X_L}{2\pi f} = \frac{41.4}{6.28 \times 60} = 0.110\text{ H}$$

Notice that the R and X_L values for the equivalent series circuit are smaller than the original R and X_L values of the parallel circuit. This must be so, since the series values are additive. This relation between series and parallel equivalent values applies to any R-L, or R-C circuit.

34-4   Resistance and capacitance in parallel.   Solution of R-C parallel circuits should present no difficulty at this time. The procedure is similar to the method discussed for R-L circuits above. The only change is that now the line current leads the line voltage by some angle less than 90°.

Example 34-3
A parallel circuit contains three branches as follows: a 350-pF capacitor, a 500-Ω resistor, and a 100-pF capacitor. The supply source is 40 V, 2000 kHz. Find:
(a) The line current.
(b) The circuit phase angle.
(c) The power dissipated.
(d) The total circuit impedance.
(e) The equivalent series circuit.

Solution

1. $$X_{C_1} = \frac{1}{2\pi f C_1} = \frac{0.159 \times 10^{12}}{2000 \times 10^3 \times 350} = 227\,\Omega$$

2. $$I_1 = \frac{E_1}{X_{C_1}} = \frac{40}{227} = 0.176\text{ A, leading by }90^\circ\text{ (pure }C\text{)}$$

3.
$$I_2 = \frac{E_2}{R} = \frac{40}{500} = 0.08 \text{ A, in phase (pure } R)$$

4.
$$X_{C_2} = \frac{1}{2\pi f C_2} = \frac{0.159 \times 10^{12}}{2000 \times 10^3 \times 100} = 795 \ \Omega$$

5.
$$I_3 = \frac{E_3}{X_{C_2}} = \frac{40}{795} = 0.0503 \text{ A, leading by } 90° \text{ (pure } C)$$

6.
$$I_T = I_1 + I_2 + I_3$$
$$= (j0.176) + (0.08) + (j0.0503) = 0.08 + j0.226$$

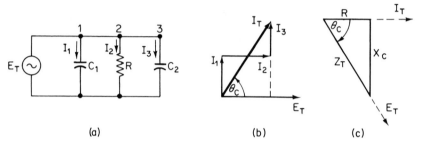

Figure 34-3   *R-C* parallel circuit.

From the phasor diagram, Fig. 34-3(b), we can see that the line current is the hypotenuse of a right triangle. One side of this triangle (in-phase component) is equal to I_2, the resistive branch current; and the other side (quadrature component) is equal to the sum of I_1 and I_3, the capacitive branch currents. Therefore,

$$I_T = \sqrt{(I_2)^2 + (I_1 + I_3)^2} = \sqrt{(0.08)^2 + (0.226)^2} = 0.240 \text{ A}$$

7. We can also see from the phasor diagram, that the tangent of the circuit phase angle is the ratio of the quadrature to the in-phase components of the line current, or

$$\theta_c = \text{arc tan} \frac{0.226}{0.08} = 70.5°$$

The line current leads the supply voltage by 70.5°, or

$$I_T = 0.240\underline{/70.5°}$$

8A.
$$P = E_T I_T \cos \theta_c = 40 \times 0.240 \times 0.334 = 3.20 \text{ W}$$

or

8B.
$$P = E_R \times I_R = 40 \times 0.08 = 3.20 \text{ W}$$

9.
$$Z_T = \frac{\dot{E}_T}{\dot{I}_T} = \frac{40\underline{/0°}}{0.240\underline{/70.5°}} = 167\underline{/-70.5°} \ \Omega$$

10. The equivalent series circuit must have an impedance of 167 Ω and a phase angle of 70.5°. Therefore,

$$R = Z_T \cos \theta_c = 167 \times 0.334 = 55.7 \, \Omega$$

$$X_C = Z_T \sin \theta_c = 167 \times 0.943 = 157.4 \, \Omega$$

$$C = \frac{1}{2\pi f \, X_C} = \frac{0.159 \times 10^{12}}{2000 \times 10^3 \times 157.4} = 505 \text{ pF}$$

[In this problem, since the individual branch currents were not required, we could have combined branch one and branch three into one branch of capacitance equal to $C_1 + C_2$ and solved the problem as a two-branch circuit (R and C_T). The method used above, although it is a little longer, does show the steps more clearly.]

34-5 Division of current: *R-C* parallel circuit.

There are times when we do not know the voltage across a parallel circuit, but we do know the current flowing into the circuit. The above technique cannot be used. Still, the principles developed can be applied in some other order. Figure 34-4 shows the phase relations in an *R-C* parallel circuit.

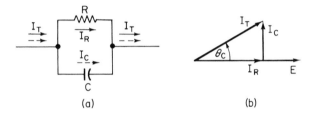

(a)   (b)

Figure 34-4   *R-C* parallel-circuit phase relations.

Let us assume that we know the value of the resistance, R; the value of reactance, X_C; and the total current, I_T. How can we proceed to find the current in each branch?

1. We know that the current divides into two components: I_R in phase with the voltage developed across the circuit, and I_C leading the voltage by 90°. But we do not know this voltage, so we cannot apply Ohm's law.

2. We do not know the impedance, Z_T, so we cannot find this voltage from $E_T = I_T Z_T$. Of course, we could first find Z_T from equation (34-2), $Z_T = \dot{Z}_1 \dot{Z}_2 / \dot{Z}_1 + \dot{Z}_2$, but there is a simpler way.

3. We know that $I_R = I_T \cos \theta_c$, but we do not know the circuit phase angle.

4. From Fig. 34-4(b) we see that the tangent of the circuit phase angle is equal to the ratio of I_C to I_R. There is the important clue. The ratio of currents must bear a definite relation to the ratio of oppositions, and we know both R and X_C. Since the units are in parallel, the voltage across each branch is the same, or

$$I_C X_C = I_R R \qquad \text{and} \qquad \frac{I_C}{I_R} = \frac{R}{X_C}$$

Since the tangent of the circuit phase angle is I_C/I_R, it is also equal to the *inverse opposition ratio* R/X_C or

$$\tan \theta_c \text{ (in an } R-X \text{ parallel circuit)} = \frac{R}{X} \qquad (34\text{-}3)$$

From this, we can get the circuit phase angle. Then, using the right-triangle relations, we solve for I_C and I_R.

Example 34-4

A current of 50 mA, 400 Hz flows through the parallel circuit shown in Fig. 34-4(a). The circuit values are 6000-Ω resistance and 0.05 μF capacitance. Find:
(a) The current through each component.
(b) The voltage across the circuit.

Solution

1. $$X_C = \frac{1}{2\pi f C} = \frac{0.159 \times 10^6}{400 \times 0.05} = 7950 \ \Omega$$

2. $$\tan \theta_c = \frac{R}{X_C} = \frac{6000}{7950}; \text{ and } \theta_c = 37.1°$$

3. $$I_R = I_T \times \cos \theta_c = 50 \times 0.798 = 39.9 \text{ mA}$$
    $$I_C = I_T \times \sin \theta_c = 50 \times 0.603 = 30.2 \text{ mA}$$

4. $$E_R = I_R \times R = 39.9 \times 10^{-3} \times 6000 = 240 \text{ V}$$
    $$E_C = I_C \times X_C = 30.2 \times 10^{-3} \times 7950 = 240 \text{ V (check)}$$

In the above example, what would be the effect of using a larger capacitor? The capacitive reactance would decrease; a greater portion of the current would flow through the capacitive branch; the total impedance would decrease; and the voltage developed across the circuit would drop.

Example 34-5

In Example 34-4, find the current division and voltage across the circuit if the capacitance is 5 μF.

Solution

1. $$X_C = \frac{1}{2\pi f C} = \frac{0.159 \times 10^6}{400 \times 5} = 79.5 \ \Omega$$

2. $$\tan \theta_c = \frac{R}{X_C} = \frac{6000}{79.5} = 75.5, \text{ and } \theta_c = 89.2°$$

3. $$I_R = I_T \cos \theta_c = 50 \times 0.013 = 0.66 \text{ mA}$$
    $$I_C = I_T \sin \theta_c = 50 \times 0.9999 = 50 \text{ mA}$$

4. $$E_C = I_C \times X_C = 50 \times 10^{-3} \times 79.5 = 3.98 \text{ V}$$

Notice that practically no current flows through the resistor, and that the voltage drop across the circuit was reduced from 240 V to less than 4 V!

34-6   Bypass capacitors.   In general, electronic circuits carry currents that are complex waves, containing a dc component and several sine-wave components of various frequencies. Very often these currents are passed through a resistor to produce a *dc voltage drop* across the resistor. If the ac components of the complex waves are also allowed to flow through the resistor, the voltage will vary depending on the magnitude and frequency of these ac components. To prevent these voltage variations, the resistor is shunted by a suitable capacitor. This capacitor is called a *bypass capacitor*. From Example 34-5 it is obvious that if a capacitor value is selected so that its reactance (compared to the resistance it is shunting) is low *for the lowest frequency component* of the complex wave, then these sine waves will pass through the capacitor, instead of the resistor. Due to the low reactance of the capacitor, the ac voltage across the parallel circuit will be negligible. Common practice dictates that the capacitor reactance should be less than one-tenth of the resistance value at the lowest frequency. Actually, the greater the capacitance, the better the bypassing action. A typical example of such a circuit is the bypass capacitor shunted across a resistor in series with the emitter (or source) leg of a transistor, for bias or bias-stability.

Another illustration of the use of a bypass capacitor is in the separation of the audio waves from a modulated carrier. From Chapter 24 we know that the modulated carrier contains the carrier frequency and audio intelligence. When this modulated carrier is put through a *detector*, sum and difference frequencies are created. The difference frequency is the desired audio signal. This complex wave of current can be passed through a resistor shunted by a suitable capacitor. This time the size of capacitor is selected so that its reactance is:

1. *High at the Audio Frequencies.* In this way, the audio currents pass through the resistor creating audio voltage drops. Since X_C is high at these frequencies, it does not bypass these frequencies, and the audio voltages developed across the resistor are high.
2. *Low at the RF Frequencies* (carrier, sidebands, and sum frequencies). These frequencies are bypassed through the capacitor and negligible RF voltages are developed across the *R-C* parallel circuit.

34-7   Resistance, pure inductance, and capacitance in parallel.
In the previous cases we have used only two types of circuit elements at a time, *R-L* or *R-C*. Since none of these circuits contained inductance in one branch and capacitance in another, we noticed that the total impedance was less than the smallest branch impedance. Now we will find that, due to the combined effect of inductive and capacitive branches, the total impedance of a circuit may become quite high. The solution of such circuits does not differ from the methods already discussed. For comparison purposes the illustrative problem used will employ the same values for *R*, *L*, and supply voltage as Example 34-2 (page 538).

Example 34-6

An inductance of 0.5 H, a resistor of 100 Ω, and a capacitor of 20 μF are connected in parallel across a 60-Hz, 120-V line. Find:

(a) The line current.
(b) The circuit phase angle.
(c) The power dissipated.
(d) The total circuit impedance.
(e) The equivalent series circuit.

Solution

1. $X_L = 2\pi f L = 2 \times \pi \times 60 \times 0.5 = 188\ \Omega$

2. $I_L = \dfrac{E_L}{X_L} = \dfrac{120}{188} = 0.638$ A, lagging by 90° (pure L) $= -j0.638$

3. $I_R = \dfrac{E_R}{R} = \dfrac{120}{100} = 1.20$ A, in phase (pure R)

4. $X_C = \dfrac{1}{2\pi f C} = \dfrac{0.159 \times 10^6}{60 \times 20} = 132.5\ \Omega$

5. $I_C = \dfrac{E}{X_C} = \dfrac{120}{132.5} = 0.906$ A, leading by 90° (pure C) $= j0.906$

6. $$I_T = \dot{I}_L + \dot{I}_R + \dot{I}_C$$

    (a) Since I_L and I_C are exactly 180° apart, the net reactive current, I_X, is the numerical difference between these two currents, or

    $$I_X = I_C - I_L = 0.906 - 0.637 = 0.269\ \text{A}$$

    leading the voltage by 90° because I_C is larger; or
    (b) Using vector algebra, we get

    $$I_X = \dot{I}_C + \dot{I}_L = (j0.906) + (-j0.637) = j0.269\ \text{A}$$

    Then

    $$I_T = \dot{I}_R + \dot{I}_X = 1.20 + j0.269\ \text{A}$$

7. Converting to polar form [or from the phasor diagram Fig. 34-5(b)] yields

    (a) I_T (magnitude) $= \sqrt{(1.20)^2 + (0.269)^2} = 1.23$ A

    (b) Circuit phase angle $=$ arc tan $\dfrac{I_X}{I_R} =$ arc tan $\dfrac{0.269}{1.20} = 12.6°$

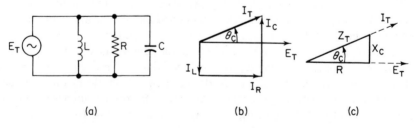

(a)   (b)   (c)

Figure 34-5 *L-C-R* parallel circuit.

$$I_T = 1.23\underline{/12.6°}\ A$$

The current leads the voltage by 12.6°.

8. Power dissipated equals

$$E_T I_T \cos \theta_c = E I_R = 120 \times 1.20 = 144\ W$$

9.
$$Z_T = \frac{E_T}{I_T} = \frac{120}{1.23} = 97.6\ \Omega$$

10. Since the line current leads the line voltage by an angle less than 90°, the circuit acts as an *R-C* series circuit. From the impedance diagram [Fig. 34-5(c)], it is obvious that the equivalent series circuit would be

(a) $Z_T = 97.6\ \Omega$

(b) $R = Z_T \cos \theta_c = 97.6 \times 0.976 = 95.2\ \Omega$

(c) $X_C = Z_T \sin \theta_c = 97.6 \times 0.218 = 21.3\ \Omega$

(d) $C = \dfrac{1}{2\pi f X_C} = \dfrac{0.159 \times 10^6}{60 \times 21.3} = 12.4\ \mu F$

A comparison of these results, and the results from Example 34-2, reveals several interesting points:

1. The power dissipated (144 W) is the same for each circuit. This should be obvious. The added capacitor dissipates no power!
2. By adding the capacitor in parallel, the line current *dropped* from 1.36 A to 1.23 A. Also, the current shifted from a lag of 27.9° to a lead of 12.6°! The capacitor not only completely counteracted the effect of the inductive branch, but made the circuit capacitive.
3. The circuit impedance is greater with the capacitor added (97.6 Ω) than it was for the previous *R-L* circuit (88.2 Ω). In this problem it is still less than the smallest branch impedance. Yet the addition of an extra branch *did increase* the impedance. Under certain conditions the impedance may actually increase to a value *much higher* than any branch impedance. (See parallel resonance, Section 34-12).

34-8 Parallel circuits—impedance branches.   Trigonometric solutions of the previously discussed parallel circuits were relatively simple because all branches were pure elements. Each branch current was either at 0° or at 90° with respect to the line voltage. Addition of these currents was readily done either by numerical addition (when the currents were in phase), or by right-triangle methods (when the currents were at 90°). However, there are circuit applications where one or more branches contain resistance as well as reactance. Such a circuit is shown in Fig. 34-6.* The phase angle of these branch currents would not be at zero or 90°. We cannot add these currents numerically or algebraically; nor can we use the Pythagorean theorem *directly*. To use a

*These are actually series–parallel circuits.

trigonometric solution, it would be necessary to:

1. Find the magnitude and phase angle of each current.
2. Resolve each current into its in-phase and quadrature components.
3. Add all the in-phase components ($\sum I_H$); and add all the quadrature components ($\sum I_V$) *algebraically*.
4. Then, to find the line current,

$$I_T = \sqrt{(\sum I_H)^2 + (\sum I_V)^2} \tag{34-4}$$

5. The phase angle of this current

$$\theta_c = \arctan \frac{\sum I_V}{\sum I_H} \tag{34-5}$$

The procedure is laborious, and care must be used to label all values carefully to prevent accidental errors.

It is in such circuits that vector-algebra solutions are preferable. Not that the calculations are any easier, but rather that the "bookkeeping" is automatic, and the chance for accidental errors or "boners" is reduced. When using vector algebra, two points should be recalled: (1) We cannot add phasors in polar form. We must first convert to rectangular form. (2) Multiplication and division of phasor quantities is much easier in polar form. So it is often necessary (or desirable) to convert back and forth between polar and rectangular forms. To illustrate, let us try an example of this type.

Example 34-7
Find the magnitude and phase angle of the line current for the circuit shown in Fig. 34-6. Also find the voltage across the capacitor.

Solution

1. $X_C = \dfrac{1}{2\pi fC} = \dfrac{0.159 \times 10^{12}}{2 \times 10^6 \times 150} = 530 \ \Omega = -j530 \ \Omega$

            $Z_1 = 200 - j530 = 566\underline{/-69.3°}$

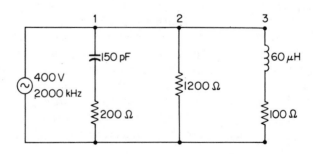

Figure 34-6 Circuit for Example 34-7.

2. Taking the supply voltage as the reference phasor, we have

$$E = 400 + j0 = 400\underline{/0°}$$

3. $$I_1 = \frac{E}{Z_1} = \frac{400\underline{/0°}}{566\underline{/-69.3°}} = 0.706\underline{/69.3°} \text{ A}$$

4. $$Z_2 = 1200 + j0 = 1200\underline{/0°}$$

$$I_2 = \frac{E}{Z_2} = \frac{400\underline{/0°}}{1200\underline{/0°}} = 0.333\underline{/0°}$$

5. $$X_L = 2\pi f L = 2 \times \pi \times 2 \times 10^6 \times 60 \times 10^{-6} = 754\ \Omega = +j754\ \Omega$$

$$Z_3 = 100 + j754 = 761\underline{/82.4°}\ \Omega$$

$$I_3 = \frac{E}{Z_3} = \frac{400\underline{/0°}}{761\underline{/82.4°}} = 0.526\underline{/-82.4°} \text{ A}$$

6. $I_T = \dot{I}_1 + \dot{I}_2 + \dot{I}_3$; but before we add these currents, we must convert them to rectangular form.

$$
\begin{aligned}
I_1 &= 0.706\underline{/69.3°} &&= 0.249 + j0.660 \\
I_2 &= 0.333\underline{/0°} &&= 0.333 + j0 \\
I_3 &= 0.526\underline{/-82.4°} &&= 0.0696 - j0.521 \\
&& I_T &= 0.652 + j0.139 \\
&& &= 0.667\underline{/12.0°} \text{ A}
\end{aligned}
$$

The line current is 0.667 A, and leads the line voltage by 12.0°.

7. The voltage across the capacitor is

$$E_C = I_1 X_C = (0.706\underline{/69.3°})(530\underline{/-90°})$$

$$= 374\underline{/-20.7°} \text{ V}$$

The capacitor voltage is 374 V, and lags the line voltage by 20.7°.

34-9 Parallel circuit impedance.   In the previous examples in this chapter, we did not solve for the impedance of the total circuit directly. Instead, using the voltage across the circuit (given or assumed value), and the impedance of each branch, we:

1. Solved for each branch current.
2. Added the branch currents vectorially.
3. Solved for Z_T by Ohm's law ($Z_T = E_T/I_T$).

We also saw that

$$\frac{1}{Z_T} = \frac{1}{\dot{Z}_1} + \frac{1}{\dot{Z}_2} + \frac{1}{\dot{Z}_3} + \cdots \qquad \text{(34-1)}$$

This equation cannot be solved by trigonometric methods. Now, using vector algebra, we can do it. The procedure involves the use of polar and rectangular forms.

1. Express each branch impedance in polar form.
2. Solve for the reciprocals, and convert to rectangular form.
3. Add these values while in rectangular form, and convert the sum back to polar form.
4. Z_T is the reciprocal of the value found in step 3.

The method, though simple, is laborious. Fortunately, most electronic circuit problems contain only two branches, or can be simplified into only two branches. We can therefore use the simplified formula for two branches:

$$Z \text{ parallel} = \frac{\dot{Z}_1 \dot{Z}_2}{\dot{Z}_1 + \dot{Z}_2} \tag{34-2}$$

Let us use this last equation to find the impedance of the parallel circuit from Example 34-2.

Example 34-8

A resistor of 100 Ω is connected in parallel with an inductor of 0.5 H across a 60-Hz 120-V line. Find the total circuit impedance.

Solution

1. $X_L = 2\pi f L = 2 \times \pi \times 60 \times 0.5 = 188 \ \Omega = +j188 \ \Omega$

2. $Z_T = \dfrac{\dot{Z}_1 \dot{Z}_2}{\dot{Z}_1 + \dot{Z}_2} = \dfrac{100 \times j188}{100 + j188} = \dfrac{18\,800 \underline{/90^\circ}}{213 \underline{/62^\circ}} = 88.3 \underline{/28^\circ} \ \Omega$

Example 34-9

Find the impedance of the three-branch parallel circuit in Example 34-6, directly from the component values. ($X_L = 188 \ \Omega$, $R = 100 \ \Omega$, $X_C = 132.5 \ \Omega$.)

Solution

$$\frac{1}{Z_T} = \frac{1}{\dot{Z}_1} + \frac{1}{\dot{Z}_2} + \frac{1}{\dot{Z}_3}$$

$$= \frac{1}{188 \underline{/90^\circ}} + \frac{1}{100 \underline{/0^\circ}} + \frac{1}{132.5 \underline{/-90^\circ}}$$

$$= (5.32 \underline{/-90^\circ} + 10 \underline{/0^\circ} + 7.55 \underline{/90^\circ}) \times 10^{-3}$$

$$= (-j5.32 + 10 + j7.55) \times 10^{-3}$$

$$= 0.010 + j0.002\,23 = 0.010\,24$$

$$Z_T = \frac{1}{0.010\,24} = 97.6 \ \Omega$$

Now let us try a problem involving the impedance of a parallel circuit with complex branches.

Example 34-10

Find the total circuit impedance, and the line current in the circuit of Fig. 34-7.

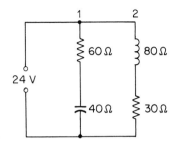

Figure 34-7 Circuit for Example 34-10.

Solution

1. $Z_1 = 60 - j40 = 72.1\underline{/-33.7°}$

2. $Z_2 = 30 + j80 = 85.4\underline{/69.4°}$

3. $Z_1 + Z_2 = 90 + j40 = 98.5\underline{/24.0°}$

4. $Z_T = \dfrac{Z_1 Z_2}{\dot{Z}_1 + \dot{Z}_2} = \dfrac{6157\underline{/35.7°}}{98.5\underline{/24.0°}} = 62.5\underline{/11.7°}\ \Omega$

5. $I_T = \dfrac{Z_T}{E_T} = \dfrac{80\underline{/0°}}{62.5\underline{/11.7°}} = 1.28\underline{/-11.7°}\ \text{A}$

The line current is 1.28 A, lagging the line voltage by 11.7°.

34-10 Series–parallel circuit impedance. Series–parallel circuits may be quite difficult to solve by trigonometric methods. For example, how
would you proceed to solve for the line current (magnitude and phase angle), in
the circuit shown in Fig. 34-8? Since the only voltage given is the line voltage,
we must find the total impedance. But we could not find Z_T by trigonometric
methods, unless we knew the current! This calls for a vector-algebra solution.
The steps are:

1. Find the equivalent impedance of the parallel branches.
2. Convert this value to rectangular form.
3. Add this value (step 2) to the series impedances (R_1L_1). This is the total
   circuit impedance (Z_T).
4. Convert Z_T to polar form and solve for I_T.

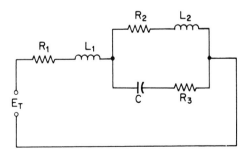

Figure 34-8 Series–parallel circuit.

Problems of this type are given in the Problems section (see Problems 24 and 25).

34-11 Conductance, susceptance, and admittance.
Up to this point, circuit solutions have been based on the property of a circuit element to *oppose* the flow of current. This gave rise to such terms as resistance, reactance, and impedance. The technique proved quite satisfactory with series circuits, and with simple parallel circuits, but as you have just seen, it can become involved when dealing with series–parallel circuits. In some circuit analyses, it may be of advantage to consider, instead, the ability of a component, or path, to *allow* current to flow. Use of vector algebra makes such an approach possible.

This idea is not new. In our studies of dc circuits, we learned that *conductance*, G (measured in siemens), was the "opposite" of resistance, in that it allowed current to flow through a circuit. We also saw that if resistors were connected in parallel, the total conductance would be the sum of the individual branch conductances, or

$$G_T = G_1 + G_2 + G_3 + \cdots \tag{34-6}$$

This concept also applies to ac circuits. However, ac circuits also have reactances that oppose the flow of current, and there must be an "opposite" to a reactance—or the ability of such components to *allow* current to flow. Because of phase differences, we cannot call this property conductance, and so we give it a new name—*susceptance*, B (also measured in siemens). As with conductances, the total susceptance in a parallel circuit is the sum of the individual branch susceptances, or

$$B_T = B_1 + B_2 + B_3 + \cdots \tag{34-7}$$

One more point—if a parallel branch contains resistance and reactance, the total branch opposition is called impedance. (where $Z = R \pm jX$). Similarly, (or "oppositely") the total "allowance" to current flow in a branch containing conductance and susceptance is called *admittance*, Y (still measured in siemens), and is equal to

$$Y = \dot{G} \pm j\dot{B} \tag{34-8}$$

This is a phasor addition, since the conductance and susceptance effects are at right angles. Also, notice the "$\pm j$" in the above equation. The "$+j$" is used for a capacitive susceptance (B_C), whereas a "$-j$" denotes an inductive susceptance (B_L).* Once we have the admittance of each branch, the total admittance for two or more branches is the sum of the admittance of each branch, or

$$Y_T = Y_1 + Y_2 + Y_3 + \cdots \tag{34-9}$$

Finally, since current in a circuit is $I = E/Z$ (using impedances), the equivalent equation using admittance values is

$$I = EY \tag{34-10}$$

---

*Notice that this is the opposite of the $+jX_L$ and $-jX_C$ we learned earlier. This reversal of signs should be obvious, since susceptances are the opposites (inverse) of reactances.

This equation can be applied to a branch (if Y is the admittance of that branch), or to the entire circuit (if Y is the total circuit admittance).

Now comes the big question: If we know the resistance and reactance of a branch, how do we calculate its conductance, susceptance, and admittance? We will explain this with the aid of a diagram. Figure 34-9(a) shows a circuit containing R and X_C in series. (This could be one branch of a series-parallel circuit.) The current in this circuit, (or branch), would be

$$I = \frac{E}{Z} = \frac{E}{R - jX_C} = E\left(\frac{1}{R - jX_C}\right)$$

Rationalizing yields

$$I = E\left(\frac{1}{R - jX_C} \cdot \frac{R + jX_C}{R + jX_C}\right) = E\left(\frac{R + jX_C}{R^2 + X_C^2}\right)$$

$$= E\left(\frac{R}{R^2 + X_C^2}\right) + E\left(j\frac{X_C}{R^2 + X_C^2}\right)$$

This is of the form

$$I = A + jB$$

where $E\left(\dfrac{R}{R^2 + X_C^2}\right)$ is the in-phase component

and $E\left(j\dfrac{X}{R^2 + X_C^2}\right)$ is the quadrature component.

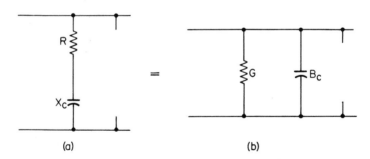

Figure 34-9   Equivalent parallel circuit.

This same current could also result from a parallel circuit with two pure-element branches (G and B_C*) as shown in Fig. 34-9(b). The current in this resistive branch is EG; the current in the capacitive branch is EB_C; and the total current is $E(G + jB_C)$. In other words, the given $R + X$ branch circuit can be replaced by an *equivalent parallel* circuit, wherein the conductance and susceptance values are

---

*The subscript C in B_C denotes a capacitive susceptance. If the circuit element were an inductor, the symbol for susceptance would be B_L.

$$G = \frac{R}{R^2 + X^2} = \frac{R}{Z^2} \qquad \text{(34-11A)}$$

$$B = \frac{X}{R^2 + X^2} = \frac{X}{Z^2} \qquad \text{(34-11B)}$$

A special case of the above general equations is obtained when a branch circuit is a pure R, pure L, or a pure C. Then, the above formulas reduce to

For a pure R branch: $\quad G = \frac{R}{R^2 + 0} = \frac{1}{R} \qquad$ (34-12A)

For a pure X branch: $\quad B = \frac{X}{0 + X^2} = \frac{1}{X} \qquad$ (34-12B)

Let us apply these ideas to some problems. We will start with the simpler case of pure-branch elements.

Example 34-11

A three-branch parallel circuit has the following branch values: $R = 200\ \Omega$; $X_L = 300\ \Omega$; and $X_C = 500\ \Omega$. Find:
(a) The conductance and susceptance of each branch.
(b) The admittance of the total circuit (in polar form).
(c) Show these relations by phasor diagram.

Solution

(a) Conductance and susceptances

Branch 1. $\qquad G_1 = \frac{1}{R} = \frac{1}{200} = 0.005\ \text{S}$

Branch 2. $\qquad B_L = \frac{1}{jX_L} = \frac{1}{300} = -j\,0.00333\ \text{S}$

Branch 3. $\qquad B_C = \frac{1}{-jX_C} = \frac{1}{500} = j0.0020\ \text{S}$

(b) Admittance

1. $\qquad B_O = \dot{B}_C + \dot{B}_L = j0.0020 - j0.003\ 33$
$$= -j0.001\ 33\ \text{S}$$

2. $\qquad Y = \dot{G} + j\dot{B} = 0.005 - j0.001\ 33\ \text{S}$

and converting to polar form, we have

$$Y = 0.005\ 17\underline{/-14.9°}\ \text{S}$$

(c) See Fig. 34-10.

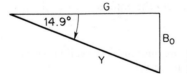

Figure 34-10 Admittance diagram for Example 34-11.

Since each of the quantities, G and B are reciprocals of R and X, it follows that Y, the phasor sum of G and B, must also be the reciprocal of Z (the phasor sum of R and X), or

$$Y = \frac{1}{\dot{Z}} \qquad (34\text{-}13)$$

Example 34-12

Find the impedance of the circuit in Example 34-11.

Solution

$$Z = \frac{1}{Y} = \frac{1}{0.005\ 17/\!-\!14.9°}$$

$$= 193/\underline{14.9°}\ \Omega$$

Since admittance is the reciprocal of impedance, the current in a circuit can be found by *multiplying* voltage by admittance.

$$I = \dot{E}\dot{Y} = E(G + jB) \qquad (34\text{-}14)$$

Example 34-13

In Example 34-11, if the line voltage is 60 V, find the line current (magnitude and phase), from the admittance value.

Solution

Using voltage as reference ($E = 60/\underline{0°}$), we get

$$I = EY = (60/\underline{0°})(0.005\ 17/\!-\!14.9°) = 0.310/\!-\!14.9°\ \text{A}$$

Now, let us try a problem with a complex branch, so we can apply equation (34-7).

Example 34-14

In Fig. 34-9(a), $R = 5000\ \Omega$, and $X_C = 3000\ \Omega$. Find the conductance, susceptance, and admittance for the equivalent parallel circuit.

Solution

1. $$G = \frac{R}{R^2 + X^2} = \frac{5000}{25 \times 10^6 + 9 \times 10^6} = 147\ \mu\text{S}$$

2. $$B_C = \frac{X_C}{R^2 + X^2} = \frac{3000}{34 \times 10^6} = 88.2\ \mu\text{S}$$

3. $$Y = G + jB_C = 147 + j88.2\ \mu\text{S}$$

To compare the two systems (impedance versus admittance methods), let us repeat Example 34-7 using the admittance technique.

Example 34-15

Find the magnitude and phase angle of the line current, for the circuit shown in Fig. 34-6.

Solution

1. In branch 1:

   (a) $X_C = \dfrac{1}{2\pi fC} = \dfrac{0.159 \times 10^{12}}{2 \times 10^6 \times 150} = 530\ \Omega$

   (b) $G_1 = \dfrac{R_1}{R_1^2 + X_1^2} = \dfrac{200}{(200)^2 + (530)^2} = 6.23 \times 10^{-4}\ S$

   (c) $B_C = \dfrac{X_C}{R_1^2 + X_1^2} = \dfrac{530}{(200)^2 + (530)^2} = 16.5 \times 10^{-4}\ S$

2. In branch 2:

$$G_2 = \frac{1}{R_2}\ (\text{pure branch}) = \frac{1}{1200} = 8.33 \times 10^{-4}\ S$$

3. In branch 3:

   (a) $X_L = 2\pi fL = 2 \times \pi \times 2 \times 10^6 \times 60 \times 10^{-6} = 754\ \Omega$

   (b) $G_3 = \dfrac{R_3}{R_3^2 + X_3^2} = \dfrac{100}{(100)^2 + (754)^2} = 1.73 \times 10^{-4}\ S$

   (c) $B_L = \dfrac{X_L}{R_3^2 + X_3^2} = \dfrac{754}{(100)^2 + (754)^2} = 13.0 \times 10^{-4}\ S$

4. Total conductance $= G_1 + G_2 + G_3$.

$$G_T = (6.23 + 8.33 + 1.73) \times 10^{-4}$$
$$= 16.3 \times 10^{-4}\ S$$

5. Total susceptance

$$B_T = B_C - B_L = (16.5 - 13.0) \times 10^{-4}$$
$$= 3.5 \times 10^{-4}\ S$$

6. $\qquad\qquad Y_T = G + jB = (16.3 + j3.50) \times 10^{-4}\ S$

   and converting to polar yields

$$Y_T = 16.7 \times 10^{-4}\underline{/12.1^\circ}\ S$$

7. Using line voltage as reference, we get

   (a) $\qquad\qquad\qquad\qquad E_T = 400\underline{/0^\circ}$

   (b) $\qquad\qquad\qquad I_T = E_T Y_T = 0.668\underline{/12.1^\circ}\ A$

34-12 Parallel resonance (pure branches). In Example 34-6 we added a capacitive branch to an *R-L* parallel circuit. The size of the capacitor added created a capacitive branch current that exceeded the inductive branch current and caused the line current to lead the line voltage. Suppose that we had used a smaller capacitor—for example, 14.1 μF. Then

$$X_C = \frac{1}{2\pi fC} = \frac{0.159 \times 10^6}{60 \times 14.1} = 188\ \Omega$$

This would make the capacitive branch reactance exactly equal to the inductive branch reactance. Under such conditions:

1. $X_C = X_L$

2. *I_T Is a Minimum and the Circuit Impedance Is a Maximum.* The reason for this can be readily explained. Since $X_L = X_C$, the reactive branch currents I_L and I_C are also equal. But we also know that these currents are 180 degrees apart. Their phasor sum is zero, and the line current would be equal to the resistive branch current alone. This is the minimum value of current, and consequently the maximum value of parallel circuit impedance. It should be noted that if the circuit contained only the inductive branch and the capacitive branch—and both branches were pure reactances—the line current would be zero, and the circuit impedance infinite!

3. *The Circuit Acts as a Pure Resistive Circuit* (line current in phase with line voltage). This should be obvious, since the phasor sum of the reactive branch currents is zero. This condition is known as *parallel resonance*. (In some texts, it is called *antiresonance*, to distinguish from the series-resonant condition.)

In the above illustration, parallel resonance was obtained by changing the capacitance value. As in the case of series circuits, resonance could also have been produced by varying the inductance value, or the applied frequency. In any parallel circuit, regardless of the values of L and C, there is always some frequency at which the circuit will be resonant. *If the reactive branches contain pure reactances*, the three conditions above will occur at the same frequency. This frequency is

$$f_0 = \frac{1}{2\pi\sqrt{LC}} \tag{34-15}$$

As the frequency is increased above the resonant value, the inductive branch reactance will increase in proportion, and the inductive branch current will decrease. Meanwhile the capacitive branch reactance will decrease, increasing the current in this branch. The two branch currents are no longer equal—the total current has increased, and is now capacitive. Obviously, the circuit acts as a capacitive circuit with the line current leading the line voltage. Since the line current has increased, the total impedance of the circuit has decreased. Conversely, if the frequency is reduced below the resonant value, the capacitive branch current decreases, the inductive branch current increases, and the line current increases. The circuit now acts as an inductive circuit with the line current lagging the line voltage. This again means that the total impedance has decreased. In other words, a parallel circuit offers a maximum impedance to currents at the resonant frequency. To all other frequency currents, the circuit offers less and less impedance, depending on how far these frequencies are removed (above or below) from the resonant frequency.

It is this characteristic of a parallel *L-C* circuit that provides us with a means of selecting one frequency from all others—as in the tuning circuit of a radio receiver. Another application is found in the "tank circuit" of a Class C amplifier in a transmitter. In this case the plate current is a complex wave containing a dc component, a fundamental frequency component, and many harmonic frequency components. By proper selection of *L-C* values, the parallel-tuned tank circuit is made resonant to the fundamental frequency component. The output will then be a pure sine wave at the fundamental frequency. We can also tune the tank circuit to the second or third harmonic frequency, and operate this stage as a frequency doubler or tripler. Again the output will be a pure sine wave—but now at the higher frequency.

34-13   Parallel resonance with impedance branches.   Although most cases of parallel resonance in electronic work are with circuits containing relatively pure reactive branches, we are sometimes faced with circuits where either the inductive branch or capacitive branch also contains some resistance. This is particularly true in the inductive branch, if the coil used has a low Q. A typical illustration is the transmitter *tank circuit* (tuning circuit), where a low Q may be necessary. Instructions for tuning such a circuit to resonance often advise the operator to tune for minimum current and then tune just beyond that point. *The line current will not be a minimum.* But if one condition of resonance is minimum line current, why do we go beyond this point? The answer is that when a parallel circuit contains an impedance branch (instead of pure reactance), the three "resonant" conditions described above *occur at three separate frequencies:*

1. The frequency at which $X_L = X_C$.
2. The frequency at which the line current is a minimum.
3. The frequency at which the parallel circuit acts as a pure resistive load.

Condition 1 is the one that applies to series-resonant circuits, regardless of the resistance of the coil. In parallel circuits, condition 3 is the most widely accepted resonant condition. However, when the Q of the coil is fairly high, the resistance in the inductive branch is negligible compared with the reactance, and all three conditions will occur at the same frequency. Since this is generally the case, resonance for the parallel circuit is considered as the frequency at which $X_L = X_C$. Still, this point must not be forgotten—*that if appreciable resistance is introduced into either reactive branch of a parallel resonant circuit, the circuit will be thrown out of resonance and must be retuned.* This point will be substantiated when studying transmitter tuning and loading techniques.

For an exaggerated case, $Q = 1$, these three conditions are shown in Fig. 34-11.

1. $X_L = X_C$. At some frequency, X_L will equal X_C. Since this condition is similar to the one discussed for series circuits, the frequency at which this condition occurs is again $f_1 = 1/(2\pi\sqrt{LC})$. But due to the resistance in the inductive branch, the impedance Z_L will be greater than X_C. Therefore, I_L will be smaller than I_C, and also it will be lagging by 45° ($Q = 1$) instead of 90°. Naturally, the quadrature components are not balanced, and the line current will lead the applied voltage. The circuit does not act as a resistive circuit, nor is the line current a minimum. Yet it does satisfy the condition $X_L = X_C$.

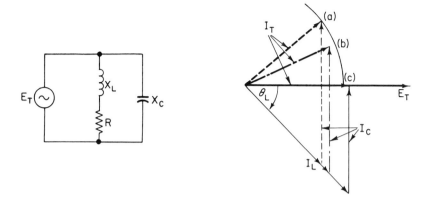

Figure 34-11 Separation of resonant conditions in a low-Q circuit due to coil resistance. (a) Frequency at which $X_L = X_C$. (b) Frequency for minimum line current. (c) Frequency for unity power factor.

2. *Minimum Line Current.* If the applied frequency is decreased, X_L and Z_L will decrease, and the current I_L will increase. The phase angle of this current was assumed to remain constant. (For a small change in frequency, and also considering the reduction of effective resistance of the coil, this assumption is permissible.) At the same time X_C will increase, thereby decreasing the capacitor branch current I_C. Both effects will cause the circuit phase angle to decrease toward zero. Meanwhile, the resultant line current will also decrease—up to a certain point. The minimum line current occurs at a frequency corresponding to

$$f_2 = \frac{1}{2\pi\sqrt{LC}}\left(\sqrt{1 - \frac{1}{4Q^2}}\right) \tag{34-16}$$

If the frequency is reduced further, the phase angle will continue to decrease, but the line current will once again start to increase. Condition (b) in Fig. 34-11 shows the situation for minimum line current.

3. *Circuit Acts as Pure Resistive Circuit.* As the frequency is reduced further, I_L will continue to increase and I_C will continue to decrease. At some point the *quadrature* component of I_L will equal the magnitude of I_C. Since the reactive currents are balanced out, the line current is in phase with the line voltage. But now, due to the increase in I_L, the line current is no longer at its minimum value. The frequency at which this unity power factor occurs is given by

$$f_3 = \frac{1}{2\pi\sqrt{LC}}\left(\sqrt{\frac{Q^2}{Q^2 + 1}}\right) \tag{34-17A}$$

or

$$f_3 = \frac{1}{2\pi\sqrt{LC}}\left(\sqrt{1 - \frac{CR^2}{L}}\right) \tag{34-17B}$$

557

These last two equations are identical—one is in terms of the coil Q, the other in terms of inductance and resistance, where R, L, and Q are all at this frequency (f_3).

If the resistive component of the inductive branch were decreased (higher Q), the angle of lag of I_L would increase. Examine Fig. 34-11 again. It is obvious that as I_L approaches 90°, the three I_C lines will fall into coincidence. Certainly for a coil Q of 10 ($\theta_L = 84.3°$) it would be impossible to distinguish the three separate conditions. With a Q of 10, also notice that the parenthetical expression in each of the above formulas becomes unity, and the equation for resonant frequency reduces to the familiar

$$f_0 = \frac{1}{2\pi\sqrt{LC}}$$

In most electronic circuits, the Q of these circuits is usually high enough so that all three conditions will occur at the same frequency. For this reason the resonant frequency is generally calculated from the simplified formula $f_0 = 1/(2\pi\sqrt{LC})$. However, it must be kept in mind that if sufficient resistance is introduced into a tank circuit (parallel-tuned circuit), and the Q drops to a low value, the resonant frequency may decrease. Remember this basic principle when you study oscillator stability or transmitter circuits. If these circuits are poorly designed and the resistance introduced into the tank circuit varies, the resonant frequency will change.

Resonant circuits, both series and parallel, will be treated in greater detail in a subsequent volume dealing with practical circuit applications.*

REVIEW QUESTIONS

1. In an ac circuit having several parallel branches:
   (a) How does the voltage across a pure resistive branch compare with the voltage across a pure capacitive branch, in magnitude? In phase? Explain.
   (b) What is the relation between the "total" voltage and each of the branch voltages?
   (c) What determines the current value in each branch?
   (d) What determines the phase relation between the current and voltage for each branch?
   (e) In making a phasor diagram for a parallel circuit, what quantity would be most suitable as the reference phasor? Why?
   (f) Can the total line current be found by numerical addition of all the branch currents? Explain.
2. In calculating the "total" impedance of an ac parallel circuit:
   (a) Can we use the *numerical* values of each branch impedance in the equation:
   $$\frac{1}{Z_T} = \frac{1}{Z_1} + \frac{1}{Z_2} + \cdots?$$ Explain.

*J. J. DeFrance, *Communications Electronics Circuits*, Holt, Rinehart and Winston, New York, 1972.

(b) Must the total impedance be less than the smallest branch impedance? Explain.

(c) When does the above relation apply?

3. (a) Is the line current in a parallel circuit always greater than the largest branch current? Explain.

(b) When does this relation apply?

4. Under what conditions can a series circuit be called equivalent to a parallel circuit?

5. (a) What combination of circuit elements should an equivalent circuit have, if the original parallel circuit contained an R branch, and an L branch?

(b) Repeat part (a) for parallel R and C branches.

(c) How will the values of the equivalent circuit components in each of parts (a) and (b) compare with the original parallel circuit values? Explain.

6. With reference to Example 34-2:

(a) To get the line current, why is it necessary to find each branch current first?

(b) In step 6, why don't we use arc tan (X/R) to find θ_c?

(c) How is this relation, arc tan (I_L/I_R), obtained?

(d) Does the equation $P = EI \cos \theta$ apply to parallel circuits?

(e) What E and I values are these?

(f) Does $P = EI_R$ give true power?

(g) How is this equation obtained?

7. With reference to Fig. 34-2(c):

(a) What is this type of diagram called?

(b) Does this type of diagram apply to a series, or a parallel circuit? Explain.

(c) Which, if any, of the values shown apply to the original parallel circuit?

(d) Which, if any, apply to the equivalent series circuit?

(e) What relation exists between the known values and the X value?

(f) Between the known values, and the value of R?

8. With reference to Example 34-3:

(a) What is the voltage across each branch? Why?

(b) In step 6 of the solution, what do the dots over the I signify?

(c) In the phasor diagram, why are I_1 and I_3 drawn directly upward?

(d) Why is I_2 drawn horizontally?

(e) In the equation for I_T (just above step 7) why are I_2 and I_3 added together numerically before squaring?

(f) In solving for θ_c (step 7), what do the numbers 0.08 and 0.226 represent? Is this a correct equation?

9. In the phasor diagram, Fig. 34-3(c):

(a) Which of the quantities, I_T, E_T, Z_T, and θ_c, belong to the original parallel circuit?

(b) Which of these quantities belong to the equivalent series circuit.

(c) At this point in the solution, which of these are known values?

(d) What do the X_C and R quantities represent?

(e) What function of Z_T corresponds to R? To X_C?

10. With reference to Fig. 34-4:

(a) Which of the quantities shown are given values?

(b) Will Z_T in this circuit be equal to $\sqrt{R^2 + X_C^2}$? Explain.

(c) Can we find Z_T directly?

     (d) Which branch will have the higher current?

     (e) What is the ratio between these currents?

     (f) From diagram (b), evaluate the tangent of angle θ_c.

     (g) What other parameters fix the value of $\tan \theta_c$?

     (h) Knowing θ_c and I_T, how can we evaluate I_R? I_C?

11. With reference to Example 34-4:

     (a) Give the equation used to find $\tan \theta_c$ in Section 32-6.

     (b) Is the equation in step 2 of the solution correct? Explain.

12. With reference to Example 34-5:

     (a) What change was made in the given data, as compared to Example 34-4?

     (b) Was the resistance value changed?

     (c) Was the total current value changed?

     (d) Why, then, is the I_R value so much lower in this problem?

     (e) Why is the E_T value also so much lower?

13. (a) What is the function of a *bypass* capacitor?

     (b) How is it connected with respect to the "other" component?

     (c) Will the capacitor in Example 34-4 serve for bypass action? Explain.

     (d) Is the capacitor in Example 34-5 satisfactory?

     (e) How is a proper value of capacitor selected?

     (f) Give a commonly used application of bypassing in electronics.

14. It is desired to separate the low-frequency component of a complex wave from a high-frequency component. State two requirements for a suitable bypass capacitor.

15. With reference to any R-L-C parallel circuit, is the total impedance always less than the smallest branch impedance? Explain.

16. In the solution to Example 34-6:

     (a) In step 2, why is 120 V used for E_L?

     (b) In step 6(a) ($I_X = 0.269$), is this a phasor or numerical solution? Explain.

17. In Fig. 34-5(b):

     (a) Why is I_L drawn downward?

     (b) Why is I_R drawn horizontally?

     (c) Why is I_C drawn upward?

     (d) What does θ_c represent?

     (e) Considering θ_c, does this circuit act as a resistive, inductive, or capacitive circuit; or what combination thereof?

18. In Fig. 34-5(c):

     (a) What does Z_T represent?

     (b) How was this value found?

     (c) What does θ_c represent?

     (d) How was this value obtained?

     (e) What does R represent, and how is its value obtained?

     (f) What does X_C represent, and how is its value obtained?

19. Comparing Example 34-2 with Example 34-6:

     (a) How do the R-branch values compare?

     (b) How do the L-branch values compare?

     (c) How do the C-branch values compare?

     (d) How do the I_T values compare?

(e) Why does the Example 34-6 circuit, with an extra branch, have a lower line current?

(f) With unequal line currents, why is the power dissipation the same?

(g) How do the total impedances compare?

(h) Why does adding a parallel branch *increase* the total impedance?

20. In a circuit with several parallel branches:
    (a) How is the current in each branch found?
    (b) Knowing these, how can the total current be found?
    (c) If each branch is a pure element, how is this addition made?
    (d) If one or more branches are not pure, is knowing the magnitude of each branch current sufficient? Explain.
    (e) How then, is the "addition" made?

21. When a parallel circuit has impedance branches:
    (a) Will the branch currents be at 0 or 90°? Explain.
    (b) Can we add these branch currents numerically? or by the Pythagorean theorem? Explain.

22. With reference to Example 34-7, Solution:
    (a) Why is Z_1 shown as $200 - j530$?
    (b) Why is E shown as $400 + j0$?
    (c) What is the phase angle of current I_1? Is it leading or lagging? Explain.
    (d) What is the phase angle of current I_2? Is this correct? Why?
    (e) In step 6, why are the currents converted to rectangular form?
    (f) In step 7, what is the phase relation of the capacitor voltage to the line voltage? How is this value obtained?

23. (a) Give the equation for the total impedance of a parallel circuit with three (or more) branches.
    (b) How is this simplified for a two-branch circuit?
    (c) If one of the branches contains an inductor of 60-Ω reactance, how must this be expressed for use in either of the above equations?

24. In equation (34-2), are the Z_1 and Z_2 values simple numbers? Explain.

25. In Example 34-8, between what two limits should the value of Z_T fall?

26. With reference to Example 34-10:
    (a) Between what two limits should the value of Z_1 fall?
    (b) Why is its polar angle a minus quantity?
    (c) Is the value 33.7° a reasonable value? Explain.
    (d) In step 3, how is the value $98.5\underline{/24°}$ obtained for $Z_1 + Z_2$?
    (e) Is this value the Z_T of the circuit?
    (f) In step 4, how is the angle 35.7° obtained for $Z_1 Z_2$?
    (g) How is the angle 11.7° obtained for Z_T?

27. With reference to Fig. 34-8, if E_T and each component value were given:
    (a) Give the steps for solving this circuit by a trigonometric method.
    (b) Give the steps for solving by vector algebra.

28. Give the name and letter symbol used to denote the opposite of:
    (a) Resistance.
    (b) Reactance.
    (c) Impedance.

29. In what type of circuit is the use of the above "opposite" quantities of possible advantage? Why?

30. In a parallel circuit, if we know the conductance and the susceptance of each branch, how do we find:
    (a) The total conductance?
    (b) The total susceptance?
    (c) The total admittance?

31. Using vector-algebra notation, what is the proper notation for:
    (a) A capacitive susceptance?
    (b) An inductive susceptance?

32. What is the unit for susceptance? for admittance? for conductance?

33. (a) If the conductance and susceptance of a circuit are known, how is its admittance evaluated?
    (b) Is this a numerical addition? Explain.

34. A circuit has a conductance of 0.4 sieman, and a susceptance of 0.6 sieman. Between what minimum and maximum limits should the admittance value fall?

35. If we know the voltage across a circuit and the admittance of the circuit, how can we find the current?

36. In a pure capacitive branch, can the branch current be found from $I = EB$? Explain.

37. In Fig. 34-9, what is the relation (if any) between diagrams (a) and (b)?

38. Equation (34-11B) gives B as $B = X/(R^2 + X^2)$. Yet Equation (34-12B) gives B as $B = 1/X$. Which is correct? Explain.

39. With reference to Fig. 34-10:
    (a) What does B_o represent?
    (b) Is this an inductive or a capacitive quantity? Explain.

40. In a parallel L-C-R circuit, with *pure* circuit elements:
    (a) What combination of circuit values will result in maximum circuit impedance? Why?
    (b) What is this condition called?
    (c) How does the line current value compare with any one branch value?
    (d) What is the value of the circuit phase angle?
    (e) What is the character of the circuit impedance?
    (f) What is the equation for finding the resonant frequency?

41. In a parallel L-C-R circuit, what is the character and relative magnitude of the total circuit impedance:
    (a) At a frequency above its resonant value? Explain.
    (b) At a frequency below its resonant value? Explain.

42. A pure inductor and capacitor are connected in parallel. What is the impedance of this circuit at resonance? Explain.

43. With reference to Fig. 34-11:
    (a) Is each branch a pure element?
    (b) What value of coil Q is used in this example?
    (c) Why is I_L drawn downward at a 45° angle?
    (d) At a frequency such that $X_L = X_C$, will I_L equal I_C? Explain.

(e) Will I_T be in phase with E_T? Explain.

(f) Will the current lead or lag E_T?

(g) In order to get $\theta_c = 0°$, should the frequency be increased or decreased? Explain.

(h) Which case in the phasor diagram represents $\theta_c = 0°$?

(i) Does $I_L = I_C$ now? Explain.

(j) Is Z_T a maximum now? Explain.

(k) Does $X_L = X_C$ now? Explain.

44. (a) Under what conditions does equation (34-15) apply?

(b) If sufficient resistance is added into the inductive branch, what happens to the frequency for resonance?

45. (a) A parallel L-C circuit of low Q is operated below its resonant frequency. As the frequency of the applied voltage is raised, which condition will occur first: $X_L = X_C$, $\theta_c = 0°$, or $I_T = $ minimum? Give the equation for this frequency.

(b) Which will occur next? Give the equation for this frequency.

(c) Which will occur last? Give the equation for this frequency.

PROBLEMS

1. A coil of 20 H is connected in parallel with a resistor of 5000 Ω across a 400-V 50-Hz supply. Find:

(a) The line current.

(b) The circuit phase angle.

(c) The power dissipated.

(d) The total impedance.

(e) The equivalent series circuit.

2. Repeat Problem 1 for $L = 80$ mH, $R = 2000 \ \Omega$, and supply of 100 V, 5000 Hz.

3. Repeat Problem 1 for $L = 30 \ \mu H$, $R = 800 \ \Omega$, and supply of 20 V, 2000 kHz.

4. A capacitor of 0.008 μF is connected in parallel with a 1000-Ω resistor across a 15-kHz, 100-V supply. Find:

(a) The line current.

(b) The circuit phase angle.

(c) The power dissipated.

(d) The total circuit impedance.

(e) The equivalent series circuit.

5. Repeat Problem 4 for a 0.0001-μF capacitor shunted by a 10 000-Ω resistor across a 20-V 500-kHz supply.

6. A transistor bias circuit consists of a 2000-Ω resistor shunted by a 2-μF capacitor. If the ac component of the current is 10 mA at 100 Hz, find:

(a) The current through the resistor.

(b) The current through the capacitor.

(c) The voltage across the combination.

7. In an R-C coupled FET amplifier, the load resistance (R_L) is 250 000 Ω. The output capacitance (transistor and circuit) shunting this resistor is 50 pF. If the ac component of the current is 0.5 mA, find:

(a) Voltage across R_L at 15 000 Hz.

(b) Voltage across R_L at 100 Hz.

(c) What is the general effect of shunt capacity on the frequency response (voltage across R_L)?

8. A complex wave, due to demodulation, contains component frequencies of 1000 kHz and 100 Hz. These components are fed to a parallel circuit consisting of a 0.5-MΩ resistor and a 100-pF capacitor. If the amplitude of the current for each frequency is 30 μA, find the rms voltage across the circuit:

(a) At the carrier frequency (1000 kHz).

(b) At the audio frequency.

9. An impedance, Z, is connected in parallel with a capacitor of 16 μF, across an 80-V, 100-Hz supply. The line current is 2.1 A, in phase with the supply voltage.

(a) Draw the phasor diagram.

(b) Find the current through the impedance Z.

(c) What are the components of this impedance?

10. A capacitor of 100 pF is connected in parallel with an inductance of 80 μH and also in parallel with a resistor of 3000 Ω across a supply voltage of 40 V, 1000 kHz. Find:

(a) The line current.

(b) The circuit phase angle.

(c) The total circuit impedance.

(d) The equivalent series circuit.

11. Repeat Problem 10, with the inductance raised to 0.4 mH.

12. A parallel circuit is connected across a 200-V 5000-Hz supply. The branches are: (1) A capacitor of 0.1 μF. (2) A resistor of 500 Ω. (3) A coil of 20 mH, and a resistor of 200 Ω. Using the rectangular coordinate method, find the line current, and express the answer in polar form. Draw the phasor diagram.

13. In each of the following cases, a commercial coil (coil has resistance) is connected in parallel with a resistor. Using a graphical method, find the coil phase angle and circuit phase angle if:

(a) $I_T = 1.2$ A, $I_R = 0.8$ A, $I_L = 0.7$ A.

(b) $I_L = 200$ mA, $I_T = 450$ mA, $I_R = 300$ mA.

(c) $I_T = 5$ A, $I_L = 3.5$ A, $I_R = 2.5$ A.

14. A coil of 5 H and 500-Ω resistance is connected in parallel with a 3500-Ω resistor across a 220-V 60-Hz supply. Find:

(a) The line current.

(b) The circuit phase angle.

(c) The power dissipated.

(d) The total circuit impedance.

(e) The equivalent series circuit.

15. Repeat Problem 13 for a 60-mH, 150-Ω coil in parallel with a 240-Ω resistor across a 30-V 2000-Hz supply.

16. A two-branch parallel circuit has a supply voltage of 120 V, 60 Hz. Branch 1 contains $L = 5$ H, in series with $R = 660$ Ω. Branch 2 contains $C = 8$ μF, in series with $R = 500$ Ω. Find the line current and circuit phase angle.

17. In Problem 16, a third branch, consisting of $R = 1000\ \Omega$, is added. Find the new line current and circuit phase angle.

18. In Fig. 34-6 the circuit values are:
   (a) Supply voltage 100 V, 400 kHz.
   (b) Branch 1: 0.0005 μF and 600 Ω.
   (c) Branch 2: 1200 Ω.
   (d) Branch 3: 1.2 mH and 300 Ω.
   Using vector algebra (either form), solve for:
   (a) The branch currents.
   (b) The line current (express in polar form).
   (c) The voltage across the coil and its phase relation to the supply voltage.
   (d) The voltage across the resistor (branch 3) and its phase relation to the supply voltage.
   (e) Draw the phasor diagram.

19. In Problem 18, if the voltage across the capacitor is 50 V, find:
   (a) The supply voltage (frequency as in Problem 18).
   (b) The current in each branch.
   (c) The total current.
   (d) Draw the phasor diagram.

20. Using vector algebra, find the total impedance of the circuit in Problem 12 *directly*.

21. Find the total impedance of the circuit in Problem 18 directly.

22. Find the total impedance of the circuit shown in Fig. 34-12 if $R_1 = 1500\ \Omega$, $X_L = 800\ \Omega$, $X_C = 2000\ \Omega$, and $R_2 = 500\ \Omega$.

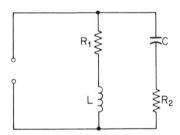

Figure 34-12

23. In Problem 22, if the line current is 80 mA, find the current in each branch.

24. In Fig. 34-8, the supply voltage is 3000 V, 50 kHz. The circuit constants are:
$$R_1 = 5000\ \Omega, \qquad R_2 = 5000\ \Omega, \qquad R_3 = 5000\ \Omega$$
$$L_1 = 50\ \text{mH}, \qquad L_2 = 5\ \text{mH}, \qquad C = 600\ \text{pF}$$
   Find:
   (a) The total impedance.
   (b) The line current.
   (c) The voltage across the parallel circuit.
   (d) The current through each branch of the parallel circuit.
   (e) Draw the phasor diagram.

25. Repeat Problem 24 for circuit values of: $E = 120$ V, 60 Hz; $R_1 = 150\ \Omega$; $R_2 = 300\ \Omega$; $R_3 = 400\ \Omega$; $L_1 = 0.5$ H; $L_2 = 1.2$ H; $C = 5.0\ \mu$F.

26. In the simple parallel circuit of Fig. 34-13, $R = 500\ \Omega$, $X_L = 500\ \Omega$, and $X_C = 200\ \Omega$. Find:

   (a) The conductance and susceptance of each branch.

   (b) The total circuit admittance.

   (c) The total circuit impedance.

   (d) Using admittance, and a line voltage of 100 V, find the line current (magnitude and phase).

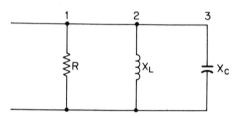

Figure 34-13

27. Repeat Problem 26 for $R = 1000\ \Omega$, $X_L = 450\ \Omega$, $X_C = 420\ \Omega$, and a line voltage of 36 V.

28. A circuit has an admittance of $0.0046\underline{/42°}$ S. Find:

   (a) The conductance and susceptance values.

   (b) The line current, if the line voltage is 90 V, at an angle of $-30°$.

29. Find the conductance, susceptance, and admittance of the circuit in Fig. 34-9(a) if $R = 250\ \Omega$ and $X_C = 500\ \Omega$.

30. Find the conductance, susceptance, and admittance of a 600-Ω resistor in series with a 500-Ω inductive reactance.

31. Solve for the line current in Problem 12 (magnitude and phase), using the conductance, susceptance, and admittance method.

32. Solve for the line current in Problem 18 (magnitude and phase), using the conductance, susceptance, and admittance method.

33. What value of inductance in Problem 10 would give a "resonant" condition $(X_L = X_C)$?

34. Using the inductance value of Problem 33, repeat Problem 10.

35. A capacitor of 0.1 μF is in parallel with a coil of 0.04 H and 200 Ω resistance (two branches). The supply voltage is 15 V, 2000 Hz. Find:

   (a) The line current.

   (b) The circuit phase angle.

   (c) The total circuit impedance.

   (d) The equivalent series circuit.

36. What value of capacitance would give $X_C = X_L$ in Problem 35.

37. Repeat Problem 35, using the value of capacitor found in Problem 36.

38. Repeat Problem 35, using the value of capacitor found in Problem 36, but dropping the applied frequency to 1840 Hz. Is this a resonant condition? Explain.

39. A coil of 10-mH inductance has a Q of 1.5. It is connected in parallel with a capacitor of 0.05 μF across a variable frequency source of 100 V. Find:

   (a) The frequency at which the two reactances are equal.

(b) The line current and circuit phase angle at this frequency.

(c) The impedance of this tank circuit.

40. Using the values of Problem 39, find the frequency that results in minimum line current.

41. Using the values of Problem 39, find the frequency that makes the circuit act as a pure resistance.

42. A coil of 0.2 H is in series with 1000-Ω resistance. The combination is in parallel with a 0.1-μF capacitor.

   (a) Find the resonant frequency. (Unity power-factor condition.)

   (b) If the resistance is neglected, how much error would have been caused?

43. A transmitter tank circuit has a Q of 50 when unloaded. Due to improper loading, the Q drops to 3. What is the percentage change in the resonant frequency (condition for pure resistive load)?

35

AC NETWORKS

Most ac circuit problems can be solved using the techniques discussed in Chapters 31 through 34. However, complex networks, and/or circuits involving more than one supply source, may require a more sophisticated attack. The network methods covered in the dc portion of this text will apply equally well to ac circuits—if we take phase angles into consideration. Obviously, vector algebra solutions will be necessary.

35-1 Kirchhoff's laws method. For use in ac circuits, Kirchhoff's voltage law can be restated as: *In any one path of a complete electrical circuit, the phasor sum of the EMFs and of each voltage drop is equal to zero.* In applying this law, it is recommended that:

1. The circuit schematic be drawn.
2. All resistances and reactances be marked, using $+j$ for X_L values, and $-j$ for X_C values.
3. One of the voltage sources be taken as reference, and all others be marked with their relative phasing.
4. Assume the most plausible direction for current flows.
5. When tracing through a resistance or reactance *in the direction of current flow*, there is a voltage drop, and the sign assigned to this quantity is negative. Conversely, when tracing in a direction *opposite to the direction of current flow*, the sign is positive.
6. Voltage sources must be treated as phasors. If they are in opposition, the phasor quantities must be subtracted.

7. Since we are adding values, all quantities (voltages and impedances) must be expressed in (or converted to) rectangular form.

Let us apply these basic steps to the simple circuit shown in Fig. 35-1.

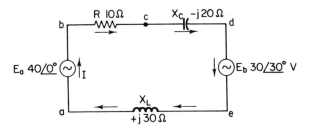

Figure 35-1   Circuit for Example 35-1.

Example 35-1

Find the current (magnitude and phase) for the circuit shown in Fig. 35-1.

Solution   Tracing through path *abcdea*, we have

1. $$E_a - IR - (-jX_cI) + E_b - (+jX_LI) = 0$$

Substituting circuit values yields

2. $$(40 + j0) - 10I + j20I + (26 + j15) - j30I = 0$$

Collecting similar terms, we get

3. $$66 + j15 - 10I - j10I = 0$$

4. $$I(10 + j10) = 66 - j15$$

5. $$I = \frac{66 + j15}{10 + j10} = \frac{67.8\underline{/12.8°}}{14.14\underline{/45°}} = 4.79\underline{/-32.2°} \text{ A}$$

The above problem, involving only one unknown, is quite simple, and would normally have been solved by finding Z_T and E_T, in conventional manner. It was used here, merely to show how to apply Kirchhoff's voltage law.

In complex networks, there will be more than one path, more than one current, and more than one unknown. This requires more than one equation, and simultaneous (or determinants) solution. The problem solution should be simplified by applying Kirchhoff's current law at the junctions, to reduce the number of unknown currents. This law can be restated as: *In any electrical network, the **phasor** sum of the currents flowing to a junction must equal the **phasor** sum of the current flowing away from that junction.*

Example 35-2

Using the T network of Fig. 35-2, find the voltage across R_3, the current through R_3, and how much of this current is delivered by each source.

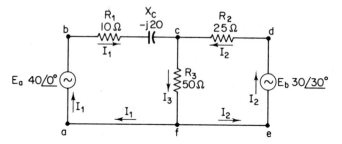

Figure 35-2 Analysis by Kirchhoff's law.

Solution

1. Applying Kirchhoff's current law at junction c, I_3 can be replaced by $(\dot{I}_1 + \dot{I}_2)$. This is a *phasor* addition.

2. Apply Kirchhoff's voltage law to path *abcfa*:

(a) $$E_a - I_1(R_1 - jX_C) - (I_1 + I_2)R_3 = 0$$

(b) $$(40 + j0) - 10I_1 + j20I_1 - 50I_1 - 50I_2 = 0$$

(c) $$40 - 60I_1 + j20I_1 - 50I_2 = 0$$

(d) $$3I_1 - jI_1 + 2.5I_2 = 2 \qquad (1)$$

3. Apply the voltage law to path *edcfe*:

(a) $$E_b - I_2R_2 - (I_1 + I_2)R_3 = 0$$

(b) $$(26 + j15) - 25I_2 - 50I_1 - 50I_2 = 0$$

(c) $$2.6 + j1.5 - 5I_1 - 7.5I_2 = 0 \qquad (2)$$

4. Multiply equation (1) by 3, and subtract equation (2):

(a)
$$9I_1 - j3I_1 + 7.5I_2 = 6$$
$$\underline{\oplus \, 5I_1 \qquad\qquad \oplus \, 7.5I_2 = \oplus \, 2.6 \oplus j1.5}$$
$$4I_1 - j3I_1 \qquad\qquad = 3.4 - j1.5$$

(b) $$I_1 = \frac{3.4 - j1.5}{4 - j3} = 0.724 + j0.168 = 0.744\underline{/13.1°} \text{ A}$$

5. Substitute for I_1 in equation (1):

(a) $$3(0.724 + j0.168) - j(0.724 + j0.168) + 2.5I_2 = 2$$
$$2.34 - j0.22 + 2.5I_2 = 2$$

(b) $$I_2 = \frac{-0.340 + j0.22}{2.5} = -0.136 + j0.088$$

$$= 0.162\underline{/147.1°} \text{ A}$$

This current, I_2, is in the second quadrant, or leads the supply voltage by 147.1°.*

*Obviously, the wrong current direction was assumed in Fig. 35-2. Reversing this current direction, we get

$$I_2 \text{ (true)} = I_2 \text{ (assumed)} - 180° = 0.162\underline{/-32.9°}$$

6. $$I_3 = I_1 + I_2 = 0.724 + j0.168 - 0.136 + j0.088$$
$$= 0.588 + j0.256 = 0.641\underline{/23.5°}\ A$$

of which 0.744 at 13.1° lead is supplied by E_a, and 0.162 at 32.9° lag is supplied by E_b.

35-2 Loop or mesh method.

This technique (as mentioned in Part 1) is almost identical to the more general Kirchhoff's law method. The only difference is with regard to the way the currents are labeled. This distinction can be seen by comparing Fig. 35-2 (Kirchhoff analysis) and Fig. 35-3 (loop analy-

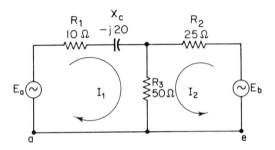

Figure 35-3 Analysis by loop method.

sis). In Fig. 35-3, two loops or paths are shown. Each loop has only one current value. Kirchhoff's voltage law is applied to each loop, in exactly the same manner as explained before. For loop 1, starting at point *a*, and tracing clockwise, the voltage equation is

$$E_a - I_1 R_1 - I_1(-jX_C) - I_1 R_3 - I_2 R_3 = 0$$

Notice that the voltage drop across R_3 has two components: $I_1 R_3$ due to the current in loop 1, and $I_2 R_3$ due to the current in loop 2. Also notice that the above equation is the same as that obtained by the general Kirchhoff's law technique. Obviously, then, the solution by either method is identical.

Let us use the loop method on a bridge-circuit problem.

Example 35-3
Find the current through the "null detector" R_D, in the bridge circuit of Fig. 35-4.

Solution In order to include each component at least once, three loops are necessary. Let us select loops 1, 2, and 3, as shown in Fig. 35-4.

1. Trace clockwise through loop 1, starting at point *c*:

(a) $$-1000I_1 - 1000I_2 - 100I_1 - (+j50)I_1 + (+j50)I_3 = 0$$

(b) $$-1100I_1 - j50I_1 - 1000I_2 + j50I_3 = 0$$

(c) Divide by 50:

$$I_1(-22 - j1) - 20I_2 + jI_3 = 0 \qquad (1)$$

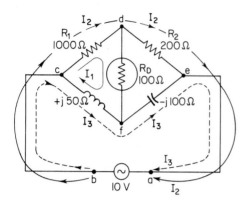

Figure 35-4 Bridge circuit analysis—loop method.

2. Trace clockwise through loop 2, starting at point a:

(a) $$10 - 1000I_2 - 1000I_1 - 200I_2 = 0$$

(b) $$5I_1 + 6I_2 + 0I_3 = 0.05 \qquad (2)$$

3. Trace clockwise through loop 3, starting at a:

(a) $$10 - (+j50)I_3 + (+j50)I_1 - (-j100)I_3 = 0$$

(b) $$10 - j50I_3 + j50I_1 + j100I_3 = 0$$

(c) $$+jI_1 + 0I_2 + jI_3 = -0.2 \qquad (3)$$

4. Solve for I_1 (through the null detector), using determinants:*

$$I_1 = \frac{\begin{vmatrix} +0 & -20 & +j1 \\ +0.05 & +6 & 0 \\ -0.2 & 0 & +j1 \end{vmatrix}}{\begin{vmatrix} (-22 - j1) & -20 & +j1 \\ +5 & +6 & 0 \\ +j1 & 0 & +j1 \end{vmatrix}} = \frac{j2.2}{12 - j32} = -0.0603 + j0.0226 \text{ A}$$

$$I_1 = -60.3 + j2.6 \text{ mA} = 64.4\underline{/159.5°} \text{ mA}$$

This current is in the second quadrant. Obviously, the current path chosen for I_1 should be reversed, and I_1 instead of leading the supply voltage by 159.5°, is more properly classified as

$$I_1 = 64.4\underline{/-20.5°} \text{ mA}$$

35-3 Nodal method. This method was described in detail in Chapter 23. It is especially suited for T or ladder networks—either dc or ac. Now, in ac circuits we must consider reactances and impedances instead of just resistances. Briefly reviewing this technique: A *node* is a junction of three (or more) elements; we will write *current* equations at one or more nodes, in terms of unknown voltages; one node is considered as the reference node, and the other

*See Appendix 5.

node potentials are evaluated with respect to this referenced node. When applied to T or ladder networks, a nodal equation reduces to the form:

node voltage × (reciprocal of each adjacent branch impedance)— (each adjacent voltage divided by the connecting element impedance)

Let us apply this technique to the T network in Example 35-2.

Example 35-4

Using the nodal method, solve for current I_3 in Fig. 35-2.

Solution   The nodes in this circuit are c and f. Selecting f as the reference node, we can write the current equations for node c. Using E_3 as the voltage between nodes c and f:

1. $E_3\left(\dfrac{1}{R_1 - jX_C} + \dfrac{1}{R_3} + \dfrac{1}{R_2}\right) - \dfrac{E_a}{R_1 - jX_C} - \dfrac{E_b}{R_2} = 0$

2. $E_3\left(\dfrac{1}{10 - j20} + \dfrac{1}{50} + \dfrac{1}{25}\right) - \dfrac{40}{10 - j20} - \dfrac{26 + j15}{25} = 0$

3. $E_3(0.02 + j0.04 + 0.02 + 0.04) - (0.8 + j1.6) - (1.04 + j0.6) = 0$

4. $E_3(0.08 + j0.04) - 0.8 - j1.6 - 1.04 - j0.6 = 0$

    $E_3 = \dfrac{1.84 + j2.2}{0.08 + j0.04} = 29.4 + j12.8 \text{ V}$

5. $I_3 = \dfrac{E_3}{R_3} = \dfrac{29.4 + j12.8}{50} = 0.588 + j0.256 \text{ A}$

    $= 0.641\underline{/23.5°} \text{ A}$

Notice that this checks very closely to the answer obtained in Example 35-2—yet it is much less work.

35-4   Superposition theorem.   This technique can be used when a network contains more than one supply source. As explained in Section 23-3, all but one of the sources are replaced by their own internal impedance.* The circuit is then solved using simple Ohm's law relations. The process is repeated using each of the supply sources as the sole supply. The actual current through any component (or portion) of the original circuit is the phasor sum of the currents obtained when each source acted individually. The previous example will be repeated using the superposition method.

Example 35-5

Using the superposition method, find current I_3 in Fig. 35-2.

Solution

1. Since no values for internal impedance of the sources are given, we will assume that they are negligible.

*If the internal impedance is negligible compared to the other circuit values, the source is replaced by a short circuit.

2. Figure 35-5(a) shows the equivalent circuit when source E_b is shorted. Notice that R_2 and R_3 are now in parallel. Their combined resistance is

(a) $$R_{2,3} = \frac{R_2 R_3}{R_2 + R_3} = \frac{25 \times 50}{75} = 16.7 \ \Omega$$

(b) The total impedance is

$$Z_T = (R_1 + R_{2,3}) - jX_C = 26.7 - j20$$

(c) $$I_T = \frac{E_a}{Z_T} = \frac{40}{26.7 - j20} = 0.958 + j0.718 \ \text{A}$$

(d) From the ratio of R_3 to R_2, the current through R_3 is only one-third of I_T, or

$$I'_3 = 0.319 + j0.239 \ \text{A}$$

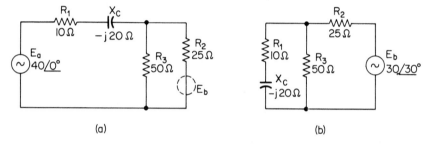

(a)   (b)

Figure 35-5   Analysis of Fig. 35-2 by superposition method.

3. Figure 35-5(b) shows the equivalent circuit when source E_a is shorted. The impedance of the parallel portion of the circuit is

(a) $$Z_{par} = \frac{(R_3)(R_1 - jX_C)}{R_3 + R_1 - jX_C} = \frac{50(10 - j20)}{60 - j20} = \frac{25 - j50}{3 - j1}$$

$$= 12.5 - j12.5$$

(b) $$Z_T = R_2 - Z_{par} = 25 + (12.5 - j12.5) = 37.5 - j12.5$$

(c) $$I_T = \frac{E_b}{Z_T} = \frac{26 + j15}{37.5 - j12.5} = 0.504 + j0.568 \ \text{A}$$

(d) Since $E_{par} = I_T Z_{par}$, and $I''_3 = E_{par}/R_3$,

$$I''_3 = \frac{I_T Z_{par}}{R_3} = \frac{(0.504 + j0.568)(12.5 - j12.5)}{50}$$

$$= 0.268 + j0.0159 \ \text{A}$$

4.   Actual $I_3 = I'_3 + I''_3 = 0.319 + j0.239 + 0.268 + j0.0159$

$$= 0.587 + j0.255 \ \text{A}$$

$$= 0.640 \underline{/23.5°} \ \text{A}$$

Notice that this answer checks very well with the previous values obtained by the Kirchhoff's law and nodal methods.

35-5 Thévenin's theorem.   As in dc circuits, Thévenin's theorem is especially valuable when we wish to analyze a complex network for various load values—and we are interested only in the load current, load voltage and load power. A typical application is in transistor circuit design, using the transistor equivalent circuit. The basic principle is quite simple. Any active network—no matter how complex—can be replaced by an *equivalent circuit* consisting of a constant-voltage source in series with an impedance, where

1. The voltage source (E_{oc}) is the *open-circuit* voltage of the original network, measured across the load terminals, *but with the load disconnected.*

2. The series impedance (Z_i) for the equivalent circuit is the *internal impedance* of the original network as seen *looking in* from the load terminals, but with all voltage sources replaced by their internal impedances. (The sources can be replaced by short circuits if their internal impedances are negligible compared to the network values.)

For comparison purposes, let us apply this technique to the circuit in Fig. 35-2.

Example 35-6

Using Thévenin's theorem and the circuit in Fig. 35-2, find the current through R_3 and the voltage across R_3.

Solution

1. R_3 is considered as the load.

2. To find the voltage (E_{oc}) for the Thévenin equivalent circuit, we must first remove the load and then calculate the voltage at the load terminals (across c-f). With R_3 removed, the circuit (Fig. 35-2) becomes a series circuit with E_b opposing E_c. [This simplified circuit is shown as Fig. 35-6(a)]. The net voltage, E_T, is therefore $\dot{E}_a - \dot{E}_b$ or

   (a) $E_T = \dot{E}_a - \dot{E}_b = (40 + j0) - (26 + j15) = 14 - j15$ V

   (b) The total circuit impedance is now

$$Z_T = \dot{R}_1 + \dot{X}_C + \dot{R}_2 = 35 - j20 \ \Omega$$

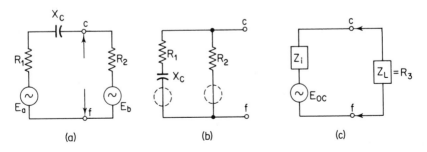

(a)   (b)   (c)

Figure 35-6 Development of the Thévenin equivalent of Fig. 35-2.

(c)
$$I_T = \frac{E_T}{Z_T} = \frac{14 - j15}{35 - j20} = 0.486 - j0.151 \text{ A}$$

(d) The voltage drop across resistor R_2 is

$$E_2 = I_T R_2 = (0.486 - j0.151) \times 25 = 12.15 - j3.77 \text{ V}$$

$$E_{c-f} = E_{oc} = E_b + E_2 = (26 + j15) + (12.15 - j3.77)$$
$$= 38.15 + j11.23 \text{ V}$$

3. To find the series impedance (Z_i), both supply sources are shorted (internal impedances assumed to be negligible). This puts R_2 in parallel with $R_1 + X_C$ [see Fig. 35-6(b)]. The series impedance is therefore

$$Z_i = \frac{(R_1 - jX_C)(R_2)}{R_1 - jX_C + R_2} = \frac{(10 - j20)(25)}{35 - j20}$$
$$= 11.5 - j7.69 \ \Omega$$

4. The equivalent Thévenin circuit is shown in Fig. 35-6(c).
5. To find the current through R_3 in Fig. 35-6(c):

$$I_3 = I_T = \frac{E_{oc}}{Z_i + R_3} = \frac{38.15 + j11.23}{(11.5 - j7.69) + 50}$$
$$= 0.588 + j0.256 \text{ A} = 0.641\underline{/23.5°} \text{ A}$$

$$E_3 = I_3 R_3 = (0.641\underline{/23.5°})(50\underline{/0°}) = 32.05\underline{/23.5°} \text{ V}$$

These answers check with the values obtained by each of the previous methods. As far as the work done to obtain the answers, there is not much choice among the various methods. The great advantage of the Thévenin method is that if we now want load current and load voltage for several other values of R_3, we need only use the simple step 5. With any of the previous methods, the full procedure would have to be repeated.

35-6 Norton's theorem.   As we saw in Section 23-5, this method is a variation of the Thévenin technique. The complex network is again replaced by an eqivalent circuit. This time, the equivalent circuit consists of a *constant-current* source in *parallel* with an impedance, where:

1. The constant-current value (I_{sc}) is the current that would flow through a short-circuited load, in the original network.
2. The shunt impedance (Z_{sh}) for the equivalent circuit is the *internal imped-ance* of the original network as seen looking in from the load terminals, but with all voltage sources replaced by their own internal impedances. (This is the same value as is used in the Thévenin method.)

Let us again use Fig. 35-2, and apply this theorem.

Example 35-7
Using Norton's theorem, and the circuit of Fig. 35-2, find the current through R_3 and the voltage across R_3.

Solution

1. R_3 is considered as the load.
2. To find the current value (I_{sc}) for the Norton equivalent circuit, the load (R_3) must be shorted. The circuit can now be drawn as in Fig. 35-7(a).

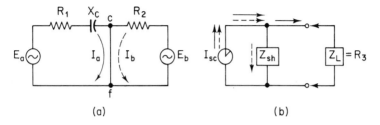

(a)   (b)

Figure 35-7   Analysis of Fig. 35-2 by Norton's theorem.

The current through the short-circuited load (I_{sc}) is obviously $I_a + I_b$, and can be readily found using the superposition theorem.

(a) With E_b shorted (internal impedance negligible),

$$I_a = \frac{E_a}{R_1 - jX_C} = \frac{40}{10 - j20} = 0.8 + j1.6 \text{ A}$$

(b) With E_a shorted,

$$I_b = \frac{E_b}{R_2} = \frac{26 + j15}{25} = 1.04 + j0.6 \text{ A}$$

(c)
$$I_{sc} = I_a + I_b = (0.8 + j1.6) + (1.04 + j0.6)$$
$$= 1.84 + j2.2 \text{ A}$$

3. To find the Norton shunt impedance, the procedure is the same as step 3 of Example 35-6, or
$$Z_{sh} = 11.5 - j7.69 \ \Omega$$

4. The Norton equivalent circuit is shown in Fig. 35-7(b). By the current ratio method, it should be obvious that the load current (I_L) is

(a)
$$I_L = I_{sc} \frac{Z_{sh}}{Z_{sh} + Z_L} = (1.84 + j2.2)\frac{11.5 - j7.69}{(11.5 - j7.69) + 50}$$
$$= 0.588 + j0.256 = 0.641\underline{/23.5°} \text{ A}$$

This is I_3 of the original circuit.

(b)
$$E_L = E_3 = I_L Z_L = (0.641\underline{/23.5°})(50\underline{/0°})$$
$$= 32.05\underline{/23.5°} \text{ V}$$

35-7 Equivalent pi and tee circuits.   Many electronic circuits, particularly filter circuits, contain impedances connected to form a π (pi) or a T network. Such circuits are shown in Fig. 35-8(a) and (b), respectively. In the "power" field, loads of this type are known as *delta* (Δ) or *wye* (Y) loads. The

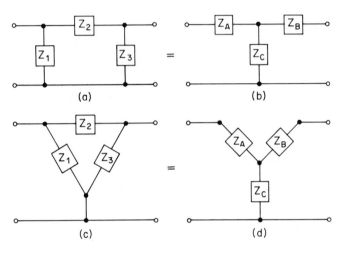

Figure 35-8 Equivalent π (or Δ) and T (or Y) networks.

delta connection [Fig. 35-8(c)] is equivalent to the π connection, but it is drawn schematically with the lower end of Z_1 and Z_3 coming to a common-point connection. The wye connection [Fig. 35-8(d)], is the same as the T connection, but is shown schematically with the horizontal arms drawn upward to simulate the letter Y. (These circuits will be discussed in more detail in chapters 37–39.) The impedances in these networks could be all resistors, all capacitors, all inductances, or any combination of R-L-C. By themselves, such circuits would not be difficult to handle. But many times the π or T circuit is only part of a more complex network. Quite often the solution of these networks can be simplified by converting the component π circuit into its equivalent T circuit, or vice versa.

Delta–wye transformations were discussed in Section 23-6 and equations were developed for converting from one type of circuit to the other. These same equations apply equally well to ac circuits—if we replace R values with Z values, and treat all values vectorially:

1. *For T to π (or Y to Δ) Conversion:*

$$Z_1 \ (\pi \text{ value}) = \frac{\dot{Z}_A \dot{Z}_B + \dot{Z}_B \dot{Z}_C + \dot{Z}_A \dot{Z}_C}{\dot{Z}_B} \qquad (35\text{-}1\text{A})$$

$$Z_2 \ (\pi \text{ value}) = \frac{\dot{Z}_A \dot{Z}_B + \dot{Z}_B \dot{Z}_C + \dot{Z}_A \dot{Z}_C}{\dot{Z}_C} \qquad (35\text{-}1\text{B})$$

$$Z_3 \ (\pi \text{ value}) = \frac{\dot{Z}_A \dot{Z}_B + \dot{Z}_B \dot{Z}_C + \dot{Z}_A \dot{Z}_C}{\dot{Z}_A} \qquad (35\text{-}1\text{C})$$

Notice that in each case the numerator is the same, while the denominator is the T impedance *which connects to the opposite line terminal.*

2. *For π to T (or Δ to Y) Conversion:*

$$Z_A \text{ (T value)} = \frac{\dot{Z}_1 \dot{Z}_2}{\dot{Z}_1 + \dot{Z}_2 + \dot{Z}_3} \tag{35-2A}$$

$$Z_B \text{ (T value)} = \frac{\dot{Z}_2 \dot{Z}_3}{\dot{Z}_1 + \dot{Z}_2 + \dot{Z}_3} \tag{35-2B}$$

$$Z_C \text{ (T value)} = \frac{\dot{Z}_1 \dot{Z}_3}{\dot{Z}_1 + \dot{Z}_2 + \dot{Z}_3} \tag{35-2C}$$

This time, the denominators are all alike, and the numerator is the *product of the two π impedances* that connect to the same terminal.

An application of the use of such conversions would apply to the solution of the ac bridge circuit shown in Fig. 35-9.

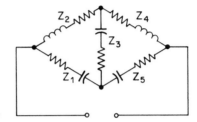

Figure 35-9 Typical ac bridge circuit.

The impedances Z_1, Z_2, Z_3 (or Z_3, Z_4, Z_5) are actually connected in a π (or delta) network. Since Z_1, Z_2, Z_3 is a π network, it can be replaced by an equivalent T circuit (Z_A, Z_B, Z_C).

Figure 35-10 shows the equivalent circuit, redrawn.

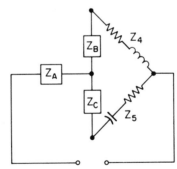

Figure 35-10 Equivalent of bridge circuit.

Notice now that:

1. Z_B and Z_4 are in series.
2. Z_C and Z_5 are in series.
3. The combination of (1) and (2) are in parallel.
4. The equivalent impedance of (3) is in series with Z_A.

Problems involving π–T conversions are included in the Problems section at the end of this chapter.

35-8 Maximum power transfer.   In Section 23-7 it was proven that maximum power is transferred from a source to a load, when the resistance of the load is equal to (or matches) the Thévenin equivalent resistance of the source and network feeding the load. In its simplest case—a source feeding a load—the load resistance should be equal to the source resistance.

The same *principle* applies to any ac circuit. However, now, the circuit may contain reactances as well as resistances. This gives rise to three possible variations:

1. *All Circuit Values Are Pure Resistances.* This situation is identical to the dc conditions analyzed in Section 23-7. Obviously then, maximum power transfer will occur when

$$R_L = R_i \qquad\qquad (35\text{-}3\text{A})$$

2. *The Source (and Connecting Network) Have Reactance as Well as Resistance, and the Load Value Is Completely Controllable.* First, it must be realized that in order to deliver power to the load, the load must contain resistance. Second, it should be recalled that the power dissipated in any resistor is proportional to the square of the current through the resistor. In turn, current is a maximum when the circuit impedance is a minimum. This condition can be achieved if the reactance of the load is made *equal and opposite* to the reactance of the source and connecting network. Now we have a pure resistive circuit— condition 1 above—and maximum power is obtained when the resistance of the load is equal to the equivalent resistance of the source and network feeding the load. For example, if the Thévenin equivalent impedance of the source and network is $20 + j30 \,\Omega$, maximum power transfer is obtained when the load impedance is $20 - j30 \,\Omega$. Such a load is said to be the **conjugate impedance**. Summarizing, when a source contains resistance and reactance, maximum power transfer occurs when

$$Z_L = \text{conjugate of } Z_i \qquad\qquad (35\text{-}3\text{B})$$

3. *R-X Source, and Pure R Load.* In some cases, the load is a pure resistance, and adjustable in value. Instead of making the load value equal to the resistance of the source, it can be proven that maximum power transfer will occur when

$$R_L = Z_i \qquad\qquad (35\text{-}3\text{C})$$

Example 35-8

A power source has an open-circuit voltage of 50 V. The internal impedance of the source plus connecting network is $100 + j80 \,\Omega$. Find:

(a) The maximum power output that this supply could deliver.

(b) The maximum power that it will deliver to a pure resistive load.

Solution

(a) For maximum power output Z_L = conjugate of Z_i:

1. $\qquad Z_L = 100 - j80$

2. $\qquad Z_T = (100 + j80) + (100 - j80) = 200 \ \Omega$

3. $\qquad I_T = \dfrac{E_T}{Z_T} = \dfrac{50}{200} = 0.25 \ \text{A}$

4. $\qquad P_o = I^2 R_L = (0.25)^2 \times 100 = 6.25 \ \text{W}$

(b) With a pure resistive load, for maximum power:

1. $\qquad R_L = Z_i = \sqrt{(100)^2 + (80)^2} = 128 \ \Omega$

2. $\qquad Z_T = Z_i + R_L = (100 + j80) + 128 = 228 + j80 = 242 \ \Omega$

3. $\qquad I_T = \dfrac{E_T}{Z_T} = \dfrac{50}{242} = 0.207 \ \text{A}$

4. $\qquad P_o = I^2 R_L = (0.207)^2 \times 128 = 5.48 \ \text{W}$

REVIEW QUESTIONS

1. **(a)** Can Kirchhoff's voltage law be applied to an ac circuit?
   (b) How does such application differ from its use in a dc circuit?
   (c) In your own words, state Kirchhoff's current law as it applies to a junction in an ac circuit.

2. With reference to Fig. 35-1:
   (a) How many paths are there?
   (b) How many unknowns?
   (c) How many equations will be needed to solve the circuit?
   (d) Why is X_C shown with a $-j$, whereas the X_L term has a $+j$?
   (e) Can the voltage value for E_b be used directly as given? Explain.

3. In the solution to Example 35-1, step 1:
   (a) Is this a current, or a voltage equation? Explain.
   (b) Explain the double negative sign with regard to the term $-(-jX_C I)$.

4. In the solution to Example 35-2:
   (a) Account for the term $-(I_1 + I_2)R_3$ in step 2(a).
   (b) How is step 2(d) obtained?
   (c) In step 4(b), how is the last answer $(0.744\underline{/13.1°})$ obtained?
   (d) In step 5(b), why is I_2 considered to have been taken in the wrong direction?
   (e) In step 5(c), how is the $-32.9°$ obtained for I_2?

5. With reference to Fig. 35-3:
   (a) When tracing through loop 1, give the term that expresses the voltage drop across R_3?
   (b) How many unknown currents are there in this circuit?
   (c) How many equations will be needed to solve this circuit?

6. In Fig. 35-4, is a fourth loop *efde* necessary? Explain.

7. With reference to the solution to Example 35-3:
   (a) In step 1(a), since we are concerned with loop 1, why are terms with I_2 and I_3 included?
   (b) In this step, what is the significance of the term $+(+j50)I_3$?
   (c) In step 3(a), this same term is now shown as $-(+j50I_3)$. Why?
   (d) How is the equation in step 3(c) obtained?

8. Define the term *node* of a circuit.

9. With reference to Fig. 35-2:
   (a) How many nodes does this circuit have?
   (b) Give their letter designation.
   (c) How many nodal equations will be needed to solve this circuit? Explain.

10. With reference to the solution of Example 35-4, step 1:
    (a) Is this a current or voltage equation.
    (b) Where in the circuit (Fig. 35-2) does this equation apply?
    (c) What does E_3 represent?
    (d) Explain the significance of each term in this equation.
    (e) For what type of circuit does such a nodal equation apply?

11. (a) How would you apply the superposition theorem to the bridge circuit of Fig. 35-4?
    (b) To what type of problem is this theorem applicable?
    (c) What is the basic principle in this method?

12. With reference to Fig. 35-5:
    (a) What does diagram (a) represent?
    (b) Explain how the total impedance of this circuit can be evaluated.
    (c) What does diagram (b) represent?
    (d) Explain how the total impedance of this circuit can be evaluated?

13. With reference to the solution to Example 35-5:
    (a) Is I_T in step 2(c) the current through R_3? Explain.
    (b) Is I'_3 in step 2(d), the true current in R_3 *of the original circuit*? Explain.
    (c) Is I''_3 in step 3(d) the true current? Explain.
    (d) In step 4, why are I'_3 and I''_3 added?

14. (a) To what types of circuits can the Thévenin theorem be applied?
    (b) What does the Thévenin equivalent circuit consist of?
    (c) How is the voltage value for this equivalent circuit evaluated?
    (d) What is this voltage value called? Give the letter symbol.
    (e) How is the series impedance value of this equivalent circuit evaluated?

15. With reference to Fig. 35-6(a):
    (a) How is this diagram obtained from Fig. 35-2?
    (b) What is the relationship of the two sources, E_a and E_b, now?
    (c) How can the current in this circuit be found?
    (d) Knowing this line current, and all component values, how can the voltage E_{cf} be evaluated?
    (e) Show another way for finding E_{cf}.
    (f) In the Thévenin equivalent circuit, what is this voltage (E_{cf}) called?

16. With reference to Fig. 35-6(b):
    (a) How is this diagram obtained from Fig. 35-2?

(b) What is the purpose of this diagram?

17. With reference to Fig. 35-6(c):
    (a) How is the value of E_{oc} affected if the Z_L value is increased? Explain.
    (b) How is the Z_t value affected if the Z_L value is decreased? Explain.
    (c) What advantage does the Thévenin method have over the previously discussed network theorems?

18. **(a)** Of what does the Norton equivalent circuit consist?
    (b) How is the source value for this equivalent circuit evaluated?

19. With reference to Fig. 35-7:
    (a) What is the purpose of diagram (a)?
    (b) Does the I_{sc} value in diagram (b) have any relation to the I_a and/or I_b value shown in diagram (a)?
    (c) Is the $(I_a + I_b)$ value equal to the current that flows in the original circuit load, R_3? Explain.
    (d) How is the Z_{sh} value in diagram (b) obtained from diagram (a)?

20. With reference to Fig. 35-8:
    (a) What is the name given to the *type* of network shown in diagram (a)? (b)? (c)? (d)?
    (b) Which of the remaining three diagrams is an exact element-by-element equivalent of network *a*.
    (c) Name two other element-by-element equivalents.
    (d) Is Z_1 in diagram (a) identical to Z_A, Z_B, or Z_C in diagram (c)? Explain.
    (e) In what respect is network *a* "equal to" network *c*?

21. It is desired to replace the T network in Fig. 35-20(b), by its equivalent π. What combination of π values would be needed to replace Z_A? Z_B? Z_C?

22. Explain why the schematic diagram, Fig. 35-10, is the equivalent of Fig. 35-9.

23. With respect to delivering maximum power to a load, in an ac circuit:
    (a) Under what condition does the relation $R_L = R_t$ apply?
    (b) What does the term *conjugate impedance* mean?
    (c) Give an example of conjugate impedances.
    (d) Under what condition does the relation $Z_L = $ conjugate of Z_t apply?
    (e) Under what condition does the relation $R_L = Z_t$ apply?

PROBLEMS

1. Using the Kirchhoff's law method, and the circuit of Fig. 35-1, find the line current for $R = 30\,\Omega$, $X_C = 40\,\Omega$, $X_L = 10\,\Omega$, $E_a = 60$ V, and $E_b = 40$ V, lagging E_a by $40°$.

2. In the schematic of Fig. 35-2, the circuit values are: $R_1 = 30\,\Omega$, $X_C = 40\,\Omega$, $R_2 = 20\,\Omega$, $R_3 = 80\,\Omega$, $E_a = 60\underline{/0°}$ V, and $E_b = 50\underline{/-36.9°}$ V. Using the Kirchhoff's law method, solve for the current in R_3.

3. Repeat Problem 2 with $X_C = 0$, and R_2 changed to an impedance value of $15 + j15\,\Omega$.

4. Using the Kirchhoff's law method and the circuit of Fig. 35-11, solve for the current through R and the voltage across R.

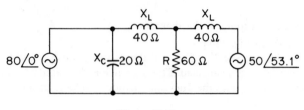

Figure 35-11

5. Using the Kirchhoff's law method and the circuit of Fig. 35-12, solve for current through R_L and the voltage across R_L.

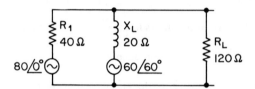

Figure 35-12

6. Using the Kirchhoff's law method and the bridged-T circuit of Fig. 35-13, solve for the current through R_L and the voltage across R_L.

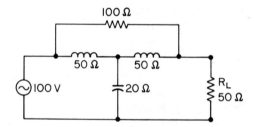

Figure 35-13

7. Repeat Problem 2 using the loop method.
8. Repeat Problem 2 using the loop method, but with $X_C = 0$, and R_2 changed to $15 + j15\ \Omega$.
9. Repeat Problem 4 using the loop method.
10. Repeat Problem 5 using the loop method.
11. Repeat Problem 6 using the loop method.
12. Repeat Problem 2 using the nodal method.
13. Repeat Problem 2 using the nodal method, but with $X_C = 0$, and R_2 changed to $15 + j15\ \Omega$.
14. Repeat Problem 4 using the nodal method.
15. Repeat Problem 5 using the nodal method.
16. Repeat Problem 2 using the superposition method.
17. Repeat Problem 4 using the superposition method.
18. Repeat Problem 5 using the superposition method.
19. Repeat Problem 2 using Thévenin's theorem.

20. Repeat Problem 4 using Thévenin's theorem.
21. Repeat Problem 5 using Thévenin's theorem.
22. Solve Example 35-3, Fig. 35-4, using Thévenin's theorem.
23. Repeat Problem 6 using Thévenin's theorem. (Save this answer for Problem 31.)
24. Repeat Problem 2 using Norton's theorem.
25. Repeat Problem 4 using Norton's theorem.
26. Repeat Problem 5 using Norton's theorem.
27. Repeat Problem 6 using Norton's theorem.
28. In Fig. 35-8(b), the circuit values are:

$Z_A = 500\ \Omega$ resistive, in series with 300 Ω capacitive reactance

$Z_B = 800\ \Omega$ resistive, in series with 1000 Ω capacitive reactance

$Z_C = 300\ \Omega$ resistive, in series with 3000 Ω inductive reactance

Find the circuit values for the equivalent π network.
29. In Fig. 35-9, find the total impedance if:

$$Z_1 = 300 - j400, \qquad Z_4 = 200 + j400$$
$$Z_2 = 100 + j80, \qquad Z_5 = 400 - j300$$
$$Z_3 = 600 - j500,$$

30. Find the maximum power that can be delivered from a 40-V source if its internal impedance is:
    (a) 50 Ω, pure resistive.
    (b) 40 Ω resistive plus 30 Ω inductive.
    (c) As in part (b), but the load is pure resistive.
31. (a) Using the circuit of Fig. 35-13, find the power delivered to the load.
    (b) What value of load would take maximum power from this circuit?
    (c) What is the power output now?

36

POWER TRANSFORMERS

In this chapter we discuss transformers used at power-line frequency (60 Hz), to "transfer" power from a source to a load. Such transformers are invariably wound on iron cores. Iron-core transformers are also used at audio frequencies to couple signal voltages from one stage to the next, or for impedance matching to a loudspeaker, or to a modulation amplifier in a transmitter. Transformers are also available with ceramic cores and with air cores. These are used at high frequencies—for example, as RF and IF transformers, to couple between stages. Since these audio and RF applications are for electronic circuits, these transformers will not be covered in this chapter, but are discussed in pertinent texts.*

36-1 Transformer principle. Basically, a power transformer consists of an iron core on which are mounted a minimum of two windings. One winding, connected to the source of power, is called the *primary*. The other winding, the *secondary*, is connected to the load, and delivers power to the load. A transformer may have more than one secondary winding, each secondary delivering power to its separate load. A basic transformer with only one secondary winding is shown in Fig. 36-1.

When the primary winding is connected to an ac source, current flows in the coil, and a magnetic field is developed around the coil. The flux path is through the iron, as shown in Fig. 36-1 by the dotted line. Since the primary is connected to an ac supply, the current and flux strength will rise to a maximum

*J. J. DeFrance, *General Electronic Circuits* (2nd ed.), Holt, Rinehart and Winston, New York, 1976; J. J. DeFrance, *Communications Electronic Circuits*, Holt, Rinehart and Winston, New York, 1972.

and fall to zero twice per cycle. In each cycle the current and flux will reverse—flowing in one direction for a half-cycle, and in the opposite direction for the other half-cycle. As the flux lines expand and collapse, they cut the primary winding. By Lenz's law, a voltage will be induced in this winding (self-induction). This voltage is in opposition to the applied voltage, and reduces the "net" primary voltage, thereby keeping the primary current at some relatively low value. In an ideal transformer, all of the primary flux reaches and cuts the secondary winding. A voltage is induced in the secondary winding.

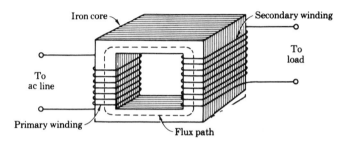

Figure 36-1   Basic transformer.

What will happen if the transformer primary is connected to a dc supply? Except for the initial surge, the primary current will be steady; no voltage will be induced in the secondary; and no counter EMF will be produced in the primary. The primary current will rise to an excessive value, and the winding will be damaged.

36-2 Voltage-turns ratio.   From our studies of induced voltage (Section 17-4), we know that the magnitude of any induced voltage depends on the number of turns in the coil (N), the strength of the magnetic field (Φ), and on the speed of cutting. For a given frequency (speed of cutting), the induced voltage in each transformer winding depends only on the number of turns in the winding. If the primary and secondary windings have the same number of turns, their induced voltages will be equal. The voltage ratio for any other turns ratio will be given by

$$\frac{E_p}{E_s} = \frac{N_p}{N_s} \tag{36-1}$$

In an ideal transformer, the resistance of the windings can be considered as zero. Also, there is no *leakage flux*; that is, all the flux produced by one winding reaches and cuts the other winding. Under these assumptions, the primary induced voltage is equal to the applied voltage, and the secondary induced voltage is equal to the output voltage. Equation (36-1) can therefore be used to represent the ratio of input to output voltage. Even in an actual transformer, the error introduced by these assumption is very small, and equation (36-1) is still valid as the ratio of input to output voltage.

When the primary and secondary have equal number of turns (a 1:1 ratio), the secondary output voltage will be the same as the applied voltage. If we need a lower output voltage, this can be obtained by using fewer turns in the secondary winding. Such a transformer is called a *step-down* transformer. Reversing the technique—using more turns in the secondary—produces a *step-up* transformer, and a higher output voltage.

Example 36-1

The generated voltage at a power plant is 22 kV. The desired transmission line voltage is 345 kV. (a) What type of transformer is needed, and (b) what should the turns ratio be?   —

Solution

(a) Obviously, a step-up transformer is needed.

(b) The turns ratio should be

$$\frac{N_s}{N_p} = \frac{E_s}{E_p} = \frac{345}{22} = = 15.68{:}1 \text{ (step-up)}$$

Example 36-2

At a switching station, it is desired to reduce the 345-kV transmission line voltage to a primary distribution voltage of 13 kV.
(a) What is the required transformer turns ratio?
(b) If the primary has 7500 turns, how many turns should the secondary have?

Solution

(a) $$\frac{N_p}{N_s} = \frac{E_p}{E_s} = \frac{345}{13} = 26.5{:}1 \text{ (step-down)}$$

(b) $$N_s = \frac{E_s}{E_p} \times N_p = \frac{13}{345} \times 7500 = 283 \text{ turns}$$

36-3   Phasor diagram relations.   In an ideal transformer (negligible resistance and negligible leakage flux), the only opposition to the flow of primary current—at no load—is the reactance X_L of the primary winding. This ideal no-load primary current is called the *magnetizing current (I_m)*, and lags the line voltage by 90°. In turn, the magnetic field strength is proportional to the magnetizing current, or—flux (ϕ) is in phase with this current. These no-load relations are shown in Fig. 36-2. Meanwhile, the secondary voltage, as with any induced voltage, must lag the flux by 90°, placing the secondary voltage 180° behind the line voltage. How much power does this ideal transformer take from

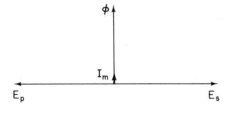

Figure 36-2 No-load, ideal transformer phase relations.

the line at no load? Since the line current lags the line voltage by 90°, the power factor is zero, and the power consumed is zero.

In an actual transformer there are some losses—hysteresis and eddy-current losses in the core—and even at no load there is some copper loss in the primary winding. To supply these losses, the actual primary no-load current cannot lag the supply voltage by a full 90°. It must consist of two components—an in-phase component ($I \cos \theta$) to supply the losses, in addition to the magnetizing current mentioned above, which becomes the quadrature component ($I \sin \theta$). *This total no-load primary current is called the **exciting current**.* However, since the losses at no load are small, the phase angle of this exciting current approaches so close to 90° that the exciting current can be considered identical to the magnetizing current I_m, shown in Fig. 36-2.

If we connect a resistive load across the secondary winding, a current I_s will flow, and this current will be in phase with the secondary voltage E_s. The secondary current in turn produces its own flux, ϕ_s. This new flux cuts the primary turns and produces a new primary voltage *in phase with the line voltage.* The resulting new component of primary current, I_1, is also in phase with the line voltage, or 180° behind I_s. These relations are shown in Fig. 36-3. Current I_1 creates another flux, ϕ_1, which is in direct opposition to the flux ϕ_s caused by the secondary current. This interaction creates a stabilizing balance wherein any secondary fluxes, ϕ_s, are counteracted by primary fluxes, ϕ_1 and neglecting losses, the *resultant flux*, ϕ_R, is identical to the no-load flux of Fig. 36-2.

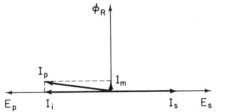

Figure 36-3 Full-load phase relations in an ideal transformer with unity power-factor load.

The total primary current I_p in Fig. 36-3 is the phasor sum of the load component, I_1, and the magnetizing current, I_m. Since the magnetizing component is much smaller than the load component, the primary current is practically in phase with the line current. (The magnetizing current as shown in Fig. 36-3 is not drawn to scale. It is an exaggerated value for plotting convenience.) The transformer takes power from the line to supply the load. Notice also in Fig. 36-3 that E_s and E_p were made equal. This would imply a 1:1 ratio. However, these phasors are generally drawn equal for convenience, regardless of the transformer ratio.

The phasor diagram in Fig. 36-3 is for a pure resistive load. Yet, in practice, particularly due to motor loads, power factors are generally lagging. How will such loads affect the phasor diagram? The principle is still the same; only the current angles will be shifted. One obvious shift is that the secondary current,

I_s, now lags the secondary voltage, E_s, by some angle θ. The primary component of this load, I_1, will still be 180° from I_s, but now it will no longer be in phase with E_p; it will lag E_p by this same angle θ. The total primary current will again be the phasor sum of I_1 and I_m, but it will lag E_p even more. If the load connected to the transformer were capacitive, the current shifts would be reversed, I_s would lead E_s; I_1 would still be 180° from I_s; but I_p, $(I_s + I_m)$, will lead E_p.

36-4 Current–turns ratio. To evaluate the current–turns relationship in a transformer, let us first consider the power relations. If a load is connected across the secondary winding, power is delivered to this load. In the general case this power will be $E_s I_s \cos \theta_s$. Assuming an ideal transformer (no losses or 100% efficiency*), the power taken from the line by the primary ($E_p I_p \cos \theta_p$) must equal the power delivered to the load. What is more, neglecting the very small magnetizing current, the primary power factor equals the secondary power factor, and

$$E_p I_p = E_s I_s$$

Solving for current ratio, we get

$$\frac{I_p}{I_s} = \frac{E_s}{E_p} \tag{36-2A}$$

and substituting turns ratio for voltage ratio yields

$$\frac{I_p}{I_s} = \frac{N_s}{N_p} \tag{36-2B}$$

Notice that *the winding with more turns has the lower current.*

Example 36-3

A transformer delivers 2500 A at 120 V. The primary voltage is 4.0 kV. Find the primary current.

Solution

1. Since this is a step-down transformer, the primary has more turns and will carry a lower current.

2. $$I_p = \frac{E_s}{E_P} \times I_s = \frac{120}{4000} \times 2500 = 75 \text{ A}$$

36-5 Transformer construction. Transformer cores are made of steel alloys carefully selected and processed so as to reduce hysteresis losses. Silicon steel alloys are quite commonly used in power transformers. In addition, the core is laminated, and each lamination is insulated from the next by an oxide and/or varnish coating. This in turn reduces eddy-current losses. To facilitate assembly of the transformer, the laminations are cut into sections and inserted through the coil openings of the prewound coils. The shape of these sections

---

*Transformers are very efficient devices. In well-designed units the efficiency at full load is in the high 90s.

depends on the core construction. There are two fundamental designs used for magnetic circuits: the *core* and the *shell* types:

1. *Core Type.* This type is characterized by one magnetic path surrounding a "window." The iron of the basic transformer shown in Fig. 36-1 is of this type. Laminations for this type of core are generally cut into L-shaped sections. As the laminations are inserted through the coil opening, every other layer is reversed so that the air gap produced at the butt joints falls at opposite corners for successive layers (see Fig. 36-4). In cheaper designs every group of three layers is reversed.

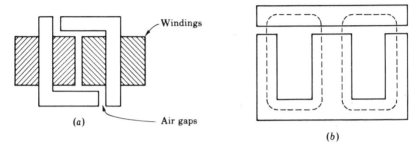

Figure 36-4   Cross sections of core-type and shell-type transformer construction.

2. *Shell Type.* The shell-type construction is characterized by two magnetic paths surrounding two "windows." This construction, shown in Fig. 36-4(b), uses E and I sections. Again, as for core-type units, successive layers (or groups of layers) are reversed so as to distribute the air-gap butt joints among the four corners. With shell-type cores only one coil form is used (containing both primary and secondary windings), and it is mounted on the "tongue" or center leg of the core. Since they are more economical, shell-type cores are more commonly used in tranformers for electronic equipment.

36-6   Grain-oriented steel.   In their research to improve core materials, steel manufacturers have developed a steel with preferred grain orientation. The sheets are cut so that the magnetic flux flows in the direction of the structural grain of the material. One such silicon steel is known by the trade name Hypersil. Grain-oriented steel has lower losses and higher permeability. To take full advantage of grain orientation, a new core construction—the type C core—became necessary. This is shown in Fig. 36-5.

The exterior appearance, size, shape, and weight of transformers differs greatly, depending on their application. For example, Fig. 36-6 shows some typical transformers used in TV or stereo receivers. These units are generally no larger than about 10 cm × 10 cm × 10 cm, and weigh (mass) no more than about 4 kg. At the other end of the scale, we have the transformers used by the utility companies. Figure 36-7 shows three single-phase pole-mounted local distribution transformers rated at 7.62 kV — 120/240 kV, 25 kVA. As you can

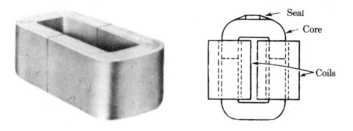

Figure 36-5 Type C grain-oriented core and its application to core-type transformers. (*Courtesy Westinghouse Corp.*)

Figure 36-6 Typical TV and stereo receiver transformers. (*Courtesy Merit Coil & Transformer Corp.*)

Figure 36-7 Pole-mounted local distribution transformers. (*Courtesy Long Island Lighting Co.*)

see from the photo, they are "a bit larger" than the TV receiver type transformers. Yet even these transformers are still rather small—compared to the three-phase unit shown in Fig. 36-8. This "monster" has a base dimension of 6.9 m $\times$ 2.5 m, and is 4.6 m high. It weighs (mass) a mere 235 metric tons! Its rating is 126 kV/69 kV, 224 MVA.

Figure 36-8   Transformer unit, Tuland Road Bank No. 1 LILCO substation. (*Courtesy Long Island Lighting Co.*)

36-7 Losses and efficiency.   In our discussion so far, we have been considering an ideal transformer, that is, one without losses. In a practical unit, however, there are some losses. The most obvious ones are the I^2R losses due to the resistance of the windings. These *copper losses* vary with the load. At no load, the secondary current is zero, and the primary current is the exciting current, which is very low. Consequently, the no-load copper losses are negligible. However, even at no load, since full voltage is applied to the primary winding, the full value of magnetizing current flows, and the full flux density exists in the core. Therefore, hysteresis and eddy-current losses are present at their full value. Since the primary is operated at full voltage regardless of the load value, these *core losses* are a fixed value, and depend on the design (material and cross section) of the core.

The efficiency of a transformer is obviously affected by the losses described above. For a given power input (P_i), the greater the losses, the less the power output (P_o) and the lower the efficiency (η, lowercase Greek letter eta), or

$$\eta = \frac{P_o}{P_i} = \frac{P_o}{P_o + \text{core losses} + \text{copper losses}} \tag{36-3}$$

Obviously, the efficiency of any transformer is zero at no load, since there is no power output. Offhand, it might seem that the efficiency should increase to higher and higher values as the load is increased indefinitely. This is not so. Remember that core losses are a fixed value, but copper losses increase as the *square* of the load current. At first, as the load increases, the power output increases, and the efficiency increases. But the copper losses, being proportional to I^2, begin to increase rapidly. A point is reached at which the increase in copper loss is greater than the increase in power output. The efficiency begins to decrease. This situation is similar to the discussion on maximum power transfer (Section 23-7), and *maximum efficiency occurs when the copper losses equal the core losses.* Transformers are generally designed so that their maximum efficiency occurs at somewhat less than full load. Efficiencies of well-designed transformers are in the high 90s.

Example 36-4

When the power input to a transfer is 300 kW, the core loss is 3.5 kW, and the copper loss is 3.75 kW. Calculate the efficiency at this load.

Solution

1.
$$P_o = P_i - \text{losses} = 300 - 3.5 - 3.75$$
$$= 292.75 \text{ kW}$$

2.
$$\eta = \frac{P_o}{P_i} = \frac{292.75}{300} \times 100 = 97.6\%$$

36-8 Alternate transformer-winding terminology. So far, we have been referring to the transformer windings as the primary and the secondary, wherein the primary is the winding connected to the power line, and the secondary is connected to the load. Now, suppose that we receive a transformer, with one winding rated at 13 kV, and the other winding rated at 69 kV. Which is the primary? Since the transformer is not connected, we cannot answer the question. Either winding could become the primary, depending on whether it will be used to step up a 13-kV supply to 69 kV, or to step down a 69-kV supply to 13 kV. However, regardless of the eventual use, the 69-kV winding is always the **high-voltage side (V_H)**, and the 13-kV winding is always the **low-voltage side (V_L)**. Using this terminology, we can rewrite equations (36-1) and (36-2) as

$$\frac{V_H}{V_L} = \frac{I_L}{I_H} = a \qquad\qquad (36\text{-}4)$$

where a is the ratio of transformation.

We said above that either side of a transformer can be used as the primary. Let us add a word of caution: *As long as the applied line voltage does not exceed the voltage rating of that winding.* For example, suppose that we had a small transformer, rated at 120/24 V. We could connect the high-voltage side (120 V) to a 120-V line, and use the transformer as a step-down unit, to supply 24 V to a transistor power supply. On the other hand, if we had a 24-V ac supply line, we could use this same transformer as a step-up unit, to obtain 120 V for use with standard line-voltage equipment. However, we *cannot* connect the low-voltage winding (24-V side) to the 120-V line and try to use this transformer to step up the 120 V to 600 V. Such a connection would exceed the voltage ratings of the windings and damage the transformer.

36-9 Evaluation of losses.   Since core losses are essentially constant, regardless of load, they can be accurately measured at no load. Rated voltage is applied to one side—whichever is more convenient and/or safer. (Usually, this is the low-voltage side.) The other side is left open, but *care must be taken so that there is no possibility of coming in contact with the terminals of this side.* A wattmeter is used to measure the power input. As noted earlier, copper losses at no load are negligible, so this wattmeter reading is a true measure of the core losses. At no load, the transformer is a highly inductive device. Therefore, a low-power-factor meter should be used for this measurement.

Copper-loss data can be readily obtained from a **short-circuit test**. One side of a transformer is short-circuited. (Usually, it is more convenient to short circuit the low-voltage side.) Then a very low voltage is applied to the other side. This voltage is slowly increased until rated current flows through the windings. Because of the shorted winding, the voltage required for this test is generally less than 5% of the rated voltage for that side. At this reduced voltage, the core losses are negligible, and the input wattmeter reading is a true measure of the copper losses for *both* windings. This is a resistive load, and a full-power-factor wattmeter should be used. A typical schematic for the copper loss test is shown in Fig. 36-9.

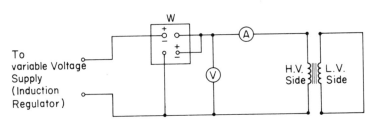

Figure 36-9   Transformer short-circuit test.

36-10 Transformer parameters—equivalent circuit.   Because of induced voltages, any winding of a transformer can be considered as a voltage source, or generator. In an ideal transformer, the voltage induced in the primary (E_p) would be equal to the applied line voltage (V_p). Also, the voltage output from the secondary (V_s) would be equal to the voltage induced in the secondary winding (E_s). But all windings have resistance, R_p for the primary, and R_s for the secondary. Using the general terminology, these would be R_H for the high-voltage winding, and R_L for the low-voltage winding. These are important transformer parameters.

Earlier in the chapter we mentioned that a small amount of the flux produced by the windings does not flow through the iron core, but "escapes" into the surrounding space. We called this the *leakage flux*. Because of this leakage, the effective, or *mutual flux*, and the induced voltages are not quite as high as they could have been. Instead of considering this reduction in voltage in terms of a reduction in flux, it is simpler to consider it as due to a reactive voltage drop (IX), due to a **leakage reactance X** *in series with the windings*. This gives rise to an X_H in the high-voltage side, and an X_L in the low-voltage side.

To evaluate the parameters above, we can use the data from a short-circuit test. The wattmeter in Fig. 36-9 measures the copper losses in both windings of the transformer. This is the copper loss at rated load if the ammeter reading shows the rated current for the high-voltage side. Let us call this value I_H. Simultaneously, the current I_L in the short-circuited low-voltage side must be more than this value (due to the transformation ratio), or

$$(1) \qquad I_L = aI_H$$

Then the copper losses can be broken down to

$$(2) \qquad P_{cu} = I_H^2 R_H + I_L^2 R_L$$

where R_H and R_L are the resistances of the high- and low-voltage windings, respectively. Replacing I_L^2 in (2) by its value from (1), we get

$$(3) \; P_{cu} = I_H^2 R_H + (aI_H)^2 = I_H^2(R_H + a^2 R_L)$$

$$= I_H^2 R_{eH}$$

$$(36\text{-}5\text{A})$$

where the term $R_{eH} = (R_H + a^2 R_L)$ is called the **equivalent resistance of both windings referred to the high side.***

So, from the wattmeter reading in Fig. 36-9, if we divide this copper loss value by the square of the ammeter reading, we get the value of R_{eH}. This is one of the parameters for the transformer equivalent circuit.

Again, referring to Fig. 36-9, the voltage V_{sH} required to force full-load current I_{sH} through the transformer winding on a short-circuit test must overcome a total impedance effect due to both windings. With the connections as shown (power fed to the high-voltage side), this impedance effect is the equivalent impedance Z_{eH}, of both windings, referred to the high side, and since

$$V_{sH} = I_{sH} Z_{eH}$$

then

$$Z_{eH} = \frac{V_{sH}}{I_{sH}} \qquad (36\text{-}7)$$

As with any impedance, Z_{eH} is made up of the resistance R_{eH}, and the reactance X_{eH}, each referred to the high-voltage side. Therefore,

---

*Similarly, had we replaced I_H^2 in (2) by its value from (1),

$$P_{cu} = I_L^2 \left(\frac{R_H}{a^2} + R_L \right) = I_L^2 R_{eL} \quad (36\text{-}5\text{B}) \qquad \text{and} \qquad \frac{R_{eH}}{R_{eL}} = a^2 \quad (36\text{-}6)$$

$$Z_{eH} = \sqrt{R_{eH}^2 + X_{eH}^2}$$

and solving for reactance,

$$X_{eH} = \sqrt{Z_{eH}^2 - R_{eH}^2}* \qquad (36\text{-}8)$$

This is the other important parameter for a transformer equivalent circuit. A simplified equivalent circuit using these parameters is shown in Fig. 36-10.

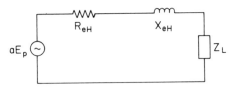

Figure 36-10 Simplified transformer-equivalent diagram (referred to the high-voltage side).

Example 36-5

A 13.2 kV — 120/240 V, 100 kVA distribution transformer is connected for a short-circuit test, with the low-voltage side shorted. The voltage applied to the high-voltage side is 420 V when the rated current of 7.58 A flows through the winding. The wattmeter reading is 1600 W. Find the equivalent resistance and reactance of this transformer, referred to the high side.

Solution

1. $$P_{cu} = I_H^2 R_{eH}$$

$$R_{eH} = \frac{P_{cu}}{I_H^2} = \frac{1600}{(7.58)^2} = 27.8 \ \Omega$$

2. $$Z_{eH} = \frac{V_{sH}}{I_{sH}} = \frac{420}{7.58} = 55.4 \ \Omega$$

3. $$X_{eH} = \sqrt{Z_{eH}^2 - R_{eH}^2} = \sqrt{(55.4)^2 - (27.8)^2}$$

$$= 47.9 \ \Omega$$

36-11 Regulation.   At no load, the secondary output voltage is the same as the induced voltage in the winding. As load is applied, there will be an IR drop and an IX drop. Therefore, the output voltage under load will differ from the no-load output voltage. Usually, the output voltage with load will be less than the no-load voltage. However, with capacitive loads, if the power factor is sufficiently low, the output voltage will actually increase with load. This effect can be seen from Fig. 36-11, which shows the phasor relations in the secondary side of a high-voltage step-up transformer. Diagram (a) is for a lagging power factor load. Starting with the full-load voltage as the reference

---

*Again, as with the resistance values, the equivalent reactance, referred to the low-voltage side, can be found from

$$\frac{X_{eH}}{X_{eL}} = a^2 \qquad (36\text{-}9)$$

phasor (V_{FL}), we must add an IR_{eH} drop (in phase with the current), and an IX_{eH} drop (leading the current by 90°), in order to get the no-load voltage, V_{NL}. As usually expected, this no-load voltage is higher than the voltage under load. Now, examine diagram (b). Again, we will use the full-load voltage as the reference. This time the load current, I_L, leads V_{FL} by some large angle θ. To get the no-load voltage, we add the IR_{eH} drop (in phase with I_L) and the IX_{eH} drop (leading I_L by 90°). *Notice that this no-load voltage is less than the full-load voltage. The output voltage increased when this capacitive load was applied.*

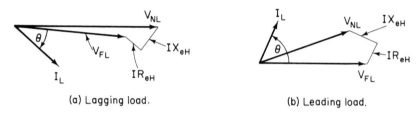

(a) Lagging load.   (b) Leading load.

Figure 36-11   Transformer regulation, phasor diagram.

The term **regulation** is used to express mathematically *the ratio of the change in voltage from no-load to full-load, divided by the full-load voltage,* or

$$\text{percent regulation} = \frac{V_{NL} - V_{FL}}{V_{FL}} \times 100 \qquad (36\text{-}10)$$

Notice that if the output voltage rises with the addition of load, the regulation will be a negative value.

Example 36-6

The transformer of Example 36-5 is used as a step-down transformer for local power distribution at 120 V. Find:
(a) The resistance and reactance referred to the low-voltage side.
(b) The full-load current at unity power factor.
(c) The no-load output voltage.
(d) The percent regulation.

Solution

(a) 1. $$a = \frac{13{,}200}{120} = 110$$

2. $$R_{eL} = \frac{R_{eH}}{a^2} = \frac{27.8}{(110)^2} = 2.30 \times 10^{-3}\ \Omega$$

3. $$X_{eL} = \frac{X_{eH}}{a^2} = \frac{47.9}{(110)^2} = 3.99 \times 10^{-3}\ \Omega$$

(b) $$I = \frac{P}{V} = \frac{100\,000}{120} = 833\ \text{A}$$

(c) $V_{NL} = V_{FL} + I(R_{eL} + jX_{eL})$

$= 120 + (833 \times 2.30 \times 10^{-3}) + j(833 \times 3.99 \times 10^{-3})$

$= 122 \text{ V}$

(d)   Regulation $= \dfrac{V_{NL} - V_{FL}}{V_{FL}} \times 100 = \dfrac{122 - 120}{120} \times 100 = 1.67\%$

36-12 Autotransformers.   If isolating the load circuit from the line is not necessary, then a single winding with a suitable tap can be used to step up or step down the line voltage. Such a device is called an *autotransformer*. Because it has only one winding, autotransformers are less expensive in first cost, and in operation, than a transformer of equivalent rating. Figure 36-12 shows how this device can be used to obtain either step-up or step-down action. In diagram (a), the line is connected across the full winding, while the lower portion is common to both the input and output sides. This is a step-down connection, and the output voltage is less than the input voltage by the same ratio as the output/input number of turns. For example, if the number of turns from the bottom (common), to the tap is one-fifth of the total, then the output voltage is one-fifth of the input voltage. Putting this relation in equation form, we get

$$\frac{E_o}{E_i} = \frac{N_o}{N_i} \qquad\qquad \textbf{(36-11)}$$

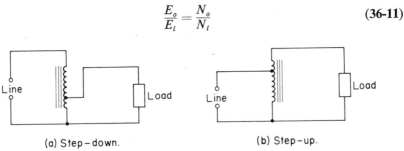

(a) Step–down.   (b) Step–up.

Figure 36-12   Autotransformer connections.

In diagram (b), the line voltage is connected to the tap, while the load is connected across the entire winding. Since the load side has more turns, the output voltage is higher than the line voltage. Equation (36-11) still applies, but this time we have a step-up action.

36-13 Transformers in a power-distribution system.   An example of the use of transformers in a power-distribution system is one used by the Long Island Lighting Company (LILCO). A typical generator voltage at the power plant is (approximately) 20 kV. For long-distance transmission, this is stepped up by transformer banks to its main-line voltage of 138 kV. When interconnections with the neighboring power company* lines are to be

*Consolidated Edison of New York.

made, a transformer substation steps up the 138 kV to 345 kV. On the other hand, for local use, the 138 kV is stepped down at distribution stations, to 69 kV and/or 13 kV. Then local underground or pole-mounted transformers step down the 69 or 13 kV to the 120/240 V for residential use. A typical transformer distribution substation is shown in Fig. 36-13.

Figure 36-13 Transformer distribution substation. (*Courtesy Long Island Lighting Co.*)

REVIEW QUESTIONS

1. Name the three basic components of a transformer.
2. (a) Which winding of a transformer is identified as the primary?
   (b) What is a secondary winding?
3. Explain briefly the principle of transformer action.
4. A transformer is connected to a dc supply at rated voltage. Explain what would happen.
5. What is meant by the term *1:1 ratio* as applied to a transformer?
6. (a) What is meant by a *step-up* transformer?
   (b) How is this feature obtained?
7. (a) What is meant by a *step-down* transformer?
   (b) How is this feature obtained?

8. Refer to Fig. 36-2.
  (a) Is this diagram for an ideal or for a practical transformer?
  (b) What is the phasor I_m called?
  (c) How does it compare with the primary current in the ideal transformer?
  (d) Why does this current lag the primary voltage by 90°?
  (e) How much power does an ideal transformer draw from the line at no load?

9. What, if any, are the losses in an actual transformer at no load?

10. (a) What is the "total" no-load primary current of a transformer called?
  (b) Does this current lag the line voltage by exactly 90°?

11. Refer to Fig. 36-3.
  (a) For what type of load does this diagram apply? How can you tell this is so?
  (b) What does the phasor I_1 represent?
  (c) What does the phasor I_p represent? How is it obtained?
  (d) Compare the length of phasors E_s and E_p. Does this imply a 1:1 transformation ratio? Explain.

12. How does the secondary current compare with the primary current:
  (a) In a step-up transformer?
  (b) In a step-down transformer?
  (c) On what does the ratio of these currents depend?

13. (a) Why is the core of a transformer laminated?
  (b) Why are laminations insulated from each other?
  (c) Why are laminations cut into sections?
  (d) Why are the laminations in each layer (or groups of layers) reversed?

14. (a) Name two fundamental core structures used for magnetic circuits.
  (b) To which type does the design shown in Fig. 36-1 belong?
  (c) In the core of Fig. 36-4(b), on which leg(s) are the windings placed?

15. (a) State two advantages of using grain-oriented steel for the core.
  (b) What type of core assembly is used with this steel?

16. (a) Name two types of losses that occur in transformers.
  (b) Which of these is considered a fixed loss? Explain.
  (c) What is the relation between the two losses when maximum efficiency occurs?
  (d) In the design of a transformer, at what load value is this maximum efficiency made to occur?
  (e) What is a typical value for the efficiency of a well-designed transformer?

17. (a) A transformer's windings are rated at 120 V and 240 V, respectively. Which is the primary?
  (b) Give another terminology that can be used to identify these windings.
  (c) What is the ratio of transformation for this transformer?
  (d) Can this transformer be used to step up a 240-V supply to 480 V? Explain.
  (e) Give two possible uses for this unit.

18. (a) A supply line feeds rated voltage to one winding of a transformer. The other side is open.
  (a) Will the transformer draw any power from the line?
  (b) If a wattmeter is used to measure this power input, of what is it a measure?
  (c) What type of wattmeter should be used for such measurements? Explain.

19. Refer to Fig. 36-9.

(a) Since the load side is shorted, wouldn't this transformer burn out? Explain.

(b) For this test, how is the current limited to the rated value?

(c) Of what is this wattmeter reading a measure?

20. (a) To what does the term *leakage flux* refer?

(b) What is the effect of this leakage flux, compared to an ideal case without leakage?

(c) Name a parameter used to account for this effect.

21. Give the letter symbols used to represent the following.

(a) The resistance of the low-voltage winding

(b) The resistance of the high-voltage winding

(c) The resistance of both windings, referred to the high-voltage side

(d) The leakage reactance of the low-voltage side.

(e) The leakage reactance of both windings, referred to the low-voltage side.

22. From what test can data be obtained to evaluate the above parameters?

23. (a) What does the term *regulation* mean?

(b) Which is better, a high or low percent regulation?

(c) What two parameters are responsible for the variation of the output voltage with load?

(d) Does the output voltage always drop with increase in load? Explain.

(e) Is the percent regulation ever a negative value? Explain.

24. Refer to Fig. 36-11.

(a) Why is I_L drawn upward in one diagram and downward in the other?

(b) What is IR_{eH}?

(c) At what angle is this voltage drop drawn?

(d) What is IX_{eH}?

(e) At what angle is this always drawn?

25. (a) What is the function of an autotransformer?

(b) How does this device differ from a "standard" transformer?

(c) Give an advantage of the autotransformer.

(d) Give a disadvantage.

PROBLEMS AND DIAGRAMS

1. The secondary winding of a transformer has 150 turns. Its primary winding, for use on a 120-V line, has 400 turns. Find the secondary voltage.

2. If a secondary voltage of 600 V were required from the transformer of Problem 1, what change in the transformer design would be necessary? Give a specific numerical answer.

3. A transformer has a primary winding of 150 turns, but only 8.3 turns in the secondary. Find the secondary voltage for an input of 120 V.

4. A 6.3-V output voltage is desired. The transformer has a 520-turn primary and will be energized from a 120-V line. How many turns should the secondary have?

5. How many primary turns should a transformer have, to get 275 V from a 1200-turn secondary, if the primary voltage is 220 V?

6. Draw a phasor diagram for an ideal transformer showing the voltage, flux, and current relations with a leading power factor load.

7. Repeat Problem 6 for a lagging power factor load.

8. The transformer of Problem 1 has a load current of 6.0 A. Find the primary current.

9. A step-down transformer has a 300-turn primary and a 15-turn secondary. If the secondary current is 6.0 A, find the current it draws from the line.

10. A step-up transformer draws 1.8 A from a 120-V line. It has a 400-turn primary and a 3300-turn secondary. Find the secondary current and voltage.

11. A transformer has a power input of 42 W. The secondary has a resistive load of 115 mA at 340 V. Find the efficiency of the transformer.

12. A transistor radio transformer supplies 30 V at 3 A, unity power factor. If its efficiency is 92%, find the primary current. (The line voltage is 120 V.)

13. A distribution transformer carries a 100 kW load. The core loss is 1.2 kW, and the copper losses at this load are 1.5 kW. Find the efficiency.

14. Find the efficiency of the transformer in Problem 13 if the load were 100 kVA at 80% power factor lagging.

15. The power input to a 300-kVA distribution transformer is 3.1 kW when rated voltage is applied to the 13-kV side and the 138-kV side is open. Then 6.9 kV is applied to the 138-kV side while the 13-kV side is shorted. The rated current of 2.7 A flows in the primary, and the power input is measured at 3.5 kW. Find:
   (a) The core loss.
   (b) The copper loss.
   (c) The efficiency at full load, unity power factor.
   (d) The efficiency at full load, 0.8 power factor lag.

16. Using the data from Problem 15, find:
   (a) The equivalent resistance, referred to the high-voltage side.
   (b) The equivalent reactance, referred to the high side.
   (*Note:* Save these answers for Problem 19.)

17. A transformer has a rated full-load output voltage of 240 V. The no-load voltage is 250 V. What is the regulation in percent?

18. A transformer with a rated full-load voltage of 13.2 kV has a regulation of 4.5%. To what value will its voltage rise at no load?

19. Using the given data from Problem 15 and the R and X values found in Problem 16, find:
   (a) The no-load voltage.
   (b) The percent regulation.

20. The transformer in Fig. 36-12(a) has a total number of turns of 200. The tap is at the fiftieth turn from the bottom. If the line voltage is 120 V, what is the load voltage?

21. The transformer of Problem 20 is reconnected as shown in Fig. 36-12(b). What is the load voltage now?

37

THREE-PHASE SYSTEMS

In Chapter 25 we saw that a sine wave of voltage was produced when a coil was rotated in a uniform magnetic field. Such a single-phase voltage source was used in the chapters that followed. Now let us suppose that the simple generator of Chapter 25 has two coils rotating in the same magnetic field. [This was done in Section 26-2 when discussing phase and time relations. Each coil produced its own sine-wave voltage. When the two coils were displaced by 90° we had two voltages 90° apart (see Fig. 26-3).] Such a source would constitute a two-phase system. This is the simplest polyphase system. Since two-phase supplies are not generally used, they will not be discussed. However, the same principle can be extended to include more coils, symmetrically spaced, producing other polyphase systems. The most common of these is the three-phase system.

37-1 Generation of three-phase supply. The generator shown in Fig. 37-1 has three coils (aa', bb', and cc') spaced 120° apart. Each of these coils generates a sine-wave voltage as explained in Chapter 25. At the instant shown, coil aa' is not cutting any lines. The voltage generated in this coil is zero. During the next half turn, terminal a will be moving down, electrons will be pulled away from this end, leaving it positive. At the same time terminal a' will be negative. Therefore, the voltage across this coil—terminal a with respect to terminal a'—starts at zero, rises to a positive maximum, and falls back to zero. As the coil continues to turn, a full sine wave of voltage is generated for each revolution. This waveshape is shown as the solid curve ($E_{aa'}$) in Fig. 37-1(b). Simultaneously, at the instant shown in Fig. 37-1(a), terminal b of coil bb' is already moving up through the magnetic field. At this instant it is negative and approaching its maximum value. Again, as the coil rotates, it will produce a sine-wave voltage, but due to its space location this wave lags the previous coil's voltage by 120°. Voltage *bb'* is shown as the long-short dash curve. Similarly, voltage *cc'*, shown as the dashed curve, lags voltage *aa'* by 240° (or leads by 120°). At the instant shown in Fig. 37-1(a), terminal *c* is moving down; its potential is positive but approaching zero.

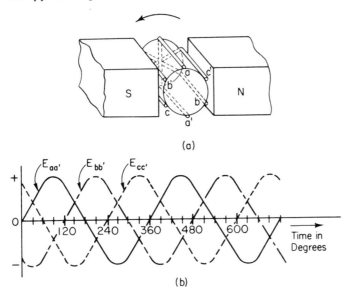

(a)

(b)

Figure 37-1   Generation of three-phase voltage supply.

The three generated voltages constitute a three-phase supply. The phase relations between these voltages are shown by the phasor diagram, Fig. 37-2.

Alternating-current generators are known as **alternators**. Their construction differs from the simple diagram shown in Fig. 37-1(a). The major difference is that the generating coils are stationary, and the field structure revolves. For a three-phase supply, three generating coils or **stator** windings are used. The windings are wound on the frame of the machine and are spaced 120 *electrical* degrees apart. The field structure is an electromagnet and is mounted on the rotating shaft of the machine. It may be a two-pole structure, or it may have any even number of poles. Where more than two poles are used, one revolution produces one cycle (360 electrical degrees) for each pair of poles. This, however, does not alter the above discussion. Details of the alternators are left to texts on machinery.

37-2   Phase sequence.   Due to the space location of the coils in a polyphase system, each coil reaches its positive maximum value at a different time. In our machine of Fig. 37-1(a), starting with coil *aa'*, the next coil to reach maximum value is *bb'*, then comes *cc'*, and then *aa'* again. This is known as the *phase sequence* of this three-phase system. If the direction of rotation of our machine were reversed, the phase sequence would have been *aa'*, *cc'*, *bb'*. No other phase sequence is possible with this machine. Instead of the positive maximum value, any other instantaneous value could be used to determine phase sequence. In general, the phase sequence of any polyphase system can be stated as the order in which the phase voltages reach a specific instantaneous value as seen by the "eye" in Fig. 37-2.

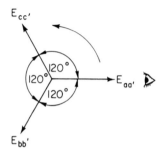

Figure 37-2   Phasor diagram of a three-phase voltage supply.

37-3   Double-subscript notation.

In the above diagrams and discussion you may have noticed the three voltages referred to by double subscripts, *aa'*, *bb'*, and *cc'*. As you will see later, the three phases of the alternator will be interconnected. Phasor addition of these voltages will be necessary. If a coil's connections are reversed with respect to another coil or a load, its instantaneous polarity is also reversed (180° phase difference). The phasor representing this coil's voltage should be shown 180° out of phase in contrast to its original direction. Some method must be used to distinguish between these two conditions. In addition, polyphase circuits generally involve several voltages and currents. When tracing through such circuits, it is important to know the instantaneous polarity of the voltages and the direction of flow of the currents encountered. Double-subscript notation will solve this problem. The first subscript is used to indicate the starting point of the coil, component, or voltage source through which we are tracing, and the second subscript is the finish point. In other words, subscript *aa'* would mean in the direction from *a* to *a'*. More specifically, with regard to voltages, the second subscript represents the reference point, so that $E_{aa'}$ means the potential at *a* with respect to *a'*.* This can be clarified by referring to Fig. 37-3. The voltage across this coil can be measured from *a'* to *a*, or from *a* to *a'*. The voltage $E_{a'a}$ is positive in polarity (*a'* is positive with respect to *a*).

Figure 37-3   Double-subscript notation.

*With regard to power applications, some texts use the first subscript as the reference point. However, in transistor circuits, voltages are specified, universally, as shown here.

Conversely, $E_{aa'}$ is negative in polarity (*a* with respect to *a'*). Obviously then

$$E_{a'a} = -E_{aa'}$$

An illustration will show the value of this system of notation. Figure 37-4(a) and (b) shows two coils of a generator, spaced 60° apart, and the phase relation between the voltages generated in their windings. If we connect these two coils in series, we have four possible output voltages, depending on which end of each coil is tied together. Using 100 V as the effective value of each coil voltage and $E_{a'a}$ as reference, let us find the output voltage for each of the four connections.

1. *Tie a and b' Together.* Tracing from the first free terminal *a'* to the other free end *b*, we pick up voltage $E_{a'a}$ and then $E_{b'b}$. Therefore, the resultant

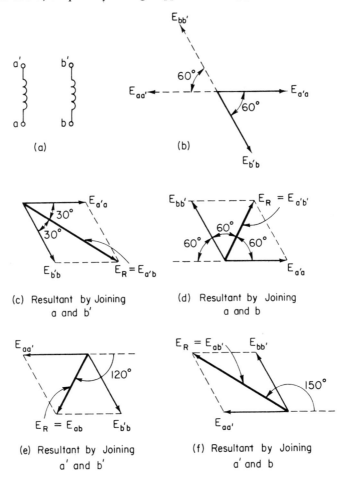

(c) Resultant by Joining
   a and b'

(d) Resultant by Joining
   a and b

(e) Resultant by Joining
   a' and b'

(f) Resultant by Joining
   a' and b

Figure 37-4   Phasor addition, showing use of double-subscript notation.

voltage is

$$E_R = \dot{E}_{a'a} + \dot{E}_{b'b} = E_{a'b}$$

This phasor relation is shown in Fig. 37-4(c). By vector algebra

$$
\begin{aligned}
E_{a'a} &= 100 + j0 \\
E_{b'b} &= \underline{\ \ 50 - j86.6} \\
E_R &= 150 - j86.6 = 173\underline{/-30°}\ \text{V}
\end{aligned}
$$

2. *Tie a and b Together.* Using the same technique—starting at terminal a' and tracing through to the other free end—we get

$$E_R = \dot{E}_{a'a} + \dot{E}_{bb'} = E_{a'b'}$$

$$
\begin{aligned}
E_{a'a} &= \ \ 100 + j0 \\
E_{bb'} &= \underline{-50 + j86.6} \\
E_R &= \ \ \ 50 + j86.6 = 100\underline{/60°}\ \text{V}
\end{aligned}
$$

This phasor addition is shown in Fig. 37-4(d).

3. *Tie a' and b' Together* [Fig. 37-4(e)].

$$E_R = \dot{E}_{aa'} + \dot{E}_{b'b} = E_{ab}$$

$$
\begin{aligned}
E_{aa'} &= -100 + j0 \\
E_{b'b} &= \underline{\ \ \ 50 - j86.6} \\
E_R &= \ \ -50 - j86.6 = 100\underline{/-120°}\ \text{V}
\end{aligned}
$$

4. *Tie a' and b Together* [Fig. 37-4(f)].

$$E_R = \dot{E}_{aa'} + \dot{E}_{bb'} = E_{ab'}$$

$$
\begin{aligned}
E_{aa'} &= -100 + j0 \\
E_{bb'} &= \underline{\ \ -50 + j86.6} \\
E_R &= -150 + j86.6 = 173\underline{/150°}\ \text{V}
\end{aligned}
$$

Notice from the above examples, that the direction in which a coil is connected can make appreciable difference not only in the magnitude but also in the phase of the resultant voltage. The double-subscript notation made circuit evaluation simple. This system of notation will be used throughout the discussion of polyphase systems.

37-4 Three-phase delta (Δ) connection.   A three-phase alternator has three sets of coils and therefore six output terminals. If six lines were used to connect between supply source and load, it would result in an inefficient distribution system because of the amount of copper required. To reduce the number of lines, the coils are interconnected at the alternator. One such method is known as the **delta** (or **mesh**) connection. This method is shown in Fig. 37-5(a). Coil end a is connected to coil end c', coil ends c to b' and b to a', to form a closed circuit. Each junction is brought out, resulting in an three-wire distribution system. But doesn't this interconnection short-circuit the alternator? It could—if the connections were improperly phased! Let us examine the phasor diagrams for the connections we are using. Figure 37-5(b) shows the phase relations among the three coil voltages. Since this is a three-phase system, these voltages are 120° apart. By connecting coil end a with coil end c', we are adding $E_{c'c}$ to $E_{a'a}$. Then when we connect coil end b' to coil end c we are adding $E_{b'b}$ to the previous two voltages. This phasor addition of the three coil voltages is shown in Fig. 37-5(c). *Notice that the resultant is zero!* Therefore, we are able to close the circuit (connect coil end b to coil end a') without creating a short circuit, and without causing any current to circulate in the delta loop.

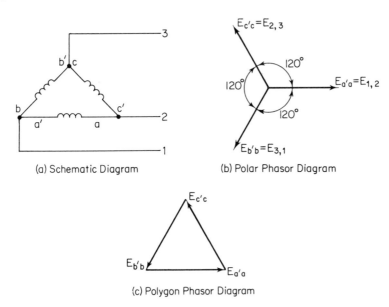

(a) Schematic Diagram   (b) Polar Phasor Diagram

(c) Polygon Phasor Diagram

Figure 37-5   Connections and voltage relations in a 3ϕ delta system.

By definition, in any polyphase circuit the voltage across an individual coil is known as the *phase* voltage, while the voltage between any two lines is called the *line* voltage. Examination of Fig. 37-5 shows that the voltage between lines 1 and 2 equals $E_{a'a}$; between lines 2 and 3 equals $E_{c'c}$; and between lines 3 and 1 equals $E_{b'b}$. In other words, *in any delta system the line voltage equals the phase voltage*, or

$$E_L = E_p \quad \text{(delta circuit)} \tag{37-1}$$

Example 37-1

A voltmeter connected across lines 1 and 2 of a three-phase delta supply (Fig. 37-5) indicates 100 V. The phase sequence is known to be $E_{a'a}$, $E_{b'b}$, $E_{c'c}$.
(a) Write the equation for each line voltage, using $E_{1,2}$ as reference.
(b) Express each voltage in polar form and in rectangular form.

Solution

(a) $\qquad\qquad E_m = \dfrac{E}{0.707} = \dfrac{100}{0.707} = 141 \text{ V}$

$$E_{1,2} = E_m \sin(\omega t + 0°) = 141 \sin \omega t$$
$$E_{3,1} = E_m \sin(\omega t - 120°) = 141 \sin(\omega t - 120°)$$
$$E_{2,3} = E_m \sin(\omega t + 120°) = 141 \sin(\omega t + 120°)$$

(b)
$$E_{1,2} = E\underline{/0°} = 100\underline{/0°} = 100 + j0 \text{ V}$$
$$E_{3,1} = E\underline{/-120°} = 100\underline{/-120°} = -50 - j86.6 \text{ V}$$
$$E_{2,3} = E\underline{/120°} = 100\underline{/120°} = -50 + j86.6 \text{ V}$$

Earlier in this discussion, we mentioned that proper phasing must be used when connecting the coils in delta. The connections shown in Fig. 37-5 were correct, but let us see what would happen if the terminals of one coil were reversed. Suppose that we had connected coil end c to b' as before, but we now connect coil end a to coil end b [see Fig. 37-6(a)]. Referring to the phasor diagram [Fig. 37-6(b)], $E_{c'c} + E_{b'b}$ would still be as shown previously. Now however, when we add $E_{aa'}$, this voltage is the reverse (180°) of $E_{a'a}$ and it would extend to the left. *The resultant voltage $\dot{E}_{c'c} + \dot{E}_{b'b} + \dot{E}_{aa'}$ is not zero, but would equal twice the phase voltage.* Coil end a' must not be connected to coil end c' as this would cause a short circuit. The circulating current within the delta may burn out the windings.

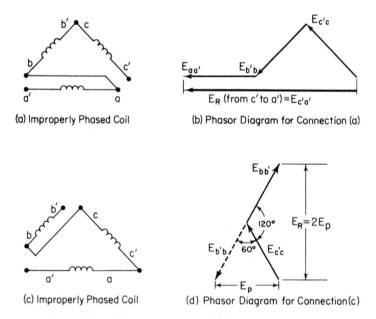

(a) Improperly Phased Coil

(b) Phasor Diagram for Connection (a)

(c) Improperly Phased Coil

(d) Phasor Diagram for Connection (c)

Figure 37-6 Effect of improper coil phasing in a delta system.

A similar situation will occur if the second coil is improperly phased. This condition is shown in Fig. 37-6(c) and (d). Notice that when two coils are properly connected the resultant voltage value is equal to E_p, whereas when one coil is reversed, the resultant value is $2E_p$.

In a commercial machine, if the two leads from each winding are brought out without any phasing identification, how can we tell if we are making the correct connections? A simple voltmeter check will solve the problem.

1. Measure the phase voltage across any one coil.
2. Connect any two coils in series and measure the voltage across their free ends. If this voltage is still equal to the phase voltage the phasing is correct; if the new reading is almost double the phase voltage ($1.73E_p$), the connection is incorrect. This effect is shown in Fig. 37-6(d). To correct this situation, merely reverse *either* coil.
3. Connect the third coil in series, and again measure the voltage across the free ends. This voltage may be zero, or twice the phase voltage. If it is zero, fine—connect the free ends together and the delta is completed. On the other hand, if the voltage is double the phase voltage, reverse the *third* coil, and now complete the delta.

37-5 Three-phase wye (Y) connection.

Another system for interconnecting the three coils of a three-phase supply is to tie one end of each coil together, and bring out the three free ends as the three line wires, to form a three-phase, three-wire system. This type of connection is known as the *wye* (*Y*) or *star* system. In addition, a fourth line—*the neutral*—is usually also brought out from the junction of the three coils. We now have a three-phase, four-wire system. This is shown in Fig. 37-7. For convenience of comparison with the previous delta system, the wye has been drawn sideways so as to keep the individual coil voltages at the same phase angles in both cases. Let us examine the voltage relations, for a phase sequence of $E_{a'a}$, $E_{b'b}$, $E_{c'c}$.

The three phase voltages, as before, are $120°$ apart. But what about the line voltages? Inspection of the circuit interconnections in Fig. 37-7(a) shows that the voltage between any two lines is the phasor sum of two of the phase voltages. For example, the potential of line 3 with respect to line 2 ($E_{3,2}$ or E_{ca}) is the phasor sum of $E_{cc'}$ ($E_{c'c}$ reversed) and $E_{a'a}$. In a similar manner we can find the remaining line voltages.

$$E_{ca} = E_{3,2} = \dot{E}_{cc'} + \dot{E}_{a'a}$$
$$E_{ab} = E_{2,1} = \dot{E}_{aa'} + \dot{E}_{b'b}$$
$$E_{bc} = E_{1,3} = \dot{E}_{bb'} + \dot{E}_{c'c}$$

These phase and line voltage relations are shown vectorially by the polygon diagram and polar diagram of Fig. 37-7(b) and (c). Notice that the phase voltage $E_{a'a}$ *leads* the line voltage E_{ca} (the line voltage with the same second subscript) by $30°$. Similarly notice that phase voltages $E_{b'b}$ and $E_{c'c}$ also lead their *respective* line voltages E_{ab} and E_{bc} by $30°$. This $30°$ phase shift between phase and line voltages is characteristic of any three-phase, four-wire wye system, or of any *balanced* three-phase, three-wire wye system.

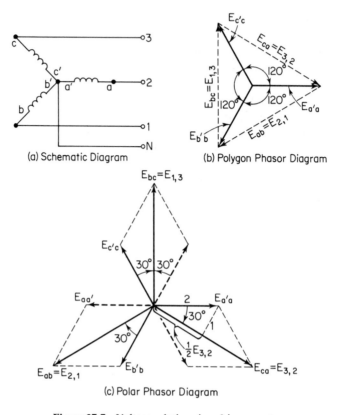

(a) Schematic Diagram

(b) Polygon Phasor Diagram

(c) Polar Phasor Diagram

Figure 37-7   Voltage relations in a 3ϕ wye system.

Let us backtrack for a moment. We obtained the above line voltages, by starting at line 3 and tracing through to line 2 ($E_{3,2}$). Then we followed in rotation, from line 2 to line 1 ($E_{2,1}$), and from line 1 back to line 3 ($E_{1,3}$). Referring to the polygon diagram, Fig. 37-7(b), we went from $E_{c'c}$ to $E_{a'a}$, $E_{a'a}$ to $E_{b'b}$, $E_{b'b}$ to $E_{c'c}$. This is the same order in which we would "see" our phase voltages.

Now there is no reason why we could not have taken our line voltages in the opposite order, that is from line 1 to line 2 to line 3. We would then get line voltages $E_{1,2}$, $E_{2,3}$, and $E_{3,1}$. For example, $E_{1,2}$ would be E_{ba} and would equal

$$E_{ba} = E_{1,2} = \dot{E}_{bb'} + \dot{E}_{a'a}$$

This line voltage is shown in Fig. 37-8. In our previous analysis of line voltages, we had

$$E_{ab} = E_{2,1} = \dot{E}_{aa'} + \dot{E}_{b'b}$$

This line voltage is also shown in Fig. 37-8.

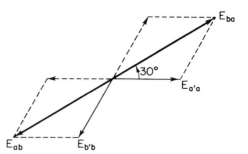

Figure 37-8 Comparison of voltage between lines 1 and 2 taken with opposite instantaneous phase polarities.

Notice that the two voltages are 180° apart. The explanation is simple. *They are the same voltage* taken with opposite ends as reference. In one case (E_{ab}) we took the *instantaneous* polarity of the component phase voltage as $E_{aa'}$ and $E_{b'b}$. In the second case (E_{ba}) we reversed the instantaneous polarity of the phases and used $E_{a'a}$ and $E_{bb'}$. Either analysis is correct, as long as we are consistent throughout the circuit. To avoid possibility of confusion, we will use the first method throughout this text (i.e., take the line voltages in the same order as the phase sequences).

Returning to the phasor diagram of Fig. 37-7, we have seen how to show the phase relation between phase and line voltages. Let us now consider the *magnitude* of the line voltage. By inspection it is obviously greater than the phase voltage. How much greater? To find this relation, notice again that each phase voltage leads its corresponding line voltage (line voltage with the same second subscript) by 30°. Now drop a perpendicular from the end of $E_{a'a}$ to $E_{3,2}$, forming a 30–60–90 triangle, and cutting $E_{3,2}$ in half. This is shown in Fig. 37-7(c). Notice that $E_{a'a}$ forms the hypotenuse of this triangle. Now in such a triangle, we know that the side opposite the 30° angle is one-half the hypotenuse. Therefore, if we assign a value of 1 to the side opposite the 30° angle, the hypotenuse $(E_{a'a})$ is 2, and by the right-triangle theorem the side adjacent $(\frac{1}{2}E_{3,2})$ is $\sqrt{2^2 - 1^2}$, or $\sqrt{3}$. But $E_{3,2}$ is twice this value, or $2\sqrt{3}$. Then

$$\frac{E_{3,2}}{E_{a'a}} = \frac{2\sqrt{3}}{2} \quad \text{and} \quad E_{3,2} = \sqrt{3}\ E_{a'a}$$

By inspection of the other line and phase voltages, it is obvious that the same relationship holds true. We can therefore generalize and say that *in a wye-connected system the line voltage leads or lags its component phase voltages by 30° and exceeds the phase voltage by $\sqrt{3}$*, or

$$E_L = \sqrt{3}\ E_p \quad \text{(wye circuit)} \tag{37-2}$$

Example 37-2

Using the phase voltage values found in Example 37-1 (for delta connection), and $E_{a'a}$ again as the reference phasor, write the equation for each line voltage when the coils are connected in wye, and express each line voltage in polar and in rectangular coordinate form.

Solution

1. From Example 37-1, we found that

$$E_{a'a} = 141 \sin \omega t$$
$$E_{b'b} = 141 \sin (\omega t - 120°)$$
$$E_{c'c} = 141 \sin (\omega t + 120°)$$

2. Each line voltage $= \sqrt{3} \, E$ phase, and lags 30° behind the adjacent phase voltage. Therefore,

(a) $E_{3,2} = \sqrt{3} \, E_{a'a}$ delayed 30° $= 244 \sin (\omega t - 30°)$

(b) $E_{2,1} = \sqrt{3} \, E_{b'b}$ delayed 30° $= 244 \sin (\omega t - 150°)$

(c) $E_{1,3} = \sqrt{3} \, E_{c'c}$ delayed 30° $= 244 \sin (\omega t + 90°)$

3. Converting to effective values and expressing in vector algebra, we get

$$E_{3,2} = 173\underline{/-30°} = 150 - j86.6 \text{ V}$$
$$E_{2,1} = 173\underline{/-150°} = -150 - j86.6 \text{ V}$$
$$E_{1,3} = 173\underline{/90°} = 0 + j173 \text{ V}$$

In a wye system, the alternator coils do not form a closed loop. Therefore, is it necessary to consider which end of each coil is connected to the neutral or common junction? Definitely yes! If we reverse a coil, the line voltages will not be 120° apart, nor will this voltage equal $\sqrt{3} \, E_p$. Again a voltmeter can be used to check proper phasing.

1. Connect two coils in series and measure the voltage across the free ends. This voltage should be $\sqrt{3} \, E_p$. If it is less, reverse one coil.

2. Now connect one end of the third coil to the common junction. Again, the voltage from the free end of this coil to the free end of each of the other two coils should be $\sqrt{3} \, E_p$. If this voltage value is not obtained, reverse the third coil connections.

37-6   Loading of a three-phase system.   We have seen so far that a three-phase alternator can be connected in delta or wye to form a three-wire distribution system. (Discussion of the four-wire wye system will be considered later.) The question now is, how is the load applied? A load can be applied across any one pair of lines to form a single-phase load; or more than one load can be connected across any combination of pairs of lines to form a polyphase load. The loads can be pure resistive, pure reactive, or any combination of resistance and reactance. The simplest type of load to analyze (and the most efficient) is the *balanced load*. Such a load is achieved when all line currents are equal and have the same phase angle with respect to the line voltages. This is the only type of load that will be discussed in this chapter.

37-7 Current relations in a balanced wye-connected load.
One method of obtaining a balanced load is to connect three identical loads in
wye. Such a load circuit is shown in Fig. 37-9(a). This load is a pure resistive
load. It is evident from the circuit diagram that the line current I_1 flows through
resistor R_1 and must be the same current as the phase current. Similarly, I_2 and
I_3 equal, respectively, the phase currents in resistors R_2 and R_3. From this we
can conclude that in a wye-connected load, the line current is the same as the
phase current.

$$I_L = I_p \quad \text{(wye circuit)} \tag{37-3}$$

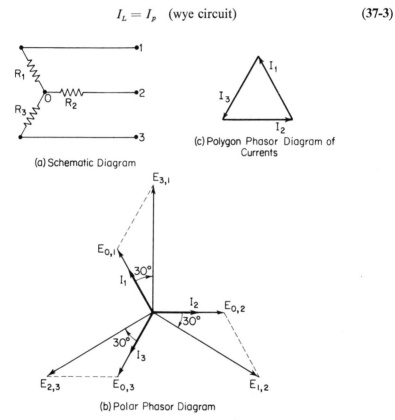

(c) Polygon Phasor Diagram of
Currents

(a) Schematic Diagram

(b) Polar Phasor Diagram

Figure 37-9   Current and voltage relations in a 3ϕ wye resistive load.

Another thought may occur to you as you look at the circuit diagram. If
all three line currents are flowing toward, or away from the free end of the
wye-connected resistors, what happens to these currents at the neutral, or junction point of the resistors? Current cannot pile up at a point. True enough, but
we have not yet considered their phase relations. From our knowledge of voltage
relations in a wye circuit, we can draw the phasor diagram for the phase and line
voltages. We also know that in any resistive circuit, the current through the

resistor is in phase with the voltage across it. But the voltage across each resistor is the phase voltage. Therefore, the line currents (which are the same as the phase currents) are in phase with the phase voltages, and are 120° apart. As can be seen in Fig. 37-9(c) the sum of these line currents is zero, and so there is no pile-up at the junction.

37-8 Phase angle in a three-phase circuit. When we studied single-phase circuits we learned that in a resistive circuit, the circuit phase angle between line voltage and line current was zero. Now examine the phasor diagram, Fig. 37-9(b). This represents a resistive circuit. What is the phase angle between line voltage and line current? That is a problem—which line current to which line voltage? Between I_1 and $E_{3,1}$, current leads by 30°; between I_1 and $F_{1,2}$ current leads by 150°. Shall we make it more confusing? Between I_1 and $E_{2,1}$ the current lags by 30°. Obviously, we can no longer speak of circuit phase angle as the angle between line current and line voltage. Now examine the phase relation between any phase current and *its phase* voltage. They are in phase, and this is a resistive circuit. This phase angle makes sense again. So in polyphase circuits *reference to phase angle, unless otherwise specified, means the angle between phase voltage and the corresponding phase current.* In a balanced load this phase angle is the same for all three phases.

Example 37-3

Three 20-Ω resistors are connected in wye across a three-phase 220-V supply. Find the line current and circuit phase angle.

Solution

1. Since the resistors are connected in wye, the voltage across each resistor is the phase voltage. In a wye system the phase voltage is *less* than the line voltage.

$$E_p = \frac{E_L}{\sqrt{3}} = \frac{220}{\sqrt{3}} = 127 \text{ V}$$

2.
$$I_p = I_L \text{ (wye)} = \frac{E_p}{R_p} = \frac{127}{20} = 6.35 \text{ A}$$

3. Since the circuit is resistive, I_p is in phase with E_p, and the phase angle is zero.

Example 37-4

Three impedances, each having a resistance of 30 Ω and an inductive reactance of 40 Ω, are connected in wye across a three-phase 208-V system. Find the line current and phase angle.

Solution

1.
$$Z_p = \sqrt{R_p^2 \times X_p^2} = \sqrt{(30)^2 + (40)^2} = 50 \ \Omega$$

2.
$$E_p = \frac{E_L}{\sqrt{3}} = \frac{208}{\sqrt{3}} = 120 \text{ V}$$

3. $$I_p = I_L \text{ (wye)} = \frac{E_p}{Z_p} = \frac{120}{50} = 2.40 \text{ A}$$

4. $$\theta_p = \text{arc tan} \frac{X_p}{R_p} = \text{arc tan} \frac{40}{30} = 53.1°$$

Let us consider now the effect of a reactive load on the phasor diagram for a wye system. The phase voltage and line voltage relations are unchanged. But the current phasors will be shifted. Referring to Fig. 37-9, an inductive load will cause the currents I_1, I_2, and I_3 to lag their respective phase voltages by some angle θ, the value of which depends on the ratio of inductive reactance to resistance of the load. Conversely, a capacitive load will cause the line currents to lead their respective phase voltages.

Example 37-5

Using a line voltage phase sequence of $E_{1,2}$, $E_{2,3}$, $E_{3,1}$, and $E_{1,2}$ as the reference, draw the phasor diagram for the currents and voltages of Example 37-4. Express each current in polar form.

Solution

1. Line voltages. Since the line voltages are 120° apart and $E_{1,2}$ is the reference voltage, $E_{2,3}$ lags by 120° and $E_{3,1}$ lags by 240° (or leads by 120°). Draw these phasors (see Fig. 37-10).

2. Phase voltages. Let us call the phase voltages $E_{0,1}$, $E_{0,2}$, and $E_{0,3}$. These phase voltages must be 120° apart, in the phase sequence as listed, and leading their respective line voltages (the line voltage with the same second subscript) by 30°. In addition, from the designation of the line voltages,

$$E_{1,2} = \dot{E}_{1,0} + \dot{E}_{0,2}$$
$$E_{2,3} = \dot{E}_{2,0} + \dot{E}_{0,3}$$
$$E_{3,1} = \dot{E}_{3,0} + \dot{E}_{0,1}$$

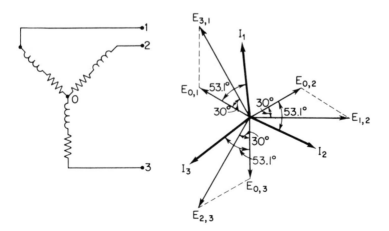

Figure 37-10   Current and voltage relations in a 3ϕ wye inductive load.

This locates our phase voltages with respect to the line voltages. Add these voltages to the phasor diagram, Fig. 37-10.

3. Line currents. In Example 37-4, the phase angle of the load was found to be 53.1°, and the load was inductive. Since the line current is also the phase current (wye load), this places each line current 53.1° lagging with respect to the *phase* voltage. Add these currents to the phasor diagram.

4. To express the currents in polar form we must know their magnitude and phase angle. From Example 37-4, we found the magnitude to be 2.40 A. Now let us use the phasor diagram (Fig. 37-10) to establish their phase.
   (a) I_2 lags $E_{0,2}$ by 53.1° and therefore lags the reference phasor ($E_{1,2}$) by 23.1°.

$$I_2 = 2.40/\!\!-\!23.1° \text{ A}$$

   (b) I_3 lags I_2 by 120°.

$$I_3 = 2.40/\!\!-\!143.1° \text{ A}$$

   (c) I_1 leads I_2 by 120°.

$$I_1 = 2.40/96.9° \text{ A}$$

Example 37-6

Solve Example 37-4, using vector algebra, and using a phase sequence of $E_{1,2}$, $E_{2,3}$, $E_{3,1}$, and $E_{1,2}$ as reference phasor.

Solution

1. To solve for currents using vector algebra, we must express all voltages and impedances in vector form. We will start with line voltages.
   (a) $E_{1,2} = 208/0° \text{ V}$
   (b) $E_{2,3} = 208/\!\!-\!120° \text{ V}$
   (c) $E_{3,1} = 208/\!\!+\!120° \text{ V}$

2. Phase voltages. Remembering that the phase voltages $= E_L/\sqrt{3}$ in magnitude and lead their respective line voltages by 30°,
   (a) $E_{0,1}$ (leading $E_{3,1}$ by 30°) $= 120/\!\!+\!150° \text{ V}$
   (b) $E_{0,2}$ (leading $E_{1,2}$ by 30°) $= 120/\!\!+\!30° \text{ V}$
   (c) $E_{0,3}$ (leading $E_{2,3}$ by 30°) $= 120/\!\!-\!90° \text{ V}$

3. Load impedances. From Example 37-4, the load impedance was found to be 50 Ω and the phase angle 53.1°. Since the load is inductive, this makes the impedances

$$Z = 50/53.1° \text{ } \Omega$$

4. Line currents.
   (a) $I_1 = \dfrac{E_{0,1}}{Z_1} = \dfrac{120/150}{50/53.1} = 2.40/96.9° \text{ A}$
   (b) $I_2 = \dfrac{E_{0,2}}{Z_2} = \dfrac{120/30}{50/53.1} = 2.40/\!\!-\!23.1° \text{ A}$

(c) $$I_3 = \frac{E_{0,3}}{Z_3} \frac{120/-90}{50/53.1} = 2.40/-143.1° \text{ A}$$

These currents check with our phasor analysis of Example 37-5.

37-9   Current relations in a balanced delta connected load.
Just as alternator coils can be connected in wye or delta, so can loads. If the impedance of each "leg" of the delta loads is the same, we will again have a balanced load. A balanced resistive delta-connected load is shown in Fig. 37-11(a). Since the load is resistive, the phase currents must be in phase with their phase voltages. But in a delta circuit, we already know that phase voltages are identical with line voltages. These phase relations are shown in Fig. 37-11(b). Now let us examine the line currents.

$$I_1 = I_{c'c} + I_{aa'}$$
$$I_2 = I_{a'a} + I_{bb'}$$
$$I_3 = I_{b'b} + I_{cc'}$$

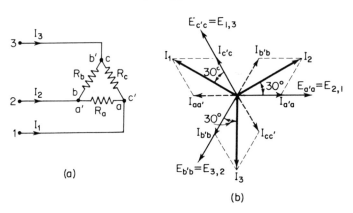

(a)

(b)

Figure 37-11   Current and voltage relations in a 3φ delta resistive load.

The line currents are greater than the phase currents, and these current phasors are similar to the voltage phasors for the wye circuit. The same method of analysis can be used to prove that

$$I_L = \sqrt{3}\, I_p \quad \text{(delta)} \qquad\qquad (37\text{-}4)$$

Example 37-7
The three resistors of Example 37-3 (20 Ω each) are reconnected in delta to the same supply line (220-V, three-phase). Find the line current and the circuit phase angle.

Solution

1. $$E_L = E_p = 220 \text{ V (delta)}$$

2. $$I_p = \frac{E_p}{R_p} = \frac{220}{20} = 11.0 \text{ A}$$

3. $I_L = \sqrt{3}\, I_p = 11\sqrt{3} = 19.0 \text{ A}$

4. Since the circuit is resistive, I_p is in phase with E_p, and the phase angle is zero.

From a comparison of Examples 37-3 and 37-7, notice that for the same supply voltage and the same resistance values, the line current in the delta load is three times the line current for the wye load.

37-10 Three-phase four-wire distribution system. Earlier in this chapter we learned that the coils of a three-phase alternator could be connected in delta or in wye. The delta connection resulted in a three-wire distribution system. The wye connection could be used as a three-wire system, or by bringing out the neutral (common terminal of the wye-connected phases), a four-wire distribution system could be obtained. This latter system (3Φ, four-wire) has definite advantages and is in common use. To explain why this is so leads us into a discussion on the selection of a suitable line voltage as related to efficiency and safety.

Power delivered to (or taken by) a load is dependent on voltage and current. For a given amount of power, the higher the line voltage, the lower the line current. From the standpoint of transmission-line efficiency, the lower the line currents, the better. The copper losses (I^2R) in the transmission lines and in the equipment itself will be reduced; or the size of wire used in the lines or equipment can be made smaller, for the same power loss; or the lines and equipment can be used at a higher power rating. On the other hand, from the standpoint of safety, it is obvious that the lower the line voltage, the less the possibility of lethal or serious shocks. As a compromise between safety and efficiency the local distribution line voltages commonly used are 120 V (approximately) for lighting and low-power equipment such as fractional horsepower motors, and 220 V (approximately) for higher-power loads. Of course, the power company can (and does) use much higher voltages (23 000 to 750 000 V) at its power stations and long-distance transmission lines, and then steps down (by transformers) to the consumer voltage level at local distribution areas.

With a three-wire distribution system, if the line voltage is 220 V, additional transformers will be needed, or the main distribution transformers must be center-tapped in order to supply the lower voltage needed for lighting loads or low-power appliances. This in turn means at least three more wires for the local distribution system. On the other hand, the four-wire distribution system, with a slight compromise, can furnish the proper voltage to either type of load. The line voltage used is 208 V. Since it is a wye-connected system, the phase voltage between any line and the neutral is 120 V:

$$E_p = \frac{E_L}{\sqrt{3}} = \frac{208}{\sqrt{3}} = 120 \text{ V}$$

Notice the flexibility available. This four-wire system can be used to supply a three-phase 208-V load by connecting the load to the three line wires;

it can be used to supply single-phase 208 V loads by connecting the loads to any pair of lines; and it can be used to supply single-phase 120-V loads by connecting the loads between any line and the neutral.

REVIEW QUESTIONS

1. With reference to Fig. 37-1:
   (a) How many coils does this generator have?
   (b) Physically, how are these coils related?
   (c) What is the effect of this physical placement?
   (d) What is such a supply called?

2. (a) What does the term "phase sequence" mean?
   (b) Starting with coil bb', what is the phase sequence for the voltages generated in Fig. 37-1?
   (c) What is the sequence of the voltages in Fig. 37-2?

3. (a) What is the purpose of specifying a coil voltage by double-subscript notation?
   (b) How does voltage $E_{nn'}$ compare with voltage $E_{n'n}$?

4. With reference to Fig. 37-3:
   (a) What is the polarity of voltage $E_{a'a}$?
   (b) What is the meaning of subscript $a'a$?
   (c) What is the polarity of voltage $E_{aa'}$? Explain.

5. In Fig. 37-4(c):
   (a) What are the given voltages?
   (b) What is their phase relation? Why?
   (c) Why is the resultant called $E_{a'b}$?
   (d) How is this condition obtained physically?

6. In Fig. 37-4(d):
   (a) How is $E_{bb'}$ obtained?
   (b) What is E_R the resultant of?
   (c) How is this condition obtained?

7. In Fig. 37-4(f):
   (a) What two voltages make up $E_{ab'}$?
   (b) How is $E_{aa'}$ obtained?
   (c) How is $E_{bb'}$ obtained?

8. With reference to Fig. 37-4(b):
   (a) Express voltage $E_{a'a}$ in vector algebra, rectangular form.
   (b) Express this voltage in polar form.
   (c) Express $E_{aa'}$ in vector algebra, both forms.
   (d) Express $E_{bb'}$ in vector algebra, polar form.
   (e) Express $E_{b'b}$ in vector algebra, polar form.
   (f) How is $E_{b'b} = 50 - j86.6$ obtained?

9. With reference to Fig. 37-5:
   (a) How many coils and leads does this generator have?
   (b) Why are the coils interconnected?
   (c) What type of interconnection is this?
   (d) Give another name for this interconnection.

10. In general, what is the distinction between the terms *phase voltage* and *line voltage*?

11. In a delta system, what is the relation between phase and line voltage? Why is this so?

12. With reference to Fig. 37-5(a), give the line voltage corresponding to each of the following phase voltages.
    (a) $E_{a'a}$
    (b) $E_{c'c}$
    (c) $E_{aa'}$
    (d) $E_{bb'}$
    (e) $E_{cc'}$
    (f) $E_{b'b}$

13. (a) What is the phasing between the line voltages of a delta-connected three-phase generator?
    (b) What is the phasing between the individual coil voltages?
    (c) What is the phase sequence for the generator in Fig. 37-5? (Use line voltages.)

14. (a) What is the phasor sum of the phase voltages for a delta connection?
    (b) Why must this be so?

15. With reference to Example 37-1, *Solution*:
    (a) Describe $E_{1,2}$ as given by the second equation in step (a).
    (b) Describe $E_{1,2}$ as given by the first equation in step (b).
    (c) Describe $E_{3,1}$ as given by the third equation in step (a).
    (d) Describe $E_{3,1}$ as given by the second equation in step (b).
    (e) Add the three line voltages in step (b). What is the resultant?

16. With reference to Fig. 37-6(c):
    (a) Which coil is improperly connected?
    (b) Which are the free coil ends now?
    (c) What is the voltage across these free ends?
    (d) Can this delta connection be completed? Explain.
    (e) How is this error corrected?
    (f) How is this error avoided when interconnecting windings?

17. (a) What type of connection is shown in Fig. 37-7(a)?
    (b) Give another name for this type of connection.
    (c) What advantage does this connection have over a delta?
    (d) When used in this manner, what else is this *output* called?
    (e) What is the phase relation between phase voltages?

18. In diagram 37-7(c):
    (a) Using phase $a'a$ as reference, what is the phase sequence?
    (b) What two voltages make up voltage $E_{1,3}$?
    (c) From the schematic (diagram a) show if this is so.
    (d) What two voltages make up voltage $E_{2,1}$?
    (e) Why is this also called E_{ab}?
    (f) How is $E_{3,2}$ obtained?
    (g) What is the phase relation between line voltages?
    (h) Compare the magnitude of the line and phase voltages.
    (i) What is the phase relation between any line voltage and its adjacent phase voltage?

19. For consistency, in what sequence should the line voltages be specified?
20. In the solution to Example 37-2, step 1:
    (a) Where does the numerical value 141 come from?
    (b) Why is $E_{a'a}$ at zero degrees?
21. In the solution to Example 37-2, step 2:
    (a) From where do the values 244 come?
    (b) Why is $E_{3,2}$ at an angle of $(\omega t - 30°)$?
    (c) Why is $E_{1,3}$ at an angle of $(\omega t + 90°)$?
22. In the solution to Example 37-2, step 3:
    (a) How are the numerical values 173 obtained?
    (b) How are the rectangular coordinate values for $E_{2,1}$ obtained?
23. (a) Since the wye connection is *not* a closed circuit, is coil phasing of any importance? Explain.
    (b) How is correct phasing established?
24. With a three-phase three-wire supply:
    (a) Can single-phase loads be used? Explain.
    (b) What is a balanced load?
25. With reference to Fig. 37-9(a):
    (a) How are the resistors connected to the supply source?
    (b) Under what condition is this a balanced load?
    (c) Where does current I_1 flow?
    (d) Trace the flow path for current I_3?
    (e) What is the relation between line and phase currents for this wye load.
    (f) If the load is unbalanced, what is the relation between line and phase currents?
26. With reference to Fig. 37-9(b):
    (a) Is this a balanced load? How can you tell?
    (b) Is the load resistive, inductive, or capacitive? How can you tell?
27. (a) In a polyphase load, to what does the term phase angle refer?
    (b) How does this apply to a balanced load?
28. In the solution of Example 37-3:
    (a) Why is it necessary to solve for E_p?
    (b) Why does $I_p = I_L$?
    (c) Why is the phase angle zero?
29. In the solution to Example 37-4:
    (a) Outline the need for each of the steps shown to find I_L.
    (b) What does θ_p represent?
30. With reference to Fig. 37-10:
    (a) What is used as the reference phasor?
    (b) Express each line voltage in vector algebra, polar form.
    (c) Why does $E_{0,2}$ lead $E_{1,2}$ by 30°?
    (d) Does this relation apply to any other phasors?
    (e) Express each phase voltage in polar form.
    (f) How is the phase angle for I_2 obtained?
    (g) Express each current in polar form.
31. With reference to the solution of Example 37-6:
    (a) How is the Z value (step 3) obtained?

(b) How is the $\underline{/96.9°}$ obtained in step 4(a)?

(c) How is the $\underline{/-23.1°}$ obtained in step 4(b)?

32. With reference to Fig. 37-11:

(a) What type of load is this?

(b) How does the line voltage compare to the phase voltage?

(c) What is the phase relation between phase current and phase voltage? Why?

(d) What does line current I_1 consist of?

(e) Why is current $I_{aa'}$ shown dotted in diagram (b)?

(f) Explain how 30° is obtained between I_1 and $I_{c'c}$?

(g) How does the magnitude of I_1 compare to $I_{c'c}$?

(h) What is the phase angle between line current and line voltage?

(i) Since this is a resistive load, explain the above phase angle value.

33. In the solution of Example 37-7, the circuit phase angle is given as zero. How does this compare with the phase relations shown in Fig. 37-11?

34. Three single-phase loads can be connected in delta, or in wye:

(a) Which connection will result in a higher line current?

(b) How much more? Why?

35. (a) If a four-wire distribution system is desired, how should the coils of a three-phase alternator be connected?

(b) What is a commonly used line voltage for such a distribution system, at the consumer level?

(c) State three types of load that can be supplied from such a system.

PROBLEMS

1. Two coils of an alternator each generate 120 V, but the phase relation between these coils is such that E_{cd} of coil 2 lags E_{ab} of coil 1 by 90°. Find the magnitude and phase angle of the four possible resultant voltages when these coils are connected in series. Use E_{ab} at zero as reference.

2. Repeat Problem 1 for a generated voltage of 80 V per coil and E_{cd} leading E_{ab} by 70°.

3. Repeat Problem 1 for a generated voltage of 120 V and E_{cd} lagging E_{ab} by 120°.

4. (a) Draw a phasor diagram for the voltages in a three-phase delta system for a phase sequence of $E_{1,2}, E_{2,3}, E_{3,1}$, and starting with $E_{1,2}$ at 90° leading.

(b) For a generated voltage of 220 V, express each line voltage in polar and rectangular form.

5. Repeat Problem 4 for a generated voltage of 200 V, phase sequence of $E_{1,2}, E_{3,1}, E_{2,3}$, and starting with $E_{1,2}$ at 0°.

6. Repeat Problem 4 for a generated voltage of 100 V, phase sequence of $E_{1,2}, E_{2,3}, E_{3,1}$, and starting with $E_{1,2}$ at 90° lagging.

7. The three phases of an alternator are to be connected in delta. The generated voltage per phase is 200 V. Describe a procedure, and give values, for obtaining the proper connections.

8. (a) Draw a phasor diagram for the phase and line voltages in a three-phase, four-wire wye system, for a phase sequence of $E_{a'a}$, $E_{b'b}$, $E_{c'c}$, and starting with $E_{a'a}$ at 90° lagging.

(b) For a generated phase voltage of 120 V, express each voltage in polar and rectangular form.

9. Repeat Problem 8 for a generated phase voltage of 100 V, phase sequence of $E_{a'a}$, $E_{c'c}$, $E_{b'b}$, and starting with $E_{a'a}$ at 0°.

10. The three phases of an alternator are to be connected in wye. The generated voltage per phase is 120 V. Describe a procedure, and give values, for obtaining the proper connections.

11. Three 50-Ω resistors are connected in wye across a 60-Hz 220-V three-phase supply. Find the line current and circuit phase angle.

12. Repeat Problem 11 for three 0.2-H inductors.

13. Three loads, each consisting of 20-Ω resistance and 30-Ω capacitive reactance, are connected in wye across a 60-Hz 208-V three-phase four-wire supply. Find the line current and circuit phase angle.

14. Repeat Problem 11 for the loads connected in delta.

15. Repeat Problem 11 for three 0.2-H inductors connected in delta.

16. Repeat Problem 13 for the loads connected in delta.

38

POWER IN THREE-PHASE SYSTEMS

In the discussion of single-phase ac circuits, it was shown that the power dissipated in such a circuit could be found from

$$P = EI \cos \theta$$

In a three-phase circuit, this same equation can be used to find the power dissipated in each phase. To keep our "bookkeeping" correct, the equation should now be rewritten as

$$P_p = E_p I_p \cos \theta_p$$

where the subscript p is used to denote phase values.

This subscript is often omitted from the $\cos \theta$ term. But you will recall that unless otherwise stated, θ refers to the phase angle between I_p and E_p. Therefore, the meaning of the equation is unchanged.

In a polyphase circuit, each phase is dissipating power (load), or delivering power (alternator). The total power for the circuit must then be the sum of the individual phase powers.

$$P_T = P_1 + P_2 + P_3 + \cdots$$

where P_1, P_2, P_3 represent the respective phase powers. This equation applies to any polyphase circuit whether balanced or unbalanced. For a *balanced* three-phase system, since each phase power is equal, we can simplify the equation above by

$$P_T = 3P_p$$

Replacing P_p by its equivalent $E_p I_p \cos \theta$, we get

$$P_T = 3E_p I_p \cos \theta \tag{38-1}$$

But in a delta system $E_p = E_L$ and $I_p = I_L/\sqrt{3}$. Substituting line values for phase values yields

$$P_T = 3E_L \frac{I_L}{\sqrt{3}} \cos\theta = \sqrt{3} \ E_L I_L \cos\theta$$

Similarly in a wye system, $E_p = E_L/\sqrt{3}$ and $I_p = I_L$. Substituting these line values for the original phase values, we have

$$P_T = 3\frac{E_L}{\sqrt{3}} I_L \cos\theta = \sqrt{3} \ E_L I_L \cos\theta$$

Notice that in *either* case, wye or delta, *for a balanced load* the equation for total power in terms of line current and line voltage is the same:

$$P_T = \sqrt{3} \ E_L I_L \cos\theta \qquad (38\text{-}2)$$

Does the $\cos\theta$ term refer to the angle between the line current and line voltage? No. Remember that there are many phase angles among the three line currents and line voltages; θ refers to the angle between the phase current and the phase voltage *of the same phase*. In a balanced circuit, this angle is the same for all three phases.

Example 38-1

A wye-connected three-phase balanced load when connected to 220-V three-phase supply draws a line current of 50 A. The power factor of the load (phase value) is 0.8 lagging. Find the total power drawn by the load.

Solution

1. Since line current and voltage are known, it makes no difference whether the load is delta- or wye-connected.
2. $P_T = \sqrt{3} \ E_L I_L \cos\theta = \sqrt{3} \times 220 \times 50 \times 0.8 = 15.24 \ \text{kW}$

Example 38-2

A delta-connected load has an impedance per phase of 20-Ω resistance and 40-Ω inductance. It is connected to a 440-V three-phase supply. Find the power taken by the load.

Solution

1. $Z_p = \sqrt{R_p^2 + X_p^2} = \sqrt{(20)^2 + (40)^2} = 44.7 \ \Omega$

2. $\cos\theta_p = \dfrac{R_p}{Z_p} = \dfrac{20}{44.7} = 0.447$

3. $E_p = E_L = 440 \ \text{V (delta)}$

4. $I_p = \dfrac{E_p}{Z_p} = \dfrac{440}{44.7} = 9.84 \ \text{A}$

5. $I_L = \sqrt{3} \ I_p = 9.84 \sqrt{3} = 17.04 \ \text{A}$

6A. $P_T = \sqrt{3} \ E_L I_L \cos\theta = \sqrt{3} \times 440 \times 17.04 \times 0.447$
$$= 5.81 \ \text{kW}$$

or

6B.
$$P_T = 3P_p = 3E_p I_p \cos \theta$$
$$= 3 \times 440 \times 9.84 \times 0.447 = 5.81 \text{ kW}$$

38-1 Measurement of power.   We have seen that power in a balanced three-phase circuit can be calculated readily from line or phase values. However, if the load is not balanced, the calculations are much more involved. In commercial applications, since the loads are often unbalanced, it is therefore desirable to measure power directly. We know how to use a wattmeter to measure power in a single-phase load (see Section 28-9). The same method could be used to measure the power in each phase. The current coil of the wattmeter should be in series with each phase, and the potential coil should be connected across each phase. In a four-wire wye system this can be done readily. Since the phase current is the same as the line current, each wattmeter current coil is connected in series with each line before tying the line to the load. Each watt-meter potential coil is connected with one end to its own current-coil line terminal, and the other end to the neutral, since this is its phase voltage. But in a three-phase balanced wye load, such as a three-phase oven, the neutral is not brought out. That does create a problem. The wattmeter connections described above become impractical if not absolutely impossible. With a delta load, the situation is even worse. Line currents and phase currents are not equal. It would be necessary to open each delta junction and insert each wattmeter current coil in series with each phase *inside* the delta junctions. This is definitely impractical. Luckily, there are other techniques for measuring power in a three-phase circuit.

38-2 The three-wattmeter method.   Here again, we are using three wattmeters, but regardless of whether the load is delta- or wye-connected, each current coil is connected in series with a *line* lead. The potential coils are connected at one end to their own current-coil line, and at the other end to a common junction (see Fig. 38-1). The potential coils actually form a wye-connected load. If the load itself were also a balanced wye-connected load, the

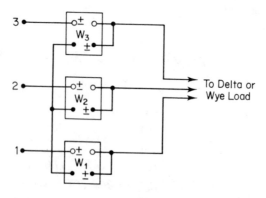

Figure 38-1   The three-wattmeter method for measuring power.

load phase-voltage would be identical with the potential coil voltage. Also, in any wye load the line current and phase current are identical. Therefore, each wattmeter indicates the phase power, and the sum of the three wattmeter readings is the total power. If the power factor of the load is changed, the phase voltage and the voltage across the potential coils are not affected. The angle of the phase current, and therefore line current, will shift with respect to phase voltage. The wattmeter readings will change, but all three wattmeter readings will still be identical. However, if the load is unbalanced, the "neutral" of the load will not correspond to the "neutral" of the potential coils. The three wattmeter readings will not be identical, but the sum of the three will still be the true total power. The same effect, unequal wattmeter readings, will also be produced if the three wattmeter potential coils do not have identical impedances, but again the correct total power will be the sum of the three wattmeter readings. Since unbalanced circuit analysis has not been covered, the proof for the effect of unbalanced loads or unbalanced potential coils cannot be given here.

If the load is a delta load, will the wattmeter connections of Fig. 38-1 give the correct power indication? To answer this we will have to use a phasor diagram. Figure 38-2(a) shows a three-phase delta resistive load being metered by the three-wattmeter method. Since the potential coils of the wattmeters form a wye load, they have been shown diagrammatically as a wye within the delta. The delta configuration and nomenclature used are the same as were used in Fig. 37-11. Therefore, the same phasor diagram can be repeated here to show current and line voltage relations. If the wattmeters are "properly" connected, each should be recording phase power $E_p I_p$. (We have simplified the problem by using resistive load which makes $\cos \theta = 1$.) For example, one meter should deflect due to the phase current $I_{a'a}$ and line voltage $E_{2,1}$. (Remember that with a delta load the line voltage is also the phase voltage.) But no wattmeter has the full delta voltage across its potential coil. Let us find these voltages. Since the potential coils are wye-connected, three *wye-phase voltages*, 120° apart, will be

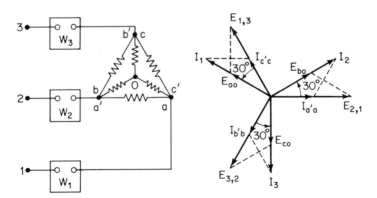

Figure 38-2   The three-wattmeter method of measuring power, used with delta load.

developed across the potential circuits. These voltages are E_{ao}, E_{bo}, and E_{co}, 30° away from the line voltages and such that

$$\dot{E}_{bo} + \dot{E}_{oa} = E_{2,1}$$

$$\dot{E}_{co} + \dot{E}_{ob} = E_{3,2}$$

$$\dot{E}_{ao} + \dot{E}_{oc} = E_{1,3}$$

This places E_{ao}, E_{bo}, and E_{co} in phase with the line currents I_1, I_2, and I_3, respectively. The deflection of the wattmeter in line 1 is due to line current I_1 and the wye-phase voltage E_{ao}. As far as magnitudes are concerned, $I_1 = \sqrt{3}\,I_p$ and $E_{ao} = E_p/\sqrt{3}$, and the product $I_1 \times E_{ao} = I_p \times E_p$. As far as phase angles are concerned, I_1 and E_{ao} are in phase and so are E_p and I_p for each phase. In other words, the power indicated by the wattmeter in line 1 is equal to the actual phase power. Similarly, the power indicated by each of the wattmeters in lines 2 and 3 is also the phase power. Therefore the connections shown will give true power indication with a three-phase delta load.

For loads of other than unity power factor the phase currents will shift with respect to their phase voltages. For example, $I_{a'a}$ will either lead or lag $E_{2,1}$ by some angle, θ, causing I_2 to shift away from E_{bo} by the same amount. The product $E_{2,1} \times I_{a'a} \cos \theta$ will still equal $E_{bo} \times I_2 \cos \theta$, and each wattmeter will still indicate the phase power. As for a wye load, if the load is unbalanced (or if the potential coils are not of identical impedance), the wattmeter readings will not be alike, but their sum will still give the true total power.

38-3 The one-wattmeter method.

It was pointed out in the above discussion, that *when the load is balanced and the three wattmeter potential coils have indentical impedances*, each wattmeter reading is alike. Each meter indicates phase power, and the total power can be obtained by multiplying any one meter reading by three. Then why use three meters? But if we remove the other two wattmeters, the potential circuit wye connection is broken! This can be remedied by using two impedances, each equal in value (*magnitude and phase*) to the impedance of the remaining wattmeter potential coil. Commercial units for this purpose are known as *Y-boxes*. They are available to duplicate the potential coils of various models of wattmeters. A Y-box for one type of meter cannot be used with a different model instrument unless the potential coil impedances are identical. The one-wattmeter method, employing a Y-box, is shown in Fig. 38-3.

Remember that this system can be used only with balanced loads.

38-4 The two-wattmeter method.

In discussing the three-wattmeter method for measuring power, it was stated that when the potential coils do not have identical impedances, the three meters do not have equal readings, but their sum is still equal to the total power. Let us carry this idea to the extreme and make the impedance of one potential circuit equal zero. For example, suppose the potential coil of W_1 in Fig. 38-1 is short-circuited. This is

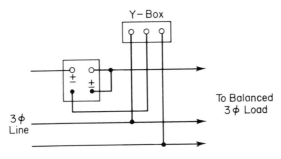

Figure 38-3 The one-wattmeter method for measuring power, using a Y-box.

the same as connecting the common junction of the wattmeters to line 1. Since there is no voltage across its potential coil, wattmeter 1 will read zero. It can therefore be removed from the circuit. This results in the two-wattmeter method for measuring power.

Figure 38-4 shows the two-wattmeter method for measuring the power taken by a balanced wye load of power factor θ. The phase and line voltage relations are shown in the usual manner, with the phase voltages leading the respective line voltages by 30°. The line currents I_1 and I_2 lag their respective phase voltages by the power factor, angle θ. The power indicated by wattmeter W_1 must be due to the torque that is produced by the current I_1 through its current coil and the voltage $E_{1,3}$ across its potential coil, with due consideration to their instantaneous values or time phase relations. In other words,

$$P_1 = E_{1,3} \times I_1 \times \cos \left. \frac{E_{1,3}}{I_1} \right]$$

The symbol $\cos \left. \dfrac{E_{1,3}}{I_1} \right]$ is used to indicate the cosine of the angle between $E_{1,3}$ and I_1. From the phasor diagram it is obvious that this angle is $(30 + \theta)$ degrees. Also $E_{1,3}$ and I_1 are the line voltage and line current, respectively. Therefore,

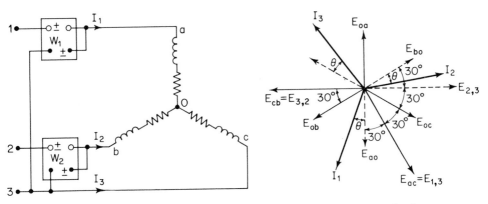

Figure 38-4 The two-wattmeter method of measuring power in a wye load.

$$P_1 = E_L I_L \cos(30 + \theta) \qquad\qquad (38\text{-}3A)$$

Similarly, the power indicated by the meter W_2 is

$$P_2 = E_{2,3} \times I_2 \times \cos\frac{E_{2,3}}{I_2}\Bigg]$$

From the phasor diagram this phase angle is seen to be $(30° - \theta)$. Again, since $E_{2,3}$ and I_2 are line values,

$$P_2 = E_L L_L \cos(30 - \theta) \qquad\qquad (38\text{-}3B)$$

In trigonometry it has been proven that the cosine of the sum of two angles α and β can be expressed as: $\cos(\alpha + \beta) = \cos\alpha\cos\beta - \sin\alpha\sin\beta$. Also, the cosine of the difference between these two angles is: $\cos(\alpha - \beta) = \cos\alpha\cos\beta + \sin\alpha\sin\beta$. Now, if we add the two wattmeter readings, meanwhile replacing the cosine terms by their trigonometric equivalents, we get

$$P_T = P_1 + P_2 = E_L I_L (\cos 30 \cos\theta - \sin 30 \sin\theta)$$
$$+ E_L I_L (\cos 30 \cos\theta + \sin 30 \sin\theta)$$

Factoring out $E_L I_L$ and simplifying, we have

$$P_T = E_L I_L (2 \cos 30 \cos\theta)$$

But $\cos 30° = 0.866$ and $2 \cos 30° = 1.732 = \sqrt{3}$. Therefore,

$$P_T = \sqrt{3}\, E_L I_L \cos\theta$$

But this is the equation for total power. In other words, the two-wattmeter method will give the true total power taken by the wye load.

Will this method measure total power in a delta-connected load? The answer is yes, it will. Analysis of a phasor diagram for a delta load will reveal that the power indicated by each wattmeter will still be $E_L L_L \cos(30 + \theta)$ for one meter, and $E_L I_L \cos(30 - \theta)$ for the other. (The proof of this is left as a problem at the end of the chapter.) As shown above, the sum of these two values equals the total power.

If the load is unbalanced, will the two-wattmeter method still measure the total power? The answer is again yes. The currents in the lines will change and the phase angles between line voltages and these currents will also change. These changes will affect the individual wattmeter indications, but their sum will still be the total power.

38-5 Effect of power factor on the two-wattmeter readings.

For the circuit shown in Fig. 38-4, the wattmeter indication for W_1 is $P_1 = E_L I_L \cos(30 + \theta)$, and for W_2 is $P_2 = E_L I_L \cos(30 - \theta)$. Had we used a leading power factor load, each current would have led its phase voltage by θ degrees, and the phase angles between line current and wattmeter potential coils would have been reversed. The reading for wattmeter W_1 would have been proportional to $\cos(30 - \theta)$, and that for wattmeter W_2 would have been proportional to cos $(30 + \theta)$. Even with a lagging power-factor load—had we put the second wattmeter (W_2) into line 3, and used line 2 as the common junction point of the wattmeter potential coils—the cosine functions would have been reversed, and again the wattmeter indications would have been W_1 proportional to cos $(30 - \theta)$ and W_2 proportional to cos $(30 + \theta)$. The important point is that in any case one wattmeter reading is $E_L I_L$ cos $(30 - \theta)$ and the other is $E_L I_L$ cos $(30 + \theta)$.

Now let us examine the effect of the power factor angle on the relative wattmeter readings.

1. At unity power factor, $\theta = 0°$, and the two cosine functions reduce to cos 30°. Both wattmeters read alike, each indicating one-half the power taken by the load.
2. For power factor angles between 0 and 30°:
   (a) $(30 + \theta)$ approaches 60°. This wattmeter reading decreases.
   (b) $(30 - \theta)$ approaches 0°. This wattmeter reading increases. It is the higher reading wattmeter.
3. For power factor angles between 30 and 60°:
   (a) $(30 + \theta)$ approaches 90°. This wattmeter reading approaches zero.
   (b) $(30 - \theta)$ increases from 0 to $-30°$. This wattmeter reading drops, but it is still the higher reading wattmeter.
4. For power factor angles between 60 and 90°:
   (a) $(30 + \theta)$ increases from 90 to 120°. The numerical value of the cosine function increases, but it is now negative. The wattmeter reading begins to increase again, *but now its indication is negative* and must be subtracted from the indication of the other wattmeter to get the total power. (It will be necessary to reverse the current coil connections in order for the meter to indicate upscale.)
   (b) $(30 - \theta)$ increases from -30 to $-60°$. This wattmeter reading drops, but (except at a power factor angle of 90°) it is still the higher reading wattmeter.
5. At zero power factor, $\theta = 90°$; $(30 + \theta)$ is 120°; $(30 - \theta)$ is $-60°$; the two cosine values are equal in magnitude, and once more the two wattmeter readings are alike. However, since the cos $(30 + \theta)$ is negative, the total power indication is zero. Such a condition, power factor angle of 90°, can be obtained only with a pure reactive load, and we know that in such a case no power is consumed by the load.

To summarize the above discussion, let us first recall that the power factor of a load is generally expressed as the cosine of the phase angle θ (either as a decimal or a percentage), rather than as the angle itself. For example, when the phase angle θ is 60°, the load has a power factor (cos θ) of 0.5 or 50%. With this in mind:

1. The wattmeter whose deflection is proportional to $(30 - \theta)$ is always positive and (except for unity or zero power factor) is always the higher-reading wattmeter.

2. The wattmeter whose deflection is proportional to $(30 + \theta)$ is positive only for power factors above 0.5, and is negative for power factors below 0.5. For a power factor of exactly 50% this wattmeter indication is zero.

38-6 Power factor from the two-wattmeter method. In the above discussion we saw that some idea of the power factor of the load could be obtained from the relative deflections of our two wattmeters. There are two methods by which this power factor can be determined accurately. One method is to take the ratio of the two wattmeter readings:

$$\frac{\text{lower WM reading}}{\text{higher WM reading}} = \frac{E_L I_L \cos (30 + \theta)}{E_L I_L \cos (30 - \theta)} = \frac{\cos (30 + \theta)}{\cos (30 - \theta)}$$

If we evaluate the ratio $\cos (30 + \theta)/\cos (30 - \theta)$ for several values of θ between 0 and 90° and the corresponding power factor $(\cos \theta)$ for each of these values of θ, we can plot a curve of wattmeter ratio versus power factor. Such a curve is shown in Fig. 38-5. Notice that this ratio is 1.0 at unity power factor since the wattmeter readings are equal; that the ratio is zero at 0.5 power factor, since the lower-reading wattmeter indicates zero; that the ratios are negative for all lower power factors; and that the ratio is -1.0 at zero power factor since the wattmeter readings are again equal but the lower reading wattmeter gives reverse indication. From this curve it is a simple matter to find the power factor of a load when we know the individual wattmeter readings. Remember, however, that this curve is applicable only for *balanced* loads when metered by the two-wattmeter method.

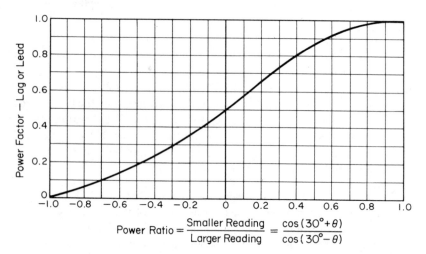

Power Ratio $= \dfrac{\text{Smaller Reading}}{\text{Larger Reading}} = \dfrac{\cos (30^\circ + \theta)}{\cos (30^\circ - \theta)}$

Figure 38-5 Relation between power factor and wattmeter ratio, for balanced 3ϕ loads.

The power factor of the load can also be found mathematically. We have already shown that the sum of the two wattmeter readings is

$$P_1 + P_2 = \sqrt{3}\ E_L I_L \cos\theta$$

Now let us take the difference between the two readings, *subtracting the smaller reading from the larger*. Calling P_2 the higher wattmeter reading, we get

$$P_2 - P_1 = E_L I_L \cos(30 - \theta) - E_L I_L \cos(30 + \theta)$$

Expanding by trigonometry yields

$$P_2 - P_1 = E_L I_L(\cos 30 \cos\theta + \sin 30 \sin\theta)$$
$$- E_L I_L(\cos 30 \cos\theta - \sin 30 \sin\theta)$$

Factoring out $E_L I_L$ and simplifying, we have

$$P_2 - P_1 = E_L I_L(2 \sin 30 \sin\theta)$$

But $\sin 30 = 0.5$, and $2 \sin 30 = 1.0$. Therefore,

$$P_2 - P_1 = E_L I_L \sin\theta$$

Now let us take the ratio of the difference between wattmeter readings over sum of wattmeter readings.

$$\frac{P_2 - P_1}{P_2 + P_1} = \frac{E_L I_L \sin\theta}{\sqrt{3}\ E_L I_L \cos\theta} = \frac{\tan\theta}{\sqrt{3}}$$

$$\tan\theta = \sqrt{3}\ \frac{P_2 - P_1}{P_2 + P_1} \qquad \text{(38-4A)}$$

or

$$\theta = \text{arc tan } \sqrt{3}\ \frac{\text{algebraic difference between WM readings}}{\text{algebraic sum of WM readings}} \qquad \text{(38-4B)}$$

Example 38-3

A three-phase balanced load is metered by the two-wattmeter method. The wattmeter readings are 4.80 kW and −1.60 kW (meter current coil reversed). Find the power factor of the load.

Solution

1. $\theta = \text{arc tan } \sqrt{3}\ \dfrac{\text{algebraic difference between WM readings}}{\text{algebraic sum of WM readings}}$

$$= \text{arc tan } \sqrt{3}\ \frac{4.80 - (-1.60)}{4.80 + (-1.60)}$$

$$= \text{arc tan } \sqrt{3}\ \frac{6.40}{3.20} = \text{arc tan } 3.46 = 73.9°$$

2. $\text{PF} = \cos\theta = \cos 73.9° = 0.277$

38-7 Wattmeter connections: polarity marks. With single-phase loads, a wattmeter will deflect backward only if it is improperly connected. This is also true when the three-wattmeter method is used to measure power in a three-phase circuit. But in the discussion above on the two-wattmeter method of measuring power, we saw that the lower-reading wattmeter will reverse its deflection if the load power factor is less than 50%. So when a wattmeter gives a reverse indication, is it because of improper connection or low power factor? To avoid this ambiguity it is necessary to pay careful attention to the polarity marks ($\pm$) on the wattmeter terminals. The procedure recommended for connecting the wattmeter for the two-wattmeter method is as follows:

1. Connect each wattmeter in the proper manner, that is, the $\pm$ terminal of the current coil on the *line* side of the line lead, and the $\pm$ terminal of the potential coil to the same line as its own current coil.
2. Now if either wattmeter deflection is backward, it is an indication of low power factor. Therefore, reverse its *current* coil connections and consider this reading a negative value.

38-8 Volt-amperes and reactive volt-amperes.

Earlier in this text, when studying single-phase circuits we saw that, due to the power factor of the load, the product of voltage and current was not the true power of the circuit (except in the special case of a pure resistive load.) This product was therefore called the *volt-amperes* (VA) of the circuit. Naturally, when either or both current and voltage are high in value, the larger unit kilovolt-amperes (kVA) is used. Can we apply this same term to a three-phase load? *Yes, but only if it is a balanced load.* Since the power in such a load is $P = \sqrt{3}\ E_L I_L \cos\theta$, if we remove the power factor term, we are left with the volt-amperes of the circuit.

$$\text{VA} = \sqrt{3}\ E_L I_L = \frac{\text{watts}}{\cos\theta}$$

Again, using the single-phase circuit for comparison, since the reactive volt-amperes (var or kvar) was the product of voltage, current, and the *sine* of the phase angle, for the balanced three-phase circuit we have

$$\text{var} = \sqrt{3}\ E_L I_L \sin\theta$$

Example 38-4

A balanced load draws 48 A from a 440-V line at 80% power factor. Find the power, volt-amperes, and reactive volt-amperes of the load.

Solution

   1. $P = \sqrt{3}\ E_L I_L \cos\theta = \sqrt{3} \times 440 \times 48 \times 0.8 = 29.3 \text{ kW}$

   2A.   Volt-amperes $= \sqrt{3}\ E_L I_L = \sqrt{3} \times 440 \times 48 = 36.6 \text{ kVA}$

       or

   2B.   Volt-amperes $= \dfrac{\text{watts}}{\cos\theta} = \dfrac{29.3 \text{ kW}}{0.8} = 36.6 \text{ kVA}$

   3. $\text{Cos }\theta = 0.8,\ \theta = 36.9°,\text{ and } \sin\theta = 0.6$

   4.   Reactive volt-amperes $=$ volt-amperes $\times \sin\theta$

                              $= 36.6 \times 0.6 = 21.9 \text{ kvar}$

38-9 Advantages of the three-phase system. By now you may be wondering why all this fuss over a three-phase distribution system? If a dual-voltage system is needed, why not use a single-phase three-wire system such as a 240/120-V three-wire supply? Offhand, it might seem that the single-phase system requiring one less wire (two wires for single supply voltage or three wires for dual supply voltage, as compared with three-phase three-wire or three-phase four-wire) would be better. But let us examine these two systems from their relative power and current relations:

$$\text{power (single-phase)} = E_L I_L \cos \theta$$
$$\text{power (three-phase)} = \sqrt{3}\ E_L I_L \cos \theta$$

For the same line voltage and line current, the three-phase system will deliver 1.73, or $\sqrt{3}$ times as much power. Therefore with only one extra wire the transmission lines can deliver almost double the power. Another way of looking at this is that for a given total load, since the line current is lower, smaller wires can be used in a three-phase system. To transmit a given amount of power with equal line loss (over a fixed distance and with equal line voltage) the three-phase system requires only 75% as much copper as a single-phase system.

Another advantage of the three-phase distribution system becomes of major importance when large motors are needed. Three-phase motors are easier to start, are simpler in construction, have higher efficiencies, require less maintenance, and have smoother torque characteristics than single-phase machines. (Discussion of machine characteristics will be left to texts on machinery.) Although a single-phase motor can be used directly on a three-phase system, the converse cannot be done without expensive conversion equipment.

REVIEW QUESTIONS

1. In a three-phase load:
   (a) How can we find the power dissipated in *one* phase?
   (b) Will this apply to a delta or wye load?
   (c) Will this apply to a balanced or unbalanced load?
   (d) If the load is balanced, how will the power per phase compare?
   (e) Does this also apply to an unbalanced load? Explain.
   (f) In a balanced load, what relation exists between total power and phase power?
   (g) Does this relation apply to an unbalanced load? Explain.

2. With reference to equation (38-2):
   (a) What do E_L and I_L represent?
   (b) Does this equation apply to delta or wye loads?
   (c) Does it apply to balanced or unbalanced loads? Explain.
   (d) What does θ represent?

3. In the statement of Example 38-1, what information is superfluous? Why?

4. In Example 38-2:
   (a) What information is superfluous?

    (b) Why is knowledge of the type of load connection now necessary?
    (c) What quantity in the power equation is unknown?
    (d) How can this quantity be obtained?
    (e) How does step 6A differ from 6B?

5. When measuring power (with wattmeters) in a three-phase load:
    (a) Is it possible to measure the individual phase powers?
    (b) Why isn't it done with wye loads?
    (c) Why isn't it done with delta loads?

6. With reference to Fig. 38-1:
    (a) How are the current coils of each wattmeter connected?
    (b) How are the potential coils connected?
    (c) What type of load do the potential coils form?

7. The meters in Fig. 38-1 are connected to a wye load:
    (a) How does the current through each meter compare with the phase currents in the load? Why?
    (b) How does the voltage across each potential coil compare with the phase voltage across each load? Why?
    (c) What does each wattmeter actually measure?
    (d) Will this apply to an unbalanced load? Explain.
    (e) If the load is balanced, how will the wattmeter readings compare? Why?
    (f) If the load is unbalanced, will the above equality apply? Explain.

8. The wattmeters in Fig. 38-1 are connected to a delta load:
    (a) What current value (line or phase) flows through each current coil?
    (b) What voltage (line or phase) is across each potential coil?
    (c) What power does each wattmeter indicate? Why?
    (d) If the load is balanced, how will the individual readings compare?
    (e) Will this system "work" with unbalanced delta loads?

9. Can one wattmeter be used to measure three-phase power:
    (a) All by itself?
    (b) What other device is needed?
    (c) How is the total three-phase power obtained?
    (d) Can this system be used with unbalanced loads? Explain.

10. With reference to Fig. 38-3:
    (a) What is a *Y-box*?
    (b) Can any Y-box be used with any wattmeter? Explain.

11. With reference to the phasor diagram, Fig. 38-4:
    (a) What type of load connection does this represent? How can you tell?
    (b) What nature of load (L, C, or R) is this? Explain.

12. With reference to the full diagram, Fig. 38-4:
    (a) What current flows through W_1?
    (b) What voltage is across its potential coil?
    (c) What is the phase angle between these two quantities?
    (d) In general terms, on what will the reading of W_1 depend?
    (e) What current flows through W_2?
    (f) Is this a line or a phase quantity?
    (g) What voltage is across its potential coil?

(h) Is this a line or a phase quantity?

(i) What is the phase angle between these two quantities?

(j) In general terms, on what will this wattmeter reading depend?

13. (a) Will the two-wattmeter system measure the total power for a delta load?

(b) In general terms, on what will each meter reading depend?

14. Will the two-wattmeter system measure total power with unbalanced loads?

15. When measuring power by the two-wattmeter method:

(a) Under what condition will the two readings be alike? Why?

(b) Under what condition will one reading be zero? Explain.

(c) One of the wattmeters deflects backward. Yet it is properly connected. Is this possible? Explain.

(d) When this happens, how is the total power obtained?

(e) When the power factor of the load is zero, will each wattmeter reading be zero? Explain.

16. (a) Explain how the curve in Fig. 38-5 can be used to evaluate the power factor of the load.

(b) Does this curve apply to unbalanced loads?

17. Give a method for evaluating power factor directly from the wattmeter readings.

18. (a) In Example 38-3, why is one reading marked as *minus* 160 kW?

(b) What does this indicate with regard to power factor?

19. (a) With single-phase loads, can a wattmeter deflect backward? Explain.

(b) How can proper connections be insured before energizing the circuit?

(c) Give two reasons why a wattmeter, in the two-wattmeter method, may deflect backward.

(d) How can you be sure which is the true reason?

20. (a) How can we obtain the volt-amperes of a balanced three-phase load?

(b) Does this also apply for an unbalanced load? Explain.

(c) How can we obtain the *vars* of a balanced three-phase load?

(d) In a "power" phasor diagram, how would the above values be related to the true power dissipated?

21. With reference to Example 38-4:

(a) How is the relation in step 2B obtained?

(b) Draw a phasor diagram to indicate the volt-amperes to power relation.

(c) From this diagram, what is the relation between watts and vars?

(d) Between watts and volt-amperes?

(e) Between volt-amperes and vars?

22. Compare a three-phase and a single-phase system with respect to:

(a) Power delivered, for the same line current and voltage.

(b) Line losses, for the same total load.

(c) Give another advantage of a three-phase supply.

PROBLEMS

1. A balanced three-phase delta load draws 120 A from a 240-V three-phase supply at a power factor of 0.74 leading. Find the power taken by the load.

2. Repeat Problem 1 if the phase current is 120 A.

3. Find the power taken by a balanced three-phase wye-connected load if the line current is 85 A, line voltage is 220 V, and the power factor of the load is 0.68 lagging.

4. Three loads, each having a resistance of 40 Ω and an inductive reactance of 30 Ω, are connected in wye across a 230-V three-phase 60-Hz supply. Find the power taken by the load.

5. Repeat Problem 4 for the loads connected in delta.

6. A balanced three-phase delta load draws 300 A from a 440-V three-phase supply. If the power taken by the load is 200 kW, find the power factor of the load.

7. A 150-hp three-phase induction motor draws 126 A from a 660-V three-phase supply at a power factor of 0.85 when operating at full load. Find the efficiency of the motor.

8. A three-phase alternator has the following rating per phase: current 60 A, voltage 230 V. Find the rated line current, line voltage, and power output at 80% power factor for:
   (a) A delta connection.
   (b) A wye connection.

9. The nameplate on a three-phase alternator gives the following ratings: phase—3; Hz—60; V—2200; A—(defaced); speed—1200 rpm; kW—400; power factor—0.8. Find the current rating for this machine.

10. A balanced three-phase load draws 60 A from a 440-V three-phase supply. The phase angle between line current and line voltage is 30°. Find the power taken by the load. (Give two possible answers.)

11. Draw a schematic diagram showing the use of three wattmeters to measure the power taken by a three-phase load. Indicate wattmeter terminal polarity marks.

12. Draw the phasor diagram for current and voltage relations in a balanced three-phase delta load, for a power factor angle of θ degrees (current leading). Prove that the power measured by any one wattmeter is the phase power.

13. Draw a schematic diagram showing the two-wattmeter method for measuring power in a three-phase load.

14. (a) Draw the phasor diagram for current and voltage relations in a balanced delta load for a power factor angle of θ degrees, lagging.
    (b) Prove that the two-wattmeter method will measure the total power taken by the load.

15. The power taken by a three-phase induction motor (delta-connected, balanced load) is checked using the two-wattmeter method. The wattmeter readings are 8.8 kW and 4.0 kW. The supply is three-phase 230-V.
    (a) Calculate the power factor of the load.
    (b) Check by finding the power factor from the curve in Fig. 38-5.
    (c) Find the line and phase current values.

16. A three-phase induction motor draws a line current of 50 A from a 220-V line at 40% power factor while starting. The two-wattmeter method is used to measure the power taken by the motor. Find the reading indicated by each wattmeter.

17. A balanced three-phase wye-connected load is connected across a 208-V three-

phase supply. The two wattmeters used to measure the power taken by the load indicate 4.6 kW and −2.4 kW.

(a) Find the power factor of the load.

(b) Find the current in each phase.

18. A balanced three-phase delta load draws a current of 95 A from a 208-V, 60-Hz three-phase supply at a power factor of 0.8 lagging. Find:

(a) The kVA of the load.

(b) The power taken by the load.

(c) The kvar of the load.

19. Repeat Problem 18 for a load drawing 120 A at a power factor of 70% lagging.

20. If the loads of Problem 18 and 19 are combined, find:

(a) The total kVA.

(b) The total power.

(c) The total kvar.

21. It is desired to raise the power factor of the load in Problem 20 to unity, by adding a three-phase delta-connected pure capactitive load.

(a) Find the reactive kVA needed.

(b) Find the capacitance value needed per phase.

22. Repeat Problem 21 if the power factor is raised to only 90%.

39

UNBALANCED THREE-PHASE LOADS

In an earlier chapter we saw that a three-phase system could be used to supply three-phase loads, or individual single-phase loads. The discussions and problems so far have all been with respect to loads having identical impedance per phase. This resulted in balanced loads. What are the chances of such a condition occurring in actual practice? Three-phase equipment, such as motors, are balanced loads, but when we add lights, appliances, and other single-phase loads on the system, the balance is lost. Homes are generally supplied with single-phase service. The power company tries to obtain a balanced load by equalizing the number of homes on each phase of its three-phase supply. Industrial consumers are generally supplied with all three lines and neutral. The internal wiring is then planned to attempt to distribute the single-phase loads equally between neutral and each of the three lines. However, any such measures, no matter how carefully planned, can prevent only drastic unbalance. This chapter deals with unbalanced circuits. The same basic principles discussed in previous chapters are employed. The problems are more tedious, and careful attention must be paid to phase relations.

39-1 Unbalanced delta loads. Whenever dissimilar loads are connected across the lines of a three-phase supply, an unbalanced delta load will result. The phase currents are the individual load currents. Since the currents in each load are not equal in magnitude and/or phase, the three line currents will be unbalanced. Each load (phase) current can be found from the line (phase) voltage and load impedance. The line currents can then be found from the vector addition of the appropriate load currents.

Example 39-1

Three loads $Z_1 = 31\,\Omega$ resistance and $59\,\Omega$ inductive reactance, $Z_2 = 30\,\Omega$ resistance and $40\,\Omega$ capacitive reactance, and $Z_3 = 80\,\Omega$ resistance and $60\,\Omega$ inductive reactance are connected one each across the lines of a 200-V three-phase supply. Find the line currents.

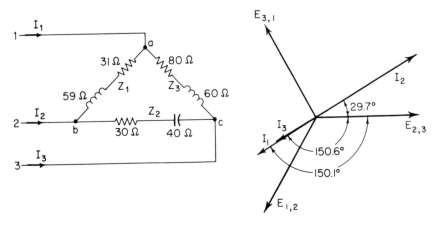

Figure 39-1   Unbalanced delta load.

Solution

1. Draw a diagram of the circuit, and a phasor diagram of the line voltages (see Fig. 39-1).

2. Express each voltage in polar form:

   (a) $$E_{1,2} = 200\underline{/-120°}$$

   (b) $$E_{2,3} = 200\underline{/0°}$$

   (c) $$E_{3,1} = 200\underline{/120°}$$

3. Express each impedance in polar form:

   (a) $$Z_1 = 31 + j59 = 66.7\underline{/62.3°}$$

   (b) $$Z_2 = 30 - j40 = 50.0\underline{/-53.1°}$$

   (c) $$Z_3 = 80 + j60 = 100\underline{/36.9°}$$

4. Solve for each phase current:

   (a) $$I_{ab} = \frac{E_{1,2}}{Z_1} = \frac{200\underline{/-120}}{66.7\underline{/62.3}} = 3.0\underline{/-182.3°}$$

   (b) $$I_{bc} = \frac{E_{2,3}}{Z_2} = \frac{200\underline{/0}}{50.0\underline{/-53.1}} = 4.0\underline{/53.1°}$$

   (c) $$I_{ca} = \frac{E_{3,1}}{Z_3} = \frac{200\underline{/120}}{100\underline{/36.9}} = 2.0\underline{/83.1°}$$

5. Convert these current values to rectangular form:

(a)
$$3.0\underline{/-182.3°} = -3.0\cos 2.3° + j3.0\sin 2.3°$$
$$I_{ab} = -3.0 + j0.12$$

(b)
$$4.0\underline{/53.1°} = 4.0\cos 53.1° + j4.0\sin 53.1°$$
$$I_{bc} = 2.40 + j3.20$$

(c)
$$2.0\underline{/83.1°} = 2.0\cos 83.1° + j2.0\sin 83.1°$$
$$I_{ca} = 0.24 + j1.99$$

6. Find the line currents:

(a)
$$I_1 = I_{ab} - I_{ca} = \begin{array}{r} -3.0 \ +j0.12 \\ \oplus 0.24 \ominus j1.99 \\ \hline -3.24 \ -j1.87 \end{array}$$
$$= 3.74\underline{/-150.1°}$$

(b)
$$I_2 = I_{bc} - I_{ab} = \begin{array}{r} 2.40 + j3.20 \\ + \\ \ominus 3.0 \ \oplus j0.12 \\ \hline 5.40 + j3.08 \end{array}$$
$$= 6.22\underline{/29.7°}$$

(c)
$$I_3 = I_{ca} - I_{bc} = \begin{array}{r} 0.24 \ +j1.98 \\ \oplus 2.40 \ominus j3.20 \\ \hline -2.16 \ -j1.22 \end{array}$$
$$= 2.48\underline{/-150.6°}$$

7. To complete the phasor diagram, add these line currents to Fig. 39-1.

39-2 Unbalanced wye load, with neutral.

In a three-phase four-wire distribution system, as described previously, the line voltage commonly used is 208 volts, and the voltage from any line to neutral is 120 V. Lighting loads and 120-V appliances will be connected between line and neutral. Except as a rare coincidence this will result in an unbalanced wye load. Again line currents will be unbalanced, and due to the unbalance, the neutral wire will also carry some current. Since the voltage between any line and neutral is fixed at $E_L/\sqrt{3}$, it is a simple matter to calculate the individual phase currents. As in any wye system, each line current will be equal to the current in the phase that it feeds. Now with an unbalanced condition, these phase currents (and line currents) are not identical, and their phasor sum does not equal zero. This phasor sum must be the current flowing in the neutral wire.

Example 39-2

A 208/120-V three-phase, four-wire system has the following loads: $Z_1 = 8\,\Omega$ resistance and $6\,\Omega$ inductive reactance, $Z_2 = 12\,\Omega$ resistance, and $Z_3 = 12\,\Omega$ resistance and $16\,\Omega$ capacitive reactance. The loads are connected between line and neutral with each load on a separate line. Find line currents and neutral current.

Solution

1. Draw a diagram of the circuit, and a phasor diagram of the phase voltages (line to neutral) (see Fig. 39-2).

2. Express each phase voltage in polar form:

(a) $E_{oa} = 120\underline{/120°}$

(b) $E_{ob} = 120\underline{/-120°}$

(c) $E_{oc} = 120\underline{/0°}$

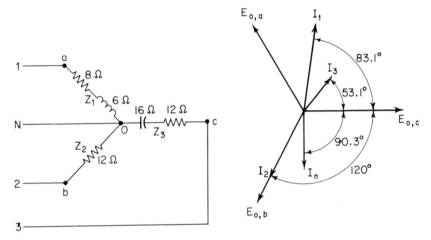

Figure 39-2   Unbalanced wye load.

3. Express each impedance in polar form:

(a) $Z_1 = 8 + j6 = 10\underline{/36.9°}$

(b) $Z_2 = 12 + j0 = 12\underline{/0°}$

(c) $Z_3 = 12 - j16 = 20\underline{/-53.1°}$

4. Solve for each phase (and line) current:

(a) $I_1 = \dfrac{E_{oa}}{Z_1} = \dfrac{120\underline{/120}}{10\underline{/36.9}} = 12\underline{/83.1°}$

(b) $I_2 = \dfrac{E_{ob}}{Z_2} = \dfrac{120\underline{/-120}}{12\underline{/0}} = 10\underline{/-120°}$

(c) $I_3 = \dfrac{E_{oc}}{Z_3} = \dfrac{120\underline{/0}}{20\underline{/-53.1}} = 6\underline{/53.1°}$

5. Draw these currents on the phasor diagram.

6. To find the neutral current, we must add the three line currents. The neutral current must then be equal and *opposite* to this sum, so that the total current

at junction will be zero. But first we must express each current in rectangular form.

(a)
$$I_1 = 12 \cos 83.1° + j12 \sin 83.1°$$
$$= 1.44 + j11.9$$

(b)
$$I_2 = -10 \cos 60° - j10 \sin 60°$$
$$= -5.0 - j8.66$$

(c)
$$I_3 = 6 \cos 53.1° + j6 \sin 53.1°$$
$$= 3.6 + j4.8$$

(d)
$$I_1 + I_2 + I_3 = \quad 1.44 + j11.9$$
$$-5.00 - j8.66$$
$$3.60 + j4.80$$
$$\overline{0.04 + j8.04}$$

(e)
$$I_n = -(0.04 + j8.04)$$
$$= 8.04\underline{/-90.3°}$$

7. Add this current to the phasor diagram.

In commercial applications with a three-phase four-wire system, the actual loading will consist of a combination of:

1. Balanced three-phase loads, wye or delta
2. Unbalanced delta loads due to single-phase loads connected across the line
3. Unbalanced wye loads due to single-phase loads connected from any line to neutral

Problems of this type are much more tedious, since each type of load must be handled separately, and the resulting line current for each type of load must then be added vectorially to get the total line currents. It would be similar to solving four problems; a balanced delta, a balanced wye (covered in Chapter 36), and the unbalanced delta and unbalanced wye as shown above.

39-3 Unbalanced wye load, no neutral.   To complete the discussion on unbalanced loads, one other type of load must be mentioned. It is the unbalanced wye load, shown above, but without interconnection between the common point of the wye load and the neutral wire of the supply line. Problems of this type are purely academic and have little practical significance. Such a load situation could arise only with a three-phase, four-wire system and single-phase loads connected between line and neutral if the connection between the supply line neutral and the "local neutral" is broken. This is really a faulty or improper operation and calls for repairs. In fact, the neutral wire in commercial operation is never fused, otherwise a blown fuse could create this situation. However, let us analyze what would happen if the neutral connection were opened.

If the load were balanced, the phasor sum of the three line currents at the common junction of load, or local neutral, would be zero. There is no current in the local neutral wire, and therefore no adverse effect is created by the break between local neutral and supply-line neutral. Meanwhile, since the wye branch impedances are identical, the voltage drops across each wye load are equal, and E_p remains $E_L/\sqrt{3}$. In other words, for a balanced load, operation remains the same with or without the neutral connection.

Now let us consider what would happen with an unbalanced load, if the tie between local neutral and supply line neutral were opened. In Example 39-2 we saw that since the phasor sum of the line currents did not equal zero, an unbalance current flows in the neutral wire. Now the neutral wire is open, and current cannot flow in an open line. But neither can current pile up at the local neutral. There is only one possible answer—the individual line currents must change so that their phasor sum becomes zero. This does not imply, however, that the currents are identical. A phasor sum of zero can be obtained with unequal line currents.

The question now is: What makes the currents change? Since the loads themselves have not changed, the voltage across each load must change. That is exactly what happens. Remember that these loads are connected from one line to neutral, and that the voltage from any line to neutral is fixed at $E_L/\sqrt{3}$ only when the neutral wire is intact. With the neutral wire open, a *floating neutral* is created, and the potential of the local neutral will depend on the impedances of the loads. The three phase voltages are unbalanced in magnitude and in phase, and it is quite possible for one phase voltage to exceed the line voltage. Such a condition is definitely undesirable. Some loads will operate inefficiently because of low voltage, while other equipment may be damaged because of overvoltage.

Since loads of this type are not normally encountered in practical applications, the solution of such problems is not discussed in this text.

REVIEW QUESTIONS

1. (a) Are three-phase loads always balanced? Explain.
   (b) What type of loads will be balanced?
   (c) What causes imbalance?
   (d) If the loads are unbalanced, how does this affect the line voltages? Explain.
   (e) How does it affect the line currents? Explain.

2. In the solution to Example 39-1:
   (a) In step 4(a), why does the phase current I_{ab}, depend only on $E_{1,2}$ and Z_1?
   (b) In step 6(a), why does $I_1 = I_{ab} - I_{ca}$?

3. In an unbalanced four-wire Y-load:
   (a) If the line voltage is 208 V, what is the phase voltage?
   (b) Will the line voltages be equal? Explain.
   (c) Will the phase voltages be equal? Explain.
   (d) Will the line currents be equal? Explain.
   (e) Will each line current equal its phase current? Explain.
4. With reference to the solution of Example 39-2:
   (a) In step 2, are these phase or line voltages?
   (b) What is the phase relation between these voltages?
   (c) Is this correct for an unbalanced load? Explain.
   (d) In step 4, is I_1 a phase or a line current?
   (e) How is the value of I_1 obtained?
   (f) In step 6, how is the value of I_n obtained?
5. (a) Does an unbalanced wye load, without a neutral, occur in commercial practice? Explain.
   (b) In such a case, would the line voltages be equal in magnitude? Explain.
   (c) Would the phase voltages be equal in magnitude? Explain.

PROBLEMS

1. A three-phase 220-V 60-Hz supply has the following single-phase loads: Z_1, 80 Ω inductive reactance across lines 1 and 2; Z_2, 60 Ω resistance across lines 2 and 3; Z_3, 100 Ω capacitive reactance across lines 3 and 1. Using a phase sequence of $E_{1,2}$, $E_{2,3}$, $E_{3,1}$, and $E_{1,2}$ as reference, calculate the three line currents (magnitude and phase). Draw the phasor diagram.

2. Repeat Problem 1 for a line voltage of 440-V and loads of $Z_1 = 20\,\Omega$ resistance and 30 Ω inductive reactance, $Z_2 = 40\,\Omega$ resistance and 50 Ω inductive reactance, $Z_3 = 10\,\Omega$ resistance and 30 Ω inductive reactance.

3. Repeat Problem 1 for a line voltage of 208 V and loads of $Z_1 = 30\,\Omega$ resistance and 40 Ω capacitive reactance, $Z_2 = 45\,\Omega$ resistive, $Z_3 = 20\,\Omega$ resistance and 15 Ω inductive reactance.

4. Repeat Problem 1 for a line voltage of 208 V and loads of $Z_1 = 26\,\Omega$ resistance, $Z_2 =$ open circuit, $Z_3 = 0.1$ H and 12 Ω resistance.

5. In each of the following cases, the supply is a three-phase four-wire 208-V 60-Hz supply. The phase sequence is $E_{1,2}$, $E_{3,1}$, $E_{2,3}$. The loads are single-phase and connected between line and neutral, with Z_1 from line 1 to neutral, Z_2 from line 2 to neutral, and Z_3 from line 3 to neutral. Find the line currents and draw the phasor diagram for:
   (a) Load values as in Problem 1.
   (b) Load values as in Problem 2.
   (c) Load values as in Problem 3.
   (d) Load values as in Problem 4.
   Note: Use $E_{1,2}$ as reference, and consider line currents flowing *out* to the line.

6. A three-phase four-wire 208-V 60-Hz supply is feeding three single-phase line-to-line loads, and three single-phase line-to-neutral loads as follows:

$Z_{1,2} = 8\,\Omega$ resistance and $10\,\Omega$ capacitive reactance

$Z_{2,3} = 4\,\Omega$ resistance and $12\,\Omega$ inductive reactance

$Z_{3,1} = 6\,\Omega$ resistance and $12\,\Omega$ inductive reactance

$Z_{0,1} = 16\,\Omega$ resistance

$Z_{0,2} = 20\,\Omega$ resistance and $16\,\Omega$ inductive reactance

$Z_{0,3} = 15\,\Omega$ resistance and $25\,\Omega$ capacitive reactance

Using a phase sequence of $E_{1,2}$, $E_{2,3}$, $E_{3,1}$, find the magnitude and phase angle of the current in each of the four lines. Draw the phasor diagram.

7. Repeat Problem 6 for

$Z_{1,2} = 20\,\Omega$ resistance and $16\,\Omega$ inductive reactance

$Z_{2,3} = 15\,\Omega$ resistance and $25\,\Omega$ capacitive reactance

$Z_{3,1} = 16\,\Omega$ resistance

$Z_{0,1} = 6\,\Omega$ resistance and $12\,\Omega$ inductive reactance

$Z_{0,2} = 8\,\Omega$ resistance and $10\,\Omega$ capacitive reactance

$Z_{0,3} = 4\,\Omega$ resistance and $12\,\Omega$ inductive reactance

APPENDIXES

EXPONENTIAL FUNCTIONS

x	e^{-x}	x	e^{-x}
0.00	1.000000	**0.50**	0.606531
0.01	0.990050	0.51	.600496
0.02	.980199	0.52	.594521
0.03	.970446	0.53	.588605
0.04	.960789	0.54	.582748
0.05	0.951229	**0.55**	0.576950
0.06	.941765	0.56	.571209
0.07	.932394	0.57	.565525
0.08	.923116	0.58	.559898
0.09	.913931	0.59	.554327
0.10	0.904837	**0.60**	0.548812
0.11	.895834	0.61	.543351
0.12	.886920	0.62	.537944
0.13	.878095	0.63	.532592
0.14	.869358	0.64	.527292
0.15	0.860708	**0.65**	0.522046
0.16	.852144	0.66	.516851
0.17	.843665	0.67	.511709
0.18	.835270	0.68	.506617
0.19	.826959	0.69	.501576
0.20	0.818731	**0.70**	0.496585
0.21	.810584	0.71	.491644
0.22	.802519	0.72	.486752
0.23	.794534	0.73	.481909
0.24	.786628	0.74	.477114
0.25	0.778801	**0.75**	0.472367
0.26	.771052	0 76	.467666
0.27	.763379	0.77	.463013
0.28	.755784	0.78	.458406
0.29	.748264	0.79	.453845
0.30	0.740818	**0.80**	0.449329
0.31	.733447	0.81	.444858
0.32	.726149	0.82	.440432
0.33	.718924	0.83	.436049
0.34	.711770	0.84	.431711
0.35	0.704688	**0.85**	0.427415
0.36	.697676	0.86	.423162
0.37	.690734	0.87	.418952
0.38	.683861	0.88	.414783
0.39	.677057	0.89	.410656
0.40	0.670320	**0.90**	0.406570
0.41	.663650	0.91	.402524
0.42	.657047	0.92	.398519
0.43	.650509	0.93	.394554
0.44	.644036	0.94	.390628
0.45	0.637628	**0.95**	0.386741
0.46	.631284	0.96	.382893
0.47	.625002	0.97	.379083
0.48	.618783	0.98	.375311
0.49	.612626	0.99	.371577
0.50	0.606531	**1.00**	0.367879

x	e^{-x}	x	e^{-x}
1.00	0.367879	**1.50**	0.223130
1.01	.364219	1.51	.220910
1.02	.360595	1.52	.218712
1.03	.357007	1.53	.216536
1.04	.353455	1.54	.214381
1.05	0.349938	**1.55**	0.212248
1.06	.346456	1.56	.210136
1.07	.343009	1.57	.208045
1.08	.339596	1.58	.205975
1.09	.336216	1.59	.203926
1.10	0.332871	**1.60**	0.201897
1.11	.329559	1.61	.199888
1.12	.326280	1.62	.197899
1.13	.323033	1.63	.195930
1.14	.319819	1.64	.193980
1.15	0.316637	**1.65**	0.192050
1.16	.313486	1.66	.190139
1.17	.310367	1.67	.188247
1.18	.307279	1.68	.186374
1.19	.304221	1.69	.184520
1.20	0.301194	**1.70**	0.182684
1.21	.298197	1.71	.180866
1.22	.295230	1.72	.179066
1.23	.292293	1.73	.177284
1.24	.289384	1.74	.175520
1.25	0.286505	**1.75**	0.173774
1.26	.283654	1.76	.172045
1.27	.280832	1.77	.170333
1.28	.278037	1.78	.168638
1.29	.275271	1.79	.166960
1.30	0.272532	**1.80**	0.165299
1.31	.269820	1.81	.163654
1.32	.267135	1.82	.162026
1.33	.264477	1.83	.160414
1.34	.261846	1.84	.158817
1.35	0.259240	**1.85**	0.157237
1.36	.256661	1.86	.155673
1.37	.254107	1.87	.154124
1.38	.251579	1.88	.152590
1.39	.249075	1.89	.151072
1.40	0.246597	**1.90**	0.149569
1.41	.244143	1.91	.148080
1.42	.241714	1.92	.146607
1.43	.239309	1.93	.145148
1.44	.236928	1.94	.143704
1.45	0.234570	**1.95**	0.142274
1.46	.232236	1.96	.140858
1.47	.229925	1.97	.139457
1.48	.227638	1.98	.138069
1.49	.225373	1.99	.136695
1.50	0.223130	**2.00**	0.135335

From "Mathematical Tables" in *Handbook of Physics*. Chemical Rubber Co., Cleveland, 1963.

x	e^{-x}	x	e^{-x}
2.00	0.135335	**2.50**	0.082085
2.01	.133989	2.51	.081268
2.02	.132655	2.52	.080460
2.03	.131336	2.53	.079659
2.04	.130029	2.54	.078866
2.05	0.128735	**2.55**	0.078082
2.06	.127454	2.56	.077305
2.07	.126186	2.57	.076536
2.08	.124930	2.58	.075774
2.09	.123687	2.59	.075020
2.10	0.122456	**2.60**	0.074274
2.11	.121238	2.61	.073535
2.12	.120032	2.62	.072803
2.13	.118837	2.63	.072078
2.14	.117655	2.64	.071361
2.15	0.116484	**2.65**	0.070651
2.16	.115325	2.66	.069948
2.17	.114178	2.67	.069252
2.18	.113042	2.68	.068563
2.19	.111917	2.69	.067881
2.20	0.110803	**2.70**	0.067206
2.21	.109701	2.71	.066537
2.22	.108609	2.72	.065875
2.23	.107528	2.73	.065219
2.24	.106459	2.74	.064570
2.25	0.105399	**2.75**	0 063928
2.26	.104350	2.76	.063292
2.27	.103312	2.77	.062662
2.28	.102284	2.78	.062039
2.29	.101266	2.79	.061421
2.30	0.100259	**2.80**	0.060810
2.31	.099261	2.81	.060205
2.32	.098274	2.82	.059606
2.33	.097296	2.83	.059013
2.34	.096328	2.84	.058426
2.35	0.095369	**2.85**	0.057844
2.36	.094420	2.86	.057269
2.37	.093481	2.87	.056699
2.38	.092551	2.88	.056135
2.39	.091630	2.89	.055576
2.40	0.090718	**2.90**	0.055023
2.41	.089815	2.91	.054476
2.42	.088922	2.92	.053934
2.43	.088037	2.93	.053397
2.44	.087161	2.94	.052866
2.45	0.086294	**2.95**	0.052340
2.46	.085435	2.96	.051819
2.47	.084585	2.97	.051303
2.48	.083743	2.98	.050793
2.49	.082910	2.99	.050287
2.50	0.082085	**3.00**	0.049787

x	e^{-x}	x	e^{-x}
3.00	0.049787	**3.50**	0.030197
3.01	.049292	3.51	.029897
3.02	.048801	3.52	.029599
3.03	.048316	3.53	.029305
3.04	.047835	3.54	.029013
3.05	0.047359	**3.55**	0.028725
3.06	.046888	3.56	.028439
3.07	.046421	3.57	.028156
3.08	.045959	3.58	.027876
3.09	.045502	3.59	.027598
3.10	0.045049	**3.60**	0.027324
3.11	.044601	3.61	.027052
3.12	.044157	3.62	.026783
3.13	.043718	3.63	.026516
3.14	.043283	3.64	.026252
3.15	0.042852	**3.65**	0.025991
3.16	.042426	3.66	.025733
3.17	.042004	3.67	.025476
3.18	.041586	3.68	.025223
3.19	.041172	3.69	.024972
3.20	0 040762	**3.70**	0.024724
3.21	.040357	3.71	.024478
3.22	.039955	3.72	.024234
3.23	.039557	3.73	.023993
3.24	.039164	3.74	.023754
3.25	0.038774	**3.75**	0.023518
3.26	.038388	3.76	.023284
3.27	.038006	3.77	.023052
3.28	.037628	3.78	.022823
3.29	.037254	3.79	.022596
3.30	0.036883	**3.80**	0.022371
3.31	.036516	3.81	.022148
3.32	.036153	3.82	.021928
3.33	.035793	3.83	.021710
3.34	.035437	3.84	.021494
3.35	0.035084	**3.85**	0.021280
3.36	.034735	3.86	.021068
3.37	.034390	3.87	.020858
3.38	.034047	3.88	.020651
3.39	.033709	3.89	.020445
3.40	0.033373	**3.90**	0.020242
3.41	.033041	3.91	.020041
3.42	.032712	3.92	.019841
3.43	.032387	3.93	.019644
3.44	.032065	3.94	.019448
3.45	0.031746	**3.95**	0.019255
3.46	.031430	3.96	.019063
3.47	.031117	3.97	.018873
3.48	.030807	3.98	.018686
3.49	.030501	3.99	.018500
3.50	0.030197	**4.00**	0.018316

x	e^{-x}	x	e^{-x}	x	e^{-x}	x	e^{-x}
4.00	0.018316	**4.50**	0.011109	**5.00**	0.006738	**5.50**	0.0040868
4.01	.018133	4.51	.010998	5.01	.006671	5.55	.0038875
4.02	.017953	4.52	.010889	5.02	.006605	5.60	.0036979
4.03	.017774	4.53	.010781	5.03	.006539	5.65	.0035175
4.04	.017597	4.54	.010673	5.04	.006474	5.70	.0033460
4.05	0.017422	**4.55**	0.010567	**5.05**	0.006409	**5.75**	0.0031828
4.06	.017249	4.56	.010462	5.06	.006346	5.80	.0030276
4.07	.017077	4.57	.010358	5.07	.006282	5.85	.0028799
4.08	.016907	4.58	.010255	5.08	.006220	5.90	.0027394
4.09	.016739	4.59	.010153	5.09	.006158	5.95	.0026058
4.10	0.016573	**4.60**	0.010052	**5.10**	0.006097	**6.00**	0.0024788
4.11	.016408	4.61	.009952	5.11	.006036	6.05	.0023579
4.12	.016245	4.62	.009853	5.12	.005976	6.10	.0022429
4.13	.016083	4.63	.009755	5.13	.005917	6.15	.0021335
4.14	.015923	4.64	.009658	5.14	.005858	6.20	.0020294
4.15	0.015764	**4.65**	0.009562	**5.15**	0.005799	**6.25**	0.0019305
4.16	.015608	4.66	.009466	5.16	.005742	6.30	.0018363
4.17	.015452	4.67	.009372	5.17	.005685	6.35	.0017467
4.18	.015299	4.68	.009279	5.18	.005628	6.40	.0016616
4.19	.015146	4.69	.009187	5.19	.005572	6.45	.0015805
4.20	0.014996	**4.70**	0.009095	**5.20**	0.005517	**6.50**	0.0015034
4.21	.014846	4.71	.009005	5.21	.005462	6.55	.0014301
4.22	.014699	4.72	.008915	5.22	.005407	6.60	.0013604
4.23	.014552	4.73	.008826	5.23	.005354	6.65	.0012940
4.24	.014408	4.74	.008739	5.24	.005300	6.70	.0012309
4.25	0.014264	**4.75**	0.008652	**5.25**	0.005248	**6.75**	0.0011709
4.26	.014122	4.76	.008566	5.26	.005195	6.80	.0011138
4.27	.013982	4.77	.008480	5.27	.005144	6.85	.0010595
4.28	.013843	4.78	.008396	5.28	.005092	6.90	.0010078
4.29	.013705	4.79	.008312	5.29	.005042	6.95	.0009586
4.30	0.013569	**4.80**	0.008230	**5.30**	0.004992	**7.00**	0.0009119
4.31	.013434	4.81	.008148	5.31	.004942	7.05	.0008674
4.32	.013300	4.82	.008067	5.32	.004893	7.10	.0008251
4.33	.013168	4.83	.007987	5.33	.004844	7.15	.0007849
4.34	.013037	4.84	.007907	5.34	.004796	7.20	.0007466
4.35	0.012907	**4.85**	0.007828	**5.35**	0.004748	**7.25**	0.0007102
4.36	.012778	4.86	.007750	5.36	.004701	7.30	.0006755
4.37	.012651	4.87	.007673	5.37	.004654	7.35	.0006426
4.38	.012525	4.88	.007597	5.38	.004608	7.40	.0006113
4.39	.012401	4 89	.007521	5.39	.004562	7.45	.0005814
4.40	0.012277	**4.90**	0.007447	**5.40**	0.004517	**7.50**	0.0005531
4.41	.012155	4.91	.007372	5.41	.004472	7.55	.0005261
4.42	.012034	4.92	.007299	5.42	.004427	7.60	.0005005
4.43	.011914	4.93	.007227	5.43	.004383	7.65	.0004760
4.44	.011796	4.94	.007155	5.44	.004339	7.70	.0004528
4.45	0.011679	**4.95**	0.007083	**5.45**	0.004296	**7.75**	0.0004307
4.46	.011562	4.96	.007013	5.46	.004254	7.80	.0004097
4.47	.011447	4.97	.006943	5.47	.004211	7.85	.0003898
4.48	.011333	4.98	.006874	5.48	.004169	7.90	.0003707
4.49	.011221	4.99	.006806	5.49	.004128	7.95	.0003527
4.50	0.011109	**5.00**	0.006738	**5.50**	0.004087	**8.00**	0.0003355

x	e^{-x}	x	e^{-x}
8.00	0.0003355	**9.00**	0.0001234
8.05	.0003191	9.05	.0001174
8.10	.0003035	9.10	.0001117
8.15	.0002887	9.15	.0001062
8.20	.0002747	9.20	.0001010
8.25	0.0002613	**9.25**	0.0000961
8.30	.0002485	9.30	.0000914
8.35	.0002364	9.35	.0000870
8.40	.0002249	9.40	.0000827
8.45	.0002139	9.45	.0000787
8.50	0.0002035	**9.50**	0.0000749
8.55	.0001935	9.55	.0000712
8.60	.0001841	9.60	.0000677
8.65	.0001751	9.65	.0000644
8.70	.0001666	9.70	.0000613
8.75	0.0001585	**9.75**	0.0000583
8.80	.0001507	9.80	.0000555
8.85	.0001434	9.85	.0000527
8.90	.0001364	9.90	.0000502
8.95	.0001297	9.95	0.0000477
9.00	0.0001234	**10.00**	0.0000454

STANDARD ANNEALED COPPER WIRE, SOLID*

American Wire Gage (B. & S.). English Units

Gage number	Diameter, mils	Cross-section		Ohms per 1,000 ft.		Ohms per mile	Pounds per 1,000 ft.
		Circular mils	Square inches	25°C. (= 77°F.)	65°C. (= 149°F.)	25°C. (= 77°F.)	
0000	460.0	212,000.0	0.166	0.0500	0.0577	0.264	641.0
000	410.0	168,000.0	0.132	0.0630	0.0727	0.333	508.0
00	365.0	133,000.0	0.105	0.0795	0.0917	0.420	403.0
0	325.0	106,000.0	0.0829	0.100	0.116	0.528	319.0
1	289.0	83,700.0	0.0657	0.126	0.146	0.665	253.0
2	258.0	66,400.0	0.0521	0.159	0.184	0.839	201.0
3	229.0	52,600.0	0.0413	0.201	0.232	1.061	159.0
4	204.0	41,700.0	0.0328	0.253	0.292	1.335	126.0
5	182.0	33,100.0	0.0260	0.319	0.369	1.685	100.0
6	162.0	26,300.0	0.0206	0.403	0.465	2.13	79.5
7	144.0	20,800.0	0.0164	0.508	0.586	2.68	63.0
8	128.0	16,500.0	0.0130	0.641	0.739	3.38	50.0
9	114.0	13,100.0	0.0103	0.808	0.932	4.27	39.6
10	102.0	10,400.0	0.00815	1.02	1.18	5.38	31.4
11	91.0	8,230 0	0.00647	1.28	1.48	6.75	24.9
12	81.0	6,530.0	0.00513	1.62	1.87	8.55	19.8
13	72.0	5,180.0	0.00407	2.04	2.36	10.77	15.7
14	64.0	4,110.0	0.00323	2.58	2.97	13.62	12.4
15	57.0	3,260.0	0.00256	3.25	3.75	17.16	9.86
16	51 0	2,580.0	0.00203	4.09	4.73	21.6	7.82
17	45.0	2,050.0	0.00161	5.16	5.96	27.2	6.20
18	40.0	1,620.0	0.00128	6.51	7.51	34.4	4.92
19	36.0	1,290.0	0.00101	8.21	9.48	43.3	3.90
20	32.0	1,020.0	0.000802	10.4	11.9	54.9	3.09
21	28.5	810.0	0.000636	13.1	15.1	69.1	2.45
22	25.3	642.0	0.000505	16.5	19.0	87.1	1.94
23	22.6	509.0	0.000400	20.8	24.0	109.8	1.54
24	20.1	404.0	0.000317	26.2	30.2	138.3	1.22
25	17.9	320.0	0.000252	33.0	38.1	174.1	0.970
26	15.9	254.0	0.000200	41.6	48.0	220.0	0.769
27	14.2	202.0	0.000158	52.5	60.6	277.0	0.610
28	12.6	160.0	0.000126	66.2	76.4	350.0	0.484
29	11.3	127.0	0.0000995	83.4	96.3	440.0	0.384
30	10.0	101.0	0.0000789	105.0	121.0	554.0	0.304
31	8.9	79.7	0.0000626	133.0	153.0	702.0	0.241
32	8.0	63.2	0.0000496	167.0	193.0	882.0	0.191
33	7.1	50.1	0.0000394	211.0	243.0	1,114.0	0.152
34	6.3	39.8	0.0000312	266.0	307.0	1,404.0	0.120
35	5.6	31.5	0.0000248	335.0	387.0	1,769.0	0.0954
36	5.0	25.0	0.0000196	423.0	488.0	2,230.0	0.0757
37	4.5	19.8	0.0000156	533.0	616.0	2,810.0	0.0600
38	4.0	15.7	0.0000123	673.0	776.0	3,550.0	0.0476
39	3.5	12.5	0.0000098	848.0	979.0	4,480.0	0.0377
40	3.1	9.9	0.0000078	1,070.0	1,230.0	5,650.0	0.0299

NOTE 1.—The *fundamental resistivity* used in calculating the tables is the International Annealed Copper Standard, viz., 0.15328 ohm (meter, gram) at 20°C. The *temperature coefficient* for this particular resistivity is $\alpha_{20} = 0.00393$, or $\alpha_0 = 0.00427$. The *density* is 8.89 g. per cubic centimeter.

NOTE 2.—The values given in the table are only for annealed copper of the standard resistivity. The user of the table must apply the proper correction for copper of any other resistivity. Hard-drawn copper may be taken as about 2.7 per cent higher resistivity than annealed copper.

NOTE 3.—Pounds per mile may be obtained by multiplying the respective values above by 5.28.

*From *Circ.* 31, U.S. Bureau of Standards.

Reproduced, by permission, from *Electrical Engineering*, Vol. 1, by C. L. Dawes. McGraw-Hill, New York, 1947.

APPENDIX 3

Allowable Carrying Capacities of Wires

National Electrical Code

A.W.G. number	Diameter of solid wires, mils	Area, circular mils	Table A, rubber insulation, amperes	Table B, varnished-cloth insulation, amperes	Table C, other insulation, amperes
18	40.3	1,624	3	.....	6
16	50.8	2,583	6	.....	10
14	64.1	4,107	15	18	20
12	80.8	6,530	20	25	30
10	101.9	10,380	25	30	35
8	128.5	16,510	35	40	50
6	162.0	26,250	50	60	70
5	181.9	33,100	55	65	80
4	204.3	41,740	70	85	90
3	229.4	52,630	80	95	100
2	257.6	66,370	90	110	125
1	289.3	83,690	100	120	150
0	325.0	105,500	125	150	200
00	364.8	133,100	150	180	225
000	409.6	167,800	175	210	275
		200,000	200	240	300
0000	460.0	211,600	225	270	325
		250,000	250	300	350
		300,000	275	330	400
		350,000	300	360	450
		400,000	325	390	500
		500,000	400	480	600
		600,000	450	540	680
		700,000	500	600	760
		750,000	525	630	800
		800,000	550	660	840
		900,000	600	720	920
		1,000,000	650	780	1,000
		1,100,000	690	830	1,080
		1,200,000	730	880	1,150
		1,300,000	770	920	1,220
		1,400,000	810	970	1,290
		1,500,000	850	1,020	1,360
		1,600,000	890	1,070	1,430
		1,700,000	930	1,120	1,490
		1,800,000	970	1,160	1,550
		1,900,000	1,010	1,210	1,610
		2,000,000	1,050	1,260	1,670

Reproduced, by permission, from *Electrical Engineering*, Vol. 1, by C. L. Dawes. McGraw-Hill, New York, 1947.

COLOR-CODING SYSTEM
FOR CAPACITORS

Capacitor rating data are often stamped directly on the unit—particularly in the larger sizes. However, certain types of capacitors are generally color-coded. The system used is similar to the Electronic Industries Association (EIA) resistor color-coding technique.

Tubular paper capacitors. This type of capacitor is coded with narrow bands, reading from left toward center. The first three bands indicate capacitance value to two significant figures—the third band being the multiplier. In this respect the coding and color values are similar to the resistor system. The fourth band designates tolerance as follows:

Green	$\pm 5\%$	Black	$\pm 20\%$
White	$\pm 10\%$	Orange	$\pm 30\%$
	Yellow	$\pm 40\%$	

Voltage ratings up to 900 volts are denoted by the fifth color band using the EIA color-coding significant figure values $\times$ 100. Ratings above 900 volts are designated by two bands (5th and 6th) again using the EIA significant figure values ($\times$ 100). For example:

(a) 1000 volts: brown band (1) and black band (0) $= 1 - 0 \times 100 = 1000$.
(b) 1600 volts: brown band (1) and blue band (6) $= 1 - 6 \times 100 = 1600$.

This system is shown in Fig. A4-1.

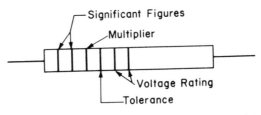

Figure A4-1   Color coding for tubular paper capacitors.

Mica capacitors. A six-dot color-code system used with mica capacitors is shown in Fig. A4-2. The upper left dot indicates whether EIA or MIL standards are being used. The lower left dot designates the "class" or "characteristic" of the capacitor, and covers temperature coefficient (ppm/C°) and

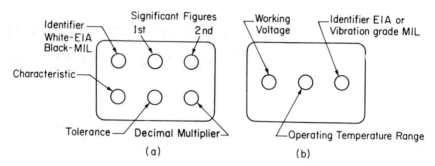

Figure A4-2 Six- and nine-dot mica capacitor color code.

capacitance drift. For this purpose, capacitors are divided into five classes (B, C, D, E, F). Details of the characteristics of each class can be found in the *EIA Standards RS-153A*, April 1964. The significance of the other dots can be seen from Table A4-1.

TABLE A4-1. **Significance of Color Designator Dots**

Dot Color	Capacitance		Tolerance (%)	Charac- terisric	Working Voltage	Operating Temperature Range(°C)
	Sig. Fig. 1 & 2	Dec. Mult.				
Black	0	1	±20	...	...	−55 to +85
Brown	1	10	± 1	B	100	...
Red	2	100	± 2	C	...	−55 to +85
Orange	3	1000	± 3	D	300	...
Yellow	4	...	...	E	...	−55 to +125
Green	5	...	± 5	F	500	...
Blue	6	...	...	...	...	−55 to +150
Violet	7	...	...	...	...	...
Gray	8	...	...	...	...	...
White	9	...	...	...	...	...
Gold	...	.1	±0.5	...	1000	...
Silver	...	.01	±10	...	...	...

A nine-dot system is also used. The first six dots are as described above. The remaining three dots are on the other side. See Fig. A4-2(b). The significance of the first two dots (left to right) is shown in Table A4-1. The third dot, if white, indicates EIA standards; any other color signifies the vibration grade under MIL standards.

Ceramic capacitors. The color-coding technique for ceramic capacitors differs somewhat, depending on whether they are of the class 1 (TC) or class 2 (high-K) type. In addition, the class 1 type may use either a five-dot or a six-dot system. The significance of the color dots for any of these systems is shown in Fig. A4-3 and Table A4-2.

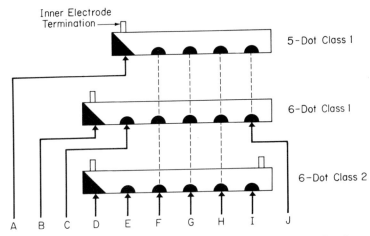

Figure A4-3 Color dot identifiers—ceramic capacitors. (Note: color bands are used on axial units).

TABLE A4-2   Ceramic Capacitor Color-Code Values

Color	A Temperature Coefficient (ppm/°C)	B Temperature Coefficient (Sig. Figure)	C Temperature Coefficient Multiplier	D Temperature Range (°C)	E Capacitance Change over Temp. Range	F 1st Significant Figure	G 2nd	H Decimal Multiplier	I Capacitance Tolerance	J Tolerance of Capacitance < 10 pF	J Tolerance of Capacitance > 10 pF
Black	0	0.0	−1	—	±2.2%	0	1	±20%	±2 pF	±20%
Brown	−33	—	−10	+10 to +85	±3.3%	1	10	—	±0.1 pF	±1%
Red	−75	1.0	−100	—	±4.7%	2	100	—	—	±2%
Orange	−150	1.5	−1000	—	±7.5%	3	1000	—	—	±3%
Yellow	−220	2.2	−10,000	—	±10%	4	10,000	+100, −0%	—	—
Green	−330	3.3	+1	—	±15%	5	—	±5%	±0.5 pF	±5%
Blue	−470	4.7	+10	—	±22%	6	—	—	—	—
Violet	−750	7.5	+100	—	+22, −33%	7	—	—	—	—
Gray	+150 to −1500	−1000 to −5200*	+1000	—	+22, −56%	8	0.01	+80, −20%	±0.25 pF	—
White	+100 to −750	—	+10,000	—	+22, −82%	9	0.1	±10%	±1 pF	±10%
Gold	—	—	—	−55 to +85	±1%	—	—	—	—	—
Silver	—	—	—	−30 to +85	±1.5%	—	—	—	—	—

DETERMINANTS

Analysis of complex electrical networks by Kirchhoff's laws, loop, or nodal methods, invariably leads to setting up two, or more, equations with two, or more, unknowns. When the coefficients of the unknown quantities are simple numbers—as in dc circuits—these simultaneous equations are quite readily solved using the elimination method. (One equation is added to, or subtracted from, another to eliminate an unknown.) This technique was used in Chapters 22 and 23. However, in ac networks containing resistors and reactors, the coefficients in the simultaneous equations will be combinations of pure and complex numbers. It is unlikely that the coefficients of the corresponding unknowns in these equations would have equal (or simple multiple) values. This makes solution by elimination quite laborious. In such cases, the *determinant* method is a more-direct technique. This appendix will show only *how to use* determinants. For the theory details, the reader is referred to any text on advanced algebra.*

Let us assmue that circuit analysis has resulted in two equations with two unknowns as follows:

$$a_1 x + b_1 y = c_1 \qquad (1)$$

$$a_2 x + b_2 y = c_2 \qquad (2)$$

These will produce a second-order determinant, where

$$x = \frac{\begin{vmatrix} c_1 & b_1 \\ c_2 & b_2 \end{vmatrix}}{\begin{vmatrix} a_1 & b_1 \\ a_2 & b_2 \end{vmatrix}}, \quad \text{and} \quad y = \frac{\begin{vmatrix} a_1 & c_1 \\ a_2 & c_2 \end{vmatrix}}{\begin{vmatrix} a_1 & b_1 \\ a_2 & b_2 \end{vmatrix}}$$

Notice that the denominators are alike, and represent the coefficients of the *unknowns* in equations (1) and (2). Also notice that in the numerator for x, the first column contains the c values instead of the x-coefficients. The second column is again the y-coefficients, as shown in equations (1) and (2). Similarly,

---

*See, for example, Mayor & Wilcox, *Comtemporary Algebra: Second Course*, Chapter 8. Prentice-Hall, Inc., Englewood Cliffs, New Jersey, 1964.

in the numerator for y, the y-coefficients (column 2) are replaced by the c values.

Setting up the determinant is the first step. Evaluation of the determinant is next. A simple rule for evaluation is to multiply the terms on diagonal lines— *adding* those that slope down to the right, and *subtracting* those that slope down to the left. Applying this rule to the numerator of the x-determinant yields

$$\begin{matrix} c_1 & b_1 \\ c_2 & b_2 \end{matrix} = (c_1 b_2) - (b_1 c_2)$$

Applying the rule to the denominator, yields

$$(a_1 b_2) - (b_1 a_2)$$

Therefore,

$$x = \frac{(c_1 b_2) - b_1 c_2}{(a_1 b_2) - (b_1 a_2)}$$

Similarly,

$$y = \frac{(a_1 c_2) - (c_1 a_2)}{(a_1 b_2) - (b_1 a_2)}$$

Now let us apply the technique to a problem yielding three equations with three unknowns:

$$a_1 x + b_1 y + c_1 z = d_1 \tag{3}$$

$$a_2 x + b_2 y + c_2 z = d_2 \tag{4}$$

$$a_3 x + b_3 y + c_3 z = d_3 \tag{5}$$

These equations will produce a third-order determinant. In setting up each determinant, it is obvious that there will be three columns. Remember that all denominators are alike, each column representing the coefficients of the unknowns in each of the three equations. Also remember, that in the numerator of the x-determinant, the column for x-coefficients is replaced by the known values (d); and the same is true for the y- and z-coefficients in the numerators of the y- and z-determinants.

The determinants will therefore be

$$x = \frac{\begin{vmatrix} d_1 & b_1 & c_1 \\ d_2 & b_2 & c_2 \\ d_3 & b_3 & c_3 \end{vmatrix}}{\begin{vmatrix} a_1 & b_1 & c_1 \\ a_2 & b_2 & c_2 \\ a_3 & b_3 & c_3 \end{vmatrix}}, \quad y = \frac{\begin{vmatrix} a_1 & d_1 & c_1 \\ a_2 & d_2 & c_2 \\ a_3 & d_3 & c_3 \end{vmatrix}}{\begin{vmatrix} a_1 & b_1 & c_1 \\ a_2 & b_2 & c_2 \\ a_3 & b_3 & c_3 \end{vmatrix}}, \quad z = \frac{\begin{vmatrix} a_1 & b_1 & d_1 \\ a_2 & b_2 & d_2 \\ a_3 & b_3 & d_3 \end{vmatrix}}{\begin{vmatrix} a_1 & b_1 & c_1 \\ a_2 & b_2 & c_2 \\ a_3 & b_3 & c_3 \end{vmatrix}}$$

These determinants are evaluated following the general rule given for the second-order determinants, i.e., multiply the terms on diagonal lines; *adding* those sloping downward to the right; and *subtracting* those sloping downward to the left. To help identify the quantities in each of the three diagonals, it is suggested that each determinant be rewritten—repeating the first two columns. For example, the numerator for the x-determinant becomes:

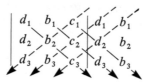

and evaluating, we get

$$(d_1 b_2 c_3) + (b_1 c_2 d_3) + (c_1 d_2 b_3) - (c_1 b_2 d_3) - (d_1 c_2 b_3) - (b_1 d_2 c_3)$$

The rest of the evaluation follows as shown before.

This *general rule* does not apply to a fourth- (or higher-) order determinant. For such cases, the reader is again referred to an advanced algebra text.

ANSWERS TO NUMERICAL PROBLEMS

Note to the classroom instructor

"There is many a slip twixt the cup and the lip."

Although these problems have been personally checked—between typographical errors, printer's errors, and just downright "boners"—it is almost impossible for a person to catch all errors, especially his own errors. Very often we see what we expect to see, rather than the actual words or numbers on the page, and errors can glide by. Since I am no longer in the classroom, these answers have not been field tested. And so, it is possible—in fact probable—that errors will be found during classroom utilization. This is bad enough, but what is more aggravating is when these errors are continued in printing after reprinting of the text.

Unfortunately, since I am not in a classroom situation, I may never become aware of such errors. I am therefore asking for your assistance. When you find (or suspect) an error, will you notify me through the publisher. I would prefer that you send me your step-by-step solution so I can pinpoint where the error was committed. I realize that I may get many responses on the same error. Yet, I would rather reply to each inquiry rather than have that error ignored and continued in perpetuum.

Chapter 0

1. (a) 46 300
   (b) 3980
   (c) 5070
   (d) 0.0629
   (e) 24.0
   (f) 3.51

2. (a) 4.63×10^4
   (b) 3.98×10^3
   (c) 5.70×10^3
   (d) 6.29×10^{-2}
   (e) 2.40×10^1
   (f) 3.51
3. (a) 36 500
   (b) 0.004 16
   (c) 722 000 000
   (d) 0.000 008 91
4. (a) 1.70×10^5
   (b) 6.79×10^{-3}
   (c) 4.09×10^3
   (d) 3.67×10^{-2}
5. (a) 2.32×10^4
   (b) 1.45×10^{-3}
   (c) 2.39×10^3
   (d) 1.69×10^{-2}
6. (a) 7.07×10^9
   (b) 1.10×10^{-5}
   (c) 2.75×10^6
   (d) 2.64×10^{-4}
7. (a) 1.32
   (b) 1.54
   (c) 3.82
   (d) 2.72

Chapter 1

1. (a) 79
   (b) 79
   (c) 118
2. (a) 92
   (b) 92
   (c) 146
3. (a) 3
   (b) 2–8–8
4. (a) 4
   (b) 2–8–18–2
5. (a) 5
   (b) 2–8–18–18–1
6. (a) 6
   (b) 2–8–18–32–18–1
7. 8; 2; 1; 1
8. (a) 10 C

(b) 2 C
(c) 1.39×10^{-2} C
10. 2500 N
11. (a) 1.58×10^{15}
    (b) 2.50×10^{15}
12. 4.24 m

Chapter 3

1. (a) 36 mils
   (b) 25 mils
   (c) 80 mils
   (d) 180 mils
   (e) 6.3 mils
   (f) 22.6 mils
2. (a) 1296 cmils
   (b) 625 cmils
   (c) 6400 cmils
   (d) 32 400 cmils
   (e) 39.7 cmils
   (f) 511 cmils
3. (a) 19 AWG
   (b) 22 AWG
   (c) 12 AWG
   (d) 5 AWG
   (e) 34 AWG
   (f) 23 AWG
4. 3250 cmils
5. 392 cmils
6. 1.27×10^6 cmils
7. 365 mils; 0.365 in
8. 40.2 mils; 0.0402 in
9. 22 AWG
10. 1 AWG
11. (a) 7.79 mm²
    (b) 1.54 mm²
    (c) 0.246 mm²
12. 2.03 S
13. 7.98 S
14. 0.285 S
15. 1.78 S
16. (a) 7.90 S
    (b) 7.68 S
    (c) 0.123 S

17. (a) 36.6 Ω
(b) 83.2 Ω
(c) 7.73 Ω
18. 0.0844 Ω
19. 0.148 in
20. 205 ft
21. 70.8 Ω
22. 0.103 Ω
23. 0.408 mm
24. 24.7 m
25. 29.4 Ω

Chapter 4

1. (a) 8.01 C
(b) 4.99 V
2. 16 V
3. (a) 4 V
(b) 10 V
(c) 22 V
(d) 12 V
4. (a) 25mA; 25 000 μA
(b) 50 μV; 0.000 05 V
(c) 0.000 25 A; 0.25 mA
(d) 0.15 MΩ
(e) 470 mV; 470 000 μV
(f) 650 000 Ω
(g) 0.000 65 A; 650 μA
(h) 0.37 mV; 0.000 37 V
(i) 10 000 Ω
(j) 3.52 MΩ
5. 1.5 A
6. 150 V
7. 275 Ω
8. 200 Ω
9. 100 V
10. 3.6 mA
11. 300 Ω
12. 81 V
13. 0.5 A
14. 800 000 μV
15. 2.32 A
16. 66.7 Ω
17. 24 V

18. 3.47 Ω; 4.46 Ω
19. 12.96 V
20. 1.11 A
21. 182 A
22. 26.4 V
23. 1714 A
24. 240 Ω
25. 8.0 A
26. 110 V
27. 400 Ω
28. 21.8 V
29. 220 V
30. 4.4 mA

Chapter 5

1. 192 W
2. 0.667 A
3. 150 V
4. (a) 48 Ω
(b) 252 W
5. (a) 4.00 W
(b) 8–16 W
7. (a) 2.5 W
(b) 112 V
(c) 22.4 mA
8. 5–10 W (nominal)
9. 160–320 W
10. 40.0 mA
11. (a) 8.58 W
(b) 15–30 W
12. 55.2 W
13. 622 W
14. 6.67 A
15. 230 V
16. 2520 W
17. (a) 0.360 kW
(b) 0.831 hp
(c) 5.63 hp
(d) 3.73 kW
(e) 2850 W
(f) 249 W
18. 23.9 A

19. 80.7%

20. 1/4 hp

21. 356 hp

22. 93.2%

23. 146 A

24. $26.54

25. $2.37

26. $0.000 037 9

27. $5.35

28. $14.07

Chapter 6

1. 535 ft

4. 475 Ω

5. 69.1 Ω

6. 52.2 Ω

7. (a) 58.9 Ω
   (b) 35.1°C

8. No. $\Delta t = 37.7°$

9. 82.2°C

10. 2563°C

Chapter 7

1. (a) 1.2 A
   (b) $IR_1 = 30$ V
      $IR_2 = 54$ V
      $IR_3 = 36$ V
   (c) $W_1 = 36$ W
      $W_2 = 64.8$ W
      $W_3 = 43.2$ W
   (d) $P_1 = 72–144$ W
      $P_2 = 130–260$ W
      $P_3 = 86.4–173$ W

2. $E_T = 200$ V
   $R_1 = 900$ Ω
   $R_2 = 500$ Ω
   $R_3 = 600$ Ω

3. (a) 60 Ω
   (b) 333–666 W

4. $R = 37.9$ Ω
   $P = 33–66$ W

5. (a) +2.4 V

(b) +14 V

(c) 11.6 V

(d) 9670 Ω

6. 1.0 A

7. (a) 110 V
   (b) 0.364 A

8. 27.2 Ω

9. (a) 195 V
   (b) 355 V

10. 583 V

11. 5.5 Ω

12. 20 Ω
   57.6 W

13. (a) 8.76 V
   (b) 221 V
   (c) 194 W

15. 127.4 V

16. 00 AWG

Chapter 8

1. (a) 20.2 Ω; (13–40 Ω)
   (b) 5.933 A
      $I_1 = 3.00$ A
      $I_2 = 1.33$ A
      $I_3 = 1.60$ A
   (c) $W_1 = 360$ W
      $W_2 = 160$ W
      $W_3 = 192$ W

2. 20 Ω

3. 16 Ω

5. 2.3 Ω

6. 3.57 A
   1.43 A

7. (a) 40.4 V
   (b) $I_1 = 1.01$ A
      $I_2 = 0.449$ A
      $I_3 = 0.539$ A

8. 62.5 Ω; 33–100 Ω

9. 9290 Ω

10. 4010 Ω

11. (a) 9.33 A
   (b) 5.60 A

12. 11.5 Ω

13. $1.0 \, \Omega$

14. $2.5 \, \text{A}$

15. $23.8 \, \text{V}$

Chapter 9

1. $I_1 = 44.9 \, \text{A}; \quad E_1 = 44.9 \, \text{V}$
$I_2 = 11.2 \, \text{A}; \quad E_2 = 10.1 \, \text{V}$
$I_3 = 33.7 \, \text{A}; \quad E_3 = 10.1 \, \text{V}$
$I_4 = I_1; \qquad E_4 = 44.9 \, \text{V}$

2. (b) $R_{AB} = 6.68 \, \Omega$
$R_{BC} = 7.79 \, \Omega$
$R_{AC} = 13.4 \, \Omega$

3. $R_x = 8 \, \Omega; \quad E_x = 24 \, \text{V}; \, I_x = 3 \, \text{A}$
$I_6 = 1 \, \text{A} \qquad E_6 = 6 \, \text{V}$
$I_4 = 1.5 \, \text{A}; \quad E_4 = 6 \, \text{V}$
$I_{60} = 0.5 \, \text{A}; \quad E_{60} = 30 \, \text{V}$
$I_{30} = 0.2 \, \text{A}; \quad E_{30} = 6 \, \text{V}$
$I_{20} = 3.5 \, \text{A}; \quad E_{20} = 70 \, \text{V}$

4. $E_1 = 0.75 \, \text{V}; \, I_1 = 1 \, \text{A}$
$R_x = 0.75 \, \Omega; \, E_x = 0.56 \, \text{V};$
$I_x = 0.75 \, \text{A}$
$E_2 = 1.69 \, \text{V}; \, I_2 = 0.19 \, \text{A}$
$E_3 = 1.69 \, \text{V}; \, I_3 = 0.56 \, \text{A}$
$E_4 = 2.25 \, \text{V}; \, I_4 = 0.25 \, \text{A}$

5. $E_T = E_{AB} = 579 \, \text{V}$

6. $I_T = 80 \, \text{mA}; \, E_T = 880 \, \text{V}$
$I_1 = 40 \, \text{mA}; \, E_1 = 200 \, \text{V}$
$I_2 = 20 \, \text{mA}; \, E_2 = 400 \, \text{V}$
$I_3 = 20 \, \text{mA}; \, E_3 = 400 \, \text{V}$
$I_4 = 40 \, \text{mA}; \, E_4 = 600 \, \text{V}$
$I_5 = 80 \, \text{mA}; \, E_5 = 200 \, \text{V}$
$I_6 = 80 \, \text{mA}; \, E_6 = 80 \, \text{V}$

7. $E_1 = 6.60 \, \text{V}; \, I_1 = 132 \, \text{mA}$
$E_2 = 32.5 \, \text{V}; \, I_2 = 81.2 \, \text{mA}$
$E_3 = 40.6 \, \text{V}; \, I_3 = 50.8 \, \text{mA}$
$E_4 = 8.12 \, \text{V}; \, I_4 = 81.2 \, \text{mA}$
$E_5 = 52.8 \, \text{V}; \, I_5 = 132 \, \text{mA}$

8. $E_T = 350 \, \text{V} \qquad R_5 = 313 \, \Omega$
$R_1 = 536 \, \Omega \quad R_{L1} = 2750 \, \Omega$
$R_2 = 625 \, \Omega \quad R_{L2} = 3330 \, \Omega$
$R_3 = 3330 \, \Omega \quad R_{L3} = 3000 \, \Omega$
$R_4 = 5000 \, \Omega \quad R_{L4} = 2500 \, \Omega$

9. $E_T = 290 \, \text{V} \qquad R_4 = 1025 \, \Omega$
$R_1 = 2630 \, \Omega \quad R_5 = 12 \, 500 \, \Omega$

$R_2 = 11 \, 100 \, \Omega \quad R_6 = 20 \, 000 \, \Omega$
$R_3 = 20 \, 000 \, \Omega \quad R_7 = 25 \, 000 \, \Omega$

10. $E_T = 275 \, \text{V} \qquad R_4 = 5000 \, \Omega$
$R_1 = 278 \, \Omega \qquad R_5 = 441 \, \Omega$
$R_2 = 1880 \, \Omega \quad R_6 = 2500 \, \Omega$
$R_3 = 1670 \, \Omega \quad R_7 = 10 \, 000 \, \Omega$

11. $I_1 = 55.0 \, \text{mA}; \quad E_1 = 33.0 \, \text{V}$
$I_2 = 116 \, \text{mA}; \quad E_2 = 23.2 \, \text{V}$
$I_3 = 110 \, \text{mA}; \quad E_3 = 30.7 \, \text{V}$
$I_4 = 58.0 \, \text{mA}; \quad E_4 = 23.2 \, \text{V}$
$I_5 = 120 \, \text{mA}; \quad E_5 = 33.0 \, \text{V}$
$I_6 = 175 \, \text{mA}; \quad E_6 = 43.8 \, \text{V}$
$I_7 = 198 \, \text{mA}; \quad E_7 = 69.3 \, \text{V}$
$I_8 = 87.7 \, \text{mA}; \quad E_8 = 30.7 \, \text{V}$
$I_T = 0.373 \, \text{A}$

12. $E_T = 170 \, \text{V}$
$E_{AB} = 47 \, \text{V}$

13. $E_T = 123.4 \, \text{V}$
$R_9 = 14 \, 900 \, \Omega$

Chapter 10

1. (a) $1270 \, \Omega; \, 3200 \, \Omega; \, 6670 \, \Omega$
(b) $3.85 \, \text{W}; \, 2.00 \, \text{W}; \, 1.50 \, \text{W}$
(c) $8–15 \, \text{W}; \, 4–8 \, \text{W}; \, 3–6 \, \text{W}$
(d) $11 \, 140 \, \Omega; \, 67–135 \, \text{W}$

2. (a) $R_1 = 160 \, \Omega; \, R_2 = 2360 \, \Omega;$
$R_3 = 2400 \, \Omega; \, R_4 = 6000 \, \Omega$
(b) $2.50 \, \text{W}; \, 7.15 \, \text{W}; \, 1.50 \, \text{W};$
$1.35 \, \text{W}$
(c) $5–10 \, \text{W}; \, 15–25 \, \text{W}; \, 5 \, \text{W}; \, 5 \, \text{W}$
(d) $10 \, 900 \, \Omega; \, 350–700 \, \text{W}$ (imprac-
tical)

3. (a) $R_1 = 100 \, \Omega; \, R_2 = 500 \, \Omega;$
$R_3 = 556 \, \Omega; \, R_4 = 3000 \, \Omega;$
$R_5 = 4000 \, \Omega$
(b) $6.25 \, \text{W}; \, 11.3 \, \text{W}; \, 4.50 \, \text{W};$
$7.50 \, \text{W}; \, 2.50 \, \text{W}$
(c) $12–25 \, \text{W}; \, 25–50 \, \text{W}; \, 10–20 \, \text{W};$
$15–30 \, \text{W}; \, 5–10 \, \text{W}$
(d) $8156 \, \Omega; \, 1000–2000 \, \text{W}$ (impractical)

4. (a) $R_1 = 318 \, \Omega; \, R_2 = 5570 \, \Omega;$
$R_3 = 5330 \, \Omega; \, R_4 = 818 \, \Omega$
(b) $3.85 \, \text{W}; \, 6.82 \, \text{W}; \, 1.20 \, \text{W};$
$9.90 \, \text{W}$

(c) 7.7–15.4 W; 13.6–27.2 W;
2.4–4.8 W; 20–40 W

(d) 12 000 Ω; 290–580 W
(impractical)

5. (a) 4500 Ω; 5500 Ω; 428 Ω
(b) 7.20 W; 2.20 W; 8.40 W
(c) 15–30 W; 5–10 W; 15–30 W
(d) 10 400 Ω; 400–800 W
(impractical)

6. (a) 318 Ω; 5570 Ω; 5330 Ω; 1500 Ω
(b) 3.85 W; 6.82 W; 1.20 W;
5.40 W
(c) 7.7–15.4 W; 13.6–27.2 W;
2.4–4.8 W; 10–20 W
(d) 12 700 Ω; 300–600 W
(impractical)

7. (a) 4500 Ω; 5500 Ω; 600 Ω
(b) 7.20 W; 2.20 W; 6.00 W
(c) 14.4–28.8 W; 4.4–8.8 W;
12–24 W
(d) 10 600 Ω; 200–400 W

Chapter 11

1. 0.2 A
3. 0.2 A
4. 0.25 Ω
5. 0.1 Ω
6. (a) $E = 3.90$ V; $R = 0.67$ Ω
(b) 0.688 A
(c) 3.44 V
7. (a) 1.4 V; 0.04 Ω
(b) 1.67 A
(c) 1.33 V
(d) 0.557 A
8. 4.44 V
9. (a) 2.10 A
(b) 0.491 A; 0.393 A
0.436 A; 0.785 A
10. (a) 4.06 Ω
(b) 16.2 V
(c) 1.78 A; 2.22 A
11. 1.45 V
12. 0.16 Ω
13. (a) 6.0 V

(b) 0.133 Ω
(c) 13.2 Ω
(d) 5.94 V
14. (a) 9.0 V
(b) 0.3 Ω
(c) 29.7 Ω
(d) 8.91 V
15. 8 cells; 2 branches of 4 each
16. (a) 57 cells
(b) 28.1 V

Chapter 15

1. (a) 4.91×10^{-4} m²
(b) 0.393 m
(c) 6.36×10^8
2. 1.06×10^6
3. 15.9 A
4. 9950 turns
5. 2.04×10^{-2} T
6. (a) 5.81×10^{-4} Wb
(b) 123 mA
7. 1.67 T
8. 432 mA
9. 162 mA
10. 9.10 A
11. 0.82 A
12. 2.13×10^{-3} Wb
13. 600 turns
14. 32 200 turns
15. 2.63 mm
16. 33.3 N
17. (a) 1.64×10^4 N
(b) 1670 kg

Chapter 16

1. (a) 5 Ω
(b) 2.22 Ω
(c) 0.408 Ω
2. (a) 980 Ω
(b) 9980 Ω
(c) 100 kΩ
(d) 250 kΩ

(f) 500 Ω/V

3. (a) 33.3 Ω; 3.03 Ω; 0.752 Ω
   (b) 9.7 kΩ; 99.7 kΩ; 399.7 kΩ;
       1000 kΩ
   (d) 2000 Ω/V

4. (a) 12 MΩ
   (b) 4 MΩ
   (c) 20 kΩ

6. (a) 65.5 V
   (b) 90.0 V
   (c) 84.7 V

7. 16.4 V

8. 9.75–10.25 A

10. 30 V; 59.7 kΩ

11. 6 MΩ; 600 Ω

12. (a) 2690 Ω
    (b) 3000 Ω
    (c) 30 Ω; 300 kΩ

13. (a) 300 Ω
    (b) 5 mA
    (c) 34.4 Ω

14. 3 Ω; 27.5 kΩ; 310 Ω

16. 428 Ω

Chapter 18

1. (a) 1 H
   (b) (1) 5 H
       (2) 0.02 H
       (3) 0.5 mH

2. (a) 20 H
   (b) 1.0 H

3. 96 μH

4. 11 turns

5. (a) 30 H
   (b) 2.46 H

6. (a) 1.7–5 H
   (b) 2.73 H

7. (a) 53–160 mH
   (b) 122 mH

8. 9.33 H

9. 8.94 mH

10. (a) 0.20 s
    (b) 533 μs

(c) 1.00 μs
(d) 0.0727 μs

11. (a) 0.14 s
    (b) add 45 Ω

12. 1.83 s

13. 48 Ω; 138 H

14. 3.67 A

15. 3.64 A

16. 1; 120 H; 0 H

17. (a) 65.6 H
    (b) 14.4 H

18. (a) 0.5
    (b) 20 H

19. (a) 35 H
    (b) 0.436
    (c) 10 H

20. (a) 250 J
    (b) 17.3 J
    (c) 80.0 J

Chapter 19

1. 20 μC

2. 50 kV

3. 1000 μF

4. (a) 0.85 μF
   (b) 0.0625 μF
   (c) 0.033–0.1 μF

5. (a) 1000 pF
   (b) 0.06 μC; 0.04 μC
   (c) 0.1 μC

6. (a) 240 pF
   (b) 0.024 μC
   (c) 0.024 μC
   (d) 40 V; 60 V

7. 667 pF

8. 50 μs

9. 0.667 μF

10. (a) 0.10 s
    (b) 2.96 mA; 126 V
    (c) 1.10 mA; 173 V
    (d) 0.5 s

11. (a) 1.08 mA; 173 V
    (b) 1.08 mA; 173 V

12. (a) 0.139 s
   (b) 0.0981 s
   (c) 0.137 s
   (d) 0.0975 s

13. 1.10 mA
   (b) 12.0 s
   (c) 282 V; 1.89 s
   (d) 282 V; 1.92 s

14. (a) 187.6 V
   (b) 187.8 V
   (c) 187.5 V

15. (b) 304 V; 33.3 μA; 60.8 μC
   (c) 0.5 s

16. (a) 0.20 s
   (b) 0.056 s
   (c) 0.179 MΩ

17. (a) 0.012 s
   (b) 1.2 mA; 240 V

18. (a) 132 μA; 26.4 V
   (b) 323 μA; 64.6 V

19. 0.0554 s

20. 0.0115 s

21. 323 μA; 64.5 V

22. 0.08 J

23. 0.54 J

Chapter 22

1. $I_1 = 5.67$ A to right
   $I_2 = 6.33$ A to left
   $I_3 = 12.0$ A down

2. $I_1 = 2.49$ A to right
   $I_2 = 1.07$ A to right
   $I_3 = 1.42$ A down

3. (a) $I_A = 124$ A; $I_B = 76$ A
   (b) 381 V

4. (a) 0.1163 km from A
   (b) 580 V

5. (a) $I_L = 8.97$ A; $V_L = 6.28$ V;
     $I_A = 16$ A; $I_B = -7.03$ A
   (b) $I_L = 30$ A; $V_L = 6.0$ V;
     $I_A = 30$ A; $I_B = 0$
   (c) $I_L = 87.3$ A; $V_L = 5.24$ V;
     $I_A = 68.2$ A; $I_B = 19.1$ A

6. (a) $I_L = 68.6$ A; $V_L = 5.49$ V;
     $I_A = 55.6$ A; $I_B = 13.0$ A
   (b) $I_L = 71.0$ A; $V_L = 5.68$ V;
     $I_A = 46.0$ A; $I_B = 25.0$ A
   (c) $I_L = 72.8$ A; $V_L = 5.82$ V;
     $I_A = 39.0$ A; $I_B = 33.8$ A

7. (a) $E_A = 110.1$ V; $E_B = 118.8$ A
   (b) $E_A = 97.0$ V; $E_B = 105.7$ A;
     $E_C = 194.7$ V

8. $I_A = 72.3$ A; $I_B = 92.2$ A;
   $I_N = 19.9$ A
   $E_A = 115.7$ V; $E_B = 110.6$ A

9. $E_A = 122$ V; $E_B = 128$ V

10. 5.48 mA

11. 12.5; 7.95 Ω

Chapter 23

1. $I_3 = 3.14$ A; $E_3 = 15.7$ V
   $I_A = 2.43$ A; $I_B = 0.714$ A

2. $E_3 = 11.43$ V; $E_4 = 8.57$ V

3. $I_L = 2.25$ A; $E_L = 67.5$ V

4. $I_L = 0.696$ A; $E_L = 41.7$ V

5. $I_L = 71.4$ mA; $E_L = 35.7$ V

6. $I_3 = 3.14$ A; $E_3 = 15.7$ V
   $I_A = 2.43$ A; $I_B = 0.715$ A

7. $E_3 = 11.45$ V; $E_4 = 8.58$ V

8. $I_L = 2.25$ A; $E_L = 67.5$ V

9. $I_3 = 3.15$ A; $E_3 = 15.75$ V
   $I_A = 2.43$ A; $I_B = 0.715$ A

10. $E_3 = 11.44$ V; $E_4 = 8.57$ V

11. $I_L = 2.25$ A; $E_L = 67.5$ V

12. $I_3 = 3.14$ A; $E_3 = 15.7$ V
   $I_A = 2.43$ A; $I_B = 0.715$ A

13. $I_L = 2.25$ A; $E_L = 67.5$ V

14. $I_L = 0.696$ A; $E_L = 41.7$ V

15. $I_L = 71.3$ mA; $R_L = 35.7$ V

16. $I_3 = 3.14$ A; $E_3 = 15.7$ V
   $I_A = 2.43$ A; $I_B = 0.715$ A

17. $I_L = 2.25$ A; $E_L = 67.5$ V

18. $I_L = 0.696$ A; $E_L = 41.7$ V

19. $I_L = 71.4$ mA; $E_L = 35.7$ V

20. (a) 42.3 mW

(b) 42.3 mW
(c) 56.3 mW

21. (a) 29.0 W
(b) 55 Ω; 29.1W

22. (a) 2.55 W
(b) 112.5 Ω; 4.24 W

Chapter 25

1. (a) 0.667
(b) 33.7°
(c) 0.555
(d) 0.832
(e) 7.21 units

2. (a) 0.643
(b) 18.7 units
(c) 14.3 units

3. (a) 9.43 units
(b) 0.530; 0.848; 0.625
(c) 32.0°

4. (a) $i = 8 \sin \theta$
(b)

θ	i
0	0
30°	4.0
60°	6.93
90°	8.0
120°	6.93
150°	4.0
180°	0.0
210°	−4.0
240°	−6.93
270°	−8.0
300°	−6.93
330°	−4.0
360°	0

6. (a) 0.454 rad
(b) 3.0 rad
(c) 47.6°
(d) 132°
(e) 180°

8. (a) 1884 rad/s
(b) 108 000 deg/s

9. (a) 377 rad/s
(b) 21 600 deg/s

10. (a) 10 Hz
(b) 25 Hz

(c) 50 Hz
(d) 2500 Hz

11. (a) 40 sin 1257t
(b) −23.5 V

12. (a) $e = 170 \sin 377t$
(b) 167 V

13. (a) 85 V; 400 Hz
(b) 250 mA; 5000 Hz
(c) 377 W; 60 Hz

14. (a) 226 V
(b) 204 V

15. (a) 8.03 A
(b) 8.91 A

16. (a) 244 V
(b) 345 V

17. (a) 54.1 mA
(b) 84.9 mA

18. $E_m = 424$ V

Chapter 26

3.

	Amp	rad/s	Hz	Phase Angle
(a)	25 V	377	60	40° lead
(b)	4 W	628	100	80° lag
(c)	0.35 A	3150	501	90° lead
(d)	120 V	628	100	180°

4. (d) 17.3; leading by 30°

7. (a) 17.3; leading by 30°
(b) 5.0 V; lagging by 130°
(c) 50 mA; lagging by 32°
(d) 78 V; leading by 23°

Chapter 27

1. $10 \sin \omega t + 5 \sin 6 \omega t$
2. $10 \sin \omega t - 5 \cos 3 \omega t$
3. $10 \sin \omega t - 5 \sin 2 \omega t$
4. $10 \sin \omega t + 5 \sin 4 \omega t$
5. $5 + 10 \sin \omega t$
6. (c) $31.8 + 50 \sin \omega t$
and $-21.2 \cos 2 \omega t$
(d) $-4.2 \cos 4 \omega t$
7. 133 V, 2nd har.

26.7 V, 4th har.
11.3 V, 6th har.

8. (a) 50 Hz
    (b) $1/15\ E_m \sin 4712\ t$
9. $-1/81\ E_m \sin 9\ \omega t$
10. $-1/8\ E_m \sin 8\ \omega t$
13. 0.20; 2.0 V
14. (b) 2.5×10^{-3}
    (c) 250 W
15. (a) 77 V
    (b) 61.5 V

Chapter 28

1. (a) 135 Ω
    (b) 150 Ω
    (c) 179 Ω
2. (a) 53 mA
4. (a) 150 W; 133 Ω
    (b) 69.4 W; 0.196 A
    (c) 160 W; 563 Ω
    (d) 1.2 W; 1333 Ω
    (e) 10.1 W; 159 V
5. 32 Ω

Chapter 29

1. (a) 318 Ω
    (b) 7.96 Ω
2. 0.0796 μF
3. 159 kHz
4. (a) 5.3 μF
    (b) 53 μF
5. (a) 166 Ω; 2.41 A
6. 11.9 μF
7. 242 pF
8. (a) Zero
    (b) 16 W
    (c) 1.74 W
    (d) 19.2 W
    (e) 0; 1.0; 0.174; 0.766

Chapter 30

1. 188 000 Ω
2. 9.95 mH
3. 1590 kHz
4. 74.6 mA
5. 44.2 mH
6. 69.8
7. 7.85 Ω
8. 8.38 Ω

Chapter 31

1. (a) $125 + j72.5$ V
    (b) $-37.6 + j0$ A
    (c) $16.7 - j18.6$ V
    (d) $0 - j85$ mA
    (e) $4.3 + j0$ A
    (f) $0 - j1200$ Ω
    (g) $0 + j850$ Ω
    (h) $600 + j0$ Ω
    (i) $0 - j350$ Ω
    (j) $600 - j1200$ Ω
    (k) $600 + j850$ Ω
    (l) $600 - j350$ Ω
2. (a) $145\underline{/30°}$ V
    (b) $37.6\underline{/180°}$ A
    (c) $25\underline{/-48°}$ V
    (d) $85\underline{/-90°}$ mA
    (e) $4.3\underline{/0°}$ A
    (f) $1200\underline{/-90°}$ Ω
    (g) $850\underline{/90°}$ Ω
    (h) $600\underline{/0°}$ Ω
    (i) $350\underline{/-90°}$ Ω
    (j) $1340\underline{/-63.4°}$ Ω
    (k) $1040\underline{/54.8°}$ Ω
    (l) $695\underline{/-30.3°}$ Ω
3. (a) $69.3 + j40$ V
    (b) $61.1 - j22.2$ mA
    (c) $161 + j344$ Ω
    (d) $0.608 + j3.45$ A
4. (a) $58.3\underline{/31°}$ mA
    (b) $46.1\underline{/-77.5°}$ V
    (c) $233.5\underline{/59°}$ Ω
    (d) $17.0\underline{/-28.1°}$ Ω

5. (a) $27.3\underline{/8.43°}$ A
    (b) $192.5\underline{/-8.97°}$ V
    (c) $76.4\underline{/46.1°}$ Ω
    (d) $111.3\underline{/0.721°}$ V

6. (a) $53.2\underline{/-65.0°}$ V
    (b) $18.6\underline{/53.7°}$ A
    (c) $30\underline{/0°}$ Ω

7. $37.5 - j50$

8. $62.5\underline{/-53.1°}$

9. $47\,500 + j145\,000$
    $153\,000\underline{/71.9°}$

10. $0.8 - j1.6$

11. $1.79\underline{/-63.4°}$

12. 585 W

13. (a) $9.27\underline{/24°}$
    (b) $20.5\underline{/-15°}$
    (c) $8.51\underline{/34.9°}$
    (d) $41.8\underline{/-15.5°}$

Chapter 32

1. 1.09 A; $0°$
    $E_1 = 21.8$ V
    $E_2 = 43.6$ V
    $E_3 = 54.5$ V

2. 18.5 mA
    $E_1 = 5.88$ V
    $E_2 = 2.35$ V
    $E_3 = 11.8$ V
    $\theta_{ckt} = 90°$

3. (a) 1.51 mA
    (b) 160 V
    (c) 200 V
    (d) 132 000 Ω

4. (a) 2.37 mA
    (b) 17.9 V; 26.8 V; 5.36 V
    (c) $\theta_{ckt} = 90°$

5. (a) 18 V
    (b) 0.143 mA
    (c) 13.4 H

6. (a) 120 mA
    (b) 72 V; 96 V
    (c) 53.1°

7. (a) 1880 Ω

    (b) 1880 Ω
    (c) 6440 Ω
    (d) 10 650 Ω
    (e) 1022 Ω
    (f) 1013 Ω

8. $37.5 - j\,50$V

9. $62.5\underline{/-53.1°}$

10. (a) 539 Ω
    (b) 0.927 A
    (c) 92.7 V; 491 V
    (d) 79.3°

11. (a) 1.51 A
    (b) 182 V; 401 V
    (c) 65.6°
    (d) 274 W

12. $R = 22.2$ Ω; $C = 34.5$ μF

13. (a) 21.1 kW
    (b) 15.8 kvars

14. (a) 0.725
    (b) 1520 vars

15. (a) 241 Ω
    (b) 0.912 A
    (c) 137 V; 172 V
    (d) 51.6°
    (e) 125 W

16. (a) 45.1 W
    (b) 367 Ω
    (c) 126 Ω; 0.457 H

17. $47.5 + j145$ V
    $153\underline{/71.9°}$ V

18. $0.80 - j1.6$ A

19. 0.115 H; 3.43

20.

	θ_C	θ_L
(a)	34.5°	76.0°
(b)	21.0°	54.0°
(c)	40.7°	68.0°

21. (a) 20 Ω; 0.107 H
    (b) 12 000 Ω; 2.56 H
    (c) 2.6 Ω; 17 mH

22. (a) 160 V; 400 V
    (b) 431 V
    (c) 68.2°
    (d) 128 W

23. (a) 1210 Ω ind; $\theta = 65.5°$

(b) 73.8 Ω cap; $\theta = 74.3°$
(c) 203 Ω ind; $\theta = 9.9°$

24.

	(a)	(b)	(c)
I_T	0.331 A	20.3 mA	1.13 A
E_R	165 V	0.406 V	226 V
E_L	436 V	11.5 V	638 V
E_C	73.1 V	12.9 V	598 V
P	54.6 W	8.24 mW	255 W

25. 21.2 V

26. 137 V

27. 25.4 V

28. (a) 67.8 V; 73.5°
(b) 14.4 V at 85.8°
(c) (1) 33.7 V
(2) 53.7°

Chapter 33

1. (a) 7110 Hz
(b) 0 Ω

2. No, $X_L > X_C$

3. Yes, $X_L = X_C$

4. (a) 0.25 A
(b) 50 V; 625 V; 625 V
(c) 0°

5. 629 kHz

6. 20 μH

7. (a) 81.1°
(b) 32.5°
(c) 17.6°
(d) 6.05°

8. (a) 8.9°
(b) 57.5°
(c) 72.4°
(d) 84.0°

9. 0.158 μF

11. (a) 32.5°
(b) 0.364°
(c) Use $C = 0.0727$ μF

12. (a) 8.43 V
(b) 10.0 V
(c) 9.98 V
(d) 10.0 V

14.

f(Hz)	$E_o(V)$
50	0.628
100	1.25
200	2.44
400	4.49
800	7.09
1600	8.96
3200	9.70
6400	9.92
12800	9.98

16. (a) 4.0 s
(b) 0.025 s
(c) 20 μs
(d) 2.0 μs

17. (a) long
(b) long
(c) medium
(d) short

18. 500 kHz and higher

19. (a) 0.5 μF and larger
(b) 0.005 μF or smaller

20. (a) 5.0 μs
(b) 500 μs
(c) 1000 Ω

21. 1.7 V

22. 9.85 V

Chapter 34

1. (a) 102 mA
(b) 38.5°
(c) 32.0 W
(d) 3910 Ω
(e) $R = 3060$ Ω, $L = 7.74$ H

2. (a) 63.9 mA
(b) 38.5°
(c) 5.0 W
(d) 1560 Ω
(e) $R = 1220$ Ω; $L = 31.0$ mH

3. (a) 58.7 mA
(b) 64.8°
(c) 0.5 W
(d) 341 Ω
(e) $R = 145$ Ω; $L = 24.5$ μH

4. (a) 125 mA

(b) 37.0°
(c) 10 W
(d) 800 Ω
(e) $R = 639\ \Omega;\ C = 0.022\ \mu F$

5. (a) 6.60 mA
   (b) 72.4°
   (c) 40 mW
   (d) 3030 Ω
   (e) $R = 916\ \Omega;\ C = 110\ pF$

6. (a) 3.69 mA
   (b) 9.29 mA
   (c) 7.38 V

7. (a) 80.8 V
   (b) 125 V

8. (a) 33.7 mV
   (b) 10.6 V

9. (b) $I_z = 2.25$ A
   (c) $R = 33.2\ \Omega;\ L = 20.2$ mH

10. (a) 56.0 mA
    (b) 76.2°
    (c) 714 Ω
    (d) 170 Ω & 110 μH

11. (a) 16.2 mA
    (b) 34.7°
    (c) 2470 Ω
    (d) 2030 Ω & 113 pF

12. $0.598/34.6°$ A

13. (a) $\theta_L = 74°;\ \theta_C = 34°$
    (b) $\theta_L = 53°;\ \theta_C = 21°$
    (c) $\theta_L = 69°;\ \theta_C = 41°$

14. (a) 143 mA
    (b) 49.9°
    (c) 20.3 W
    (d) 1540 Ω
    (e) $R = 991\ \Omega;\ L = 3.12$ H

15. (a) 138 mA
    (b) 16.0°
    (c) 3.98 W
    (d) 217 Ω
    (e) $R = 209\ \Omega;\ L = 4.76$ mH

16. 194 mA; 16.1° lead

17. 311 mA; 9.93° lead

18. (a) $I_1 = 100/53°$ mA
    $I_2 = 83.3/0°$ mA
    $I_3 = 33.0/-84.3°$ mA

(b) $154/17.8°$ mA
(c) $99.5/5.7°$ V
(d) $9.9/-84.3°$ V

19. (a) 62.6 V
    (b) 62.9 mA; 52.2 mA; 20.7 mA
    (c) $96.8/17.8°$ mA

20. $336/-34.5°$ Ω

21. $649/-17.9°$ Ω

22. $1503/-16.9°$ Ω

23. $70.7/-28.1°$ mA
    $58.2/76°$ mA

24. (a) $17.4/60.6°$ kΩ
    (b) $172/-60.6°$ mA
    (c) $615/-69.4°$ V
    (d) $117/-86.8°$ mA;
    $84.4/-22.7°$ mA

25. (a) $710/22.9°$ Ω
    (b) $169/-22.9°$ mA
    (c) $86.5/-13.1°$ V
    (d) $159/-69.6°$ mA;
    $130/39.9°$ mA

26. (a) $0.002 + j0$ S;
    $0 - j0.002$ S;
    $0 + j0.005$ S
    (b) $0.003\ 61/56.3°$ S
    (c) $277/-56.3$ Ω
    (d) $0.361/56.3$ A

27. (a) $0.001 + j0$ S;
    $0 - j0.002\ 22$ S;
    $0 + j0.002\ 38$ S
    (b) $0.001\ 01/9.09°$ S
    (c) $990/-9.09°$ Ω
    (d) $36.4/9.09$ mA

28. (a) $G = 3.42 \times 10^{-3}$ S
    $B = 3.08 \times 10^{-3}$ S
    (b) $414/72°$ mA

29. $G = 8.0 \times 10^{-4}$ S
    $B = 16.0 \times 10^{-4}$ S
    $Y = 1.79 \times 10^{-3}/63.4°$ S

30. $G = 9.84 \times 10^{-4}$ S
    $B = 8.19 \times 10^{-4}$ S
    $Y = 12.8 \times 10^{-4}/-39.8$ S

31. $0.598/34.6°$ A

32. $0.154/17.8°$ A

33. 253 μH

34. (a) 13.3 mA
(b) 0°
(c) $R = 3000\ \Omega$
(d) $R = 3000\ \Omega$

35. (a) 12.3 mA
(b) 33.7°
(c) 1220 Ω
(d) 1014 Ω; 53.8 mH

36. 0.158 μF

37. (a) 11.0 mA
(b) 21.9°
(c) 1360 Ω
(d) 1260 Ω; 0.157 μF

38. (a) 11.84 mA
(b) 0.339°
(c) 1267 Ω
(d) $R = 1267\ \Omega$; $C = 11.5\ \mu$F
(e) Yes, practically.

39. (a) 7110 Hz
(b) 124 mA; 33.8°
(c) 806 Ω

40. 6700 Hz

41. 5920 Hz

42. (a) 795 Hz
(b) 29.3%

43. 5.11%

14. $0.925\underline{/1.53°}$ A; $55.5\underline{/1.53°}$ V
15. $0.672\underline{/38.2°}$ A; $80.6\underline{/38.2°}$ V
16. $0.454\underline{/-23.5°}$ A
17. $0.925\underline{/1.55°}$ A; $55.5\underline{/1.55°}$ V
18. $0.671\underline{/38.2°}$ A; $80.5\underline{/38.2°}$ V
19. $0.455\underline{/-23.4°}$ A
20. $0.925\underline{/1.55°}$ A; $55.5\underline{/1.55°}$ V
21. $0.671\underline{/38.2°}$ A; $80.5\underline{/38.2°}$ V
22. $64.3\underline{/-20.5°}$ mA
23. $1.23\underline{/139°}$ A; $61.3\underline{/139°}$ V
24. $0.454\underline{/-23.4°}$ A
25. $0.926\underline{/1.55°}$ A; $55.6\underline{/1.55°}$ V
26. $0.672\underline{/38.2°}$ A; $80.6\underline{/38.2°}$ V
27. $1.23\underline{/139°}$ A; $61.5\underline{/139°}$ V
28. $z_1 = 4050\underline{/83.8°}\ \Omega$
$z_2 = 1720\underline{/-51.8°}\ \Omega$
$z_3 = 8900\underline{/63.5°}\ \Omega$
29. $493 + j212\ \Omega$
30. (a) 8.0 W
(b) 10.0 W
(c) 8.88 W
31. (a) 74.4 W
(b) $2.72 - j16.3\ \Omega$
(c) 919 W

Chapter 35

1. $2.22\underline{/29.2°}$
2. $0.454\underline{/-23.5°}$ A
3. $0.480\underline{/-22.1°}$ A
4. $0.925\underline{/1.55°}$ A; $55.5\underline{/1.55°}$ V
5. $0.671\underline{/38.2°}$ A; $80.5\underline{/38.2°}$ V
6. $1.23\underline{/139°}$ A; $61.5\underline{/139°}$ V
7. $0.455\underline{/-23.3°}$ A
8. $0.477\underline{/-22.8°}$ A
9. $0.926\underline{/1.60°}$ A; $55.6\underline{/1.60°}$ V
10. $0.671\underline{/38.2°}$ A; $80.5\underline{/38.2°}$ V
11. $1.23\underline{/139°}$ A; $61.5\underline{/139°}$ V
12. $0.455\underline{/-23.3°}$ A
13. $0.481\underline{/-22.2°}$ A

Chapter 36

1. 45.0 V
2. 2000 turns
3. 6.64 V
4. 27.3 turns
5. 960 turns
8. 2.25 A
9. 0.30 A
10. 0.218 A; 990 V
11. 93.1%
12. 0.815 A
13. 97.4%
14. 96.7%
15. (a) 3.1 kW
(b) 3.5 kW

   (c) 97.8%
   (d) 97.3%
16. (a) 480 Ω
   (b) 2510 Ω
17. 4.17%
18. 13.8 kV
19. (a) 139.5 kV
   (b) 1.06%
20. 30 V
21. 160 V

$$E_{ba} = 173\underline{/-30°} = 150 - j86.6$$
$$E_{ac} = 173\underline{/-150°} = -150° - j86.6$$
$$E_{cb} = 173\underline{/90°} = 0 + j173$$

11. 2.54 A at zero deg.
12. 1.68 A lagging 90°
13. 3.33 A leading 56.3°
14. 7.62 A at zero deg.
15. 5.05 A lagging 90°
16. 10.6 A leading 56.3°

Chapter 37

1. $E_{ac} = 170\underline{/45°}$
  $E_{bc} = 170\underline{/135°}$
  $E_{bd} = 170\underline{/-135°}$
  $E_{ad} = 170\underline{/-45°}$

2. $E_{ad} = 131\underline{/35°}$
  $E_{bd} = 91.8\underline{/125°}$
  $E_{ac} = 91.8\underline{/-55°}$
  $E_{bc} = 131\underline{/-145°}$

3. $E_{ac} = 208\underline{/30°}$
  $E_{bc} = 120\underline{/120°}$
  $E_{bd} = 208\underline{/-150°}$
  $E_{ad} = 120\underline{/-60°}$

4. $E_{1,2} = 220\underline{/90°} = 0 + j220$
  $E_{2,3} = 220\underline{/-30°} = 190.5 - j110$
  $E_{3,1} = 220\underline{/-150°}$
    $= -190.5 - j110$

5. $E_{1,2} = 200\underline{/0°} = 200 + j0$
  $E_{3,1} = 200\underline{/-120°}$
    $= -100 - j173.2$
  $E_{2,3} = 200\underline{/120°} = -100 + j173.2$

6. $E_{1,2} = 100\underline{/-90°} = 0 - j100$
  $E_{2,3} = 100\underline{/150°} = -86.6 + j50$
  $E_{3,1} = 100\underline{/30°} = 86.6 + j50$

8. (b) $E_{a'a} = 120\underline{/-90°} = 0 - j120$
  $E_{b'b} = 120\underline{/150°} = -104 + j60$
  $E_{c'c} = 120\underline{/30°} = 104 + j60$
  $E_{bc} = 208\underline{/0°} = 208 + j0$
  $E_{ca} = 208\underline{/-120°} = -104 - j180$
  $E_{ab} = 208\underline{/120°} = -104 + j180$

9. (b) $E_{a'a} = 100\underline{/0°} = 100 + j0$
  $E_{c'c} = 100\underline{/-120°} = -50 - j86.6$
  $E_{b'b} = 100\underline{/120°} = -50 + j86.6$

Chapter 38

1. 36.9 kW
2. 63.9 kW
3. 22.0 kW
4. 848 W
5. 2540 W
6. 0.875
7. 91.4%
8. (a) 104 A; 230 V; 33.1 kW
  (b) 60 A; 398 V; 33.1 kW
9. 131 A
10. 45.72 or 22.86 kW
15. (a) 0.839
  (b) ≅ 0.84
  (c) 38.3 A; 22.1 A
16. −1226 W & 8850 W
17. (a) 0.179
  (b) 34.1 A
18. (a) 34.2 kVA
  (b) 27.4 kW
  (c) 20.5 kvar
19. (a) 43.2 kVA
  (b) 30.3 kW
  (c) 30.9 kvar
20. (a) 77.4 kVA
  (b) 57.7 kW
  (c) 51.4 kvar
21. (a) 51.4 kvar
  (b) 1049 μF
22. (a) 23.6 kvar
  (b) 482 μF

Chapter 39

1. $I_1 = 2.52\underline{/-40.9°}$
   $I_2 = 1.88\underline{/-167°}$
   $I_3 = 2.08\underline{/91.9°}$
2. $I_1 = 20.6\underline{/-96.8°}$
   $I_2 = 17.6\underline{/140°}$
   $I_3 = 18.5\underline{/30.2°}$
3. $I_1 = 5.15\underline{/-73.1°}$
   $I_2 = 8.77\underline{/-123°}$
   $I_3 = 12.7\underline{/74.9°}$
4. $I_1 = 5.90\underline{/-41.1°}$
   $I_2 = 8\underline{/180°}$
   $I_3 = 5.26\underline{/47.6°}$
5. (a) $I_1 = 1.5\underline{/-60°}$
       $I_2 = 2\underline{/150°}$
       $I_3 = 1.2\underline{/0°}$
       $I_N = 0.372\underline{/-53.7°}$
   (b) $I_1 = 3.33\underline{/-26.3°}$

$I_2 = 1.88\underline{/98.7°}$
$I_3 = 3.80\underline{/-161.6°}$
$I_N = 1.22\underline{/-138°}$
(c) $I_1 = 2.4\underline{/83.1°}$
    $I_2 = 2.67\underline{/150°}$
    $I_3 = 4.80\underline{/-127°}$
    $I_N = 4.91\underline{/-179°}$
(d) $I_1 = 4.61\underline{/30°}$
    $I_2 = $ zero
    $I_3 = 3.03\underline{/-162.3°}$
    $I_N = 1.76\underline{/51.0°}$

6. $I_1 = 9.06\underline{/154°}$
   $I_2 = 34.4\underline{/26.3°}$
   $I_3 = 28.0\underline{/-139°}$
   $I_N = 1.90\underline{/29.0°}$
7. $I_1 = 28.1\underline{/116°}$
   $I_2 = 11.3\underline{/67.5°}$
   $I_3 = 20.4\underline{/-87.5°}$
   $I_N = 16.9\underline{/115°}$

INDEX